The growth and
functioning of leaves

AF679292

# The growth and functioning of leaves

*Proceedings of a Symposium held prior to the Thirteenth International Botanical Congress at the University of Sydney 18–20 August 1981*

EDITED BY
J. E. DALE AND F. L. MILTHORPE

CAMBRIDGE UNIVERSITY PRESS
CAMBRIDGE
LONDON NEW YORK NEW ROCHELLE
MELBOURNE SYDNEY

CAMBRIDGE UNIVERSITY PRESS
Cambridge, New York, Melbourne, Madrid, Cape Town,
Singapore, São Paulo, Delhi, Tokyo, Mexico City

Cambridge University Press
The Edinburgh Building, Cambridge CB2 8RU, UK

Published in the United States of America by Cambridge University Press, New York

www.cambridge.org
Information on this title: www.cambridge.org/9780521279284

© Cambridge University Press 1983

This publication is in copyright. Subject to statutory exception
and to the provisions of relevant collective licensing agreements,
no reproduction of any part may take place without the written
permission of Cambridge University Press.

First published 1983
First paperback edition 2011

*A catalogue record for this publication is available from the British Library*

*Library of Congress Catalogue Card Number: 82–4377*

ISBN 978-0-521-23761-1 Hardback
ISBN 978-0-521-27928-4 Paperback

Cambridge University Press has no responsibility for the persistence or
accuracy of URLs for external or third-party internet websites referred to in
this publication, and does not guarantee that any content on such websites is,
or will remain, accurate or appropriate.

# Contents

# Participants

(Names in italic indicate principal contributors and chairmen-recorders)

| | |
|---|---|
| Dr H. Y. Adamson | School of Biological Sciences, Macquarie University, North Ryde, NSW 2113, Australia |
| Dr D. Aspinall | Waite Agricultural Research Institute, Department of Plant Physiology, Glen Osmond, SA 5064, Australia |
| *Dr N. R. Baker* | Department of Biological Sciences, University of Essex, Wivenhoe Park, Colchester, Essex CO4 35Q, UK |
| *Dr E. W. R. Barlow* | School of Biological Sciences, Macquarie University, North Ryde, NSW 2113, Australia |
| Dr A. J. E. van Bel | Department of Botany, University of Utrecht, Utrecht, The Netherlands |
| Dr F. W. Bentrup | Institüt für Biologie, Universität Tübingen, D-7400 Tübingen, West Germany |
| Professor G. Bernier | Department of Botany, University of Liège, Bat. B22, Sart Tilman, B-4000 Liège, Belgium |
| Dr P. V. Biscoe | Broom's Barn Experiment Station, Higham, Bury St Edmunds, Suffolk IP28 6NP, UK |
| Dr W. M. Blacklow | Department of Agronomy, University of Western Australia, Nedlands, WA 6009, Australia |
| Dr K. Bradford | Department of Vegetable Crops, University of California, Davis, CA 95616 USA |
| Dr C. Brady | Plant Physiology Group, CSIRO, School of Biological Sciences, Macquarie University, North Ryde, NSW 2113, Australia |
| Professor M. M. Caldwell | Department of Range Science, College of |

| | |
|---|---|
| | Natural Resources, Utah State University, Logan, UT 84321, USA |
| Ms J. Campbell | Agricultural Research Centre, Forest Road, Orange, NSW 2800, Australia |
| *Dr D. A. Charles-Edwards* | CSIRO, The Cunningham Laboratory, Mill Road, St Lucia, Qld 4067, Australia |
| Professor R. E. Cleland | Department of Botany, University of Washington, Seattle, WA 98195, USA |
| Dr B. Clough | AIMS, PMB No. 3, Townsville MS0, Qld 4180, Australia |
| Dr D. J. Connor | School of Agriculture, La Trobe University, Bundoora, Vic. 3083, Australia |
| *Dr J. E. Dale* | Department of Botany, The King's Buildings, University of Edinburgh, Mayfield Road, Edinburgh EH9 3JH, UK |
| Dr W. Day | Physics Department, Rothamsted Experimental Station, Harpenden, Herts. AL5 2JQ, UK |
| Associate Professor N. G. Dengler | Department of Botany, University of Toronto, Toronto, Ont. M5S 1A1, Canada |
| *Professor J. Elston* | Department of Plant Sciences, University of Leeds, Leeds LS2 9JT, UK |
| Professor E. Epstein | Department of Land, Air and Water Resources, Hoagland Hall, Davis, CA 95616, USA |
| *Professor R. O. Erickson* | Department of Biology, University of Pennsylvania, Philadelphia, PA 19063, USA |
| *Mr M. G. Erwee* | Department of Agronomy & Horticultural Science, University of Sydney, NSW 2006, Australia |
| Dr C. Field | School of Life Sciences, NSW Institute of Technology, PO Box 123, Broadway, NSW 2007, Australia |
| *Professor D. Geiger* | Department of Biology, University of Dayton, Dayton, OH 45469, USA |
| Dr R. M. Gifford | CSIRO Division of Plant Industry, PO Box 1600, Canberra City, ACT 2601, Australia |
| *Dr P. B. Goodwin* | Department of Agronomy and Horticultural Science, University of Sydney, NSW 2006, Australia |
| *Dr A. J. Gordon* | Grassland Research Institute, Hurley, Maidenhead, Berks. SL6 5LR, UK |
| Dr P. B. Green | Department of Biological Sciences, Stanford University, Stanford, CA 94305, USA |

| | |
|---|---|
| Professor R. H. Hageman | College of Agriculture, University of Illinois at Urbana-Champaign, Urbana, IL 61801, USA |
| Dr A. J. Hall | Department de Ecologia, Facultad de Agronomia, Av. San Martin 4453, 1417 Capital, Argentina |
| Dr C. Harte | Institüt für Entwicklungs Physiologie, Universität zu Köln, D-5000 Köln 41, Federal Republic of Germany |
| Dr C. Hecht-Buchholtz | Institute of Crop Research, Technische Universität, Berlin, Federal Republic of Germany |
| *Dr J. D. Hesketh* | S212 Turner Hall – AR, SEA, USDA, University of Illinois, Urbana, IL 61801, USA |
| Dr Lim Ho | Glasshouse Crops Research Institute, Rustington, Littlehampton, West Sussex, BN16 3PU, UK |
| Dr M. Howard | Botany Department, University of Massachusetts, Amherst, MA 01003, USA |
| Dr R. Huffaker | Department of Agronomy, University of California, Davis, CA 95616, USA |
| Professor P. G. Jarvis | Department of Forestry and Natural Resources, University of Edinburgh, Darwin Building, The King's Buildings, Mayfield Road, Edinburgh EH9 3JH, UK |
| Mr B. Jenka | Institüt für Pflanzenbau, ETH-Zentrum CH-8092 Zürich, Switzerland |
| *Dr G. I. Jenkins* | Department of Plant Sciences, University of Leeds, Leeds LS2 9JT, UK |
| Dr D. Kemp | Department of Agriculture, Agricultural Research Centre, Forest Road, Orange, NSW 2800, Australia |
| *Dr R. W. King* | CSIRO Division of Plant Industry, PO Box 1600, Canberra City, ACT 2601, Australia |
| Dr E. J. M. Kirby | Plant Breeding Institute, Maris Lane, Trumpington, Cambridge CB2 2LQ, UK |
| Dr M. Kluge | Botanisches Institüt, Fachbereich Biologie, Technische Hochschule, Schnittspahnstrasse 3–5, D-6100 Darmstadt, West Germany |
| Dr R. Lainson | School of Life Sciences, NSW Institute of Technology, Westbourne Street, Gore Hill, NSW 2065, Australia |
| *Dr E. M. Larson* | AR, SEA, USDA, S-212 Turner Hall, University of Illinois, Urbana, IL 61801, USA |

| | |
|---|---|
| *Dr P. R. Larson* | United States Department of Agriculture, Forest Service, Forestry Sciences Laboratory, PO Box 898, Rhinelander, WI 54501, USA |
| Dr U. H. K. Law | School of Biological Sciences, University of Sydney, NSW 2006, Australia |
| *Professor R. M. Leech* | Department of Biology, University of York, Heslington, York YO1 5DD, UK |
| Dr B. Legge | CSIRO Division of Environmental Mechanics, PO Box 821, Canberra, ACT 2601, Australia |
| Dr A. Lindenmayer | Theoretical Biology Group, University of Utrecht, Utrecht 3508 TB, The Netherlands |
| Professor J. F. Loneragan | School of Environmental and Life Sciences, Murdoch University, Murdoch, WA 6150, Australia |
| Ms M. Lowman | School of Biological Sciences, University of Sydney, Macleay Building A12, Sydney, NSW 2006, Australia |
| *Dr M. M. Ludlow* | Division of Tropical Crops and Pastures, CSIRO, The Cunningham Laboratory, Mill Road, St Lucia, Qld 4067, Australia |
| *Dr U. Lüttge* | Institute of Botany, Technische Hochschule, 61 Darmstadt, Schnittspahnstr 3, Darmstadt, West Germany |
| *Dr R. F. Lyndon* | Department of Botany, The King's Buildings, University of Edinburgh, Mayfield Road, Edinburgh EH9 3JH, UK |
| Mr G. McDonald | Department of Agronomy, University of Sydney, NSW 2006, Australia |
| Dr J. Marc | School of Botany, University of NSW, PO Box 1, Kensington, NSW 2033, Australia |
| Dr R. Maksymowych | Department of Biology, Villanova University, Villanova, PA 19085, USA |
| Dr R. D. Meicenheimer | Department of Botany, Washington State University, Pullman, WA 99164, USA |
| Dr N. J. Mendhan | Department of Agricultural Science, University of Tasmania, PO Box 252C, GPO, Hobart 7001, Australia |
| *Professor F. L. Milthorpe* | School of Biological Sciences, Macquarie University, North Ryde, NSW 2113, Australia |
| *Professor J. L. Monteith* | School of Agriculture, University of Nottingham, Sutton Bonington, Loughborough LE12 5RD, UK |
| *Dr J. Moorby* | Agricultural Research Council, 160 Great |

| | |
|---|---|
| | Portland Street, London W1N 6DT, UK |
| Dr J. M. Morgan | Agricultural Research Centre, Tamworth, NSW 2340, Australia |
| Mr G. C. Morrison | 34 Leura Avenue, Wahroonga 2076, Australia |
| Dr R. C. Muchow | CSIRO Division of Tropical Crops & Pastures, Kimberley Research Station, Kununurra, WA 6743, Australia |
| Dr R. Munns | CSIRO Division of Plant Industry, PO Box 1600, Canberra City, ACT 2601, Australia |
| Dr G. J. Murtagh | Agricultural Research Centre, Wollongbar, NSW 2480, Australia |
| Mr R. Nable | School of Environmental and Life Sciences, Murdoch University, Murdoch, WA 6150, Australia |
| Professor Y. S. Nasyrov | Institute of Plant Physiology & Biophysics, Tadjik Academy of Sciences, Dushanbe 734630, Tadjikistan, USSR |
| Dr C. E. Offler | Department of Biological Sciences, The University of Newcastle, NSW 2308, Australia |
| Dr J. Palmer | School of Botany, University of NSW, PO Box 1, Kensington, NSW 2033, Australia |
| Dr J. W. Patrick | Department of Biological Sciences, The University of Newcastle, Newcastle, NSW 2308, Australia |
| Dr J. M. Peacock | ICRISAT, Patancheru PO, Andhra Pradesh 502324, India |
| Dr C. J. Pearson | Department of Agronomy, University of Sydney, Sydney, NSW 2006, Australia |
| *Dr D. B. Peters* | AR, SEA, USDA, S-212 Turner Hall, University of Illinois, Urbana, IL 61801, USA |
| Dr K. A. Platt-Aloia | Department of Botany and Plant Sciences, University of California, Riverside, CA 92521, USA |
| Miss S. P. Radford | Department of Agronomy & Horticultural Science, The University of Sydney, NSW 2006, Sydney, Australia |
| Dr A. S. Raghavendra | Department of Botany, Sri Venkateswara University, Tirupati, 517502, Andhra Pradesh, India |
| Dr J. T. Ritchie | USDA Grassland, Soil & Water Research Laboratory, Temple, Texas 76501, USA |
| Dr T. Sachs | The Hebrew University of Jerusalem, Department of Botany, Jerusalem, Israel |

| | |
|---|---|
| Professor E.-D. Schulze | Lehrstuhl für Pflanzenokölogie, Universität Bayreuth, 8580 Bayreuth, West Germany |
| Dr L. Schultz | Intermountain Herbarium, Utah University, Logan, UT 84321, USA |
| Dr R. Scott Poethig | Curtis Hall, University of Missouri, Columbia, MO 65211, USA |
| *Dr W. K. Silk* | Department of Land, Air and Water Resources, University of California, Davis, CA 95616, USA |
| Dr J. H. Silsbury | Department of Agronomy, Waite Agricultural Research Institute, Glen Osmond, SA 5064, Australia |
| Dr J. H. J. Spiertz | Centre for Agrobiological Research, PO Box 14, 6700AA Wageningen, The Netherlands |
| Dr Z. Starck | Warsaw Agricultural University, Department of Plant Biology, Rakowiecka str. 26/30, 02-528 Warsaw, Poland |
| Dr B. T. Steer | CSIRO Division of Irrigation Research, Private Bag, Griffith, NSW 2680, Australia |
| *Professor T. A. Steeves* | Department of Biology, University of Saskatchewan, Saskatoon S7N OWO, Canada |
| *Dr S. E. Taylor* | Department of Plant and Soil Biology, University of California, Berkeley, CA 94720, USA |
| *Dr N. Terry* | Department of Plant and Soil Biology, University of California, Berkeley, CA 94720, USA |
| *Dr D. J. Tucker* | Glasshouse Crops Research Institute, Worthing Road, Littlehampton, West Sussex BN16 3PU, UK |
| Dr N. C. Turner | CSIRO Division of Plant Industry, PO Box 1600, Canberra City, ACT 2601, Australia |
| *Dr D. Vince-Prue* | Glasshouse Crops Research Institute, Worthing Road, Littlehampton, West Sussex BN16 3PU, UK |
| *Dr L. J. Waldron* | Department of Plant & Soil Biology, University of California, Berkeley, CA 94720, USA |
| Dr I. F. Wardlaw | CSIRO Division of Plant Industry, PO Box 1600, Canberra City, ACT 2601, Australia |
| Mr I. J. Warrington | DSIR, Plant Physiology Division, Palmerston North, New Zealand |
| Dr D. P. Webb | Canadian Forestry Service, Great Lakes Forest Research Centre, PO Box 490, Sault Ste. Marie, Ontario P6A 5M7, Canada |

| | |
|---|---|
| Dr R. Welch | USDA Plant, Soil and Nutrition Laboratory, Cornell University, Ithaca, NY 14853, USA |
| Professor F. Wightman | Department of Biology, Carleton University, Ottawa K1S 5B6, Canada |
| Dr R. F. Williams | Botany Department, Australian National University, PO Box 4, Canberra City, ACT 2601, Australia |
| Dr D. Wilson | The University of Wales, Welsh Plant Breeding Station, Plas Gogerddan, Aberystwyth SY23 3EB, UK |
| Professor G. L. Wilson | Department of Agriculture, University of Queensland, St. Lucia, Qld 4067, Australia |
| *Professor H. W. Woolhouse* | John Innes Institute, Colney Lane, Norwich NR4 7UH, UK |

# Preface

In 1956 the Easter School of the University of Nottingham School of Agriculture was devoted to the topic of leaf growth, and the published volume that resulted has been widely referred to for many years. However, in the quarter of a century that has passed there has been no major meeting devoted exclusively to a consideration of the growth, development and functioning of leaves, despite the fact that leaf production and performance constitute the biological basis for all agriculture. It therefore seemed appropriate to use the occasion of the Thirteenth International Botanical Congress at Sydney in August 1981 to promote a pre-Congress Symposium on this topic. This meeting brought together experts from a range of related areas – anatomists, morphologists, laboratory- and field-oriented physiologists, biochemists and biologically oriented mathematicians – to talk about the current understanding of leaves. In the event this multidisciplinary strategy, reflecting as it does the highly successful practical approach to many biological problems, proved enormously stimulating to all who took part. It is hoped that this published account of the papers and summaries of the ensuing discussions will give an indication of the substantial progress that has been made since 1956 in our understanding of leaves, as well as indicating some of the difficult problems that remain.

Space and time restricted the cover to the basic understanding concerning growth and functioning of leaves; apart from the final chapter, it was not possible to deal with the significance of these issues in the field situation at a level which some may have thought desirable. These matters and the complex interrelations with the soil and aerial environments are well covered in texts on crop physiology; we trust therefore that this volume will complement others with a broader canvas and fill a long-felt need.

In planning this book contributors were asked to write accounts of their subject suitable for senior undergraduates and postgraduate workers requiring an insight into the system; a basic knowledge of plant structure, physiology and biochemistry was to be assumed and authors were asked to survey the field rather than to give an account of their latest research. In this way it was hoped that the book will be useful to a wide range of workers in botanical and agricultural disciplines. In passing, we also hope that this volume will not have to wait into the twenty-first century for its successor – if so, the current editors will certainly not be participating.

It is a pleasure to thank the various authors whose work this is, and the chairmen who took charge of the various sessions at the Symposium and who prepared at very short notice the summaries of the discussions following the presentations. We also thank the University of Sydney for hospitality in the busy pre-Congress period. Macquarie University also gave important support and one of us gratefully acknowledges receipt of a Visiting Fellowship which allowed work on this book to proceed without distractions. Cambridge University Press made generous financial provision towards the cost of getting contributors to Australia.

Finally, and formally, we thank all those authors and publishers who have allowed copyright material to be used in this book.

J. E. DALE AND F. L. MILTHORPE

# PART I

## Initiation and early growth

An important feature of true leaves is that irrespective of final shape or size they have similar origins, arising as lateral superficial outgrowths, called *primordia*, at predictable positions on the flanks of the *apical dome*. True, the shape and size of the apical dome can vary widely, as in the examples shown below which are by no means extreme, but these differences give no indication of the form or size of the leaf to be produced. The three-dimensional nature of the *shoot apex*, properly considered as the apical dome plus the primordia of all unexpanded leaves and the associated stem tissues, is seen in the reconstructions from serial sections of apices of *Lupinus angustifolius* and *Triticum aestivum* made by R. F. Williams (1975 – *The Shoot Apex and Leaf Growth*. Cambridge University Press).

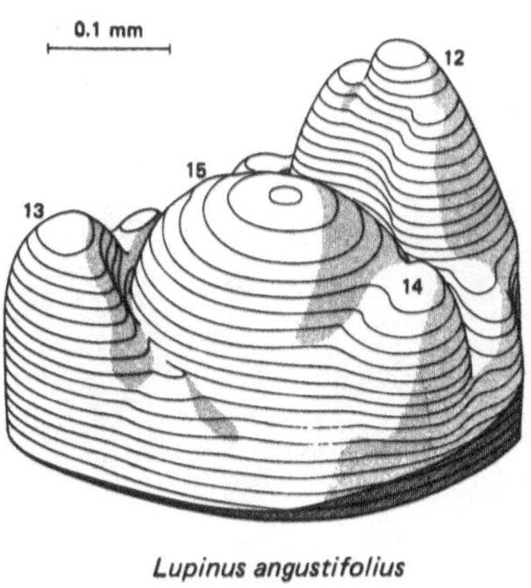


*Lupinus angustifolius*
DAY 7

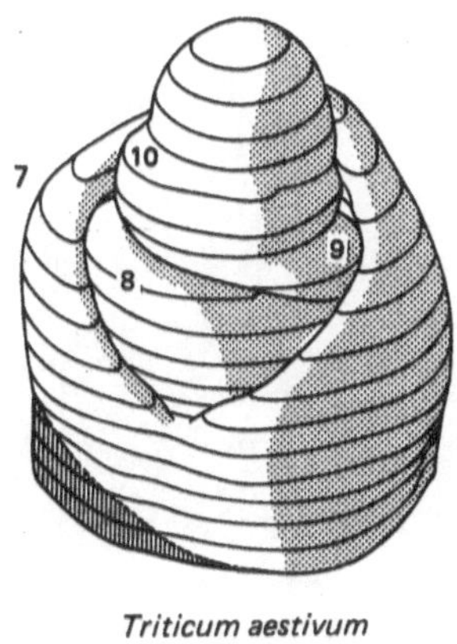


*Triticum aestivum*
DAY 22

The diagrams show the regular arrangement of primordia at the apex, the early development of leaflet initials in *Lupinus* (compare primordia 12 and 13), and the lateral extension of the base of older primordia (e.g. primordium 7) in wheat (*Triticum*).

The chapters in this section are concerned, first, with the events that occur prior to and during the initiation of a primordium where local changes in cell growth and metabolism within the *plastochron*, the time interval between the initiation of successive primordia, are important. Interest in the patterns and arrangements of leaves at the apex has a long history and in Chapter 2 the importance of embryonic vascular tissue, the procambium, to the siting of primordia is explored, while the mathematical basis of leaf arrangement, *phyllotaxis*, is considered in Chapter 3. A further mathematically based analysis of the dynamics of apex growth and primordium development follows in Chapter 4. In addition to producing leaf primordia, the shoot apex also produces floral primordia of diverse kinds, as well as primordia of bracts and *cataphylls* (scale leaves). The transition from leaf production to the production of other types of primordia is considered in Chapter 5 where the importance of study at the molecular level is clearly demonstrated.

# 1

# The mechanism of leaf initiation

R. F. LYNDON

## Introduction

The whole of the leafy shoot originates from the activity of the meristem at the shoot apex, a minute part of the plant usually about 100–300 $\mu$m in diameter and usually consisting of no more than a few hundred cells. The leaves are formed below the extreme tip of the apex, on the flanks of the apical dome which is that region of the shoot meristem distal to the youngest primordium. The apical dome is free of primordia until a terminal flower is formed. A major change in the functioning of the shoot apex is when it makes the transition from vegetative to reproductive growth. Here we will be concerned only with vegetative growth and in particular with the initiation of leaves. Various aspects of leaf initiation and the structure of the shoot apex are reviewed by Cutter (1965), Gifford & Corson (1971), Lyndon (1973, 1976, 1977), Halperin (1978) and Green (1980).

## Problems of leaf initiation

The process of leaf initiation poses two questions: (1) how is a leaf primordium formed and (2) where is it formed? The latter is really the question of phyllotaxis and is considered in Chapter 3. Here we will confine ourselves to the question of how a primordium is formed, although the question of where it is formed inevitably arises since the position is a function of the size of the primordium and the area of the apex over which the events of primordial initiation occur.

The problems of leaf initiation are:

(1) What is the mechanism of leaf primordium initiation?

(2) How is apical structure related to leaf initiation?
(3) How is apical structure maintained?

The structure of the apex is considered first, then how leaves are initiated, and lastly how structure and functioning may be linked.

## The structure of the shoot apex: the basis for leaf initiation

### *The cellular pattern*

The apical dome in many plants is radially symmetrical, as best seen in those plants, such as conifers or the lupin, in which leaves are initiated in a spiral or helix. Its structure is less symmetrical where the apical dome is small relative to the primordia as, for example, in *Clethra* (Hara, 1971) or in plants with decussate or distichous phyllotaxis. In pteridophytes with a large apical cell which cuts off cells on three faces, radial symmetry may be replaced by trilateral symmetry. But even where the sectors of cells so generated are clearly distinguishable there is no correlation between the geometry of the cell sectors in the apex and the positions of leaf initiation (Golub & Wetmore, 1948; Hébant-Mauri, 1975).

A pattern seen in many angiosperm apices is the tunica-corpus arrangement of the cells in which the outer layers of cells (the tunica) show only anticlinal divisions (Clowes, 1961; Esau, 1965). It is most marked in dicotyledons, and in *Acorus*, for example, up to seven tunica layers may be discerned (Kaplan, 1970). However, in grasses the tunica typically consists only of the epidermis (Esau, 1965) and in gymnosperms there often does not appear to be a distinct tunica; instead, the epidermis often shows periclinal divisions and so contributes cells to the inner layers. The number of tunica layers can vary from plant to plant of the same species and also varies according to the stage of the plastochron (Clowes, 1961; Lyndon, 1976; Agarwal & Puri, 1977). In the grasses *Saccharum* and *Erianthus* it varies according to the age and the size of the apex, periclinal divisions occurring frequently in the epidermis of the apical dome of young plants and those with small apices (Thielke, 1965). In *Silene* plants subjected to severe nutrient deficiency, the apices are small, the number of tunica layers is reduced, and periclinal divisions may be seen in the epidermis but the apices function apparently normally (R. F. Lyndon, unpublished). Periclinal divisions in the epidermis of the apical dome have been recorded as occurring rarely in a number of dicotyledons and monocotyledons (Popham, 1963;

Stewart & Dermen, 1970) but may well be less rare than is commonly supposed.

Because of the limited applicability of the tunica-corpus concept apices have been classified according to more complex schemes (Popham, 1951; Newman, 1965; Cutter, 1971). Since all apices, with or without apical cells and with or without tunicas, seem to function similarly in producing leaves on the flanks of the apex but not on the apical dome itself, the tunica-corpus arrangement and the cellular pattern in the apical dome cannot be crucial to leaf initiation. This emphasizes that the differences are likely to be at the molecular level and are not dependent on the cellular arrangements we happen to see; moreover, there are no apparent ultrastructural differences between tunica and corpus cells (Mauseth, 1981a). Models indicate that the visible cellular configurations in the apex could result from the physical constraints and pressures resulting from the shape of the apex and its derivatives (Hejnowicz, 1955; Niklas & Mauseth, 1980). The important point then becomes how this shape is maintained.

### *Cytochemical zonation*

A common, perhaps universal, feature of active vegetative shoot meristems is the occurrence of a zonation, revealed histochemically, of less dense staining at the primordia-free summit of the apex (central zone) and deeper staining on the flanks (peripheral zone) where the leaves are initiated. These regions correspond to the '*méristème d'attente*' and '*anneau initial*' respectively of the French school of anatomists (Buvat, 1952; Nougarède, 1967). This zonation reflects the greater concentrations of nucleic acids, especially RNA, and proteins, seen at the ultrastructural level as a greater concentration of ribosomes, in the cells on the flanks than in those of the central zone (Gifford & Stewart, 1967; Nougarède, 1967; Cecich, 1977). Other differences in ultrastructure have been looked for but have not been found (Lyndon & Robertson, 1976; Mauseth, 1981a). Non-ribosomal proteins may often be in greater concentration on the flanks of the apex than in the central zone (Jacqmard, 1978; Fosket & Miksche, 1966; Riding & Gifford, 1973) but this is not always so (Evans & Berg, 1972; Mia & Pathak, 1968; Vanden Born, 1963; Thielke, 1965).

The central zone can often be distinguished because its nuclei stain for DNA more diffusely than the nuclei of the peripheral zone. This is not because the DNA content per nucleus is any less but because the nuclei are larger. In pea, for instance, nuclei with the 2C content of DNA are two to three times the volume of 2C nuclei in the peripheral zone (Lyndon, 1973; Toupiol, 1976). In contrast, in the cactus *Echinocereus*, the nuclei in the central zone are smaller than in the peripheral zone but the central zone cells are more vacuolate (Mauseth, 1981a) and this apparently leads to the lower concentration of proteins and nucleic acids.

The cytochemical zonation pattern is independent of the tunica-corpus arrangement or cell pattern since the differential staining of central and peripheral regions occurs not only in angiosperms with diverse tunica-corpus arrangements, but also in gymnosperms with mantle-core patterns and pteridophytes with single apical cells. In different species, the cell size can be larger or smaller in the central zone than on the flanks of the apex (Lyndon, 1972a), so cell size, too, seems to be independent of the cytochemical zonation. The universality of the cytochemical zonation in vegetative apices suggests that it may be essential for apical functioning and leaf initiation.

### *Ultrastructure*

Although ultrastructural differences have been looked for, none (except for ribosome concentration) have been found which are correlated with the various cyto-histological patterns. The ultrastructural differences within the apex are related to the differentiation of the cells (Lyndon & Robertson, 1976; Mauseth, 1981b). As cells differentiate into pith in the pith-rib meristem the volume per cell of all organelles except the nucleus increases (Table 1.1) and the cells become much more vacuolate; the number of mitochondria per cell increases but plastid number remains the same as in the apical dome. The developing leaf axils consist of smaller cells with more densely-packed organelles, fewer mitochondria and less endoplasmic reticulum. In the young leaf primordium, although the volume of organelles is very much the same as in the rest of the apex, there are greater numbers of plastids, mitochondria and dictyosomes per cell. The initiation of the primordium is marked by plastid replication becoming faster than cell replication and this could be regarded as one of the first signs of leaf differentiation (Lyndon & Robertson, 1976). There is a similar major development of chloroplasts in the cortical chlorenchyma of the cactus *Echinocereus*, which does not have leaves (Mauseth, 1981b).

In other respects the ultrastructure of the shoot meristem appears to be unremarkable. No changes in ultrastructure during a plastochron have been detected (Lyndon & Robertson, 1976; Mauseth, 1981a) although there is an abrupt increase in the number of plastids containing starch half a plastochron before leaf initiation (Lyndon & Robertson, 1976). However, this is a metabolic rather than an ultrastructural change and serves to emphasize that the controls for leaf initiation are almost certainly at the molecular level and that the precise cellular or ultrastructural organization of the apex is irrelevant to the zonation.

### *Gradient of growth rate*

It has been inferred from the cytochemical zonation and from mitotic indices that there is a gradient of growth and cell division rates from a minimum at the summit to a maximum on the flanks of the apex where the leaves are initiated (Nougarède, 1967). In plants in which direct experimental measurements have been made there is a two- to three-fold difference in division rates between the summit and the flanks (Lyndon, 1973) but in some plants (e.g. *Helianthus* and *Nicotiana*), the difference may be much greater and the central zone relatively less active (Steeves, Hicks, Naylor & Rennie, 1969; Sussex & Rosenthal, 1973). How this growth gradient is maintained is not known. It means that changes in form will tend to take place on the flanks of the apex because for most of the plastochron this is where the growth rate is fastest.

Table 1.1. *Ultrastructure of the pea shoot apex (Mit, mitochondria; Pla, plastids; Dic, dictyosomes; ER, endoplasmic reticulum; Vac, vacuoles)*

| | Volume per cell ($\mu m^3$) | | | | | Number per cell | | | |
|---|---|---|---|---|---|---|---|---|---|
| Region | Mit | Pla | Dic | ER | Vac | Mit | Pla | Dic | Vac |
| Central zone | 26 | 23 | 3.7 | 25 | 1.0 | 58 | 11 | 15 | 1.0 |
| Peripheral zone ($I_1$) | 26 | 21 | 7.7 | 32 | 0.6 | 75 | 14 | 16 | 0.5 |
| Leaf primordium | 24 | 25 | 5.6 | 34 | 1.8 | 85 | 16 | 23 | 2.2 |
| Incipient pith | 31 | 47 | 7.8 | 43 | 11.9 | 83 | 10 | 20 | 2.9 |
| Axillary cells of $P_2$ | 14 | 20 | 5.4 | 28 | 2.6 | 51 | 9 | 13 | 1.4 |

$P_2$, the second youngest primordium.
After Lyndon & Robertson (1976).

### *Polarity in the apical dome*

A consequence of the gradient in growth rate, with a minimum at the summit, is that if the apex is domed, as it is in most plants, then the growth of the apical surface cannot be isotropic since this would require maximum growth rate at the summit and minimum on the flanks (Green, Erickson & Richmond, 1970). In fact growth is polarized predominantly longitudinally, or radially in a flatter apex, and is most obvious in the peripheral surface cells (Hara, 1962, 1971, 1975, 1980; Lyndon, 1976). Such polarity is not only compatible with the maintenance of domed apex but also with the extreme polarization of growth and the rapid longitudinal extension of the axis below the apical dome. Although such a longitudinal, or radial, polarization would not in theory be obligatory for a flat disc-like apex, in the flat apices of *Clethra* and *Ginkgo* radial files of cells are nevertheless a distinctive feature of the apical surface (Hara, 1971, 1980). In the newly regenerated *Graptopetalum* apex there appear to be two axes of longitudinal polarization normal to each other, corresponding to the axes on which the decussate pairs of leaves are initiated (Green & Lang, 1981).

### *The significant structural features*

Three features seem to be universal to vegetative shoot apices: the cytochemical zonation of central and peripheral zones which reflects differences in the concentration of proteins and nucleic acids; the gradient of cell division and growth rates which are least at the summit and greatest on the flanks of the apex; and the radial or longitudinal polarity of growth in the apical dome. All three features suggest a lower rate of growth and metabolism at the summit of the apex than on the flanks. Labelling patterns, such as those after supplying radioactive precursors of RNA, show little difference between different parts of the apex (West & Gunckel, 1968; Lyndon, 1972b) but these data alone are not reliable indications of relative metabolic rates.

The patterns of cytochemical staining and of growth rates are similar but not necessarily coincident. In pea, for example, the central zone is smaller than the region of lowest division rate (Lyndon, 1970a, 1972a). They may, however, be functionally correlated since an increase in growth rate in the central zone at the moment of leaf initiation is accompanied by a reduction in size of the central zone (Lyndon, 1968, 1970a). Also, in *Cosmos* an increase in the mitotic index in the tunica of the central zone is correlated with a decrease in the size of the paler staining central zone (Molder & Owens, 1972).

The tendency for the central zone to disappear at the time of leaf initiation and for its growth rate to increase to approximately that of the peripheral parts of the apical dome suggests that the maintenance of gradients of growth rate and cytochemical zonation is not essential for leaf initiation. However, if it is the longitudinal polarity of the apical dome which is the important feature, then the growth rate gradient and zonation would be necessary consequences rather than vice versa. A temporary lessening of the degree of anisotropic growth at leaf initiation would be compatible with temporary disappearance of the growth rate gradient and the associated cytochemical zonation.

Without a predominantly longitudinal polarization the growth rate at the base of a domed apex would be zero, or at least minimal, and rapid longitudinal extension of the supporting axis would have to be brought about by a combination of renewed growth and a newly imposed longitudinal polarity at the base of the apical dome. It seems therefore most reasonable to suppose that the important feature of the structure of the apical dome is the longitudinal polarization and that the gradient of growth rate and the associated cytochemical zonation are necessarily associated with it. Since the growth rate is least at the summit of the apical dome the cytochemical zonation might simply be the visible evidence of the difference in concentration of the protein-synthesizing apparatus bringing this about. It has yet to be shown, however, that the rates of growth and cell division are governed by the concentration of ribosomes in the cells.

How the polarity of the apex arises and is maintained is unknown. If a control function is in some way exercised by the central zone, perhaps as the source of a growth substance, it should be possible to demonstrate this by suitable experimentation.

## The events of leaf initiation

In angiosperms the first visible sign of leaf initiation is the occurrence of periclinal divisions in the tunica, which in grasses is the epidermis. In some species of conifers these periclinal divisions may be seen in the epidermis as well as in the underlying cells (Fahn, 1967). It is frequently assumed that for a leaf primordium to be formed there must arise a growth centre where growth is faster than elsewhere in the apex.

We shall see that both types of change – in the direction of growth and in the rate of growth – are involved in leaf initiation.

### *Changes in the rates of growth and cell division*

Although a faster growth rate at the site of primordium initiation might be expected, when direct measurements of the rates of cell division in the different parts of the apex of clover and pea were made only a relatively small difference was found between the site at which a leaf was about to be initiated ($I_1$) and the opposite flank of the apex ($I_2$) (Denne, 1966; Lyndon, 1970a). Even in the young primordium the rate of cell division and growth did not appear to be very different from that of most of the cells of the apical dome (Lyndon, 1970a). The only direct measurements of the changes in the rate of cell division throughout the plastochron and during primordium initiation are those for the pea, on which this account will be based (Lyndon, 1970a, b, 1971, 1982).

The rate of cell division increased in the $I_2$ region at the end of a plastochron (Table 1.2), was maintained at this higher rate as the cells were displaced into the $I_1$ region in the next plastochron and then increased at initiation of the primordium (Lyndon, 1970a). Two-thirds of a plastochron after its initiation, when its axil was formed, the division rate of the primordium fell again to the lower value originally found in $I_2$.

The values for rates of cell division were supplemented by maps showing the distribution of colchicine metaphases throughout the apex (Fig. 1.1). On the basis of such maps, Hussey (1972) pointed out that, in both pea and tomato, there was indeed a growth centre of cells with a higher rate of division in the $I_1$ region, and localized at the site of

Table 1.2. *Rates of cell division in cells as they pass from $I_2$ to $I_1$ and into the primordium; (1) = first 30 h and (2) = subsequent 16 h of each plastochron*

| | $I_2$ | | $I_1$ | | Primordium | |
|---|---|---|---|---|---|---|
| Region | (1) | (2) | (1) | (2) | (1) | (2) |
| Rate of cell division (% cells per h) | 1.9 | 2.7 | 2.5 | 2.5 | 2.9 | 2.0 |

After Lyndon (1970a).

initiation of the future primordium. Although this growth centre was first apparent at the base of the $I_2$ region and persisted into the base of the $I_1$ region, in the incipient primordium itself there was no longer a region of high division rate and the procambium had differentiated (Lyndon, 1970a). In pea, other regions of high division rate which were present at the sides of the apex, where leaves were never initiated, were the forerunners of the procambium, well before it became visibly

Fig. 1.1. Rates of cell division as indicated by the distribution of colchicine metaphases in (*a*) a 30 μm thick longitudinal median section and (*b*) a 40 μm thick transverse section at the base of the apical dome. Both diagrams represent the pea apex towards the end of a plastochron. The youngest primordium is at the left in both cases; the next primordium is about to be initiated in the $I_1$ region, on the right.

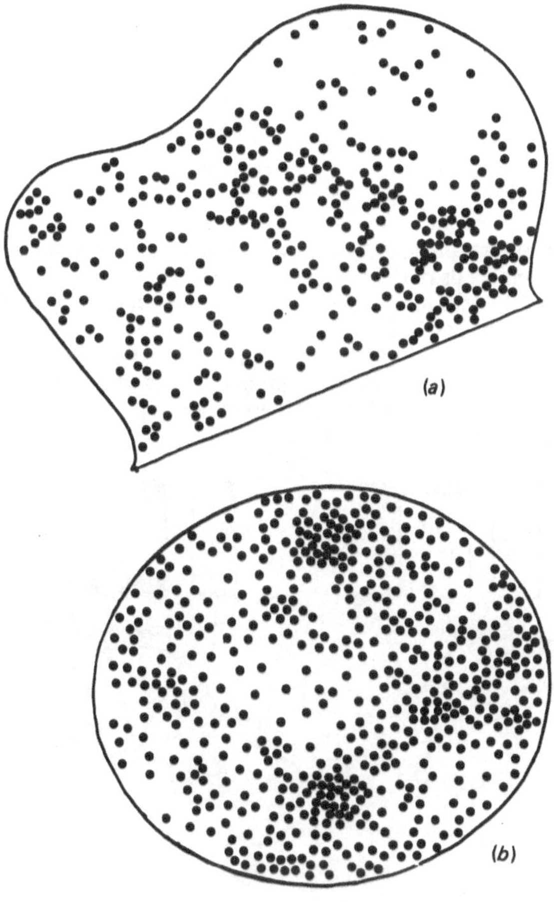

differentiated. It can therefore be argued that the regions of high division rate ('growth centres') are regions in which procambium differentiation is beginning, one and a half plastochrons before leaf initiation, and may not be directly concerned with leaf initiation (cf. Chapter 2).

### *Changes in the orientation of growth and cell division*

An event which seems to be more directly associated with the emergence of a primordium (i.e. the formation of a bulge on the apical flank) is the change in orientation of growth and the occurrence of periclinal divisions in the $I_1$ region of the apical dome half a plastochron before primordium emergence (Fig. 1.2). In the $I_1$ region in the first part of the plastochron only one mitotic figure out of 65 (from 30 apical domes) was a periclinal division whereas in the second part of the plastochron periclinal divisions were seen consistently (22 out of 111 mitotic figures from 36 apical domes). The increase in the proportion of periclinal divisions was quite abrupt at 16 hours before leaf initiation (Lyndon, 1970b). Divisions in the epidermis were always anticlinal but,

Fig. 1.2. Proportions of anticlinal and periclinal cell divisions in the non-epidermal cells of the pea apical dome (*a*) during the 30 h just after initiation of the leaf primordium shown to the left and (*b*) during the subsequent 16 h, just before initiation of the next primordium at the $I_1$ region, at the right. Stippled, anticlinal divisions with new cell walls normal to the section; white, anticlinal divisions with new cell walls in the plane of the section; black, periclinal divisions. (Data from Lyndon, 1972a.)

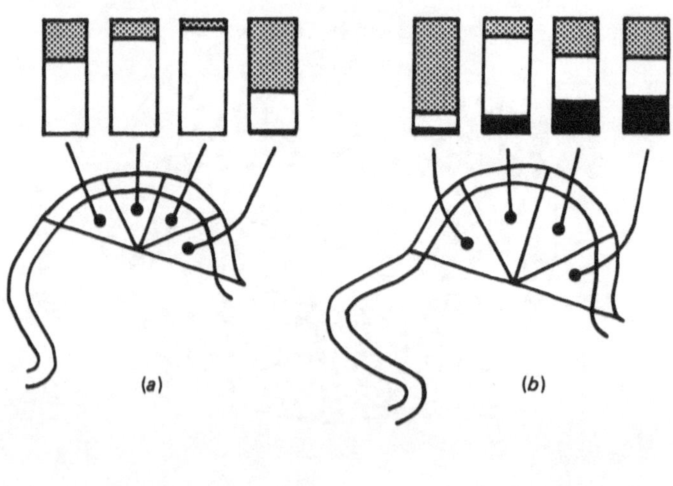

when orientations of mitotic figures in other tissues were examined, it was found that the proportions of divisions in the three planes of space into which the data were analysed were equal. There was therefore no preferred orientation, a state which persisted until primordium emergence (Lyndon, 1972a). The occurrence of periclinal divisions is therefore not because this orientation is imposed on the cells but because the restriction on periclinal divisions is lifted. Moreover, the periclinal divisions were not restricted to the tunica but seemed to be more frequent in the corpus soon after they first appeared. This has been noted for other species (Tepfer, 1960; Rickson, 1969; Lyndon, 1972a; Sistrunk & Tucker, 1974). Since this change in growth orientation occurred without a change in the growth rate of the $I_1$ region as a whole, it was concluded that the primary event in primordium initiation, which took place half a plastochron before primordium emergence, was a change in the plane of growth rather than a change in the rate of growth.

### *Rates of growth and cell division in the median part of the incipient primordium*

Closer examination shows the situation to be more complicated. When the division rates in the median and lateral parts of the incipient primordium and of the $I_1$ and $I_2$ regions were examined separately, it was found that the rate of division increased by about 30% in the midline of the primordium at initiation (Lyndon, 1982) but half a plastochron later the rate had fallen and become lower than in most other parts of the apex (Table 1.3). This more pronounced increase in division rate in the midline of the new primordium occurred 16 hours

Table 1.3. *Rates of cell division (% cells per h) in the median and lateral regions of the apex in the first 30 h (1) and the subsequent 16 h (2) of the plastochron*

| | | $I_2$ | | $I_1$ | | Primordium | |
|---|---|---|---|---|---|---|---|
| | | (1) | (2) | (1) | (2) | (1) | (2) |
| Median 30μm | Epidermis | 0.3 | 1.4 | 2.3 | 2.1 | 3.6 | 1.4 |
| | Underlying cells | 1.0 | 2.6 | 3.2 | 2.8 | 3.6 | 2.0 |
| Lateral regions | Epidermis | 1.1 | 2.2 | 2.5 | 2.4 | 2.9 | 2.1 |
| | Underlying cells | 3.0 | 3.4 | 2.4 | 2.6 | 2.9 | 2.1 |

After Lyndon (1982).

after the change in orientation of growth so that the two changes were not simultaneous. Despite the change in plane of growth in the underlying cells, the apex as a whole does not change shape substantially until the moment of primordium appearance.

### *Growth of the epidermis and the non-epidermal cells*

Comparison of the rates of growth with the changes in shape of the apex suggested that during the first part of the plastochron there was a restraint such that growth within the apical dome led to a bulging into the tissues proximal to it (Lyndon, 1970a). A good candidate for a restraining layer would be the outer layers of the apex, perhaps the epidermis. Although it was concluded that the epidermis was acting like a skin, responding to the growth of the tissues within it (Lyndon, 1971), this view was influenced by the apparent deep-seated nature of the morphogenetic events, i.e. the occurrence of periclinal divisions which appeared to originate deep in the corpus and did not extend to the epidermis. If instead we regard the localized growth of the epidermis as being essential for primordium initiation, then the increase in growth rate in the epidermis and the underlying cells in the midline of the primordium at initiation becomes crucial, and we can tentatively assign a controlling role to the processes which allow this increased growth rate.

The orientations of growth of the epidermis and the underlying cells differ (Table 1.4). During the second part of the plastochron the subepidermal cells in $I_1$ show random orientation of planes of division

Table 1.4. *Percentages of mitotic spindles orientated towards (T) or not towards (N) the tip of the primordium or putative primordium; the orientation in the epidermis is adopted by the underlying cells in the primordium*

| | $I_1$ | | Primordium | |
|---|---|---|---|---|
| | T | N | T | N |
| Epidermis | 50 | 50 | 53 | 47 |
| Underlying cells | 38 | 62 | 57 | 43 |

After Lyndon (1982).

(Lyndon, 1972a). In the epidermis, however, the longitudinal orientation of growth which becomes established at the transition from $I_2$ to $I_1$ is largely maintained throughout $I_1$ and into the primordium and this polarity is adopted by the underlying cells in the developing primordium (Lyndon, 1982). This suggests a controlling role for the epidermis in the establishment of polarity in the young primordium. If the epidermis is the restraining layer preventing outgrowth of the apex then the immediate event resulting in primordium initiation would be an increase in growth rate of the epidermis.

Two independent observations are relevant and striking. The first is that in irradiated wheat seedlings the first stages of primordium initiation occurred in the absence of cell division and consisted of the swelling of the epidermal cells to form a bulge (Foard, 1971). The second is the overriding control of primordium initiation shown by the epidermis in a *Camellia* chimera (Stewart, Meyer & Dermen, 1972). The parents were *C. sasanqua* which shows normal flower development, and *C. japonica*, in which the flowers have sepals and petals but no stamens or carpels. In the chimera with a *sasanqua* epidermis and a *japonica* interior, stamens and carpels were formed. Whether or not stamens and carpels were initiated therefore depended solely on the genetic constitution of the epidermis suggesting that the control of primordium initiation is exercised by the epidermis.

The possible importance of epidermal control is also suggested by the observations of the reorientation of microtubules and cellulose microfibrils in the epidermal cells of the succulent *Graptopetalum* before primordium initiation (Hardham, Green & Lang, 1980) and before any cellular rearrangement or reorientation of growth direction can be seen in the underlying cells (Green & Lang, 1981). However, it is not known whether reorganization of the microtubules and cellulose microfibrils had also taken place in the underlying cells, or whether the orientation of their mitotic spindles had changed, and so it is not clear whether the reorientation of growth direction in *Graptopetalum* is confined to the epidermis. In pea, the reorientation of growth planes certainly occcurs in the underlying cells before there is any evidence of the primordium as a bump (Lyndon, 1970b, 1972a), and in fact it is the epidermal cells which are conservative in their polarity (Lyndon, 1982). Pea and *Graptopetalum* differ, however, in that in the former the apex was already established and had been functioning for about four plastochrons after germination whereas in *Graptopetalum* the regenerating apex was being organized at the same time as it was initiating its first leaves. It is therefore not entirely clear whether the epidermal reorganization seen in *Graptopetalum* was concerned primarily with leaf initiation or whether it might be more to do with the reorganization of the apex itself.

The reorientation of growth axes on the sides of the primordium, stressed by Green (1980), may be more obvious in *Graptopetalum* in which the leaves are more-or-less centric, than in pea in which the primordium, which is essentially a crescent, gives rise to a laminar leaf (Lyndon, 1982).

## The cellular basis for leaf initiation: changes during the plastochron

We may distinguish two sets of events during a plastochron: those which occur before primordium emergence and which are therefore perhaps requirements for initiation but are not directly concerned with the actual change in shape of the apex, and those which are the immediate cause of the change in shape of the apex to give the bulge which is the primordium. There may be a third set of events, which so far have not been identified and which cause determination (i.e. fix the fate) of the primordium and its parts; these will be considered later.

### *The first set of events: before leaf initiation*

Those events occurring half a plastochron before primordium initiation in the pea are:

(i) The randomization of the growth axis in the subepidermal cells of $I_1$.
(ii) A decrease of starch in $I_1$ and an increase on the opposite side of the apex.
(iii) Increase in division rate in $I_2$ and decrease in division rate in the youngest existing primordium.
(iv) A transient increase in division rate in the central zone.
(v) The formation of a distinct axil in the youngest existing primordium.

Since the randomization of the division planes in $I_1$ takes place 16 hours before any visible change in the shape of the apical dome at this point (Lyndon, 1970a) it presumably cannot be a response to removal of physical restraint. If it were, then it is not clear why other parts of the apical dome should not also show the same randomization of division planes, which they do not (Lyndon, 1982). This points to the randomization being the result of localized chemical changes in the cellular environment of $I_1$, which may also be responsible for causing the changes in starch accumulation in the plastids (Lyndon & Robertson, 1976) and the transient increase in division rate in the central zone (Lyndon, 1970a).

The axil of the youngest primordium also forms at this time, so that an axillary distance (Hussey, 1971) becomes measurable. It is an event which determines the diameter of the apical dome by fixing the position of the axillary cells at a more-or-less set distance from the opposite axil. If this did not occur the apical dome and primordium would continue growing as a single unit with the axillary cells being carried upwards passively by the growth of the tissues beneath them.

Whether the increase in division rate in $I_2$ and the decrease in the primordium at this time are also linked events is less certain.

The possibility then is that 30 hours after the initiation of the last primordium and 16 hours before the initiation of the next there is a change in concentration of a chemical substance at the apex which has these diverse effects. The different concentrations of starch on the opposite sides of the apex are consistent with these arising from the same stimulus as causes randomization of the plane of division in $I_1$ but not in $I_2$. Since the position of the next leaf is governed by the positions of the existing leaves (Richards, 1951) the different effects of the stimulus on the opposite sides of the apex may well result from some influence from the existing leaf primordia.

### *The second set of events: at leaf initiation*

The second set of events is those occurring at primordium emergence. In pea these are:

(i) Increase in division rate in cells (epidermal and subepidermal) in the midline of the incipient primordium.
(ii) Beginning of outward growth of the primordial bulge.
(iii) Transient increase in division rate and reduction in size of the central zone.

All can be accounted for by an increase in the rate of growth and cell division, the significance of which is easier to comprehend for the emerging primordium than it is for the central zone. In order for the primordium to bulge out there must be a more rapid localized growth of the epidermis since this grows only by increase in area.

It is pertinent here to consider what might be meant in physiological terms by 'the size of a primordium at initiation'. This may be regarded as the area of the apical surface and the associated cells beneath which contribute to form the primordial bulge. This area, relative to the apical area, is the principal determinant of phyllotaxis (Richards, 1951). The size of the incipient primordium could be described by two parameters: (1) the size of the region of the apical dome in which randomization of cell division planes occurs, this being probably larger than the primordial size (Lyndon, 1972a); and (2) the area over which the properties of the surface change so that a bulge is formed. This will be determined by the area of influence of whatever increases division rate in the midline of the incipient primordium and delimits the area of the epidermis that will grow faster to accommodate the bulge.

If a substance such as auxin, which could cause wall loosening in the epidermis and so allow the tissues to bulge, were produced in the inner tissues it would have to diffuse outwards to reach sufficient concentration at the epidermis at the time of primordium initiation for this event to occur. Is the randomization of division planes in $I_1$, 16 hours before this, the moment at which the localized activation of an enzyme system for synthesis of auxin, or some other morphogen, begins? If so, this might account for the apparent spread of periclinal divisions to the tunica from the inner tissues. But such divisions are found in the tunica at least 10 hours before primordium initiation and it would be necessary to make the further postulate that the threshold concentration of morphogen for changing the plane of division would have to be quite low and much higher for increased wall growth in the epidermis.

Such a hypothetical scheme would link the events occurring half a plastochron before leaf initiation and those occurring at leaf initiation, but raises the questions of whether a single substance could possibly be responsible for all events and whether a delay of 16 hours between events in the underlying cells and the epidermis could plausibly be accounted for by the need for different threshold concentrations.

The temporary reduction in size of the central zone (presumably by a change in the degree of hydration of the cells) and the simultaneous and also temporary disappearance of the gradient of growth rate in the apex suggests that whatever conditions maintain the cytological features of the central zone also maintain its low growth rate and that these conditions are changed transiently when a leaf is initiated. But whether the events in the incipient leaf control those in the central zone, or vice versa, is a matter for speculation.

**Structure of the apical dome in relation to leaf structure**

In an established apex, such as that of the pea, the dome has an essentially radial structure with predominantly longitudinal polarity which corresponds to the requirements for a domed apex with minimum growth rate at the summit (Green, Erickson & Richmond, 1970). This polarity then may determine the polarity of growth in the leaves through the influence of the epidermis (Lyndon, 1982). However, an established radial structure of this kind is not essential since in *Graptopetalum* the polarity may have to change at the initiation of successive leaf pairs (Hardham, Green & Lang, 1980). If this is so, there are two possibilities: one is that this lack of radial symmetry may be related to the decussate phyllotaxis. The other is that the reorganization of polarity is specifically linked to and imposed by the processes of leaf initiation. The leaves of *Graptopetalum* are centric or nearly so, whereas pea leaves are strongly dorsiventral and it is possible that the degree of dorsiventrality of a leaf may therefore be a function of the degree of polarity already existing in the apical dome at the site of leaf initiation.

It is not clear exactly what are the properties of the apex which make a leaf dorsiventral. Experiments on potato apices showed that when a potential leaf site ($I_1$) was isolated from the summit of the apical dome by cuts the resulting organ which developed was centric (either a leaf or a bud) instead of being dorsiventral (Sussex, 1951). This suggested that some influence, presumably chemical, from the apical summit was essential to cause dorsiventrality in the developing primordium. However, these results could not be repeated with a different cultivar of potato, although they could be repeated in *Epilobium* (Snow & Snow, 1959). Such experiments, and those in which the apical summit was destroyed by a prick but the meristems continued to produce dorsiventral leaves (Wardlaw, 1950; Sussex, 1954), suggests that induction from the summit is not needed to make the leaves dorsiventral. Snow & Snow (1959) suggested that 'possibly in these species the tissues of the apical cone may become longitudinally polarized in their microstructure as they pass down from the summit of the apex to the leaf-forming zone so that the leaves arising from such polarized tissues are inherently dorsiventral from the start'. In view of what we now know this seems a very pertinent comment. A repetition of the experiments on potato apices in conjunction with a detailed examination of the surface structure could perhaps lead to an understanding of why contradictory results were obtained.

The dorsiventrality of the leaf may also, at least in part, owe its origin to the difference in growth rate between the adaxial and abaxial surfaces of the leaf which can be traced back to the difference in growth rates between the upper and lower parts of the $I_1$ region (Lyndon, 1970a, 1976; Hussey, 1972). The abaxial surface of the leaf characteristically grows quicker, resulting in the upwards growth of the young primordium and the resulting enshrouding of the apex by the growing leaves. These various considerations show the need to investigate in more detail the relationship between the structure of the apex and the form of the leaves which are initiated on it.

### Determination of the leaf

There are probably several steps in the determination of a leaf of which it is possible to identify the following:

(i) The determination of a site on the apical surface as a primordium. This would mark the fixing of the position of a leaf.

(ii) The determination of the nature of this primordium as a bud, leaf or other organ (e.g. as a centric organ, or as a spine).

(iii) Determination of the parts of the primordium as axis, lamina, leaflets or stipules.

Surgical experiments on apices have shown that the position of the next primordium may be altered until half a plastochron before its emergence. Incisions made in the apex during the half plastochron before initiation resulted in the formation of bracts rather than leaves (Snow & Snow, 1933). These experiments therefore suggest that the position and the nature of the primordium both become determined at the same time, about half a plastochron before emergence. In ferns, however, the primordium is not committed to being a leaf until about three plastochrons after its initiation, until which time it could transform into either a bud or a leaf (Cutter, 1954). When the $I_1$ regions of potato apices were isolated by cuts they developed into leaves showing that this region was already committed as a leaf (Sussex, 1951). The upper and lower surfaces of the pea leaf may also be determined as such at the time of initiation. This is suggested by pea mutants for waxiness and leaf shape which show that the lower surface of the leaf corresponds in its expression of certain genes to the whole of the normal stipule but that the gene expression on the upper surface of the leaf may differ from that on the lower (Marx, 1977). The determination process (whatever it is) continues until one plastochron after initiation. By this time the excision of parts of the leaf does not allow regeneration and so the form of the leaf has become fixed (Sachs, 1969).

The process of determination for a leaf may therefore extend over one and a half plastochrons, or longer in the ferns. Processes of determination continue in the leaf with the differentiation of the various tissues in the mature leaf. The maturation of the leaf progresses from the tip downwards (see Chapter 6). This process may start at the moment of leaf initiation when the epidermal cells at the tip of the incipient primordium begin to divide more slowly than the other cells (Lyndon, 1982), and prompts the interesting speculation that perhaps the epidermis is the site where leaf maturation begins.

## In conclusion

It is easy to forget that the leaf is not a single but a dual structure – a leaf with a bud in its axil. The visible formation of the axil half a plastochron after leaf initiation may well be preceded by the determination of these cells as axillary long before this. The axil cannot be explained simply as a group of cells which grow more slowly. Apart from having the potential to grow into a new shoot with its own apical meristem, their significance for leaf initiation is that they remain in a particular spatial relationship to the opposite side of the apical dome. It may well be more pertinent to concentrate on what happens in the axillary cells and on the flanks of the primordium in order to understand the mechanisms controlling the changing shape of the apical system which we call leaf initiation.

## References

Agarwal, R. M. & Puri, V. (1977). Ontogenetic studies in some important timber trees of India. I. Shoot apex organization and leaf development in *Dalbergia sissoo*. *Phytomorphology*, **27**, 296–302.

Buvat, R. (1952). Structure, évolution et fonctionnement du méristème apical de quelques dicotylédones. *Annales Sciences Naturelles (Botanique) Series II*, **13**, 199–300.

Cecich, R. A. (1977). An electron microscopic evaluation of cytohistological zonation in the shoot apical meristem of *Pinus banksiana*. *American Journal of Botany*, **64**, 1263–71.

Clowes, F. A. L. (1961). *Apical Meristems*, 217 pp. Oxford: Blackwell Scientific Publications.

Cutter, E. G. (1954). Experimental induction of buds from fern leaf primordia. *Nature*, **173**, 440–1.

Cutter, E. G. (1965). Recent experimental studies of the shoot apex and shoot morphogenesis. *Botanical Review*, **31**, 7–113.

Cutter, E. G. (1971). *Plant Anatomy: Experiment and Interpretation. Part 2. Organs*, 343 pp. London: Arnold.

Denne, M. P. (1966). Morphological changes in the shoot apex of *Trifolium repens L*. 1. Changes in the vegetative apex during the plastochron. *New Zealand Journal of Botany*, **4**, 300–14.

Esau, K. (1965). *Plant Anatomy*, 2nd edn, 767 pp. New York & London: Wiley.

Evans, L. S. & Berg, A. R. (1972). Qualitative histochemistry of the shoot apex of *Triticum*. *Canadian Journal of Botany*, **50**, 241–4.

Fahn, A. (1967). *Plant Anatomy*. 534 pp. Oxford: Pergamon.

Foard, D. E. (1971). The initial protrusion of a leaf primordium can form without concurrent periclinal cell divisions. *Canadian Journal of Botany*, **49**, 1601–3.

Fosket, D. E. & Miksche, J. P. (1966). A histochemical study of the seedling shoot apical meristem of *Pinus lambertiana*. *American Journal of Botany*, **53**, 694–702.

Gifford, E. M. & Corson, G. E. (1971). The shoot apex in seed plants. *Botanical Review*, **37**, 143–229.

Gifford, E. M. & Stewart, K. D. (1967). Ultrastructure of the shoot apex of *Chenopodium album* and certain other seed plants. *Journal of Cell Biology*, **33**, 131–42.

Golub, S. J. & Wetmore, R H. (1948). Studies of development in the vegetative shoot of *Equisetum arvense* L. I. The shoot apex. *American Journal of Botany*, **35**, 755–67.

Green, P. B. (1980). Organogenesis – a biophysical view. *Annual Review of Plant Physiology*, **31**, 51–82.

Green, P. B., Erickson, R. O. & Richmond, P. A. (1970). On the physical basis of cell morphogenesis. *Annals New York Academy of Sciences*, **175**, 712–31.

Green, P. B. & Lang, J. M. (1981). Toward a biophysical theory of organogenesis: Birefringence observations on regenerating leaves in the succulent, *Graptopetalum paraguayense* E. Walther. *Planta*, **151**, 413–26.

Halperin, W. (1978). Organogenesis at the shoot apex. *Annual Review of Plant Physiology*, **29**, 239–62.

Hara, N. (1962). Structure and seasonal activity of the vegetative shoot apex of *Daphne pseudo-mezereum*. *Botanical Gazette*, **124**, 30–42.

Hara, N. (1971). Structure of the vegetative shoot apex of *Clethra barbinervis*. I. Superficial and transectional views. *Botanical Magazine Tokyo*, **84**, 8–17.

Hara, N. (1975). Structure of the vegetative shoot apex of *Cassiope lycopodioides*. *Botanical Magazine Tokyo*, **88**, 89–101.

Hara, N. (1980). Morphological study on early ontogeny of the *Ginkgo* leaf. *Botanical Magazine Tokyo*, **93**, 1–12.

Hardham, A. R., Green, P. B. & Lang, J. M. (1980). Reorganization of cortical microtubules and cellulose deposition during leaf formation in *Graptopetalum paraguayense*. *Planta*, **149**, 181–95.

Hébant-Mauri, R. (1975). Apical segmentation and leaf initiation in the tree fern *Dicksonia squarrosa*. *Canadian Journal of Botany*, **53**, 764–72.

Hejnowicz, Z. (1955). Growth distribution and cell arrangement in apical meristems. *Acta Societatis Botanicorum Poloniae*, **24**, 583–608.

Hussey, G. (1971). *In vitro* growth of vegetative tomato shoot apices. *Journal of Experimental Botany*, **22**, 688–701.

Hussey, G. (1972). The mode of origin of a leaf primordium in the shoot apex of the pea (*Pisum sativum*). *Journal of Experimental Botany*, **23**, 675–82.

Jacqmard, A. (1978). Histochemical localization of enzyme activity during floral evocation in the shoot apical meristem of *Sinapis alba*. *Protoplasma*, **94**, 315–24.

Kaplan, D. R. (1970). Comparative foliar histogenesis in *Acorus calamus* and its bearing on the phyllode theory of monocotyledonous leaves. *American Journal of Botany*, **57**, 331–61.

Lyndon, R. F. (1968). Changes in volume and cell number in the different regions of the shoot apex of *Pisum* during a single plastochron. *Annals of Botany*, **32,** 371–90.
Lyndon, R. F. (1970a). Rates of cell division in the shoot apical meristem of *Pisum*. *Annals of Botany*, **34,** 1–17.
Lyndon, R. F. (1970b). Planes of cell division and growth in the shoot apex of *Pisum*. *Annals of Botany*, **34,** 19–28.
Lyndon, R. F. (1971). Growth of the surface and inner parts of the pea shoot apical meristem during leaf initiation. *Annals of Botany*, **35,** 263–70.
Lyndon, R. F. (1972a). Leaf formation and growth at the shoot apical meristem. *Physiologie Végétale*, **10,** 209–22.
Lyndon, R. F. (1972b). Nucleic acid synthesis in the pea shoot apex. In *Nucleic Acids and Proteins in Higher Plants*, ed. G. L. Farkas. Symposia Biologica Hungarica, **13,** pp. 345–53.
Lyndon, R. F. (1973). The cell cycle in the shoot apex. In *The Cell Cycle in Development and Differentiation*, ed. M. Balls & F. S. Billett, pp. 167–83. Cambridge University Press.
Lyndon, R. F. (1976). The shoot apex. In *Cell Division in Higher Plants*, ed. M. M. Yeoman, pp. 285–314. London: Academic Press.
Lyndon, R. F. (1977). The shoot apical meristem. In *The Physiology of the Garden Pea*, ed. J. F. Sutcliffe & J. S. Pate, pp. 183–211. London: Academic Press.
Lyndon, R. F. (1982). Changes in polarity of growth during leaf initiation in the pea. *Annals of Botany*, **49,** 281–90.
Lyndon, R. F. & Robertson, E. S. (1976). The quantitative ultrastructure of the pea shoot apex in relation to leaf initiation. *Protoplasma*, **87,** 387–402.
Marx, G. A. (1977). A genetic syndrome affecting leaf development in *Pisum*. *American Journal of Botany*, **64,** 273–7.
Mauseth, J. D. (1981a). A morphometric study of the ultrastructure of *Echinocereus engelmannii* (Cactaceae). II. The mature, zonate shoot apical meristem. *American Journal of Botany*, **68,** 96–100.
Mauseth, J. D. (1981b). A morphometric study of the ultrastructure of *Echinocereus engelmannii* (Cactaceae). III. Subapical and mature tissues. *American Journal of Botany*, **68,** 531–4.
Mia, A. J. & Pathak, S. M. (1968). A histochemical study of the shoot apical meristem of *Rauwolfia* with reference to differentiation of sclereids. *Canadian Journal of Botany*, **46,** 115–20.
Molder, M. & Owens, J. N. (1972). Ontogeny and histochemistry of the vegetative apex of *Cosmos bipinnatus* 'Sensation'. *Canadian Journal of Botany*, **50,** 1171–84.
Newman, I. V. (1965). Pattern in the meristems of vascular plants. III. Pursuing the patterns in the apical meristem where no cell is a permanent cell. *Journal of the Linnaean Society (Botany)*, **59,** 185–214.
Niklas, K. J. & Mauseth, J. D. (1980). Simulations of cell dimensions in shoot apical meristems: implications concerning zonate apices. *American Journal of Botany*, **67,** 715–32.
Nougarède, A. (1967). Experimental cytology of the shoot apical cells during vegetative growth and flowering. *International Review of Cytology*, **21,** 203–351.
Popham, R. A. (1951). Principal types of vegetative shoot apex organization in vascular plants. *Ohio Journal of Science*, **51,** 249–70.
Popham, R. A. (1963). Developmental studies of flowering. In *Meristems and Differentiation*, Brookhaven Symposia in Biology Number 16, pp. 138–56.
Richards, F. J. (1951). Phyllotaxis: its quantitative expression and relation to growth in the apex. *Philosophical Transactions of the Royal Society*, **B235,** 509–64.
Rickson, F. R. (1969). Developmental aspects of the shoot apex, leaf and Beltian bodies of *Acacia cornigera*. *American Journal of Botany*, **56,** 195–200.

Riding, R. T. & Gifford, E. M. (1973). Histochemical changes occurring at the seedling shoot apex of *Pinus radiata*. *Canadian Journal of Botany*, **51**, 501–12.
Sachs, T. (1969). Regeneration experiments on the determination of the form of leaves. *Israel Journal of Botany*, **18**, 21–30.
Sistrunk, D. R. & Tucker, S. C. (1974). Leaf development in *Doxantha unguis-cati* (Bignoniaceae). *American Journal of Botany*, **61**, 938–46.
Snow, M. & Snow, R. (1933). Experiments on phyllotaxis. II. The effect of displacing a primordium. *Philosophical Transactions of the Royal Society*, **B222**, 353–400.
Snow, M. & Snow, R. (1959). The dorsiventrality of leaf primordia. *New Phytologist*, **58**, 188–207.
Steeves, T. A., Hicks, M. A., Naylor, J. M. & Rennie, P. (1969). Analytical studies on the shoot apex of *Helianthus annuus*. *Canadian Journal of Botany*, **47**, 1367–75.
Stewart, R. N. & Dermen, H. (1970). Determination of number and mitotic activity of shoot initial cells by analysis of mericlinal chimeras. *American Journal of Botany*, **57**, 816–26.
Stewart, R. N., Meyer, F. G. & Dermen, H. (1972). *Camellia* + 'Daisy Eagleson', a graft chimera of *Camellia sasanqua* and *C. japonica*. *American Journal of Botany*, **59**, 515–24.
Sussex, I. M. (1951). Experiments on the cause of dorsiventrality in leaves. *Nature*, **167**, 651–2.
Sussex, I. M. (1954). Experiments on the cause of dorsiventrality in leaves. *Nature*, **174**, 351–2.
Sussex, I. M. & Rosenthal, D. (1973). Differential $^3$H-thymidine labelling of nuclei in the shoot apical meristem of *Nicotiana*. *Botanical Gazette*, **134**, 295–301.
Tepfer, S. S. (1960). The shoot apex and early leaf development in *Clematis*. *American Journal of Botany*, **47**, 655–64.
Thielke, C. (1965). Strukturwechsel und Enzymmuster am Sprossscheitel einiger Gräser. *Planta*, **66**, 310–19.
Toupiol, D. (1976). Mis en place de l'inhibition du bourgeon cotylédonaire chez le *Pisum sativum* L.: étude morphologique et cytophotométrique. *Comptes Rendus Hebdomadaires de l'Académie des Sciences, Paris*, **D282**, 281–4.
Vanden Born, W. H. (1963). Histochemical studies of enzyme distribution in shoot tips of white spruce (*Picea glauca* (Moench) Voss). *Canadian Journal of Botany*, **41**, 1509–28.
Wardlaw, C. W. (1950). The comparative investigation of apices of vascular plants by experimental methods. *Philosophical Transactions of the Royal Society*, **B234**, 583–604.
West, W. C. & Gunckel, J. E. (1968). Histochemical studies of the shoot of *Brachychiton*. II. RNA and protein. *Phytomorphology*, **18**, 283–93.

# 2

# Primary vascularization and the siting of primordia

P. R. LARSON

## Introduction

Botanists often make a distinction between vascular and non-vascular plants. This distinction implies that the vascular system performs a major role in the life of higher plants. And, indeed it does. The main body of most higher plants consists of an elaborate vascular system that permeates almost every organ and tissue of the plant. The vasculature serves as an extensive transport system, and blockage or disruption of this system can either kill or debilitate the affected plant part. Surprisingly, however, we are relatively uninformed about this complex circulatory system when viewed in its entirety. We know much about specific aspects of vascular structure and function. But no concerted effort has ever been made either to describe or to comprehend how this system develops, how it is organized, or how it functions throughout development in any one species of plant.

The procambium is the precursor of the vascular system. By demarcating the prospective vascular system in meristematic tissues it determines vascular organization. And by serving as a template for primary vascular differentiation it directs vascular function. Development and organization of the procambium is therefore fundamental to development and organization of the mature vascular system. Yet, the procambium is perhaps the most neglected and least understood phase of vascular development. Despite many descriptive studies of the procambium, we have relatively meagre knowledge of how the procambium develops, how it is organized, and how it functions. We know even less about possible regulatory roles that the procambial system might perform in plant development.

It is obvious that the vascular system is essential to the development and functioning of leaves. It is perhaps less obvious that prior knowledge of stem vasculature is essential to understanding leaf vasculature. A consideration of the procambial system that serves both stem and leaf is therefore germane to an inquiry into leaf growth and function.

Highly relevant and significant research on procambial development has been discussed in previous reviews (Esau, 1954, 1965). However, to maintain a semblance of continuity, much of the discussion will be confined to a single species, *Populus deltoides*. Procambial development in *P. deltoides* is typical of that in most dicotyledons investigated, and its vasculature may be considered as a representative model system.

## Procambial system of stem

### *Leaf trace concept*

The procambial systems of most higher plants develop in a very orderly and systematic manner, usually conforming to some phyllotactic pattern. Although the contact parastichy notation (1,2; 2,3; 3,5, etc.) is often used to describe the arrangement of primordia on the shoot apex (Richards, 1951), divergence fractions (1/2, 2/5, 3/8, etc.) best describe organization of the procambial system. For a given phyllotactic order, the denominator of the divergence fraction is either equal to or a multiple of the number of main bundles, the sympodial bundles, that traverse the stem. The sympodial bundles are also equivalent to orthostichies in procambial systems (Larson, 1977).

In some species, the sympodial bundles traverse the stem indefinitely with leaf traces simply diverging from each sympodial bundle at regular intervals (Dormer, 1954; Ezelarab & Dormer, 1963; Devadas & Beck, 1971). In other species, the sympodial bundles consist of interconnected leaf traces of finite length (Alexandrov & Alexandrova, 1929; Esau, 1943a; Larson, 1975a). When one leaf trace departs from the procambial cylinder, a new one diverges from it and continues upward to serve the next younger primordium on the orthostichy. The divergent trace also perpetuates the sympodium. At this time it is not known whether these sympodial patterns represent real taxonomic differences or differences in interpretation. If real taxonomic differences exist, the formulation of hypotheses for both procambial organization and the siting of primordia will become more complicated.

According to the leaf trace concept, a leaf trace extends from its point of divergence on either a sympodial bundle or a parent trace to a leaf base (Esau, 1965). The bundles do not, however, terminate at the leaf base. They extend into the node, through the petiole, and finally into the lamina to form the ramifying vein system. The leaf trace system is therefore a functional as well as a developmental continuum between stem and leaf. Because of this continuum, the entire vascular cylinder may be considered an aggregation of leaf traces with every trace in the vascular cylinder serving a leaf at some position on the stem.

The uniformity of the leaf trace system and its symmetrical organization according to a phyllotactic pattern can be readily visualized in a seedling stem of *P. deltoides*. An entire seedling was serially sectioned and each leaf trace in the stem vascular cylinder was identified and related to the leaf it served (Larson, 1979). Leaf traces in the lower stem region bearing secondary leaves conformed to a 2/5 phyllotaxis (Figs. 2.1–2.2). A two-dimensional plot of the leaf-trace system (Fig. 2.3) shows that it is composed of five sympodia, with each sympodium consisting of a series of perpetuating central leaf traces, or C-traces. Each C-trace terminates in a leaf and a new one diverges from it to serve the leaf situated on the next orthostichy above. Right and left lateral traces diverge from the C-traces at regular intervals consistent with the phyllotactic pattern of leaf display. There was no evidence of cauline bundles that indefinitely traversed the stem and the entire vascular cylinder conformed to the leaf trace concept.

### *Acropetal development of procambial system*

Anatomists generally agree that the procambial system progresses acropetally and in continuity with existing procambium in the shoots of most higher plants (Esau, 1954, 1965; Cutter, 1971). This acropetally progressing system consists of procambial leaf traces that have united with their respective primordia.

In some species procambial strands have also been observed. Procambial strands are incipient leaf traces that are still developing toward the prospective sites of their primordia. They develop acropetally in uncommitted tissue of the residual meristem, a derivative tissue of the apical meristem that occurs between the prospective pith and cortical regions of the stem (Esau, 1954).

The structure of procambial strands verifies their acropetal and continuous development. New procambial strands that diverge from parent traces in the main axis and develop acropetally are both larger

Figs. 2.1–2.2. Cross-section of young *Populus deltoides* stem with 2/5 vascular phyllotaxis. The leaf traces were identified by analysis of serial sections. C, central trace; R, right trace; L, left trace. Trace 5Ls is a basipetal bundle split in which subsidiary bundles are split from the parent trace when they encounter an exiting trace three nodes below.

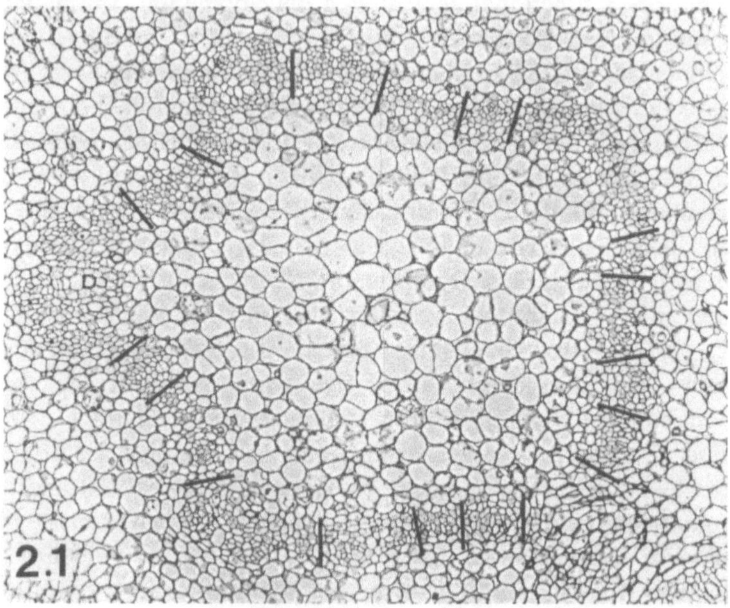


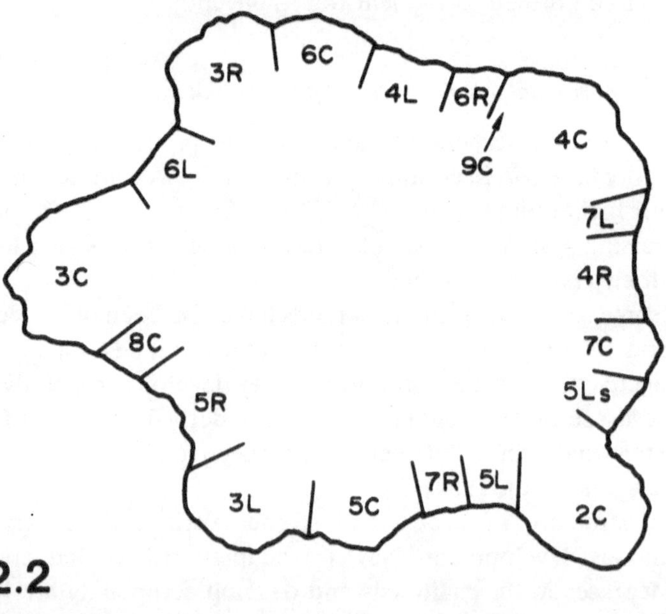


2.2

and more differentiated at their points of divergence than at their acropetal fronts. For example, in *Linum* (Esau, 1943a), *Lupinus* (O'Neill, 1961), and *Populus* (Larson, 1975a), protophloem has been observed maturing acropetally in the lower levels of procambial strands while the upper levels were still developing in residual meristem (Figs. 2.4–2.6). Procambial strands enlarge by cell division within the strand and by acquisition of new cells from the residual meristem.

When followed acropetally through serial sections, the continuum of procambial strands fades out within the residual meristem. This does not imply that the procambial strands terminate but that they can no longer be anatomically distinguished from cells of the residual meristem. The subapical region in which the procambial strands develop was previously designated the 'meristematic ring', a term that generated considerable controversy. Kostytschew (1924), for example, believed

Fig. 2.3. Diagrammatic representation of 2/5 vascular phyllotaxis. The vascular system has been displayed as if it were unrolled from the stem and laid flat. ×, central traces; closed triangles, right traces; open triangles, left traces. The horizontal arrows indicate the position of Fig. 2.1 in the phyllotactic array. Each of the three traces serving a leaf arises from a different sympodium. C-traces are the perpetuating members of the phyllotactic system; e.g. 6C is the parent trace for 11C and also for 8R and 9L. The leaves at nodes 6 and 1 are antecedent leaves for the primordium at node 11. Note that this is an open system. Subsidiary bundles that develop basipetally in each leaf trace unite the system by a series of bundle splits or phloem bridges.

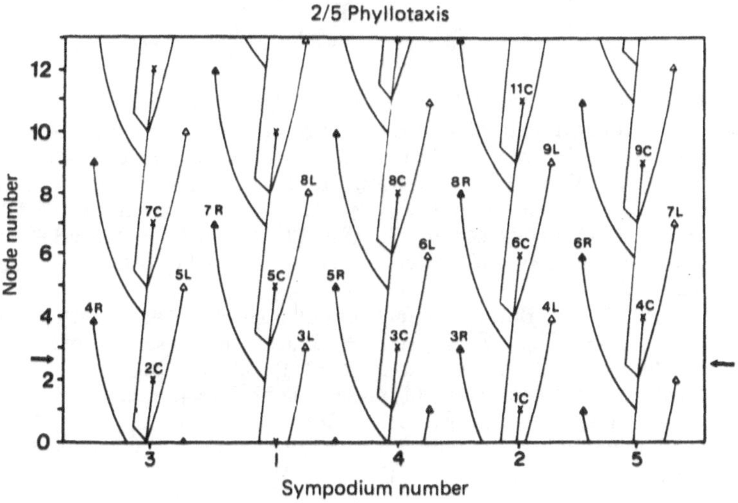

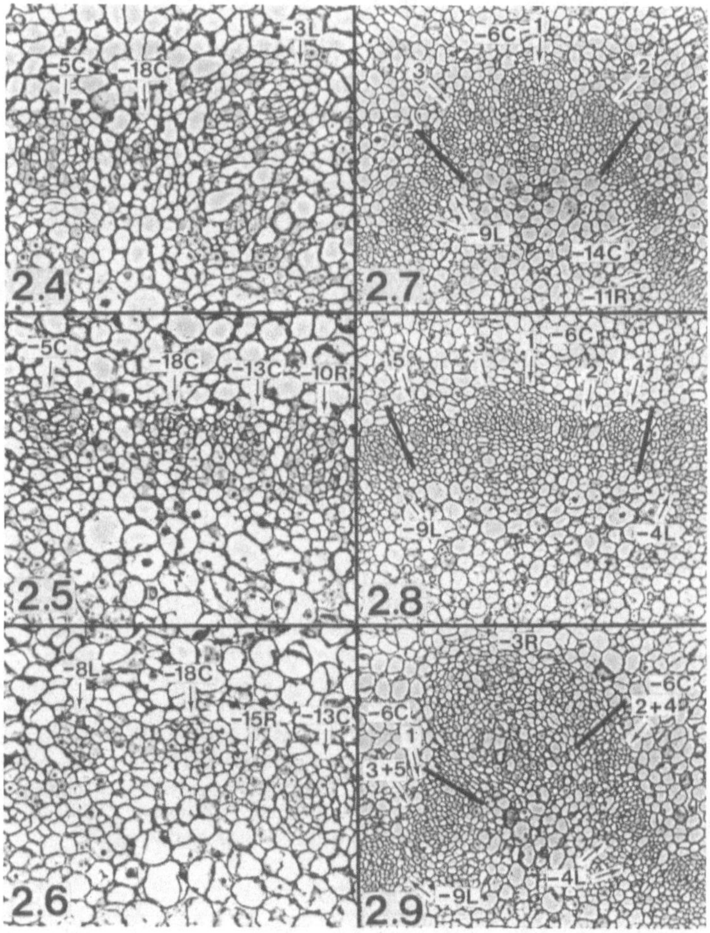


Figs. 2.4–2.6. Acropetal course of development of procambial strand −18C from origin to fade-out. Levels refer to Fig. 2.12. Fig. 2.4, 1040 μm level. Strand −18C immediately after divergence from parent trace −5C. Trace −3L is separating from vascular cylinder before exit. Fig. 2.5, 686 μm level. Strand −18C lying between parent trace −5C and −13C following exit of −3L. Fig. 2.6, 262 μm level. Strand −18C just before fade-out in residual meristem. It now lies between −8L and −15R. 204×.

Figs. 2.7–2.9. Basipetal course of development of subsidiary bundles in trace −6C. Fig. 2.7, 376 μm level. Trace −6C consists of three bundles as it exits the vascular cylinder; the original procambial bundle that developed acropetally (1) and two subsidiary bundles that differentiated basipetally (2 & 3). 128×. Fig. 2.8, 492 μm level. At level of maximum development, subsidiary bundles 2 & 3 have divided to form bundles 4 & 5. 139×. Fig. 2.9, 918 μm level. The outermost that the meristematic ring was the source of all procambium. Helm (1931), to the contrary, argued that what appeared to be a ring was a histological artifact and that it was actually an aggregation of individual procambial bundles in the process of differentiation. Analysis of thin (1–2 μm) serial sections usually supports the latter interpretation. What appears to be a meristematic ring in conventional microscopic preparations is in reality the front of acropetally developing procambial strands.

This view is not shared by all investigators, some of whom believe that procambial development may be discontinuous. For example, McArthur & Steeves (1972) observed a ring of 'provascular' tissue immediately beneath the apex of *Geum chiloense* in which bidirectionally differentiating procambial strands were presumably initiated. Schnettker (1976) described a similar process of bidirectional differentiation from an 'Urmeristeme' in *Clematis vitalba*. Although it is conceivable that some species may possess the equivalent of a ring meristem, an acropetally discontinuous procambium has seldom been observed when serial sections have been critically analysed.

### *Organization of the procambial system*

The formation of a highly ramified procambial system does not arise by chance nor does it occur randomly. On the contrary, it develops according to a precise phyllotaxis conforming to that of the leaves the procambial traces will eventually serve. Most vascular plants progress through one to several phyllotactic transitions before attaining a mature state (Allard, 1942; Pulawska, 1965; Heimans, 1978; Larson, 1980a). In dicotyledons, the first primary leaf (or leaf pair) usually occurs midway between the oppositely situated cotyledonary pair and in monocotyledons opposite the single cotyledon. Subsequent leaves are situated either singly, or in pairs, or in whorls depending on the phyllotaxis characteristic of the plant. Because every leaf must be vascularized, the procambial system also conforms to the phyllotactic pattern and to the changes that occur in that pattern during ontogeny. The association between leaf and vascular phyllotaxis is most easily observed during development of the procambium and the earliest stages of primary vascular differentiation. During later stages, subsidiary bundles and

Caption to Figs 2.7–2.9 (*cont.*)
basipetal bundles of trace −6C (2+4) are split off from the parent trace (1 & 3+5) when they encounter exiting trace −3R three nodes below. 128×. (From Larson, 1975a.)

other changes in vascular development can completely mask the phyllotactic pattern of the procambium. Observations based on cleared preparations of leaves or stem segments must be interpreted with particular care, because these require differentiated vascular elements.

*P. deltoides* is one of a few plants in which organization of the procambial system has been investigated in detail from the seedling to an advanced state (Larson 1975a, 1977, 1979, 1980a). In a newly germinated seedling, two embryonic bundles bifurcate to serve the two cotyledons. These cotyledonary traces conform to a 1/2 phyllotaxis. Helical phyllotaxis begins with the primary leaf traces, arranged in a 1/3 phyllotaxis, followed by the first secondary leaves in a 2/5 phyllotaxis. As new leaves are produced and the plant increases in size and age, the procambial system usually progresses to a 3/8 and finally a 5/13 phyllotaxis, which is presumably the stable phyllotactic order for this species. These transitions occur at relatively consistent stages of plant development, suggesting that they are somehow programmed in the plant's development.

### *Procambial template system*

The acropetal procambial strands block out the prospective vasculature in the residual meristem, and the organized procambial system thus formed serves as a template for further vascularization. In most higher plants, protophloem is the first vascular tissue to differentiate (Esau, 1954). It differentiates acropetally in the phloic procambium of each leaf trace in continuity with more mature protophloem below. Protoxylem is usually initiated at a site within or near the primordium base. From this site, protoxylem differentiates acropetally in the node and basipetally in the xylary procambium of each leaf trace; its differentiation is therefore discontinuous.

While the protophloem is differentiating acropetally a cambium-like zone of cells begins to form deep within the trace internal to the protophloem (Fig. 2.10). It, too, differentiates acropetally in the leaf trace in continuity with more mature cambium-like cells below. Thus, each procambial trace serves as a template for further vascularization. Protophloem, protoxylem, and the cambium-like cells all differentiate from procambial initials, but they do so at different times and at different locations within the procambial trace.

In *P. deltoides*, the cambium-like cells have been referred to as metacambium because the daughter cells differentiate as metaphloem and metaxylem (Larson, 1976). The metacambium is followed by the secondary cambium which also differentiates acropetally and continuously; its daughter cells differentiate as secondary phloem and xylem (Larson, 1980b). Thus, the developmental continuum, procambium→metacambium→cambium, serves as the template for all subsequent vascularization. Although modified in several ways, this pattern is typical of that found in most higher plants. As Esau (1943b) has pointed out, the procambium is related to primary tissues as cambium is related to secondary ones.

Thus, every leaf trace progresses through the same sequence of events during its development and maturation. Both the trace and the leaf it serves depend on products derived either from antecedent leaves below or from stored reserves for much of their development. Each leaf via its traces is therefore first an importer and then an exporter in the developmental sequence. These import and export functions are closely correlated with the structure of each leaf trace as it proceeds through the procambium→cambium continuum.

### *Basipetal subsidiary trace bundles and procambial integration*

Vascular systems have been described as either open or closed, depending on whether adjacent sympodia of bundles appear to be independent or united (Dormer, 1972). Although small herbaceous plants with simple vasculature may conceivably possess an open system, it is difficult to visualize such a system satisfying the functional requirements of large plants with many leaves. The accuracy of many vascular diagrams in the literature must be questioned because they were reconstructed from either tissue clearings or short segments of stem.

An example of an open system is illustrated in Fig. 2.3. The only interconnections between adjacent sympodia are the three traces serving each leaf. Obviously, such a system cannot properly fulfil the functional requirements of a *Populus* plant. Further integration of this system is accomplished by subsidiary bundles that differentiate basipetally in each leaf trace. These bundles develop according to rather precise rules as follows.

The procambial strands that develop acropetally – three per leaf – do so as individual strands referred to as original procambial strands or bundles (Fig. 2.10). Approximately three to four plastochrons after primordium establishment, new procambial bundles differentiate basipetally from the primordium base. These subsidiary bundles differentiate in the interfascicular residual meristem with the original procambial bundle of each trace serving as a template for their descent. A typical leaf trace will thereafter consist of at least three bundles, the original procambial strand that developed acropetally and two subsidiary bundles that differentiated basipetally (Fig. 2.7). The subsidiary bundles may, however, subdivide further depending on the vigour of the leaf or the plant (Fig. 2.8). C-traces serving leaves of vigorous *P. deltoides* plants may contribute six to eight subsidiary bundles to the vascular cylinder of the subtending internode but lateral traces seldom contribute more than four to six bundles.

Integration of the sympodial system is by bundle splits. During descent, the basipetally developing subsidiary bundles encounter exiting traces and one, or sometimes two, of these bundles are split from their parent trace (Fig. 2.9). For example, in the stem region of *P. deltoides* conforming to 5/13 phyllotaxis, bundles of C-traces are always split by exiting R-traces three nodes below: bundles of L-traces are always split by exiting C-traces three nodes below (Fig. 2.11). The

Fig. 2.10. Sequence of cambial and vascular development in sympodial series of leaf traces 1C→16C. The cambial continuum develops upward with the procambium blocking out pattern and serving as a template for all subsequent vascularization. The phloem continuum differentiates acropetally whereas the xylem continuum differentiates basipetally. The sequence (*a*)–(*d*) presents an approximation of the time and position of each vascular event (from Larson, 1980a).

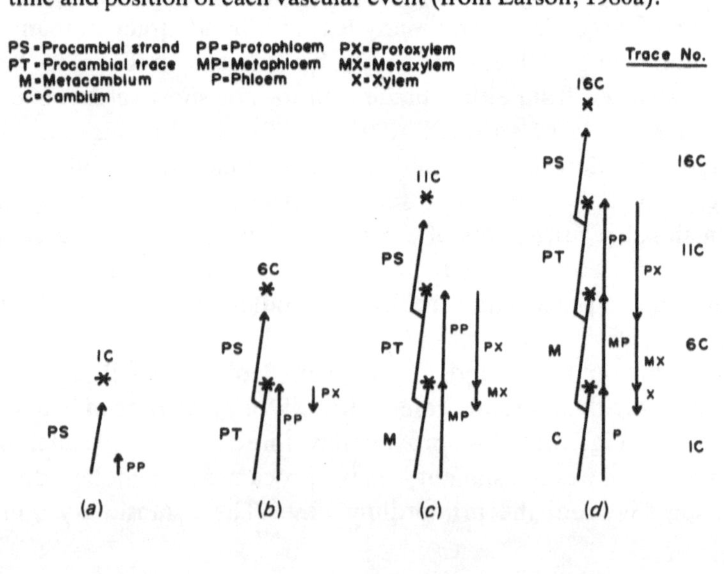

basipetal bundles and their bundle splits unite the otherwise open procambial system into a fully integrated and functional whole.

Basipetally developing subsidiary bundles merge with the original procambial bundles of their respective traces several internodes below their nodes of origin. In a similar manner, bundle splits merge with bundles of the traces that split them. Bundles also occasionally merge with subsidiary bundles of adjacent traces. In either case a new sequence of trace development begins following the initiation of subsidiary bundles. For example, a procambial trace is larger at its base than at its terminus when it first unites with a new primordium. But the initiation and development of subsidiary bundles reverse this process, and thereafter the trace is largest at the primordium base and it tapers downward.

Subsidiary bundles are not unique to *Populus*. In *Carya*, Foster (1935) observed procambial strands that differentiated bidirectionally

Fig. 2.11. Sequence of events during development of basipetal subsidiary bundles and bundle splits in *Populus deltoides* plants with 5/13 vascular phyllotaxis. (*a*) An original procambial C-strand develops acropetally toward the site of its primordium, ×. (*b*) About three to four plastochrons after primordium establishment subsidiary bundles differentiate basipetally from the primordium base using the original procambial bundle (that developed acropetally) as a template in their descent. (*c*) During their descent, one or two of the outermost subsidiary bundles are split from the main C-trace when they encounter an exiting right trace (solid triangle) three nodes below. (*d*) Subsidiary bundles are produced in all three traces serving a leaf, and bundle splits occur systematically to integrate the entire vascular system. For example, in 5/13 phyllotaxis, subsidiary bundles from a C-trace are always split by an exiting R-trace three nodes below, and those from an L-trace (open triangle) are always split by an exiting C-trace three nodes below. R-traces are never split, and L-traces never split another trace.

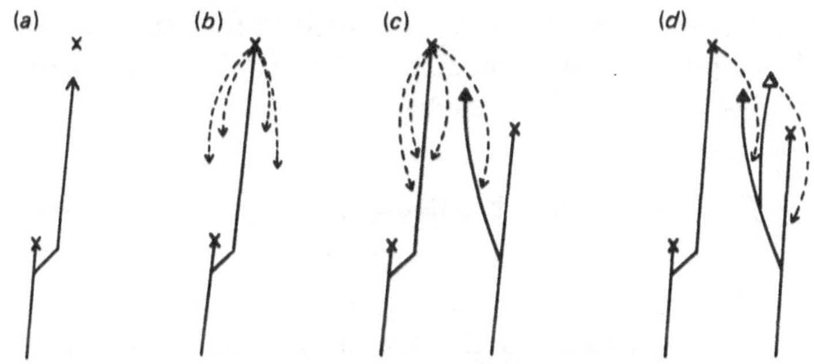

from the bases of primordia that had attained heights of 250 to 300 $\mu$m. Alexandrov & Alexandrova (1929) described bundle splits in the primary vasculature of *Helianthus* very similar to those of *Populus*. *Helianthus* leaves are also served by three traces. As each trace enters the cylinder it splits a bundle from a leaf situated three nodes higher in the stem. The bundle splits then merge with the trace that split them. These authors also stress the fact that traces are larger at the leaf base than four internodes down the stem because of the mergers of trace bundles.

Reference is seldom made to subsidiary bundles or their equivalents in botanical literature, perhaps because few sufficiently detailed anatomical studies have been made of the primary vascular system. However, many references can be found to various categories of accessory bundles (Devadas & Beck, 1971), acropetal blind bundles (Crooks, 1933), and basipetal blind bundles (Dormer, 1954). Unrecognized basipetal bundles may also account for the numerous reports that established primordia or leaves are essential for the development of their traces (Young, 1954; Wardlaw, 1965). Most botanical texts contain photomicrographs and camera lucida drawings of transverse stem sections showing obvious subsidiary bundles in traces leaving the vascular cylinder. Similar transverse sections of primary vascular cylinders show far more discrete bundles than are required to serve individual leaves. These excess bundles are probably subsidiary bundles.

A basipetal trace system analogous to that of the subsidiary trace bundles in dicotyledons may also exist in monocotyledons. Zimmermann & Tomlinson (1972) described how the vasculature of palms is organized in an inner and outer system of bundles. In describing a similar vascular system in *Andropogon*, Maze (1977) suggested that the inner systems developed acropetally but the outer system of cortical bundles developed basipetally. The interrelations of the inner and outer bundle systems of monocotyledons have recently been clarified by Bell (1980). However, further work is required on the comparative anatomy and development of the acropetal and basipetal bundle systems of both monocotyledons and dicotyledons.

### Procambial strand hypothesis

In previous sections, it was deliberately implied that procambial strands develop acropetally in advance of the primordia they are destined to serve. Although documented reports of precocious procambial strands are not numerous, these relate to species across a broad range of taxonomic groups. For example, precocious procambial strands have been observed in angiosperms (White, 1955) and gymnosperms (Crafts, 1943; Gunckel & Wetmore, 1946), in dicotyledons (Reeve, 1942) and monocotyledons (Maze, 1977), in species with helical (Larson, 1975a) and decussate (Shushan & Johnson, 1955) phyllotaxis, and in species with multilacunar (Priestley, Scott & Gillett, 1935) and unilacunar nodes (Girolami, 1953, 1954). Although most reports are of only one to several precocious procambial strands, as many as six or more have been observed in certain species such as *Pseudotsuga taxifolia* (Sterling, 1947) and *P. deltoides* (Larson, 1975a). The acropetal and continuous development of the procambial system, and the presence of precocious strands in particular, have suggested to some anatomists that the strands might be involved in the initiation of new primordia at the shoot apex. The most intriguing question is: does the procambium perform a regulatory role in primordia initiation in addition to demarcating the prospective vascular system?

Many hypotheses have been proposed by both theorists and experimentalists to account for the origin of new primordia, but most of these hypotheses concern events within the shoot apex independent of the procambial system (Wardlaw, 1965; Steeves & Sussex, 1972; Halperin, 1978). Bolle (1939) proposed what he considered a new theory of phyllotaxis based on earlier observations of Nägeli and Szabo (cited in Bolle, 1939). Bolle postulated that plant vascular systems develop according to certain phyllotactic laws. For example, the vasculature of a dicotyledon with helical phyllotaxis would begin with the two traces serving the cotyledons; it would then increase to higher phyllotactic orders by repeated bifurcations. According to Bolle, the phyllotactic organization of the vasculature would be essentially unaltered by and independent of contact parastichies, divergence angles, and the wanderings of foliar orthostichies. To account for the siting of new primordia, Bolle postulated an 'Antriebe', implying that a stimulus propagated acropetally by the advancing vascular bundles might be responsible for the initiation of primordia.

Bolle's work has been overlooked in the literature, possibly because of its inaccessibility, although Jean (1978) has considered it in his discussions of hierarchical systems. The procambial strand hypothesis has been neither pursued vigorously nor subjected to rigorous experimental study. Consequently, it has not been seriously considered by either plant morphologists or theorists, and it has been seldom mentioned as an alternative hypothesis in plant morphology texts. However, plant anatomists have continued to provide supporting evidence and the procambial strand hypothesis therefore remains a viable option.

### *Evidence from* Populus deltoides

Recent work on *P. deltoides* lends support to the procambial strand hypothesis (Larson, 1975a). Each *P. deltoides* leaf is served by three traces, a central (C) trace and right (R) and left (L) lateral traces (Fig. 2.3). The leaf traces of older plants are organized in a sympodial system with C-traces comprising the 13 sympodia in a 5/13 phyllotaxy and with R and L traces systematically diverging from each C-trace in a helicoid configuration. The sympodia of *P. deltoides* are not cauline indefinitely traversing the stem as visualized by Ezelarab & Dormer (1963). Rather, each C-trace terminates in a primordium and a sympodium must therefore be regenerated by the initiation of a new C-trace divergence. Thus, the divergence of a new procambial strand initiates a new histogenic event. The fact that sympodia are finite is important to the procambial strand hypothesis, because it provides for a procambial system that can both progress and regress to higher and lower phyllotactic orders, respectively. Phyllotactic regressions have been observed in *Nicotiana* (Allard, 1942) and *Betula* (Heimans, 1978).

Several features observed in development and organization of the procambial system of *P. deltoides* support the procambial strand hypothesis. First, C-traces are the perpetuating members of the procambial system. In one carefully analysed terminal bud, six precocious procambial C-strands, −16C through −21C, were observed in the residual meristem (Fig. 2.12). Each C-strand arose in its turn in the ontogenetic sequence, and advanced acropetally according both to the state of development of and to the time of divergence from its parent trace. For example, protophloem, like the procambium, differentiates acropetally and continuously from existing protophloem, and a new C-strand does not diverge until differentiated protophloem extends throughout the parent trace. Thus, the parent trace must apparently attain a certain functional state of development before giving rise to a new C-strand. As to time of divergence, although several C-strands were simultaneously advancing toward the apex, the one destined to serve the next primordium to arise (−16C) was larger and better developed than younger ones because it diverged earlier at a lower level in the procambial system. Protophloem had differentiated acropetally for a considerable distance in the oldest C-strands whereas the youngest one remained procambial (cf. −15C and −21C). These events suggest that advance and perpetuation of the procambial system require products from older leaves, the antecedent leaves, situated at lower levels in the phyllotactic array. They also suggest that divergence of new C-strands and their acropetal advance are somehow precisely controlled, or programmed, within the vascular system.

Secondly, the C-strand always preceded the lateral strands serving a new primordium both in time of divergence and in subsequent development. Although both the right and left lateral strands diverged from their parent traces several plastochrons in advance of the primordium they served, they did not unite with the primordium until after its initiation. For example, −13R diverged from parent trace −11C and −14L diverged from −13R. Trace −13R had already united with its primordium, but −14L was still progressing acropetally in residual meristem toward the site of its primordium which was already present on the apex. Nonetheless, both lateral traces diverged at the proper

Fig. 2.12. Diagrammatic display of *Populus deltoides* vascular system with 5/13 dextrorse phyllotaxis based on quantitative data obtained from serially sectioned bud. Each point at which a strand or trace diverges from a parent trace and a trace departs from the vascular cylinder is plotted logarithmically relative to its distance below the apical summit and to Leaf Plastochron Index (see Chapter 6). (From Larson, 1975a.)

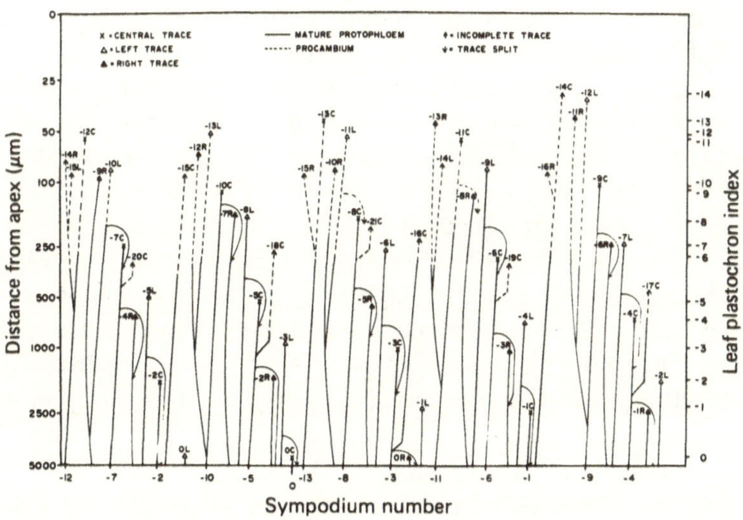

time and correct position in the phyllotactic array to serve the new primordium when it became established.

Each of the three traces serving a new primordium also arose on a different sympodium and, consequently, from a different parent trace served by a different series of antecedent leaves (e.g. −13L from −10C, −13C from OC, −13L from −11C). These facts again suggest a precisely programmed procambial system that advances acropetally and continuously under the regulatory influence of a more mature system below. Such precise synchrony is difficult to account for if the primordium is initiated under the influence of the apex independent of the vascular traces destined to serve it.

As a third line of evidence, it can be demonstrated that both C-strands and lateral strands are guided in their development to the apex. When new C-strands diverge from parent traces they advance acropetally in uncommitted interfascicular tissue of the residual meristem. The new C-strands are then guided acropetally first by older adjacent traces serving newly emergent primordia and finally by later-arising lateral strands. For example, strand −16C was bounded first by −1L and −3C, then by −6L and −13R, and finally by −11L and −13R, traces that were all further advanced and that differentiated before −16C. Strand −16C can therefore terminate at only one position on the apex and its terminus is dictated by its position in the array of leaf traces. Thus, the phyllotactic system assures that the oldest C-strand will arise at the apex at the precise time and in the correct position to serve the next primordium to arise. Moreover, because of its position in the phyllotactic array, the new C-strand will advance, quite naturally, immediately beneath the first available space on the apex.

The leaf trace system requires that one trace or strand cannot cross another. If they were to cross, internal phloem would be produced, a condition that does not occur in *Populus*. Thus, C-strands must either ascend between or be guided in their ascent by adjacent traces. During their ascent, new C-strands maintain their individuality and they do not merge with adjacent traces. Sachs (1969) has demonstrated that vascular bundles emanating from sources high in auxin tend to repel one another while those from sources relatively low in auxin tend to be attracted and to merge. A similar situation may occur at the advancing front of the procambial system, although the source may not necessarily produce auxin. C-strands advancing under the influence of vigorous antecedent leaves do so in a strictly polar manner as discrete bundles. In contrast, basipetal trace bundles, originating from new primordia that are presumably weak sources, anastomose with bundles from other traces several nodes below their points of origin.

Finally, as noted earlier, the procambial system of *P. deltoides* progresses through a series of phyllotactic transitions before attaining the stable 5/13 order. Because of the way in which the procambial system is propagated acropetally, each of these phyllotactic transitions must occur during procambial development. It is extremely unlikely that phyllotactic transitions arise independently both in the procambial system and in new primordia and then join. If they do, we must account for some well-coordinated and highly synchronized regulatory process that can function between the acropetally developing vascular system and the apex.

Higher orders of phyllotaxis necessitate a larger apex to accommodate more primordia. This problem has been approached theoretically by assuming a change in the packing order of primordia (Erickson, 1973). However, a phyllotactic transition occurring in the procambial system should also result in an increase in apical size because more procambial strands are progressing toward their prospective primordial sites (Larson, 1980a). Fosket (1968) provided evidence that factors outside the bud of *Catalpa* were responsible for changes within the apex, resulting in a shift in phyllotaxis.

### *Uncertainties regarding the hypothesis*

#### *Recognition of procambial front*

The procambial strand hypothesis has never been systematically investigated. Consequently, supporting evidence has been obtained indirectly, primarily from anatomical studies of vascular development. Many questions therefore arise that not only disclose uncertainties regarding the procambial strand hypothesis but also reveal our inadequate knowledge of how the procambial system develops and functions. Several of the more important questions will be considered.

For example, if procambial strands are involved in the siting of new primordia, why aren't procambial cells evident immediately below the prospective primordial site? Perhaps they are, but it is beyond our present technical abilities either to detect or to recognize them. Few details are known of procambial ultrastructure (Catesson, 1974), and no distinguishing characteristics have been observed in advance of morphological differentiation even in culture systems (Phillips, 1980). Moreover, at the light microscope level, it is seldom possible to recognize anatomically a single procambial cell in either the transverse or longitudinal plane. Groups or islands of procambial cells can be recognized by their dense staining and small size relative to adjacent cells of the residual meristem in the transverse plane and by their elongated appearance in the longitudinal plane. However, elongation of procambial cells is a secondary event. Procambial cells do not elongate by intrusive tip growth but by accommodative symplastic growth in response to expansion of pith or ground tissue cells during early internodal elongation. Because internodal elongation occurs in the subapical region of the shoot (Loy, 1977), there is little opportunity for procambial cells to elongate immediately beneath the prospective primordial site.

The criteria by which we recognize procambial cells imply anatomical differentiation. Nonetheless, biochemical events must precede anatomical differentiation and these events should be detectable in advance of the recognizable procambial front. Few attempts have been made to biochemically characterize procambium, particularly in the shoot apex. Vanden Born (1963) demonstrated enzyme activity associated with young vascular strands in *Picea* apices. Acid phosphatase activity was localized in a broad band extending from the apex to the subapical tissue with the highest concentration occurring in provascular tissue and at the bases of new needle primordia. High levels of esterase activity were found only in young vascular tissues. Evans & Berg (1972) also obtained evidence that specific enzymes were associated with differentiating vascular tissue in apices of *Triticum*. Although the results of these histochemical studies are highly suggestive, they do not provide definitive evidence that the biochemical events actually preceded anatomical differentiation. McLean & Gahan (1970), however, detected increased levels of enzyme activity that reflected distinctive changes in cellular metabolism preceding morphological differentiation in roots of *Vicia*. Procambial cells destined to become protophloem sieve elements were characterized by high levels of $\beta$-glycerophosphatase, acid naphthol AS-BI phosphatase, and indoxyl esterase. More recently, Khavkin, Markov & Misharin (1980) reported evidence for proteins specific to differentiating vascular elements in *Zea* based on immunodiffusion tests.

There is little doubt that biochemical differentiation in cells will eventually be detected in advance of morphological differentiation. Although such information may help us recognize procambial strands immediately beneath a prospective primordial site, it will not provide decisive evidence that the procambial strands induce primordia. This problem of biochemical differentiation is not unique to the procambial strand hypothesis. All of the apical hypotheses are beset with equally formidable tasks of identifying the fields of inhibition (Veen & Lindenmayer, 1977) and morphogen gradients (Thornley, 1975) essential to the hypothetical models.

Similarly, studies of cell division have not been particularly helpful in revealing patterns of procambial differentiation in the shoot apex. Lyndon (1976) concluded that data from several studies were consistent in showing that division rates of summit cells were approximately one-half that of peripheral cells. Although a plate of rapidly dividing cells often extends down the sides of the apex to link with procambium, the procambium is not necessarily distinguishable from adjacent tissues. Shininger (1979) also emphasized the absence of good cytological and biochemical evidence for identifying procambium.

Our inability to recognize the procambial front either cytologically or biochemically presents problems for all hypotheses of primordia initiation. Whether primordia are initiated independently by events in the apical region or induced by advancing procambial strands, they must be served eventually by an organized procambial system.

How prevalent are precocious procambial strands? Carefully conducted analytical studies covering a broad range of species are required to answer this question. Does the procambium simply keep pace with growth of the advancing apex as suggested by Clowes (1961), or do procambial strands actually advance within the interfascicular residual meristem? The latter process implies some degree of differentiation. Can the transition from one meristematic condition (residual meristem) to another (procambium) be considered a form of differentiation as suggested by Larson (1975a), or should the procambium be considered as ordinary meristematic cells (Shininger, 1979)? Obviously our knowledge of procambial development in the subapical region is extremely limited. Sophisticated research from interacting disciplines is required to clarify the developmental events in both the subapical and the peripheral regions of the apex.

### *Procambial transport*

If procambial strands are involved in the siting of new primordia, how might they exert their influence? The most logical interpretation would be that they provide a transport pathway. Although the procambial strands may not exert a direct morphogenetic influence, they may transport the essential nutrients and hormones that do. The procambial strands differentiate in a polar gradient; in their ascent they are guided to the apex in such a manner that each new strand arises in its turn in the first available space on the apex (Larson, 1975a). This region is presumably competent to initiate a primordium once it is suitably induced by a stimulus delivered by its respective procambial strand. At this time, we do not have good evidence that procambial cells can translocate, although indications are that they do (Greenwood & Goldsmith, 1970; Fry & Wangermann, 1976). The procambium presents the paradoxical situation of cells translocating to other cells while they are themselves differentiating. An analogous situation occurs in intercalary meristems and in differentiating cambia.

If the existence of precocious procambial strands is denied or if the differentiating procambial strands are considered incapable of serving as transport pathways, then it becomes difficult to account for nutrition of the apex. The extent of this problem can be illustrated by analysis of the procambial system of leaf traces. For example, in a 5/13 phyllotactic system a daughter C-trace traverses at least 13 to 15 internodes from its point of divergence on a parent C-trace to its departure in a leaf primordium (Fig. 2.12); (see also Sterling, 1947 and Pulawska, 1965). If precocious procambial strands do not exist, then the initiation of each new primordium would require either that a procambial trace differentiates through 13 to 15 internodes in one plastochron or that the procambial strands differentiate basipetally from the primordium site. The former situation seems highly improbable and the basipetal procambial strands necessary for the latter possibility have never been observed or verified. Alternatively, acropetally developing procambial strands have been observed in many species as noted previously. For a 5/13 phyllotactic system, however, many precocious procambial strands must be present and suitably differentiated to serve the prospective primordia. The number of such strands would necessarily vary with the order of phyllotaxy, a 2/5 order requiring fewer strands than a 5/13. Consequently, when one considers the distance that a procambial strand must traverse, particularly in a high-order phyllotactic system, then it appears essential that the strand diverge well in advance of the primordium it will eventually serve. Moreover, if procambial strands diverge in advance of the primordium, then the phyllotactic transitions that occur during ontogeny must also be initiated in the procambial system independent of the apex.

Other nutritional problems arise if precocious procambial strands exist but do not translocate. It is conceivable that preferred diffusion pathways could operate by mechanisms proposed by Meinhardt (1976, 1978). However, it is questionable if oriented diffusion pathways could function for more than a few cell diameters (Mitchison, 1980). Sussex (1963) has demonstrated that cytoplasmic connections are necessary for nutrition of the apex, but he did not determine whether they were procambial. Nonetheless, diffusion pathways either within or leading to the apical region must at some point be provided by a source of vascular tissue (Brown, 1976, Nicholls, 1978). Procambial strands are logically situated to serve this function.

Analysis of the procambial system in the subapical region shows that it consists of divergent traces and convergent strands (Fig. 2.13). The divergent procambial traces serve the newly differentiated and enlarging primordia on the apex. These traces are continuous with more mature vascular tissues below. Although these traces are entirely procambial in the upper levels, protophloem is differentiating acropetally in the lower levels (see Fig. 2.12). The convergent strands represent precocious procambial strands differentiating acropetally toward their prospective primordial sites; i.e. those strands that have faded out in the so-called meristematic ring. Each of these strands will become a divergent procambial trace when the primordium it serves is initiated.

Fig. 2.13. The procambial system consists both of divergent procambial traces (large arrows) serving established primordia and convergent procambial strands (small arrows) differentiating toward their prospective primordial sites. When a new primordium is initiated, the oldest procambial strand will diverge as a procambial trace and a new convergent procambial strand will differentiate acropetally to replace it.

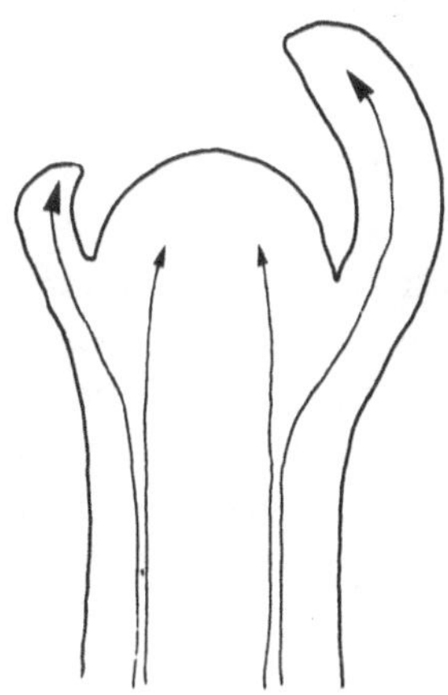

The divergent traces serving established primordia are presumably translocating even though protophloem has not yet differentiated throughout the trace. However, it is highly questionable if these traces would divert assimilates either to the apical summit or to other prospective primordial sites. Each divergent C-trace is a polarly oriented, discrete tissue system serving a specific primordium developing as a concentrated metabolic sink. It is therefore highly unlikely that this transport system would release metabolites to serve other less demanding sinks in the apex.

The convergent procambial strands, on the contrary, would be ideally positioned to deliver nutrients to the apical summit as well as to the prospective primordial sites. Not only are these strands in continuity with parent traces below, but the more advanced strands possess well-developed protophloem throughout their lower levels, providing access to assimilates produced by exporting leaves below. According to the procambial strand hypothesis, the principal metabolic sink for each of these advancing C-strands would be its prospective primordial site. Before primordium establishment, an advancing procambial strand may possess no specific metabolic sink other than its own terminus. However, because of its position in the phyllotactic array, the most advanced procambial strand and its associated antecedent leaf(s) will be at different stages of development than less-developed procambial strands and their antecedent leaf(s). Consequently, metabolites translocated by the most advanced procambial strand might also be expected to differ both quantitatively and qualitatively from those translocated by less-developed procambial strands. It is suggested that these quantitative and qualitative changes in metabolites delivered to the apex, together with the precise orientation of each procambial strand beneath its prospective primordial site, may be responsible for the induction of new leaf primordia.

Protophloem cannot fulfil the functional requirements of the apical and subapical regions of the shoot without the participation of the procambial strands. Protophloem maturation in procambial strands usually terminates 100–300 $\mu$m beneath the apex (Esau, 1943a; Pulawska, 1965; Larson, 1975a). It does not extend into the developing primordium until several to many plastochrons after its establishment depending on the order of phyllotaxis. In the convergent procambial strands, protophloem either has not differentiated or differentiation extends acropetally for only a short distance.

*Source of leaf-forming stimulus*

If the procambial strands do indeed promote primordia initiation by serving as a transport pathway, then the regulatory influence must arise elsewhere in the plant. The procambial system develops by a series of events with each new event requiring an essential prior event. Prior events presumably occur both in the parent leaf traces and in the leaves that they serve. For example, a new procambial strand diverges from a parent trace serving an older primordium, which in turn diverges from another parent trace serving a still older antecedent leaf below. Not only must the procambial strand be sufficiently developed to transport the inductive stimulus, but the antecedent leaf (or leaves) must also attain critical developmental states.

Each antecedent leaf progresses through a developmental sequence related to its position in the phyllotactic array. Examination of this developmental sequence both morphologically and physiologically suggests that events occurring during physiological maturation of the antecedent leaves may influence events occurring during development of their daughter primordia and the procambial traces that unite them (Larson, 1975b).

The leaf-forming stimulus need not be a specific morphogen. A combination of essential nutrients and hormones exported by the antecedent leaf(s) could conceivably provide the necessary stimulus. Under all conditions conducive to vegetative growth, the leaf-forming stimulus would be transported to the leaf initiation sites via the procambial strands. However, if an antecedent leaf(s) was suitably induced by a specific set of conditions then the stimulus would be altered to a flower-forming stimulus. Again, this stimulus need not be florigen or any other specific substance, but it would differ from the leaf-forming stimulus. Alternatively, if an antecedent leaf(s) was suitably induced by yet another set of conditions, then a dormancy-inducing stimulus would be transported to the apex. Thus, the procambial strand hypothesis can account for the major growth stages exhibited by a plant during ontogeny – vegetative, reproductive or dormancy. The procambial strand hypothesis is obviously a whole-plant hypothesis. And, when viewed in its entirety, the procambial strand hypothesis has the potential for clarifying many events that occur both during primordia initiation and subsequent organ development.

## Conclusion

Strong arguments can be advanced for the procambial strand hypothesis but counter-arguments can also be advanced for each of the apical hypotheses. Uncertainties exist in all current hypotheses because of the large gaps in our knowledge of events in the apical and subapical regions of the shoot. Such uncertainties have led several investigators to adopt a compromising view that the initiation of primordia and the arrival of procambial strands at the initiation site are either simultaneous events or complementary aspects of a single process (Philipson, 1949; Esau, 1954; Catesson, 1964; Schnettker, 1976). As well as intellectually unsatisfying, such a compromising view cannot at present be factually argued. One can, of course, adopt the more extreme conclusion of Barsch-Gollnau, Ritterbusch & Mohr (1980) that not only is a convincing theory relating geometric and physiological parameters unavailable, but one is not even in sight.

The author would like to thank Dr R. V. Jean for providing him with a copy of the paper by Bolle (1939).

## References

Alexandrov, W. G. & Alexandrova, O. G. (1929). Über die Struktur verschiedener Abschnitte ein und desselben Bündels und den Bau von Bündeln verschiedener Internodien des Sonnenblumenstengels. *Planta*, **8**, 456–86.

Allard, H. A. (1942). Some aspects of the phyllotaxy of tobacco. *Journal of Agricultural Research*, **64**, 49–55.

Barsch-Gollnau, S., Ritterbusch, A. R. & Mohr, H. (1980). Photomorphogenesis and phyllotaxis during vegetative growth in *Sinapis alba* and *Xanthium strumarium. Plant, Cell and Environment*, **3**, 363–70.

Bell, A. (1980). The vascular pattern of a rhizomatous ginger (*Alpina speciosa* L. Zingiberaceae). I. The aerial axis and its development. *Annals of Botany*, **46**, 203–12.

Bolle, F. (1939). Theorie der Blattstellung. In *Botanischer Verein der Provinz Brandenburg*, pp. 153–91. Berlin.

Brown, R. (1976). Significance of division in higher plants. In *Cell Division in Higher Plants*, ed. M. M. Yeoman, pp. 3–46. London: Academic Press.

Catesson, A. M. (1964). Origine, fonctionnement et variations cytologiques saisonnières du cambium de l'*Acer pseudoplatanus* L. *Annales des Sciences Naturelles, Botanique et Biologie Végétale*, Série 12, **5**, 229–498.

Catesson, A. M. (1974). Cambial cells. In *Dynamic Aspects of Plant Ultrastructure*, ed. A. W. Robards, pp. 358–90. New York: McGraw-Hill.

Clowes, F. A. L. (1961). *Apical Meristems*, 217 pp. Oxford: Blackwell Scientific.

Crafts, A. S. (1943). Vascular differentiation in the shoot apex of *Sequoia sempervirens. American Journal of Botany*, **30**, 110–21.

Crooks, D. M. (1933). Histological and regenerative studies on the flax seedling. *Botanical Gazette*, **95**, 209–39.

Cutter, E. G. (1971). *Plant Anatomy: Experiment and Interpretation*, Part 2, *Organs*, 343 pp. London: Edward Arnold.
Devadas, C. & Beck, C. B. (1971). Development and morphology of stelar components in the stems of some members of the Leguminosae and Rosaceae. *American Journal of Botany*, **58,** 432–46.
Dormer, K. J. (1954). The acadian type of vascular system and some of its derivatives. I. Introduction, Menispermaceae and Lardizabalaceae, Berberidaceae. *New Phytologist*, **53,** 301–11.
Dormer, K. J. (1972). *Shoot Organization in Vascular Plants*, 240 pp. London: Chapman & Hall.
Erickson, R. O. (1973). Tubular packing of spheres in biological fine structure. *Science*, **181,** 705–16.
Esau, K. (1943a). Vascular differentiation in the vegetative shoot of *Linum*. II. The first phloem and xylem. *American Journal of Botany*, **30,** 248–55.
Esau, K. (1943b). Origin and development of primary vascular tissues in seed plants. *Botanical Review*, **9,** 125–206.
Esau, K. (1954). Primary vascularization in plants. *Biological Reviews of the Cambridge Philosophical Society*, **29,** 49–86.
Esau, K. (1965). *Vascular Differentiation in Plants*, 160 pp. New York: Holt, Rinehart & Winston.
Evans, L. S. & Berg, A. R. (1972). Qualitative histochemistry in the shoot apex of *Triticum*. *Canadian Journal of Botany*, **50,** 241–4.
Ezelarab, G. E. & Dormer, K. J. (1963). The organization of the primary vascular system in Ranunculaceae. *Annals of Botany*, **27,** 23–38.
Fosket, E. B. (1968). The relation of age and bud break to the determination of phyllotaxy in *Catalpa speciosa*. *American Journal of Botany*, **55,** 894–9.
Foster, A. S. (1935). A histogenetic study of foliar determination in *Carya buckleyi* var. *arkansana*. *American Journal of Botany*, **22,** 88–147.
Fry, S. C. & Wangermann, E. (1976). Polar transport of auxin through embryos. *New Phytologist*, **77,** 313–17.
Girolami, G. (1953). Relation between phyllotaxis and primary vascular organization in *Linum*. *American Journal of Botany*, **40,** 618–25.
Girolami, G. (1954). Leaf histogenesis in *Linum usitatissimum*. *American Journal of Botany*, **41,** 264–73.
Greenwood, M. S. & Goldsmith, M. H. M. (1970). Polar transport and accumulation of indole-3-acetic acid during root regeneration by *Pinus lambertiana* embryos. *Planta*, **95,** 297–313.
Gunckel, J. E. & Wetmore, R. H. (1946). Studies of development in long shoots and short shoots of *Ginkgo biloba* L. I. The origin and pattern of development of the cortex, pith and procambium. *American Journal of Botany*, **33,** 285–95.
Halperin, W. (1978). Organogenesis at the shoot apex. *Annual Review of Plant Physiology*, **29,** 239–62.
Heimans, J. (1978). Problems of phyllotaxis. *Proceedings Koninklijke Nederlandse Akademie von Wetenschappen*, Series C, **81,** 91–8.
Helm, J. (1931). Untersuchungen über die Differenzierung der Sprossscheitelmeristeme von Dikotylen unter besonderer Berücksichtung des Procambiums. *Planta*, **15,** 105–91.
Jean, R. V. (1978). Growth and entropy: Phylogenism in phyllotaxis. *Journal of Theoretical Biology*, **71,** 639–60.
Khavkin, E. E., Markov, E. Y. & Misharin, S. I. (1980). Evidence for proteins specific for vascular elements in intact and cultured tissues and cells of maize. *Planta*, **148,** 116–23.
Kostytschew, S. (1924). Der Bau und das Dickenwachstum der Dicotylenstämme. *Beihefte zum Botanischen Centralblatt*, **40,** 295–350.

Larson, P. R. (1975a). Development and organization of the primary vascular system in *Populus deltoides* according to phyllotaxy. *American Journal of Botany*, **62**, 1084–99.
Larson, P. R. (1975b). The leaf-cambium relation and some prospects for genetic improvement. In *Tree Physiology and Yield Improvement*, ed. M. G. R. Cannell & F. T. Last, pp. 261–82. New York: Academic Press.
Larson, P. R. (1976). Procambium vs. cambium and protoxylem vs. metaxylem in *Populus deltoides* seedlings. *American Journal of Botany*, **63**, 1332–48.
Larson, P. R. (1977). Phyllotactic transitions in the vascular system of *Populus deltoides* Bartr. as determined by $14^{C}$ labeling. *Planta*, **134**, 241–9.
Larson, P. R. (1979). Establishment of the vascular system in seedlings of *Populus deltoides* Bartr. *American Journal of Botany*, **66**, 452–62.
Larson, P. R. (1980a). Interrelations between phyllotaxis, leaf development and the primary–secondary vascular transition in *Populus deltoides*. *Annals of Botany*, **46**, 757–69.
Larson, P. R. (1980b). Control of vascularization by developing leaves. In *Control of Shoot Growth in Trees*, ed. C. H. A. Little, pp. 157–72. Proceedings International Union of Forest Research Organizations, Fredericton, New Brunswick, Canada, July 20–24, 1980.
Loy, J. B. (1977). Hormonal regulation of cell division in the primary elongating meristems of shoots. In *Mechanisms and Control of Cell Division*, ed. T. L. Rost & E. M. Gifford, pp. 92–111. Stroudsburg: Dowden, Hutchinson & Ross.
Lyndon, R. F. (1976). The shoot apex. In *Cell Division in Higher Plants*, ed. M. M. Yeoman, pp. 285–314. New York: Academic Press.
McArthur, I. C. S. & Steeves, T. A. (1972). An experimental study of vascular differentiation in *Geum chiloense* Balbis. *Botanical Gazette*, **133**, 276–87.
McLean, J. & Gahan, P. B. (1970). The distribution of acid phosphatases and esterases in differentiating roots of *Vicia faba*. *Histochemie*, **24**, 41–9.
Maze, J. (1977). The vascular system of the inflorescence axis of *Andropogon gerardii* (Gramineae) and its bearing on concepts of monocotyledon vascular tissue. *American Journal of Botany*, **64**, 504–15.
Meinhardt, H. (1976). Morphogenesis of lines and nets. *Differentiation*, **6**, 117–23.
Meinhardt, H. (1978). Space-dependent cell determination under the control of a morphogen gradient. *Journal of Theoretical Biology*, **74**, 307–21.
Mitchison, G. J. (1980). A model for vein formation in higher plants. *Proceedings of the Royal Society of London*, **B207**, 79–109.
Nicholls, P. B. (1978). Response of barley shoot apices to application of gibberellic acid: initial response pattern. *Australian Journal of Plant Physiology*, **5**, 311–19.
O'Neill, T. B. (1961). Primary vascular organization of *Lupinus* shoot. *Botanical Gazette*, **123**, 1–9.
Philipson, W. R. (1949). The ontogeny of the shoot apex in dicotyledons. *Biological Reviews of the Cambridge Philosophical Society*, **24**, 21–50.
Phillips, R. (1980). Cytodifferentiation. In *Perspectives in Plant Cell and Tissue Culture, Part A., International Review of Cytology* Supplement **11A**, 55–70.
Priestley, J. H., Scott, L. I. & Gillett, E. C. (1935). The development of the shoot in *Alstroemeria* and the unit of shoot growth in monocotyledons. *Annals of Botany*, **49**, 161–79.
Pulawska, Z. (1965). Correlations in the development of the leaves and leaf traces in the shoot of *Actinidia arguta* Planch. *Acta Societatis Botanicorum Poloniae*, **34**, 697–712.
Reeve, R. M. (1942). Structure and growth of the vegetative shoot apex of *Garrya elliptica* Dougl. *American Journal of Botany*, **29**, 697–711.
Richards, F. J. (1951). Phyllotaxis: Its quantitative expression and relation to growth in the apex. *Philosophical Transactions of the Royal Society of London*, **B235**, 509–64.

Sachs, T. (1969). Polarity and the induction of organized vascular tissues. *Annals of Botany*, **33,** 263–75.

Schnettker, M. (1976). Anlage und Differenzierung des Prokambiums im Sprossscheitel von *Clematis vitalba* (Ranunculaceae). *Plant Systematics and Evolution*, **125,** 59–75.

Shininger, T. L. (1979). The control of vascular development. *Annual Review of Plant Physiology*, **30,** 313–37.

Shushan, S. & Johnson, M. A. (1955). The shoot apex and leaf of *Dianthus caryophyllus* L. *Bulletin of the Torrey Botanical Club*, **82,** 262–83.

Steeves, T. A. & Sussex, I. M. (1972). *Patterns in Plant Development*, 302 pp. Englewood Cliffs, New Jersey: Prentice-Hall.

Sterling, C. (1947). Organization of the shoot of *Pseudotsuga taxifolia* (Lamb.) Britt. II. Vascularization. *American Journal of Botany*, **34,** 272–80.

Sussex, I. M. (1963). The permanence of meristems: Developmental organizers or reactors to exogenous stimuli? *Brookhaven Symposium in Biology*, **16**, 1–12.

Thornley, J. H. M. (1975). Phyllotaxis. I. A mechanistic model. *Annals of Botany*, **39,** 491–507.

Vanden Born, W. H. (1963). Histochemical studies of enzyme distribution in the shoot-tips of white spruce. *Canadian Journal of Botany*, **41,** 1509–27.

Veen, A. H. & Lindenmayer, A. (1977). Diffusion mechanism for phyllotaxis. *Plant Physiology*, **60,** 127–39.

Wardlaw, C. W. (1965). The organization of the shoot apex. In *Encyclopedia of Plant Physiology 15*, pp. 996–1076. Berlin: Springer-Verlag.

White, D. J. B. (1955). The architecture of the stem apex and the origin and development of the axillary buds in seedlings of *Acer pseudoplatanus* L. *Annals of Botany*, **19,** 437–49.

Young, B. S. (1954). The effects of leaf primordia on differentiation in the stem. *New Phytologist*, **53,** 445–60.

Zimmermann, M. H. & Tomlinson, P. B. (1972). The vascular system of monocotyledonous stems. *Botanical Gazette*, **133,** 141–55.

# 3

# The geometry of phyllotaxis

R. O. ERICKSON

## Introduction

The remarkable regularity in the arrangement of lateral organs (leaves on a stem, leaf primordia in a growing shoot apex, cone scales on a cone axis, florets or achenes in a composite head, etc.), which most higher plants exhibit, has attracted the interest of many authors. This subject, phyllotaxis, has an ancient and extensive literature which, it seems to me, reached a culmination at the beginning of this century with the publication of Church's (1904) and van Iterson's (1907) monographs. The older work of Bonnet, Schimper and Braun, the Bravais brothers, Hofmeister, Airy, Delpino, Schwendener and others is cited in these treatises. Both Church and van Iterson constructed geometrical models of leaf arrangements, Church largely on the basis of orthogonally intersecting spirals, and van Iterson principally of regularly arranged contiguous circles or other figures. I will review these classical geometrical models, particularly those of van Iterson, since I believe that they can have great utility in empirical studies. They can form a guide for the analysis of descriptive data and experimental results, allowing one to put his results in more objective and general terms than would be possible without a model. This also seems worthwhile since the classical work is old, relatively inaccessible, and poorly known to most botanists.

Following Church's and van Iterson's work, Schoute (1913) proposed a hypothesis that existing leaf primordia secrete a diffusible substance inhibiting the formation of new leaves in their vicinity, and that leaf initiation can only occur at a distance from existing leaves. Schüepp (1923 and later) constructed many models of phyllotactic patterns. Richards (1948, 1951) described phyllotactic models which have much in common with Church's, emphasizing their relation to growth processes in the shoot apex, and proposed the use of the plastochron ratio and a phyllotactic index in describing patterns which are found in plants. The subject of phyllotaxis has also attracted the attention of mathematicians, notably Turing (1959) and Coxeter (1961). The beginnings of experimental investigation of phyllotaxis have been made by several investigators. M. and R. Snow (1935), for example, made diagonal incisions in the shoot apices of *Epilobium hirsutum* plants, which normally exhibit decussate phyllotaxis, and found that the majority of regenerated apices had a spiral leaf arrangement.

Quite recently a number of theoretical considerations of phyllotaxis have appeared – undertaken to account for the regularity and stability of the process of leaf production at the apex, and also for the remarkable fact that the numbers of spirals which can be traced through a phyllotactic pattern are predominantly integers of the Fibonacci sequence. Adler (1974, 1975, 1977) theorizes that contact pressures between leaf primordia lead to a continual readjustment of their positions by maximizing the minimum distances between them. He identifies the conditions under which this 'maximin' principle leads to Fibonacci phyllotaxis, and asserts that this model can be tested in actual plant development. Roberts (1977) has extended Adler's analysis to apparently aberrant patterns, such as semidecussate phyllotaxis. It is interesting that in Williams's (1975) detailed reconstructions of shoot apices in several plants, he has reached the conclusion that the growth and form of young leaves is largely determined by the physical constraint to which they are subjected. He is careful, however, not to identify this constraint with contact pressure, and it is doubtful that his data provide a test of Adler's theory.

Other studies are explicit formulations of Schoute's (1913) theory of a diffusible inhibitor of leaf initiation. In the studies by Hellendoorn & Lindenmayer (1974) and Veen & Lindenmayer (1977), a cylindrical surface was represented with an array of locations in the memory of a computer, at each of which a concentration of the hypothetical inhibitor could be stored. Initial high concentrations were specified at one end of the cylinder (the shoot apex) and at one or more locations to represent leaf primordia. A program was written to compute new concentrations at each location, on the basis of current concentrations, a growth simulation, finite difference approximations of diffusion in the surface, continued high concentrations at the apex and leaf primordia, and decay of the inhibitor. After a number of iterations, when the concentration fell below a threshold value at some location, a new primordium was defined there. This program succeeded in simulating a number of phyllotactic patterns, and the results were related to the dimensions, growth rate, and frequency of leaf initiation in actual shoot apices. Thornley (1975) simulated leaf initiation in a linear model assuming the diffusion in one dimension circumferentially of an inhibitor of leaf initiation, a morphogen, from sources at the apex and existing primordia. Under certain conditions his solutions approximate the Fibonacci angle. Young (1978) has carried out a similar simulation assuming the diffusion of an inhibitor in the surface of a cylindrical model. Mitchison (1977) proposed that an inhibitor mechanism can be formally equivalent to the contact pressure theory of Adler (1974). Assuming steep concentration gradients about primordia, threshold contours play the role of 'contact circles' within which leaf initiation is prohibited. With computer simulations he finds that a range of Fibonacci contacts can be established with good approximation of the Fibonacci angle. Roberts (1978) discusses the relationship between Adler's and Mitchison's models. Richter & Schranner (1978) also present an inhibition theory arguing that the Fibonacci angle can be deduced on a very simple basis from a proportionality in inhibitory strengths of the three youngest existing leaves at the site of initiation of a new primordium, $i_3 : i_2 = i_2 : i_1$. The simplicity of their argument is appealing although they do not make a detailed analysis.

These recent theoretical works have certainly thrown new light on the subject of leaf arrangement. The occurrence of the Fibonacci numbers and of the Fibonacci angle, which is an expression of the golden mean, has often been regarded as a great mystery or as a mathematical curiosity. It is now clear that it has a solid theoretical basis, whether or not the details of a particular theory are correct. However, there are many arbitrary aspects and approximations in each of the theoretical models, and I feel that there is a great need for empirical studies. The contact pressure which Adler hypothesizes has not been demonstrated, and the inhibitor of leaf initiation discussed by other authors is purely hypothetical. One of the difficulties in empirical studies of leaf initiation is in the small size and relative inaccessibility of the shoot apex, enclosed as it usually is by the young leaves of the bud. Usually rather extensive dissection is required to expose it to view, and it is dubious if its further growth is normal. Much of our knowledge of its structure is therefore indirect, based on inferences from sectioned material, or dissections viewed by stereo light microscopy or scanning electron microscopy. It is likely to continue to be so. I believe that a knowledge of the classical geometrical models can be helpful in making inferences about the activity of the shoot apex from such observations, and will illustrate this with some of our work, and the recent studies of Meicenheimer.

## Geometric models

The basic property of a phyllotactic pattern is undoubtedly that the lateral organs (leaves, cone scales, etc.) are arranged in obvious ranks or *parastichies*. There have been three ways of regarding these patterns: as lattices, as networks of intersecting parastichies, and as arrays of figures (e.g. circles) which are in contact with each other. In the first, the leaves or other lateral organs are represented by points, and the pattern represented as a *lattice*, or symmetrical two-dimensional array of points. Interest then centres on the symmetry properties of various lattices, just as in crystallography the symmetries of three-dimensional lattices are studied. The second model, represented by Church (1904, 1920) and Richards (1948, 1951), visualizes leaves as existing at the intersections or in the interstices of two sets of intersecting helices or spirals. Church supposed that these curves intersect at 90° and apparently saw some significance in the resemblance of such sets of *orthogonal parastichies* to the orthogonal lines of equal action in a spiral vortex or in an electromagnetic field. Richards extended Church's constructions to include models in which parastichies intersect at angles other than 90° and in which the divergence departs from the 'ideal' angle. A third model, elaborated by van Iterson (1907), is of a surface on which circles, or other figures, are drawn so as to be tangent to other circles, ranks of such circles constituting *contact parastichies*. These three views are not wholly distinct from each other. The symmetry properties of a lattice are shared by a diagram of intersecting parastichies, or of ranks of circles in proximity to each other. While Church's emphasis was on orthogonal parastichies, he also drew circles or 'quasi-circles' within the four-sided areas delimited by them. Van Iterson's equations are quite explicit in relating the symmetry of point sets to patterns of contiguous figures, and he specifies those sets of contact parastichies which intersect orthogonally.

Theoretical patterns have usually been constructed on the surface of a cylinder, or about a central point in a plane, but other geometries, such as the surface of a cone or of the three-dimensional packing of spheres, are possible and have been studied. Church concerned himself mainly with the plane model in which the parastichies are represented as equiangular spirals, arguing that phyllotaxis is properly studied in the primary growth zone, the shoot apex, where the pattern originates, and that mature patterns, such as those on cylindrical stems, are distorted by secondary growth inequalities. Van Iterson discussed cylindrical and plane models, as well as those on a cone, and the conformal projection of one pattern into another, in an admirably comprehensive way. In this discussion, I will recast van Iterson's equations, and those of Church and Richards, into a somewhat more general and unified form. I will redefine certain parameters so as to be more consistent with familiar morphological notions and to be amenable to certain growth interpretations.

### *Lattices and parastichies on a cylindrical surface*

Some phyllotactic patterns which are found in plants may be regarded as approximately cylindrical, and one may take as a model of such a pattern a symmetrical array of points placed on a circular cylindrical surface. Such an array can be generated from a single point by symmetry operations or isometries (Coxeter, 1961), also called congruences (Weyl, 1952). The most general isometry in this context is the screw displacement, whose components are translation along the axis of the cylinder and rotation about this axis. Van Iterson tabulates six cases of symmetrical lattices, generated by these three isometries, and proposes that for discussion of phyllotaxis, they be grouped into two classes, *simple* lattices, in which a horizontal circle passes through only a single point, and *conjugate* lattices, in which two or more points lie on a circle circumscribing the cylinder. Simple lattices correspond to the 'alternate' leaf arrangements of plant taxonomy, in which there is a single leaf attached to the stem at each node, and conjugate patterns to 'opposite' and 'whorled' arrangements, with two or more leaves attached at each node.

In a simple lattice on a cylinder, the points may be connected by helical lines as shown in Fig. 3.1. There is an indefinite number of such lines, just as in an orchard the trees are aligned in rows in many directions. If the points are numbered in order of their increasing distance along the axis of the cylinder, lines can be found which connect them in serial order. The shortest of these, the 1-parastichy, is often called the generative helix. In Fig. 3.1, a right-hand generative helix is shown, but it could as well be a left-hand helix. It may be characterized by two parameters, the angular displacement, $\alpha$, and the vertical distance, $h$, from one point to the next in serial order, corresponding to the rotation about the axis and translation parallel to the axis, which constitute a single screw displacement of the pattern. The angle, $\alpha$, will be referred to as the *divergence angle*, or principal divergence, consistent with van Iterson's notation, and the terminology of virtually all authors. To provide a scale for $h$, the radius of the circular cylinder will be taken as unity. Many of van Iterson's equations are written in terms of $b$, the radius of a circle as a fraction of the circumference of the cylinder in the contiguous circle model (see below), and some of them will be quoted. However, it seems preferable to use the parameter, $h$, since it corresponds to the intuitive notion of internode length, and leads to simpler formulae. It is related to the pitch of the helix, as that term is understood in machine-shop practice, except that the pitch of a screw is the translation corresponding to rotation through a full circle. Clearly, $h = (\text{pitch}) \cdot \alpha/2\pi$. It may be noted that Turing (1959) proposed to consider three parameters, $J$ for jugacy, $\lambda$ for vertical displacement, and $\alpha$, the divergence angle. Coxeter (1961) describes the pattern of a pineapple fruit in terms of divergence angle, $\tau$, and height, $h$, as a fraction of the circumference of the cylinder.

In addition to the generative helix, there are other parastichies connecting successive points which differ in serial number by integers greater than one. If $m$ is such an integer, there will be many helices

Fig. 3.1 (*a*) Lattice of points on the surface of a cylinder. (*b*) The same cylinder unrolled. Points are numbered in a single sequence along the generative helix. A 3-parastichy is shown as a dashed line, 5- and 8-parastichies as solid lines.

(*a*) (*b*)

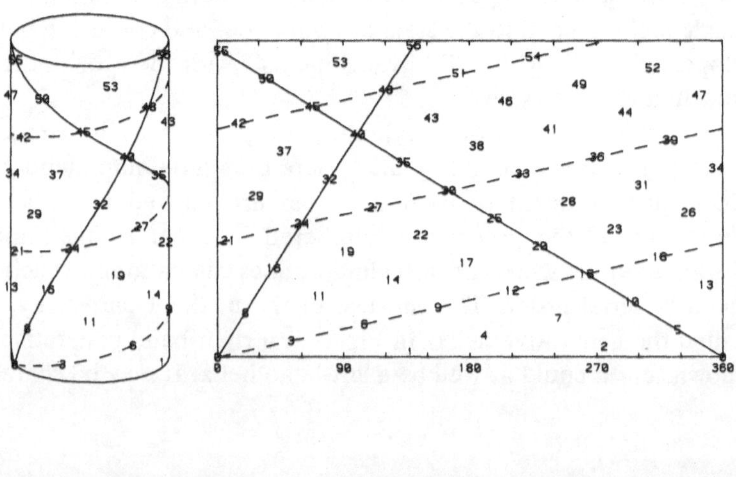

connecting points which differ in serial number by $m$. Considering again only the shortest, that is, the steepest of these, there is a set of $m$ parastichies, each of which connects points differing in serial number by $m$. For instance, one may trace two 2-parastichies and a set of five congruent 5-parastichies.

The properties of an $m$-parastichy may be considered in terms of a screw displacement along the 1-parastichy, consisting of a rotation through the angle $m\alpha$, and a translation of length $mh$. Taking a point 0 as origin, the generative helix from 0 to $m$ may extend one or more times around the cylinder, and it is convenient, instead of $m\alpha$, to define the *secondary divergence*, $\delta_m$, as the smaller angle, to the right or left, from point 0 to $m$. This will be less than or equal to 180°, or $\pi$ radians. Its relation to $\alpha$ is

$$\delta_m = m\alpha - \Delta_m 2\pi \tag{3.1}$$

where $\Delta_m$ is an integer which van Iterson termed the *encyclic number*. It is the number of turns around the cylinder, rounded upward or downward to the nearest integer, which the 1-parastichy describes between point 0 and point $m$. Coxeter (1961) expresses essentially the same relationship in terms of multiples of a circle, rather than in degrees or radians. The secondary divergence is then the fractional part, and the encyclic number the integral part of the divergence along the generative helix.

Of the innumerable ranks of points which can be traced in a lattice, a limited number are subjectively more evident than others. In particular, one's eye tends to see two series of parastichies running in opposite directions, and it seems probable that these evident parastichies are those of intermediate pitch, which run in opposite directions and intersect approximately at right angles, or perhaps which connect points with their nearest neighbours. Citing these evident parastichies is a convenient way of designating a pattern, e.g. a (2, 3) pattern, or (34, 55) or $(m, n)$. For conjugate patterns, the jugacy, for which I suggest $k$, may be placed before the parenthesis, $k(m, n)$. Thus, the frequently occurring decussate pattern might be designated 2(1, 1). This designation is not appropriate for stems in which there has been great elongation of the internodes, since the evident parastichies do not intersect orthogonally, and nearest neighbours are not easily evident. And the frequently used fractional notation, e.g. 2/5, 3/8, is not very informative. It is sometimes said that the numerator indicates the number of orthostichies, and the denominator the number of parastichies, implying incorrectly in most cases that the divergence angle is a rational fraction of a circle. I would suggest that with many mature stems it would be sufficient to note that the leaf arrangement is helical (or decussate, etc.), and to indicate the value of $h$. For a helical arrangement the divergence angle might also be cited if it can be estimated. In common woody dicots, internodes often vary greatly in length, but $h$ is often in the range of 5 to 50.

The evident parastichies of a pattern can be related to its divergence, $\alpha$, its vertical spacing, $h$, and jugacy, $k$. To do this for a simple pattern, $k = 1$, consider, in addition to the set of $m$-parastichies, another along which points differ in serial number by $n$, the set of $n$-parastichies. It will be assumed that $n > m$ and has no common divisor with it. (The patterns (0, 1) and (1, 1) must be considered separately.) The $n$-parastichy is characterized by a translation $nh$, a secondary divergence, $\delta_n$, and an encyclic number, $\Delta_n$, the latter two related to the principal divergence by

$$\delta_n = n\alpha - \Delta_n 2\pi \tag{3.2}$$

This equation holds simultaneously with equation (3.1), with the same sign convention for $\delta_n$ as for $\delta_m$. If equation (3.1) is multiplied by $n$ and equation (3.2) by $m$, and the term $mn\alpha$ is eliminated, the following relation between $\delta_m$ and $\delta_n$ results

$$n\delta_m - m\delta_n = (m\Delta_n - n\Delta_m)2\pi \tag{3.3}$$

The $m$- and $n$-parastichies intersect at the point $mn$, and the secondary divergence from point 0 to point $mn$ may be designated $\delta_{mn}$. As van Iterson has shown, $m\delta_n = \delta_{mn}$. A variety of dispositions of intersecting sets of parastichies is possible, with $\delta_m$ and $\delta_n$ both positive angles, both negative, or of opposite sign, and with various angles of intersection. Van Iterson has shown that for all arrangements $n\delta_m = \delta_{mn} \pm 2\pi$. Substituting these values for $m\delta_n$ and $n\delta_m$ into equation (3.3) gives

$$n\Delta_m - m\Delta_n = \pm 1 \tag{3.4}$$

The difference is positive when the $m$-helices are homodromous with, i.e. run in the same direction as, the $l$-helix, negative when the $m$- and $l$-parastichies are antidromous, i.e. have the opposite sense. This convention will be followed in subsequent equations. Using equation (3.4), equation (3.3) can also be solved to give

$$m\delta_n - n\delta_m = \pm 2\pi \tag{3.5}$$

For given values, $m$ and $n$, the smallest positive integers, $\Delta_m$ and $\Delta_n$, which satisfy equation (3.4) are the encyclic numbers for the pattern $(m, n)$. Often $\Delta_m$ and $\Delta_n$, corresponding to given $m$ and $n$, can be found easily by trial, and there is a systematic method of solution with continued fractions. (If the fraction $m/n$ is expanded as a finite continued fraction with numerators equal to 1, and this is truncated by omitting the last term, the truncated fraction is equal to $\Delta_m/\Delta_n$. As an example, find $\Delta_m$, $\Delta_n$, given $m = 5$, $n = 12$. First expand, $\frac{5}{12} = \frac{1}{2+}\frac{1}{2+}\frac{1}{2}$. Then truncate, and evaluate the resulting fraction, $\frac{1}{2+}\frac{1}{2} = \frac{2}{5}$. Therefore, $\Delta_m = 2$, $\Delta_n = 5$. Actually the original fraction can be expanded further, $\frac{5}{12} = \frac{1}{2+}\frac{1}{2+}\frac{1}{1+}\frac{1}{1}$, but the expansion should be terminated with the first integral denominator (van Iterson, 1907).)

Of the many possible patterns of intersecting parastichies, only those simple patterns in which the $m$- and $n$-parastichies run in opposite directions, that is are antidromous to each other, will be discussed in detail. It follows from equations (3.1) and (3.2) that for these patterns, $\alpha$ lies between $2\pi\Delta_m/m$ and $2\pi\Delta_n/n$. These are the limits for $\alpha$, within which a pattern exhibiting the evident parastichies, $m$ and $n$, can exist. Some examples of these limits are listed in Table 3.1 (in angular degrees), with corresponding values of $m$, $n$, $\Delta_m$ and $\Delta_n$. These limits can also be seen graphically in Fig. 3.5. In general, they become narrower as $m$ and $n$ increase.

### *Lattices and parastichies in a plane and on a conical surface*

Planar models can also be constructed to resemble patterns such as those of leaf primordia about a growing shoot apex seen from above, of the fruits in a sunflower head and of the leaves of many succulent rosette plants. These examples have the property that the lateral organs are symmetrically placed about a central point. Arrays of points with this property may be generated by rotation about the centre through a specified angle, $\alpha$, accompanied by dilatation, that is, increasing the radial distance to a point, by a specified factor, $a$. These correspond to rotation and translation in the cylindrical case. The general symmetry operation in the plane is a similarity, as compared to the isometry, or congruence, in the case of a cylindrical surface, and it can be defined as a motion of a point along an equiangular spiral. As before, the resulting lattices may be grouped into simple or conjugate lattices, depending on whether only one, or whether two or more points may be found at a given distance from the centre.

Fig. 3.2. Lattices in a plane, in which points are numbered in a single sequence along an equiangular spiral, representing the generative spiral. At the left, 3-, 5- and 8-parastichies (equiangular spirals) are drawn, and at the right, 5-, 8- and 13-parastichies.

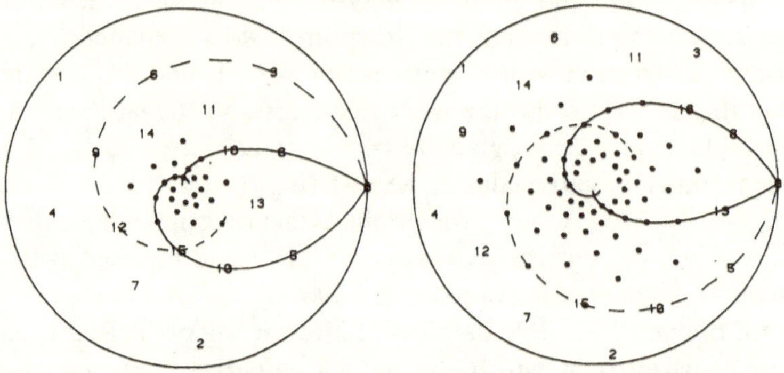

Table 3.1. *Patterns of orthogonal parastichies, and of contiguous circles on the surface of a cylinder, each circle tangent to four neighbours. Parastichy numbers are given by the symbol k(m, n) and encyclic numbers by $\Delta_m$, $\Delta_n$. The range of divergence angles, α, within which each pattern of orthogonal parastichies can exist is shown in columns 3 and 4. Columns 5 and 6 give the divergence α and vertical spacing h for orthogonal intersection of parastichies drawn through the centres of contiguous circles, from simultaneous solution of equations (3.7) and (3.13)*

| $k(m, n)$ | $\Delta_m, \Delta_n$ | Angular range $\alpha$ | | Orthogonal contact parastichies | |
|---|---|---|---|---|---|
| | | | | $\alpha$ | $h$ |
| (1, 1) | 0, 1 | 0° | 360° | 180° | 3.14593 |
| (1, 2) | 0, 1 | 0 | 180 | 144 | 1.25664 |
| (2, 3) | 1, 1 | 180 | 120 | 138.4615 | 0.48332 |
| (3, 5) | 1, 2 | 120 | 144 | 137.6471 | 0.18480 |
| (5, 8) | 2, 3 | 144 | 135 | 137.5281 | 0.07060 |
| (8, 13) | 3, 5 | 135 | 138.4615 | 137.5107 | 0.02697 |
| (13, 21) | 5, 8 | 138.4615 | 137.1429 | 137.5082 | 0.01300 |
| (21, 34) | 8, 13 | 137.1429 | 137.6471 | 137.5078 | 0.00393 |
| (1, 3) | 0, 1 | 0 | 120 | 108 | 0.62832 |
| (3, 4) | 1, 1 | 120 | 90 | 100.8 | 0.25133 |
| (4, 7) | 1, 2 | 90 | 102.8571 | 99.6923 | 0.09666 |
| (7, 11) | 2, 3 | 102.8571 | 98.1818 | 99.5294 | 0.03696 |
| (11, 18) | 3, 5 | 98.1818 | 100 | 99.5056 | 0.01412 |
| (18, 29) | 5, 8 | 100 | 99.3103 | 99.5021 | 0.00539 |
| 2(1, 1) | 0, 1 | 0 | 180 | 90 | 1.57080 |
| 2(1, 2) | 0, 1 | 0 | 90 | 72 | 0.62832 |
| 2(2, 3) | 1, 1 | 90 | 60 | 69.2307 | 0.24166 |

In such a lattice, ranks of points may be connected by equiangular spirals, which will be called parastichies, just as in the cylindrical model the parastichies are represented by helices. In Fig. 3.2, such lattices are illustrated, with a number of the spirals drawn in.

The relationships between parastichy numbers, $m$, $n$, and encyclic numbers, $\Delta_m$, $\Delta_n$, the relationship between the principal divergence, $\alpha$, and the secondary divergences, $\delta_m$, $\delta_n$, and the angular limits for a given pattern, $(m, n)$, are identical with those in the cylindrical case, as van Iterson has shown in detail. Equations (3.1) to (3.5) and the numerical examples of Table 3.1 then apply to both cases. The relationship of the dilatation factor, $a$, to these parameters is worked out below.

Much the same can be said of lattices on a conical surface, except that the general symmetry operation, a similarity, is more complex, consisting of a rotation, dilatation and translation. The lines connecting ranks of points, or parastichies, are taken as conical loxodromes, as illustrated in Fig. 3.3, and the general symmetry operation can be considered as a motion of a point along this curve. An additional parameter must be considered, the apical angle of the cone, $\psi$. A point on the cone may be located in a cylindrical coordinate system ($\varphi$, $r$, $z$), or alternatively, by giving its angular position, $\varphi$, and its distance from the vertex, $R$, where $R = r/\sin(\psi/2)$. The numerical and angular relationships expressed in equations (3.1) to (3.5) apply to this model as well, as van Iterson has shown. Other parameters, and their relationships are discussed below.

Fig. 3.3. (*a*) Lattice on the surface of a right circular cone. (*b*) The same cone cut along a generator and unrolled. Points can be numbered in a single sequence along the generative spiral, a conical loxodrome. 3-, 5- and 8-parastichies, which are also loxodromes, are drawn.

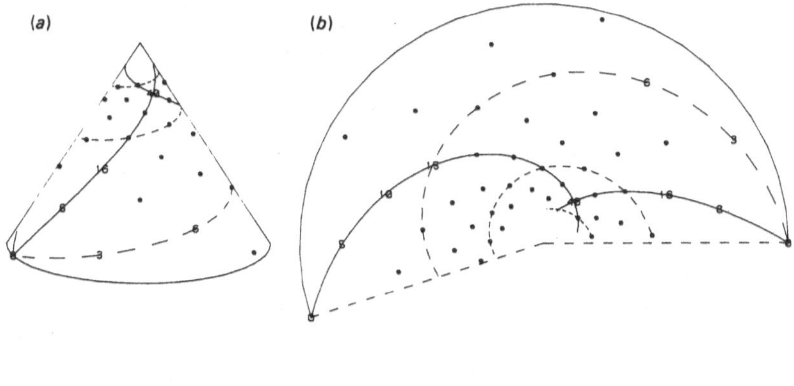

### *Angle of intersection of parastichies on a cylinder; orthogonal parastichies*

I suggested above that a phyllotactic pattern might be specified by citing those sets of parastichies which are most evident, and that they may be the two which intersect most nearly at right angles. To designate a pattern in this way, an equation will be found relating the vertical spacing, $h$, to the divergence, $\alpha$, for given values of $m$, $n$; or relating $h$ to the secondary divergences, $\delta_m$, $\delta_n$, which is equivalent. This will be done only for simple systems, of jugacy $k=1$.

An expression for the angle of intersection of the $m$-, $n$-parastichies can be found using relationships in the triangle of Fig. 3.4, which is a portion of an unrolled cylindrical lattice with $m$-, $n$-parastichies drawn in. This is the visible opposed parastichy triangle of Adler (1974). The base of the triangle is the circumference of the cylinder through point 0, and the other two sides are $m$-, $n$-parastichies through point 0 and point $mn$. Since $h$ is the vertical displacement between successive points on the 1-helix, the altitude of the triangle is $mnh$. The length of the base is $2\pi r$, and it is divided into two segments, $\pm n\delta_m r$ and $\mp m\delta_n r$ (equation 3.5). Taking the radius of the cylinder as unity, the base is $2\pi$ and its segments are $\pm n\delta_m$, $\mp m\delta_n$, regarding $\delta_m$, $\delta_n$ as measures of arc length. When the $m$- and 1-parastichies are homodromous, the upper sign in each expression applies, when antidromous, the lower. To find the angle of intersection, $\varepsilon$, note the angles $\beta$ and $\gamma$ which the parastichies make with the altitude and whose sum is $\varepsilon=\beta+\gamma$. Then, using

Fig. 3.4. Diagram of a portion of a point lattice on an unrolled cylindrical surface. An 'opposed parastichy triangle' is shown, in which the base is a circumference of the cylinder of length $2\pi$. The other two sides are $m$- and $n$-parastichies, $\delta_m$ and $\delta_n$ are secondary divergence angles.

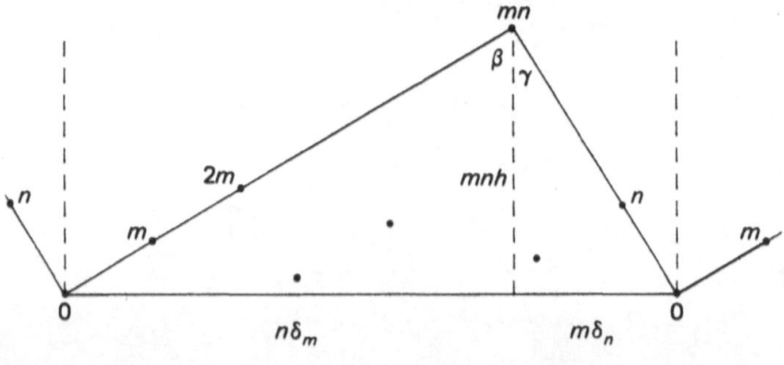

$\tan \beta = \delta_m/mh$ and $\tan \gamma = \delta_n/nh$, the trigonometric identity for the tangent of the sum of two angles, and equation (3.5)

$$mnh^2 + 2\pi h(\cot \varepsilon) + \delta_m\delta_n = 0 \tag{3.6}$$

The condition for orthogonal intersection of the $m$-, $n$-parastichies is that $\cot \varepsilon = 0$. Substituting this into equation (3.6)

$$h = \sqrt{(-\delta_m\delta_n/mn)} \tag{3.7}$$

To relate $h$ to $\alpha$ directly, for orthogonal intersection, substitute the values of $\delta_m$, $\delta_n$ from equations (3.1) and (3.2) into equation (3.7)

$$h^2 = -\alpha^2 + 2\pi\alpha \cdot (m\Delta_n + n\Delta_m)/mn - 4\pi^2 \cdot \Delta_m\Delta_n/mn \tag{3.8}$$

Equation (3.7) seems preferable for numerical exemplification of various patterns, as in Table 3.1 and Fig. 3.5, where one can first find $\delta_m$, $\delta_n$, then solve equation (3.7). Equation (3.6) can be solved for sets of parastichies intersecting at other angles than 90° ($\frac{1}{2}\pi$), and a few solutions are plotted in Fig. 3.6 for the (2, 3) and (3, 5) patterns. For a given divergence, $\alpha$, an acute angle of intersection leads to greater vertical spacing, $h$, of the points, and an obtuse angle to less.

Fig. 3.5. Parameters of possible simple, or unijugate, patterns of orthogonal parastichies which can be drawn on the surface of a cylinder; $h$, the vertical spacing of points, and $\alpha$, the divergence angle. Symbols of the form $(m, n)$ designate the patterns. For conjugate patterns $k(m, n)$, where $k>1$, other diagrams can be drawn. This diagram also applies to simple patterns of orthogonal spirals in the plane, if the vertical axis is taken to represent ln $a$, where $a$ is the plastochron ratio.

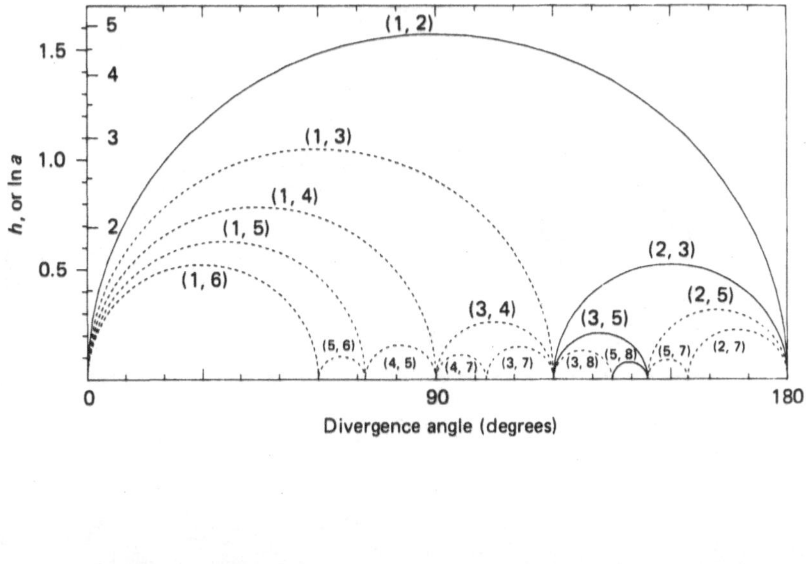

Figures 3.5 and 3.6 show the range of values of $\alpha$ and $h$, for which various patterns of orthogonal parastichies $(m, n)$ can exist. The $(1,2)$ pattern can be drawn with divergence angles from 0° to 180°, with a maximum value of $h = \frac{1}{2}\pi = 1.571$, at $\alpha = 90°$. The $(2,3)$ pattern is possible between 120° and 180°. The $(3,5)$, $(5,8)$, and higher order patterns have successively narrower angular limits, and smaller vertical spacings. The $(1,1)$ pattern, which is not shown, exists at any value of $\alpha$, and has a maximum value of $h = \pi$ at 180°. Equations (3.6) and (3.7) are indeterminate for the pattern $(0,1)$.

### *Angle of intersection of parastichies in the plane model; orthogonal parastichies*

In the plane model with points symmetrically placed around a central point, the parastichies may be represented by equiangular spirals, corresponding to the symmetry operation of a rotation and dilatation, which characterizes the model. In polar coordinates, $(r, \varphi)$,

Fig. 3.6. Parameters of (2, 3) and (3, 5) patterns of parastichies or helices on a cylinder, which intersect at 90° (solid line), or at 75° and 105° (dashed lines). The same diagram applies to (2, 3) and (3, 5) patterns of intersecting equiangular spiral parastichies in the plane.

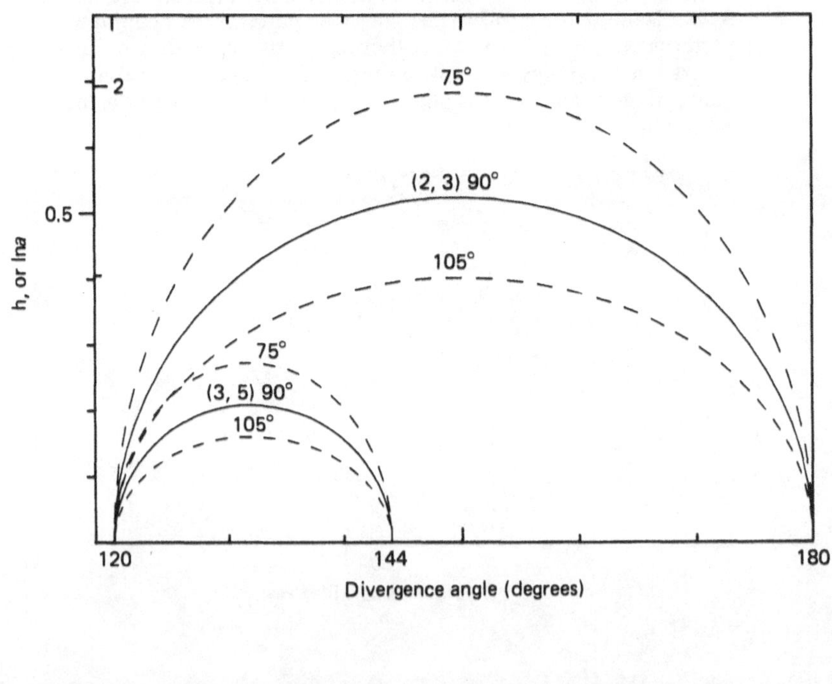

the equation of an equiangular spiral is $r = Ce^{q\varphi}$, where $q$ is the cotangent of the angle between the spiral and the radius vector to each point. If this equation is taken to represent the 1-parastichy, two successive lattice points along it, $(r_0, \varphi_0)$ and $(r_1, \varphi_1)$ are related by the dilatation factor, $r_0/r_1$. This will be symbolized $a$, and will be taken to be greater than one, in agreement with the plastochron ratio of Richards (1951), so that it is the reciprocal of van Iterson's factor, $a$. Using the equation above, $a$ may be written $r_0/r_1 = e^{q(\varphi_0 - \varphi_1)}$, where $(\varphi_0 - \varphi_1)$ is the angular divergence, $\alpha$, so that $a = e^{q\alpha}$. These are illustrated in Fig. 3.7.

The intersecting sets of $m$-, $n$-parastichies which characterize a given pattern are also equiangular spirals, with greater pitch than the 1-parastichy. A single displacement along the $m$-parastichy, from point 0 to point $m$, is equivalent to $m$ displacements along the 1-parastichy, the ratio of radii being $r_0/r_1 = a^m$, and the angular divergence, as in the cylindrical model, being $\delta_m = m\alpha - \Delta_m 2\pi$ (equation 3.1). If $\beta$ is the angle between the tangent to the spiral, and the radius, $r_0$, the relation between successive points along the $m$-parastichy is $a^m = e^{(\cot\beta)\delta_m}$, and by the same argument $a^n = e^{(\cot\gamma)\delta_n}$. Taking logarithms of these equations and rearranging, $\tan\beta = \delta_m/m\ln a$, and $\tan\gamma = \delta_n/n\ln a$. The angle of intersection of the $m$- and $n$-parastichies is $\beta + \gamma$, symbolized $\varepsilon$ as above. Using equation (3.5) and a trigonometric identity

$$mn(\ln a)^2 + 2\pi(\ln a)(\cot\varepsilon) + \delta_m\delta_n = 0 \tag{3.9}$$

Fig. 3.7. Diagram of $m$- and $n$-parastichies in the plane which intersects orthogonally. Radii to points 0, $m$ and $n$ are shown, and secondary divergence angles $\delta_m$, $\delta_n$.

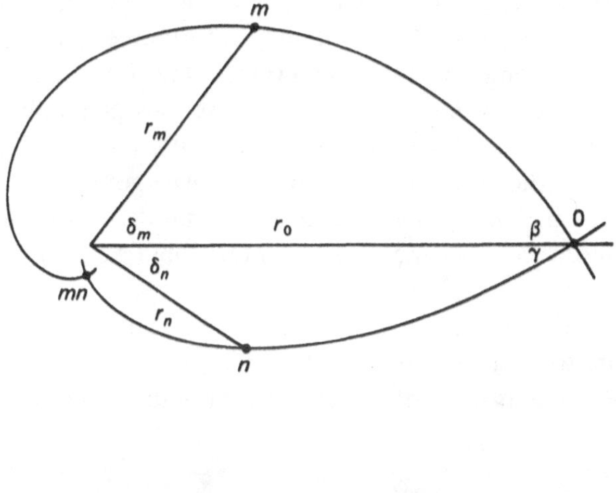

This equation is identical to equation (3.6) if $\ln a$ is substituted for $h$, a result which is not easily apparent in either van Iterson's (1907) or Richards' (1948) work.

Just as in the cylindrical case, the condition for orthogonal intersection is obtained by substituting $\cot \varepsilon = 0$ into equation (3.9)

$$\ln a = \sqrt{(-\delta_m \delta_n / mn)} \tag{3.10}$$

This equation is given by Richards (1948, 1951). His term, the plastochron ratio $a$ in these equations, is useful for a growing system since it is the ratio by which a leaf primordium is displaced radially from the shoot apex during one plastochron in steady growth. Van Iterson (1907) also considered orthogonal intersection of parastichies in systems of contiguous circles (see below).

The numerical and graphical illustrations of equation (3.7), which were given in Table 3.1 and Figs 3.5 and 3.6, apply equally to equation (3.10) with the additional point that values of $a$, the plastochron ratio, may be noted.

This equivalence of the equations for intersecting parastichies in the cylinder and the plane can also be obtained by considering the plane model as a conformal projection, or mapping, of the cylindrical model, such that the coordinates of each point on a cylinder of unit radius, ($z$, $\theta$), and the corresponding point in polar coordinates, ($r$, $\varphi$), are related by $r = \ln z$, $\theta = \varphi$. The origins and directions of the $r$- and $z$-axes can be chosen so as to be consistent with the numbering of points in Figs 3.1 and 3.4.

### *Intersection of parastichies on a cone*

Patterns of intersecting sets of parastichies on a conical surface may be considered as projections of cylindrical or planar patterns, as van Iterson (1907) has discussed in detail. Alternatively, the cone may be viewed as cut along one of its generators, and unrolled into a flat sheet, a sector of a circular surface. The equations given above apply with slight modification. A right circular cone is illustrated in Fig. 3.3, in which the angle at the vertex between two opposite generators is $\psi$. The 3-, 5- and 8-parastichies have been drawn in the form of conical loxodromes on the cone surface, and as interrupted equiangular spirals on the plane sector. Corresponding to $r$ in the plane model, one can consider $R$, the distance from the vertex of the cone, or centre of the circular sector, to a given point, with $r = R \sin \frac{1}{2}\psi$. The factor $\sin \frac{1}{2}\psi$, symbolized $N$, appears in all of the equations relating to conical models.

If $\alpha$ is the principal divergence about the axis of the cone, the divergence in the unrolled circular sector is

$$\alpha' = \alpha \sin \tfrac{1}{2}\psi = N\alpha$$

and for secondary divergences, $\delta_m' = N\delta_m$, $\delta_n' = N\delta_n$. The relationships between parastichy numbers, $m$, $n$, and encyclic numbers, $\Delta_m$, $\Delta_n$, given in equations (3.1) to (3.5) apply unaltered to the conical model, so that substituting $\alpha'$, $\delta_m'$ and $\delta_n'$ for $\alpha$, $\delta_m$ and $\delta_n$ in the subsequent equations gives corresponding equations for the conical model. For example, the relation between the plastochron ratio, $a$, the secondary divergences and the parastichy intersection angle, is a modification of equation (3.9)

$$mn(\ln a)^2 + 2\pi(\ln a)(\cot \varepsilon) + N^2 \cdot \delta_m \delta_n = 0 \qquad (3.11)$$

and the equation specifying orthogonal intersection of parastichies on a cone, similar to equation (3.10) is

$$\ln a = N \cdot \sqrt{(-\delta_m \delta_n / mn)} \qquad (3.12)$$

### *Contiguous circles, cylinder model*

An alternative basis for designating the evident ranks of points in a phyllotactic pattern is to consider the neighbours of a given point. A set of $m$-parastichies may be chosen which connects each point with its nearest neighbour, so that a point 0 will have as its nearest neighbours, $m$ and $-m$. A second set of parastichies may connect point 0 with its next nearest, or equidistant, neighbours, $n$ and $-n$. If the distances from a point to its next neighbours along the $m$-, $n$-parastichies are equal, the surface is divided into rhombuses, or, if the intersection angle is 90°, into squares, as illustrated by van Iterson (1960). If the parastichies intersect at 60° or 120°, the shorter diagonal of each rhombus is equal to its side, each point is equidistant from six neighbours, and it is appropriate to designate the pattern by citing three rather than two parastichy numbers.

When a point is equidistant from four (or six) of its neighbours, particularly suggestive diagrams may be constructed by drawing patterns of circles on the unrolled cylindrical surface, with the points as centres and diameter equal to the distance between points. Each circle is tangent to four (or six) others, the points of tangency lying on the $m$-, $n$-parastichies (or also on a third set of parastichies, in the 60° or 120° case).

Van Iterson's (1907) discussion is largely based on this model of contiguous circles, or other figures, on a surface, and I will reproduce his derivations but with some change in definition of the parameters.

In Fig. 3.8, a portion of an unrolled cylinder on which such a pattern of contiguous circles has been drawn, each circle is tangent to four others. The diameter of each circle, as a fraction of the radius of the cylinder, is $d$, and equals the distance between points of the lattice. Van Iterson used $b$ to symbolize the ratio of the diameter of each circle to the circumference of the cylinder, so that $d = 2\pi b$. As in Fig. 3.4 an opposed parastichy triangle has been constructed, with the circumference of the cylinder through point 0 as its base and, as its other two sides, the $m$-, $n$-parastichies passing through points 0 and $mn$. The perpendicular to the base from point $mn$ divides this triangle into two right triangles, in which $(n\delta_m)^2 + (mnh)^2 = (nd)^2$ and $(m\delta_n)^2 + (mnh)^2 = (md)^2$. On eliminating $d^2$ and rearranging

$$h = \sqrt{[(\delta_m^2 - \delta_n^2)/(n^2 - m^2)]} \tag{3.13}$$

This is the counterpart of equation (3.7) which gives $h$ for orthogonal parastichies. It can also be put in a form similar to equation (3.8) giving $h$ as a function of $\alpha$, but the form given is probably more convenient for numerical evaluation.

The diameter of the circles, $d$, may be found by eliminating $(mnh)^2$ between the same two equations

$$d = \sqrt{[(n^2\delta_m^2 - m^2\delta_n^2)/(n^2 - m^2)]} \tag{3.14}$$

Fig. 3.8. Diagram of opposed parastichy triangle on the surface of a cylinder, in which points of the lattice are equidistant along $m$- and $n$-parastichies. Circles can be drawn, centred on the points, such that they will be tangent along the $m$- and $n$-parastichies.

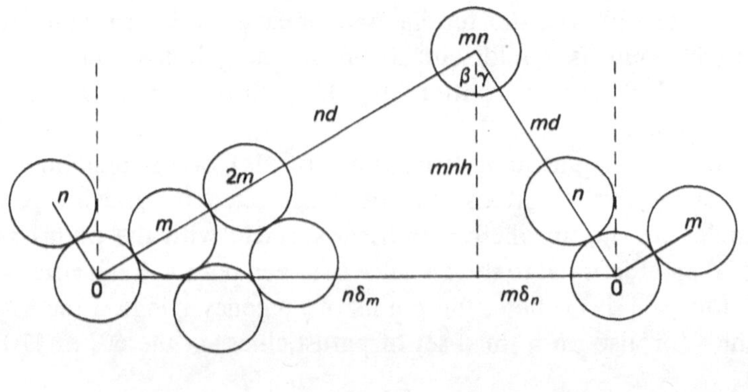

This differs from van Iterson's $b$ by the factor $2\pi$.

In solving equations (3.13) and (3.14) for representative values of $h$ versus $\alpha$, a complication arises, in that the range of values of $\alpha$ within which a given pattern of contiguous circles can exist in narrower than that given above – $(\Delta_m/m) \cdot 360°$ and $(\Delta_n/n) \cdot 360°$ in Table 3.1. Since van Iterson (1907) discussed this matter thoroughly, I will only illustrate the relationship graphically and recast his equation for consistency with others in this chapter. Consider first the pattern $(m, n)$ in which each circle is tangent to four others (Fig. 3.9*b*). For concreteness, the illustration is of a (3, 5) pattern in which $\alpha = 138°$, $\delta_3 = 54°$, $\delta_5 = -30°$, $h = 0.1959$, $d = 1.1107$, in accordance with equations (3.13) and (3.14). Now, if $h$ were decreased, it would be necessary to decrease $\alpha$ and the diameter of the circles, to maintain tangency along the $m$-, $n$-helices, that is, to maintain the contact of circle 0 with 3 and 5. If $h$ is progressively decreased in this manner, circle 8 approaches 0, and eventually comes into contact with it (Fig. 3.9*c*). There are now three parastichies along which circles are tangent, and one may say that a new pattern has been generated, $(m, n, m+n)$ or specifically, (3, 5, 8). Starting again with Fig. 3.9*b*, if $h$ were increased progressively in such a way as to maintain tangency of circle 0 with circles 3 and 5, circle 2 would approach and finally touch circle 0, and a second new pattern would appear, $(n-m, m, n)$, or (2, 3, 5). It is natural to speak of the $(n-m, m, n)$ and $(m, n, m+n)$ patterns as exhibiting triple contacts, and the $(m, n)$ pattern as having double contacts, and to regard the former as limiting patterns between which the $(m, n)$ pattern can exist.

It is then of interest to evaluate $\alpha$ ($\delta_m$ and $\delta_n$), $h$ and $d$ for triple contact patterns. The diameter, $d$, can be readily evaluated for the pattern $(m, n, m+n)$ using an equation for the base of the triangle of Fig. 3.8 in terms of the other two sides and their included angle

$$(2\pi)^2 = (nd)^2 + (md)^2 - 2(nd)(md)\cos(2\pi/3)$$

or, on simplification

$$d = 2\pi/\sqrt{(m^2 + mn + n^2)} \tag{3.15}$$

Rearrangement of equation (3.14) and use of equation (3.5) yields expressions for $\delta_m$, $\delta_n$

$$\delta_m = \pi \cdot (m + 2n)/(m^2 + mn + n^2);$$
$$\delta_n = \pi \cdot (2m + n)/(m^2 + mn + n^2) \tag{3.16}$$

Fig. 3.9. Representative patterns of tangent circles drawn on an unrolled cylindrical surface. (*a*) and (*c*) are triple-contact patterns, (2, 3, 5) and (3, 5, 8) in which each circle is tangent to six neighbours. (*b*) and (*d*) are double-contact patterns (3, 5) and (5, 8) in which each circle has four tangent neighbours.

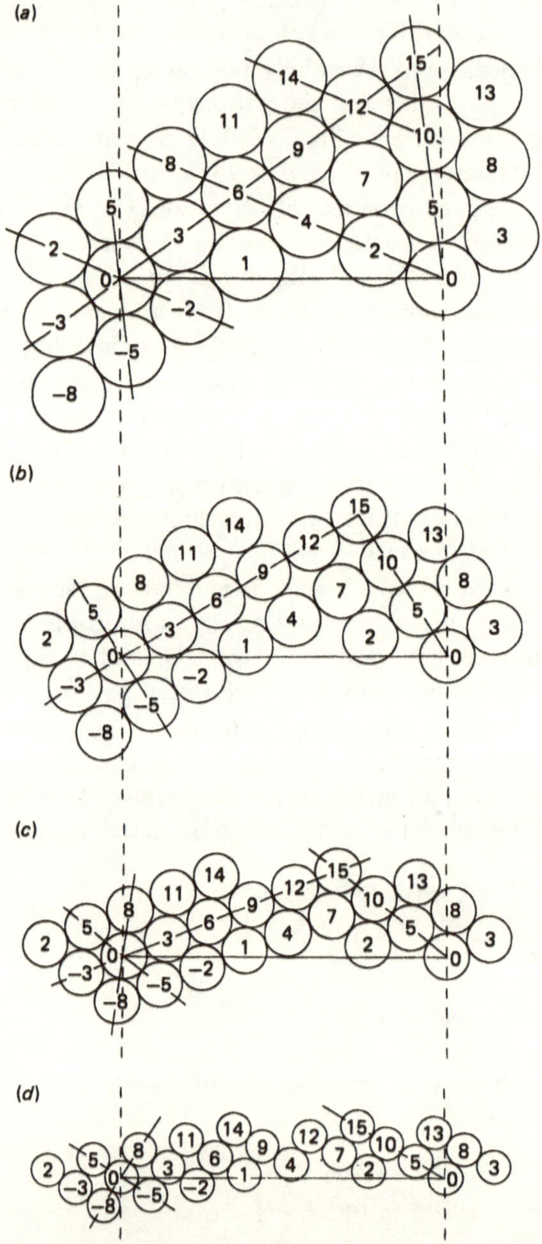

and these lead to expressions for $\alpha$. If the $m$-helix is homodromous to the generative helix

$$\alpha = \pi \cdot [2\Delta_m/m + (m + 2n)/(m^2 + mn + n^2)] \qquad (3.17a)$$

If it is antidromous

$$\alpha = \pi \cdot [2\Delta_n/n + (2m + n)/(m^2 + mn + n^2)] \qquad (3.17b)$$

The vertical spacing of the lattice points is

$$h = \pi\sqrt{3}/(m^2 + mn + n^2) \qquad (3.18)$$

Similar derivations can be made for the pattern $(n - m, m, n)$, but equations (3.15) to (3.18) can be made to serve for all simple triple-contact patterns of circles on the cylinder. Angular limits for patterns of contiguous circles are listed in Table 3.2 for comparison with those for patterns of intersecting parastichies.

Equation (3.13) has been solved for a number of patterns $(m, n)$ and these are plotted in Fig. 3.10 (solid line) as $h$ against $\alpha$. Van Iterson's graphs of $b$ against $\alpha$ are similar. The possible values of $h$ and $\alpha$ for a given pattern $(m, n)$ are represented by an arc extending from one

Table 3.2 *Triple-contact patterns of contiguous circles, on a cylindrical surface, and in the plane. The divergence angles α listed are limiting divergences for double-contact patterns. For example, the (5, 8) pattern of contiguous circles on the cylinder can be constructed with any divergence α between 135.9184° and 138.1395°*

| | Cylinder | | Plane | | |
|---|---|---|---|---|---|
| $k(m, m + n, n)$ | $\alpha$ | $h$ | $\alpha$ | $a$ | $\ln a$ |
| (0, 1, 1) | 180° | 5.44139 | | | |
| (1, 1, 2) | 180 | 1.81380 | | | |
| (1, 2, 3) | 128.5714 | 0.77734 | 128.1727° | 2.89005 | 1.06128 |
| (2, 3, 5) | 142.1053 | 0.28639 | 142.1147 | 1.36986 | 0.31471 |
| (3, 5, 8) | 135.9184 | 0.11105 | 135.4178 | 1.12184 | 0.11497 |
| (5, 8, 13) | 138.1395 | 0.04218 | 138.1398 | 1.04365 | 0.04273 |
| (8, 13, 21) | 137.2700 | 0.01615 | 137.2700 | 1.01636 | 0.01623 |
| (13, 21, 34) | 137.5991 | 0.00616 | 137.5991 | 1.00620 | 0.00617 |
| (1, 3, 4) | 96.9231 | 0.41857 | 96.8723 | 1.62328 | 0.48445 |
| (3, 4, 7) | 102.1622 | 0.14706 | 102.1630 | 1.16654 | 0.15404 |
| (4, 7, 11) | 98.7097 | 0.05851 | 98.7096 | 1.06138 | 0.05957 |
| 2(1, 2, 3) | 64.2857 | 0.38867 | 64.2778 | 1.51235 | 0.41366 |
| 2(2, 3, 5) | 71.0526 | 0.14319 | 71.0529 | 1.15765 | 0.14639 |
| 2(3, 5, 8) | 67.9592 | 0.05552 | 67.9592 | 1.05759 | 0.05600 |

triple-contact point to another. When the $m$-helix is homodromous, that is, runs in the same direction as the generative helix, the arc extends downward to the left, and when antidromous, downward to the right. For instance, the (3,5) pattern of contiguous circles extends from the point designated (2,3,5) with $\alpha = 142.105°$, $h = 1.4415$, to point (3,5,8) at $\alpha = 135.918°$, $h = 0.8976$.

The conditions under which contact parastichies intersect orthogonally can also be found. These would correspond to the points of

Fig. 3.10. Parameters of possible simple patterns of contiguous circles drawn on the surface of a cylinder (solid lines), or in the plane (dashed lines). For the cylindrical patterns of contact parastichies, the vertical spacing, $h$, is plotted against divergence angle, $\alpha$. For the plane patterns, the plastochron ratio, $a$, or its logarithm, ln $a$, is plotted against $\alpha$. The arcs represent the parameters of double contact patterns, $(m, n)$. The points of intersection of three arcs represent triple-contact patterns, $(m, n, m+n)$. Similar diagrams can be drawn for conjugate patterns, $k(m, n)$ and $k(m, n, m+n)$, where $k>1$. Similar diagrams can also be drawn for patterns of contact parastichies on a conical surface.

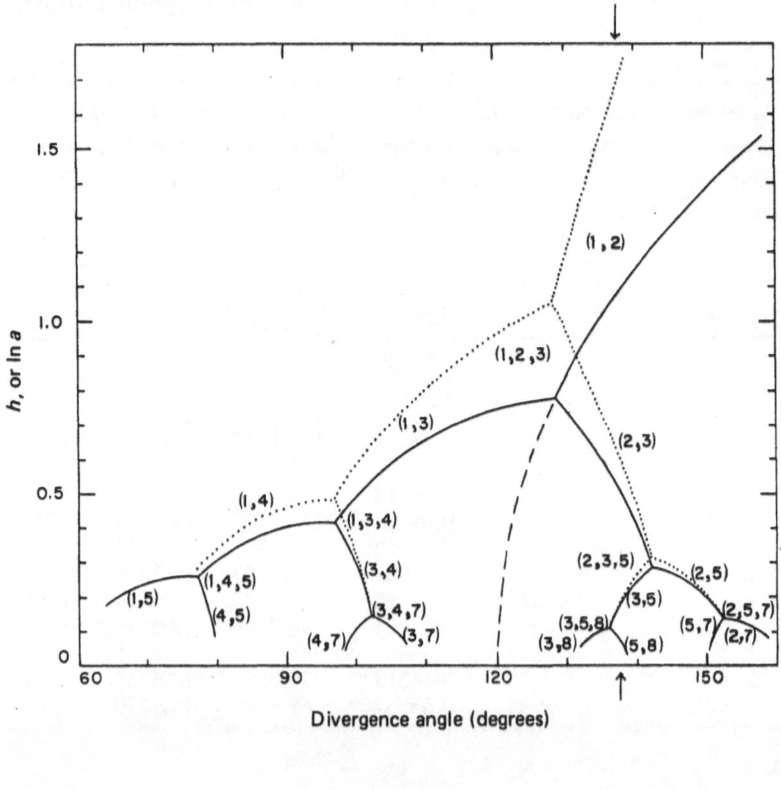

intersection of the arcs of Fig. 3.10 and those of Fig. 3.5, and one might contemplate solving equation (3.7) simultaneously with equation (3.13). It is more direct to consider that in this case Fig. 3.8 represents a right triangle with the hypotenuse $2\pi$ as base. Then $(nd)^2 + (md)^2 = (2\pi)^2$, so that $d = 2\pi/(m^2 + n^2)$.

The perpendicular from $mn$ to the hypotenuse divides the triangle into two triangles which are similar to the whole, so that $nd/2\pi = mnh/md$, and substitution from the last equation gives $h = 2\pi/(m^2 + n^2)$. Another proportionality between the sides of the triangles, $n\delta_m/nd = nd/2\pi$, and substitution for $d$, gives the secondary divergence, $\delta_m = 2\pi \cdot n/(m^2 + n^2)$. In the same way, $\delta_n = -2\pi \cdot m/(m^2 + n^2)$. These two equations apply in the homodromous case; in the antidromous case, the signs are changed. An equation can also be found for $a$, when contact parastichies on the cylinder intersect orthogonally. Van Iterson has tabulated some values of $b$ and $a$ to illustrate these relationships.

Models have been proposed in which other figures than circles are drawn on a cylindrical surface. Van Iterson discussed briefly the geometry of patterns of contiguous ellipses, referring to older work of Schwendener. Coxeter (1961) describes a pattern which consists of Dirichlet regions, or Voronoi polygons, about the points of a lattice, whose sides are the perpendicular bisectors of the lines between neighbouring points. In Coxeter's model, there is a marked similarity between the hexagons which he constructed and the surface outlines of segments of the fruit of pineapple, *Ananas comosus*, which he illustrated. I have proposed that Van Iterson's models of contiguous spheres, whose centres lie on a cylindrical surface, be considered as models for the tubular packing of the protein subunits of such biological ultastructures as bacterial flagella, microtubules and virus capsids (Erickson, 1973a,b). Harris and I (1980) have made an extensive analysis of the properties of such structures, largely from a crystallographic point of view. The possibility of contraction of a tubular packing from one form to another is suggestive of several aspects of the structure and function of tubules.

### *Contiguous circles, plane model*

In discussing orthogonal parastichies in the plane, it was noted that the plane model with spiral parastichies could be considered as a conformal mapping of the cylindrical model of helical parastichies. The essential identity of equations (3.7) and (3.10) is a consequence of the way in which the projection was defined. The same conformal projection, where $a$ is unchanged and $h$ of the cylinder is replaced by $\ln a$, can be considered for the pattern of equidistant points, or contiguous circles. One may immediately write, similar to equation (3.13)

$$\ln a = \sqrt{[(\delta_m^2 - \delta_n^2)/(n^2 - m^2)]}$$

which relates the ratio of distances from the centre of two successive points, $a$, and the divergence angle, $\alpha$, represented of course by $\delta_m$, $\delta_n$. While the projection preserves angular relationships, the distances between points, taken as equal in the cylinder model, are not equal in the projection, and the circles which were drawn with lattice points as centres are distorted. In Church's study, the consequences of this projection were studied, and a general equation for 'quasi-circles', into which the circles of the cylinder model are transformed, was given (Church, 1904, appendix by Hayes). The equation given above applies to a model of contiguous quasi-circles in the plane, as well as the illustrative values of Table 3.1, and the solid line curves of Fig. 3.10. Some qualification of equations (3.14) and (3.15) for $d$ is necessary but other equations of the previous section apply.

Van Iterson (1907) described an alternative plane model, in which contiguous circles are drawn about points of a lattice which lie along equiangular spirals, and I will follow his derivation. In Fig. 3.11 the $m$- and $n$-spirals which pass through point 0 have been drawn, and two triangles constructed by drawing radii, $r_0$, $r_m$, $r_{2m}$, and the chords connecting point 0 to $m$ and $m$ to $2m$. The angle at the pole in each triangle is $\delta_m$, and it is bisected in each case by the dashed line, dividing the chords into the line segments $s_1$ and $s_2$, $s_3$ and $s_4$. For the moment disregard the circles, since it must be shown that they can be drawn.

As in the discussion of intersecting spiral parastichies, $r_0 = r_m a^m$, $r_m = r_{2m} a^m$. Since the two triangles with vertices, 0, $m$, $2m$, are similar, the chord lengths are in the same ratio, $(s_1 + s_2) = a^m\,(s_3 + s_4)$. The bisector of angle $\delta_m$ in each triangle divides the opposite side such that $s_1/r_0 = s_2/r_m$, and $s_3/r_m = s_4/r_{2m}$. Also $s_2/r_m = (s_1 + s_2)/(r_0 + r_m)$, and $s_3/r_m = (s_3 + s_4)/(r_m + r_{2m})$. On substituting $a^{-m}(s_1 + s_2)$ for $(s_3 + s_4)$, $a^{-m}r_0$ for $r_m$, and $a^{-m}r_m$ for $r_{2m}$ in the right side of the last equation, it follows that $s_2 = s_3$, so that a circle can be drawn with point $m$ as centre and $s_2$ as radius, passing through points P and Q, at which the bisectors of $\delta_m$ intersect the two chords. The same relationships hold for other triangles constructed with radii and chords to other points on the

$m$-spiral, so that, for instance, a circle may be drawn with 0 as centre, $s_1$ as radius, another with $2m$ as centre and $s_4$ as radius. These circles will pass through points P and Q, and are clearly tangent to circle $m$ at these points. Hence, series of contiguous circles may be drawn corresponding to the set of $m$-parastichies. By the same argument, series of contiguous circles corresponding to the set of $n$-parastichies may be drawn.

To find the conditions under which each circle has points of tangency with other circles along both the $m$- and $n$-parastichies, $s_1$ will be equated with $s_5$ (Fig. 3.11). In this case circle 0 will pass through points P and R, the points of tangency with circles $m$, $n$. For the chord between points 0 and $m$

$$s_1 + s_2 = \sqrt{(r_0^2 + r_m^2 - 2r_0 r_m \cos \delta_m)}$$

or recalling equation (3.1), and the relationships, $r_m = a^{-m} r_0$, $r_{2m} = a^{-2m} \cdot r_0$, $s_1/r_0 = s_2/r_m = (s_1 + s_2)/(r_0 + r_m)$, one can write

$$s_1/r_0 = \sqrt{(1 + a^{2m} - 2a^m \cos m\alpha)}/(1 + a^m) \tag{3.19}$$

Similarly for the chord segment $s_5$

$$s_5/r_0 = \sqrt{(1 + a^{2n} - 2a^n \cos n\alpha)}/(1 + a^n) \tag{3.20}$$

Fig. 3.11. Diagram of a portion of a simple contact parastichy pattern in a plane, in which circles are in contact along $m$- and $n$-parastichies. Radii, $r_0$, $r_m$, $r_n$, and secondary divergence angles, $\delta_m$ and $\delta_n$, are shown.

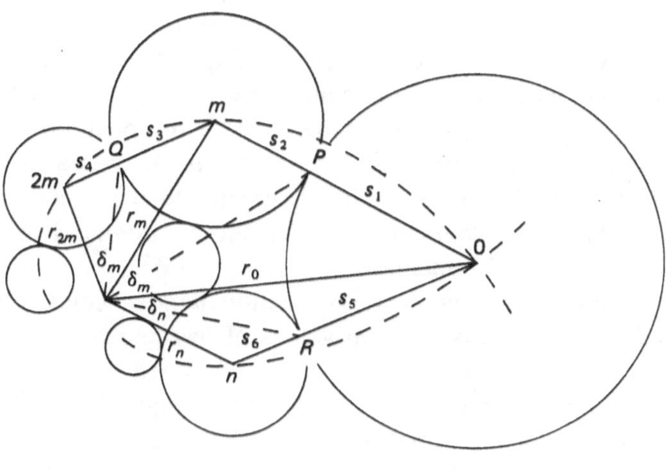

Equating these two expressions gives

$$(1 + a^{2m} - 2a^m \cos m\alpha)/(1 + a^{2n} - 2a^n \cos n\alpha) = (1 + a^m)^2/(1 + a^n)^2$$

This can be rearranged and factored to give

$$(1 + \cos m\alpha)/(1 + \cos n\alpha) = (a^{-m} + 2 + a^m)/(a^{-n} + 2 + a^n)$$

Extracting the square root of each term and using identities for the cosine of a half angle, and for the hyperbolic cosine, gives the following for the relationship between the plastochron ratio, $a$, and the divergence, $\alpha$

$$\cosh \tfrac{1}{2}m \ln a/\cosh \tfrac{1}{2}n \ln a = \pm \cos \tfrac{1}{2}m\alpha/\cos \tfrac{1}{2}n\alpha \qquad (3.21)$$

Van Iterson's (1907) equation 40 is equivalent to this, i.e.

$$\cos \tfrac{1}{2}m\alpha/\cos \tfrac{1}{2}n\alpha = \pm\sqrt{(a^{n-m})} \cdot (1 + a^m)/(1 + a^n)$$

These implicit equations can be solved by successive approximation methods. We have written computer programs incorporating the interval-halving, or the Newton-Raphson, methods to do this. Some representative values of $a$ and $\alpha$ are listed in Table 3.2 and plotted in Fig. 3.10 (dotted lines).

Just as in the case of tangent circles on a cylindrical surface, the arc representing possible values of $a$ and $\alpha$ for a pattern $(m, n)$ extends from the triple-contact point $(n - m, m, n)$ to the point $(m, n, m + n)$. In principle, the coordinates of these triple-contact points may be found by writing equation (3.21) for $(m, n)$ and for $(n, m + n)$, then solving the two expressions simultaneously. But the algebraic approach does not yield simple expressions, so that estimation of representative values is again done by computer approximation, using second degree interpolation. Values of $a$ and $\alpha$ for triple-contact patterns are given in Table 3.2 for comparison with $h$ and $\alpha$ for the cylindrical model. Note that the angular limits for a given pattern are wider in the plane than on the cylinder, and that the value of $\ln a$ is greater than the corresponding $h$-value of the cylindrical model, but that these differences become less pronounced as $m$ and $n$ increase.

The conditions for orthogonal intersection of parastichies in the plane model have been discussed. The values of $\ln a$ and $\alpha$ for orthogonal intersection of contact parastichies are those in which equations (3.21) and (3.10) hold. This problem does not yield a simple algebraic solution, so that a computer interpolation method has again been used to find representative values. Again, there are differences between the cylinder and plane models, which become negligible for higher-order patterns.

In order to draw patterns of contiguous circles such as Fig. 3.12, it is necessary to find the diameter of each circle, at each distance, $r$, from the origin. The radius of a circle has been discussed, so that the required equation is a slight modification of equation (3.19) or (3.20)

$$d = 2r\sqrt{(1 + a^{2m} - 2a^m \cos m\alpha)}/(1 + a^m) \qquad (3.22)$$

Fig. 3.12. Representative patterns of contiguous circles in a plane. (*a*) and (*c*) are the triple-contact patterns, (2, 3, 5) and (3, 5, 8). (*b*) is the double-contact pattern (3, 5).

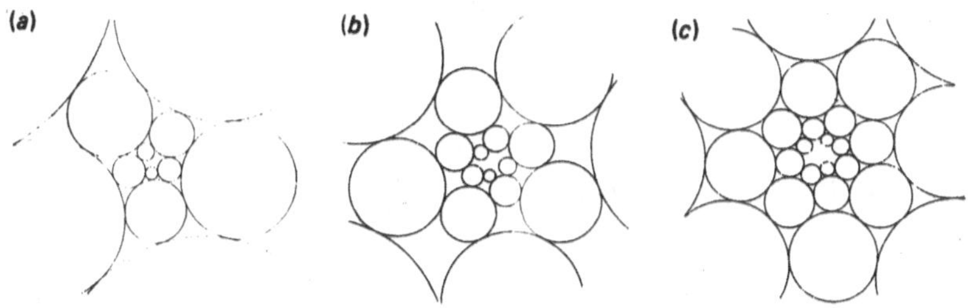


Table 3.3. *Patterns of contiguous circles and contiguous folioids in the plane. Equations (3.21) and (3.23) have been solved to find the plastochron ratio* a *for 'ideal' divergence angles* $\alpha = (3 - \sqrt{5})\pi$, $(5 - \sqrt{5})\pi/5$, $(3 - \sqrt{5})\pi/2$. *For the folioid patterns,* $N = \frac{1}{4}$, $\psi = 30°$

| | | Contiguous circles | | Contiguous folioids | |
|---|---|---|---|---|---|
| $k(m, n)$ | $\alpha$ | $a$ | $\ln a$ | $a$ | $\ln a$ |
| (1, 2) | 137.5078° | 5.43112 | 1.69215 | 1.31969 | 0.27740 |
| (2, 3) | | 1.84260 | 0.61118 | 1.14054 | 0.13150 |
| (3, 5) | | 1.20843 | 0.18932 | 1.04621 | 0.04518 |
| (5, 8) | | 1.07540 | 0.07269 | 1.01801 | 0.01785 |
| (8, 13) | | 1.02742 | 0.02705 | 1.00674 | 0.00672 |
| (13, 21) | | 1.01040 | 0.01034 | 1.00258 | 0.00258 |
| (21, 34) | | 1.00394 | 0.00394 | 1.00098 | 0.00098 |
| (1, 3) | 99.5016 | 1.76290 | 0.56696 | 1.12942 | 0.12171 |
| (3, 4) | | 1.41933 | 0.35018 | 1.08346 | 0.08016 |
| (4, 7) | | 1.09749 | 0.09302 | 1.02299 | 0.02273 |
| (7, 11) | | 1.03911 | 0.03837 | 1.00955 | 0.00950 |
| (11, 18) | | 1.01413 | 0.01403 | 1.00350 | 0.00350 |
| (18, 29) | | 1.00544 | 0.00542 | 1.00135 | 0.00135 |
| 2(1, 2) | 68.7539 | 1.80243 | 0.58914 | 1.14658 | 0.13678 |
| 2(2, 3) | | 1.31015 | 0.27014 | 1.06758 | 0.06533 |
| 2(3, 5) | | 1.09545 | 0.09117 | 1.02279 | 0.02254 |

Note that for given $m$, $n$, $a$, $\alpha$, the diameter of each circle is directly proportional to its distance from the centre. Diagrams such as Fig. 3.12 are drawn efficiently by a computer programmed to solve the equations and operate an $x$-, $y$-plotter.

Van Iterson's definition of $b$ in this model is in terms of the angle under which the circle is seen from the centre, $\omega$, as a fraction of the circle, $b = \omega/2\pi$, and again adapting equation (3.19) or (3.20)

$$\sin \tfrac{1}{2}\omega = d/2r = \sqrt{(1 + a^{2m} - 2a^m \cos m\alpha)}/(1 + a^m) = \sqrt{(1 + a^{2n} - 2a^n \cos n\alpha)}/(1 + a^n)$$

On rearrangement this is

$$b = \cos^{-1}(2 \cos \tfrac{1}{2}m\alpha/\cosh \tfrac{1}{2}m \ln a)/\pi$$

I would propose that the plastochron ratio, $a$, or $\ln a$, be used as a general parameter to characterize contiguous circle patterns in the plane, and that it should not be necessary to consider $d$ or $b$ except in the practical context of drawing a diagram.

### *Contiguous circles on a conical surface*

Patterns of contiguous circles can also be drawn on a conical surface, since it, like the cylinder, can be cut along a generator and unrolled into a flat sheet. In the discussion above of patterns of intersecting parastichies on a conical surface, it was found that the equations relating to a centric plane surface are modified only by inclusion of the factor $N = \sin \tfrac{1}{2}\psi$. The same is true of patterns of contiguous circles on a cone. However, it is desirable to use secondary divergences, $\delta_m$, $\delta_n$, rather than $\alpha$. With this modification of equation (3.21), the applicable equation is

$$\cosh \tfrac{1}{2}m \ln a/\cosh \tfrac{1}{2}n \ln a = \pm \cos \tfrac{1}{2}N\delta_m/\cos \tfrac{1}{2}N\delta_n \qquad (3.23)$$

Values of $a$ and $\alpha$ for triple-contact patterns, and for orthogonal intersection of contact parastichies on the cone can be found by the same numerical approximation methods which are used for solving contiguous circle patterns in the plane.

### *Contiguous folioids, plane model*

The vertical projection of the conical pattern of contiguous circles, onto the plane through point 0, perpendicular to the axis of the cone, is itself an interesting pattern, since the circles on the conical surface are transformed into figures which van Iterson termed folioids. These have a suggestive similarity to the outline of a transverse section of an actual leaf primordium. Equation (3.23) serves to define the parameters $a$, $\alpha$, for double- and triple-contact patterns of folioids. Van Iterson presented a general equation for a folioid but, as a matter of computing strategy, one can easily evaluate the polar coordinates of a circle centred on each lattice point, then adjust the angle coordinate to convert the circle into a folioid. A computer, interfaced to an $x$-, $y$-plotter, has been used to draw diagrams such as those in Fig. 3.13.

**The Fibonacci sequence**

It was seen above that in a cylindrical pattern $(m, n)$ of contiguous circles, decreasing $h$ while maintaining the contacts leads to the appearance of a new contact set, $m + n$, and increasing $h$ leads to the appearance of $n - m$ contacts. Alternatively, without regard to the model of circles, if a point in the lattice, 0, has as its nearest neighbours, $m$ and $n$, then its next nearest neighbours are $n - m$, and $m + n$. These four numbers, in order of magnitude, form a part of a sequence whose rule of formation is that each member, after the first two, is the sum of the two preceding members. A famous sequence with this property is the Fibonacci sequence, 0, 1, 1, 2, 3, 5, 8, 13 . . ., which arises from the initial numbers, 0, 1, or 1, 1, or 1, 2. These numbers have, of course, occupied an exceedingly prominent place in the literature of phyllotaxis. Other sequences can be formed with other initial numbers and the same

Fig. 3.13. Representative patterns of contiguous folioids in the plane, (1, 2), (2, 3) and (3, 5). Alternatively, these can be considered as projections into the plane of patterns of contiguous circles drawn on the surface of a cone.

rule of formation. The many remarkable properties of these numbers, their relation to the golden mean and to continued fractions, their role in optimization algorithms, etc., have been discussed by many authors. I will not attempt to review this large subject.

It will be seen by studying Fig. 3.10, that the curves describing the parameters $h$ (or ln $a$) and $\alpha$ for various patterns fall readily into sequences of this type. It will also be useful to consider Fig. 3.9 further, recalling that the triple-contact pattern $(3, 5, 8)$, Fig. 3.9*c*, was derived from the $(3, 5)$ pattern, Fig. 3.9*b*, by decreasing $h$. If $h$ is decreased further, at the same time increasing $\alpha$ in accordance with equation (3.13), the 3-contacts are broken, resulting in $(5, 8)$, Fig. 3.9*d*. If, on the other hand, $\alpha$ is decreased as $h$ is decreased, the 8-contacts are broken, resulting in $(3, 8)$. Similarly, having followed the transition from $(3, 5)$ to $(2, 3, 5)$ on increase of $h$, a further increase of $h$ with decrease of $\alpha$ breaks the 5-contacts, giving $(2, 3)$ (not shown), and altering $h$, $\alpha$, so as to break the 3-contacts leads to $(2, 5)$. It is suggested that these transitions also be checked in Fig. 3.10.

Thus, decreasing $h$, with corresponding adjustments of $\alpha$, leads to higher order patterns, i.e. larger $m$, $n$. These are represented by descending arcs which intersect at triple-contact points. If, in following these arcs downward, one chooses at each intersection to follow the arc with the steeper slope, $dh/d\alpha$, the arcs chosen represent patterns designated by adjacent members of the Fibonacci sequence, or a sequence with the same rule of formation. If one chooses, at an intersection, to follow the arc with the lesser slope, one initiates a new sequence. For instance, starting with the $(1, 1)$ arc (not shown in Fig. 3.10), it may be followed to the intersection $(1, 1, 2)$ at $\alpha = \pi$, $h = 1.8138$. The steeper arc descending from this point is $(1, 2)$, and continuing in this manner, one would choose the $(2, 3)$, $(5, 8)$, $(8, 13)$, etc., patterns. These represent the Fibonacci sequence, or as van Iterson termed it, the *Hauptreihe*. If, at the $(1, 2, 3)$ intersection, one had followed the arc with lesser slope, $(1, 3)$, one would have chosen numbers which initiate the Lucas sequence, 1, 3, 4, 7, 11, 18 . . ., which van Iterson termed the first of the *Nebenreihen*. All of the possible simple patterns can be organized into sequences in this manner.

These relationships may be significant in explaining the preponderance of Fibonacci patterns in plants, namely, that transitions from one pattern to another can be accomplished by changing the spacing of leaves with minimum adjustment of the divergence angle, as long as one follows such an additive sequence. Departing from the sequence requires a greater angular adjustment for a given change of vertical or radial spacing.

Other authors have discussed these relationships, e.g. van Iterson (1907) and Adler (1974). In the context of the tubular packing of spheres, Harris and I (1980) have studied similar relationships and have termed these changes in pattern 'continuous contraction'. We have shown that for simple packings, $k = 1$, $h$ changes monotonically with change in the radius of the cylinder if one follows a sequence of descending arcs, representing the Fibonacci, or other additive, sequence. If one departs from such a sequence, $h$ changes non-monotonically. That is, $h$ must increase slightly before it can decrease. This may indicate that contractions of tubular packings which follow the Fibonacci sequence are favoured energetically.

It is well known that many plants change the phyllotactic pattern of leaf primordia at the shoot apex as they grow. Young seedlings exhibit a lower order pattern, and the larger apex of an older shoot has a higher order pattern. I think it doubtful that this transition from a lower to a higher order pattern in the growth of a shoot apex occurs in the manner suggested by the discussion above, that is, by way of intermediate triple-contact patterns. This would require considerable shifts in the divergence angle. Williams (1975) has provided admirably complete studies of changing phyllotaxis in several species. For *Linum usitatissimum*, he has plotted the divergence angle against primordium number for plants of different ages. There are considerable fluctuations in the angle, and an eventual approach to the Fibonacci angle, 137.5°. He also determined the plastochron ratio and the phyllotaxis index (Richards, 1948) for these plants. But it is not clear whether the measured divergence angles correlate with the plastochron ratio, as predicted by Fig. 3.10. Clearly, additional analysis and empirical study are in order.

### Application

I believe that the geometrical models described here can have great value for empirical studies of phyllotaxis and growth of the shoot apex. However, there are very few studies in which sufficiently detailed data have been obtained to make a close comparison with the models possible or useful. But Williams (1975) has provided a lucid account of the growth of the shoot apex and young leaves of about 12 angiosperm species, illustrated with remarkable reconstructions from serial sections. There are many growth data. The phyllotactic aspects of his studies are presented in terms of measured divergence angles, the plastochron ratio, and Richards's phyllotaxis index. He also presents several graphical models of circles drawn in the plane. Clearly, the attention to phyllotactic parameters, and modelling, have added much of value to Williams's studies and simulations.

Schwabe (1971) experimentally altered the phyllotaxis of *Chrysanthemum* cuttings by treatment with triiodobenzoic acid. He reported a change in the average divergence angle from 136° in control apices to 158° in treated apices. He analysed the placement of leaf primordia in terms of Richards's phyllotaxis index, 2.23 in controls, 1.96 in treated apices, from which I calculate plastochron ratios of 1.47 and 1.65. Comparing these data with Fig. 3.10, and considering that the leaf primordia could be better modelled with folioids than with circles, suggests that the control apices might be designated (2, 3) and the treated ones (1, 1).

Maksymowych & Maksymowych (1973) have found that a single treatment of vegetatively growing *Xanthium* plants with gibberellic acid produces an immediate, striking and long-lasting effect on their morphology. There is an increase in the rate of growth in height of the plants, a change of leaf shape, and an acceleration of the frequency of leaf initiation by a factor of 1.8. A striking aspect of this effect is a change in the pattern of leaf arrangement as seen in sections of the apex (Fig. 3.14). We undertook a detailed analysis of these sections, largely in terms of the model of folioids in the plane (Maksymowych & Erickson, 1977). In comparing control and treated plants we found no significant difference in the divergence angle. But its average value, $\alpha = 139.36°$, interestingly deviates significantly from the often assumed angle of 137.51°. We proposed a method for objectively estimating the plastochron ratio, based on measurements of chord lengths between primordia as seen in the sections. Plastochron ratios and divergence angles are plotted in Fig. 3.15 for control and experimental plants at several times after the treatment. The figure also includes arcs representing the theoretical parameters of the orthogonal parastichy model, and that of contiguous circles. Numerical fitting of the data to the models showed that the control apices were best represented by a (2, 3) model of contiguous folioids, and the treated plants by a similar (3, 5) model (dashed lines). The analysis yielded estimates of the folioid parameter, $N$, from which it was concluded that the cone taken to represent the surface of the shoot apex was flatter after treatment, $\psi = 92.1°$, than in the control apices, $\psi = 78.1°$. Data on the duration of the plastochron were also obtained; from these we concluded that the rate of radial expansion of the apex did not differ significantly between the two groups. Hence, the effect appeared to be purely geometrical, on the relative placement of new primordia.

Fig. 3.14. (B) Transverse sections of a normal shoot apex of *Xanthium* and (A) of an apex collected 28 days after treatment of the plant with gibberellic acid. The treatment has altered the phyllotactic arrangement from (2, 3) in the untreated plant to (3, 5) after treatment.

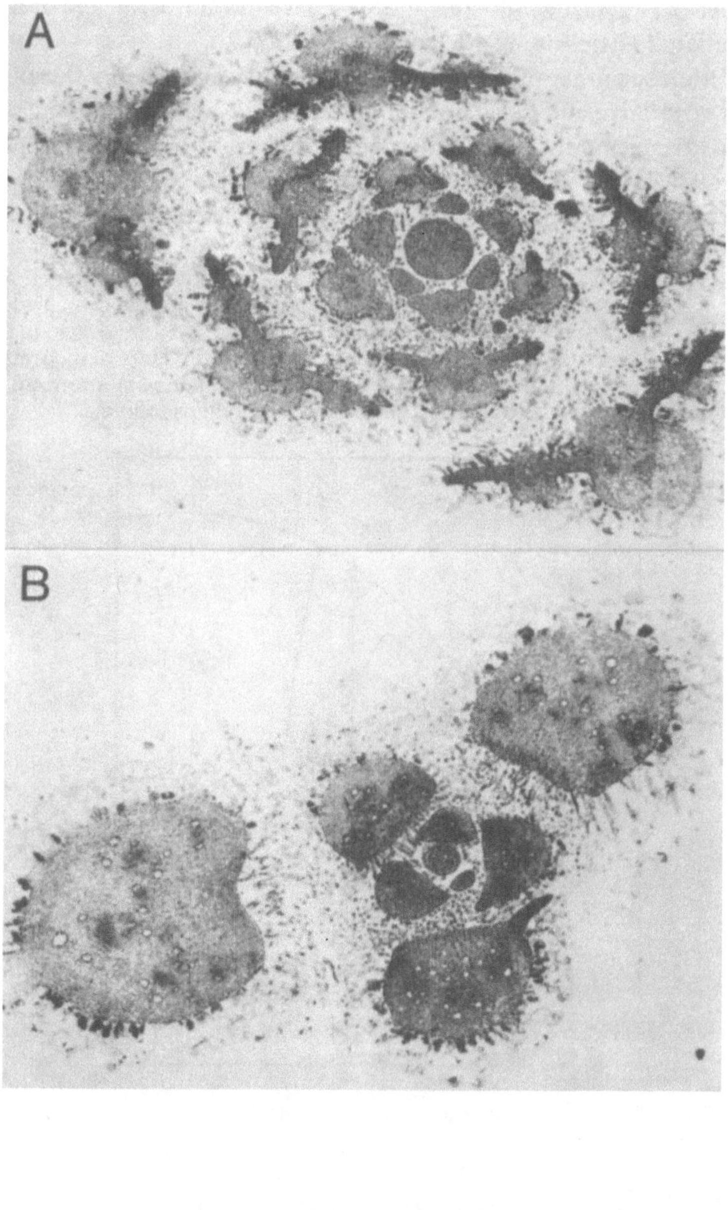

There are many indications that gibberellic acid may play some role in floral induction. To learn if there is a similar effect of floral induction on the *Xanthium* apex to that of gibberellic acid treatment, apices collected during photoperiodic induction were analysed by the methods described above. During the first five days of long-night treatment there was a reduction of the plastochron ratio, and acceleration of the rate of leaf initiation, nearly identical to that resulting from gibberellic acid treatment. This effect was transient, however, since the apex was then reorganized into the inflorescence primordium and florets were initiated (Erickson & Meicenheimer, 1977).

Meicenheimer (1979) has carried out a commendably thorough study of vegetative and floral growth in *Ranunculus*, correlating the phyllotactic arrangement of vegetative and floral parts with their initiation and growth rates. The acceleration of the rate of initiation of floral primordia as compared with leaves and bracts is striking. It is interesting

Fig. 3.15. Phyllotactic parameters estimated from untreated *Xanthium* apices (circles), and gibberellic acid treated apices (triangles). Arcs corresponding to those in Fig. 3.5 and Fig. 3.10 are drawn to allow comparison of these estimates with the parameters of theoretical models. The dashed lines represent parameters of patterns of contiguous circles on a cone, or of contiguous folioids.

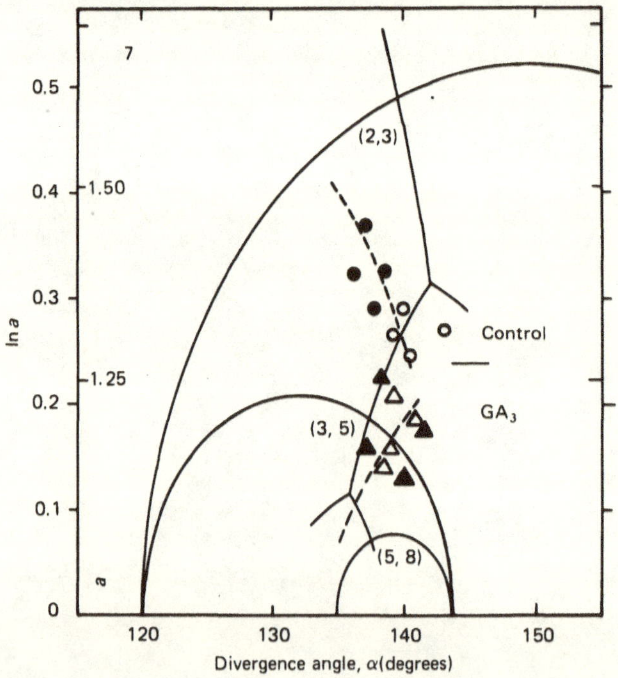

that parastichies connecting vegetative leaves could be traced without interruption through the insertions of floral parts, including the apparent whorl of five petals and the closely packed stamens and carpels. Meicenheimer's work is distinguished by his skilled use of scanning electron micrographs for precise measurement, and his analysis in terms of phyllotactic and growth parameters.

Meicenheimer has made a similar analysis of the growth and phyllotaxis of *Epilobium*, preliminary to his experimental work on altering the phyllotaxis of vegetative shoot apices with chemical agents (Meicenheimer, 1981). He dissected shoot apices so as to expose the youngest bijugate pair of primordia. An auxin transport inhibitor, naphthylphthalamic acid, and an antagonist of auxin activity, chlorophenoxyisobutyric acid, were used. A small amount of the agent in lanolin paste was applied to one of the youngest pair of primordia. This was followed by angular shifts in the positioning of new primordia, resulting in a change from bijugate to spiral phyllotaxis. From analysis of morphological and growth data he found that, with both chemicals, the relative plastochron rates of radial ($\ln a$) and vertical growth of induced spiral shoot apices is about one-half that of bijugate controls. The same is true of the relative plastochron rate of leaf elongation. Since the duration of the plastochron is also about one-half, one concludes that the treatments led to no significant change in growth rates. The effect of the treatments therefore is only on the relative positioning of new primordia, again a purely geometric effect.

## References


Adler, I. (1974). A model of contact pressure in phyllotaxis. *Journal of Theoretical Biology*, **45**, 1–79.

Adler, I. (1975). A model of space filling in phyllotaxis. *Journal of Theoretical Biology*, **53**, 435–44.

Adler, I. (1977). The consequences of contact pressure in phyllotaxis. *Journal of Theoretical Biology*, **65**, 29–77.

Church, A. H. (1904). *On the Relation of Phyllotaxis to Mechanical Laws*. London: Williams & Norgate.

Church, A. H. (1920). *On the Interpretation of Phenomena of Phyllotaxis*. Botanical Memoirs, No. 6. London: Oxford University Press. (Reprinted by Hafner Publishing Co., New York.)

Coxeter, H. S. M. (1961). *Introduction to Geometry*. New York: John Wiley & Sons.

Erickson, R. O. (1973a). Tubular packing of spheres in biological fine structure. *Science*, **181**, 705–16.

Erickson, R. O. (1973b). Geometry of tubular ultrastructures. *Brookhaven Symposium in Biology*, **25**, 111–28.

Erickson, R. O. & Meicenheimer, R. D. (1977). Photoperiod induced change in phyllotaxis in *Xanthium*. *American Journal of Botany*, **64**, 981–8.

Harris, W. F. & Erickson, R. O. (1980). Tubular arrays of spheres: geometry, continuous and discontinuous contraction, and the role of moving dislocations. *Journal of Theoretical Biology*, **83**, 215–46.
Hellendoorn, P. H. & Lindenmayer, A. (1974). Phyllotaxis in *Bryophyllum tubiflorum*: morphogenetic studies and computer simulations. *Acta Botanica Néerlandica*, **23**, 473–92.
Maksymowych, R. & Erickson, R. O. (1977). Phyllotactic change induced by gibberellic acid in *Xanthium* shoot apices. *American Journal of Botany*, **64**, 33–44.
Maksymowych, R. & Maksymowych, A. B. (1973). Induction of morphogenetic changes and acceleration of leaf initiation by gibberellic acid in *Xanthium pennsylvanicum*. *American Journal of Botany*, **60**, 901–6.
Meicenheimer, R. D. (1979). Relationships between shoot growth and changing phyllotaxy in *Ranunculus*. *American Journal of Botany*, **66**, 557–69.
Meicenheimer, R. D. (1981). Changes in *Epilobium* phyllotaxy induced by *N*-1-naphthylphthalamic acid and α-4-chlorophenoxyisobutyric acid. *American Journal of Botany*, **68**, 1139–54.
Mitchison, G. J. (1977). Phyllotaxis and the Fibonacci series. *Science*, **196**, 270–5.
Richards, F. J. (1948). The geometry of phyllotaxis and its origin. *Symposium of Society for Experimental Biology*, **2**, 217–44.
Richards, F. J. (1951). Phyllotaxis: Its quantitative expression and relation to growth in the apex. *Philosophical Transactions of the Royal Society of London*, **B235**, 509–64.
Richter, P. H. & Schranner, R. (1978). Leaf arrangement. Geometry, morphogenesis, and classification. *Naturwissenschaften*, **65**, 319–27.
Roberts, D. W. (1977). A contact pressure model for semi-decussate and related phyllotaxis. *Journal of Theoretical Biology*, **68**, 583–97.
Roberts, D. W. (1978). The origin of Fibonacci phyllotaxis – an analysis of Adler's contact pressure model and Mitchison's expanding apex model. *Journal of Theoretical Biology*, **74**, 217–33.
Schoute, J. C. (1913). Beiträge zur Blattstellungslehre. *Récueil des Travaux Botaniques Néerlandais*, **10**, 153–324.
Schüepp, O. (1923). Konstruktionen zur Blattstellungstheorie. I. *Berichte der Deutsche Botanische Gesellschaft*, **41**, 255–62.
Schwabe, W. (1971). Chemical modification of phyllotaxis and its implications. In: *Control Mechanisms of Growth and Differentiation*. Symposium of Society for Experimental Biology, XXV, 301–22.
Snow, M. & Snow, R. (1935). Experiments on phyllotaxis. III. Diagonal splits through decussate apices. *Philosophical Transactions of the Royal Society of London*, **B225**, 63–94.
Thornley, J. H. M. (1975). Phyllotaxis. I. A mechanistic model. II. A description in terms of intersecting logarithmic spirals. *Annals of Botany*, **39**, 493–524.
Turing, A. M. (1959). In *Alan M. Turing, Mathematician and Scientist*, by S. Turing. Cambridge: Heffer.
Van Iterson, G. (1907). *Mathematische und mikroskopish-anatomische Studien über Blattstellungen*. Jena: Gustav Fischer.
Van Iterson, G. (1960). New studies on phyllotaxis. *Proceedings of Koninklijke Nederlandse Akademie van Wetenschappen Ser. C*, **63**, 137–50.
Veen, A. H. & Lindenmayer, A. (1977). Diffusion mechanism for phyllotaxis, theoretical, physico-chemical and computer study. *Plant Physiology*, **60**, 127–39.
Weyl, H. (1952). *Symmetry*. Princeton University Press.
Williams, R. F. (1975). *The Shoot Apex and Leaf Growth*. Cambridge University Press.
Young, D. A. (1978). On the diffusion theory of phyllotaxis. *Journal of Theoretical Biology*, **71**, 421–32.

# 4

# Kinematic analysis of leaf expansion

WENDY K. SILK

## Introduction

The term 'kinematics' is used by fluid dynamicists to refer to motion or deformation in a fluid or continuum without reference to underlying forces or causes. Mathematical expressions used for the kinematics of the flow field provide a quantitative, succinct description of the phenomenology of the flow. Experimentally determined evaluation of these expressions can be used in turn to test hypotheses of force-deformation, energy-deformation and other derived relationships.

The relationship between plant growth and fluid flow becomes apparent when we consider the organization of growing regions, particularly apices, plumular hooks, and vascular cambia. The apical dome, for instance, is a structure which may change slowly in time and which may even, for some time periods, appear unchanging. Anatomists recognize regions of large and small cell size and slow and rapid cell division (see Chapter 1). Histologists can identify regions by their staining properties. But the tissue elements which comprise the structure are displaced through these regions and experience continuing change. An ivy branch provides another illustration of the analogy between plant development and fluid flow. If the branch is examined at successive plastochrons, it may appear to have a 'steady' (time-invariant) form. The youngest expanded leaf appears on a given day to be identical to the youngest expanded leaf two days later. But if we mark the youngest expanded leaf and examine the same leaf two days later, we observe that it develops to resemble its older neighbour at the time of the original observation. The constant form of the branch is produced by a succession of leaves, each of which enlarges, develops pigment, and acquires carbohydrate as it is displaced from the apex. As plant morphologists have long recognized, the structure of the branch implies a developmental history. One can predict the fate of a young leaf by looking at leaves progressively farther from the apex. Conversely, one can infer the history of an older leaf from the characteristics of leaves closer to the apex.

The existence of steady organs composed of changing elements is reminiscent of fluid structures, such as waterfalls, fountains, and boat wakes, structures which do not change but whose elements flow through them. Indeed, the concepts and methods of classical fluid dynamics prove a powerful tool for the analysis of developmental processes.

A particularly relevant fluid-dynamical concept is the distinction between spatial and material specifications of variables. In the ivy branch example, we could describe leaf size in spatial terms by measuring leaves encountered at different positions on the stem and by plotting leaf size against distance from the tip. An equally valid way to specify leaf size would be to choose a 'material' (real) element, a particular leaf, and to measure its size as it is displaced to various distances from the tip. Since the form of the branch is steady (at the same plastochron stage), the spatial and material specifications of leaf size will be identical. Suppose, however, that the branch changes from the juvenile to the adult form. Now successive leaves no longer undergo the same sequence of morphological and physiological events. The form of the branch is nonsteady (changing in time), and the spatial and material specifications of developmental variables no longer coincide.

The apical dome can also be specified in either spatial or material terms. Growth field variables, i.e. functions of position in the apical dome and of time, can be described by values at distances from the centre of the apical dome. Alternatively, the structure can be described by the properties of the material elements, the cells or cell components, as they are displaced through space and time. In botanical terms the distinction is that between 'site specific' (spatial) or 'cell specific' (material) developmental patterns (Green, 1976). The botanical terminology is familiar and therefore easy to use, but it must be remembered that variables should be specified at instants in time and points (infinitesimally small increments) of space. The term 'cell specific' is usable only if the group of cells under consideration does not become very large.

A related idea is the distinction between local and material rates of change. Again, the ivy branch provides an illustration. Leaf length at a position, say 3 cm from the tip, may not change in time. But the leaf which is momentarily located 3 cm from the tip is growing; and, as it is displaced to 4 cm from the tip, it increases in length. It is characteristic of plants that, if an apex is chosen as origin of the coordinate system, the material rates of change in stem and root elements are often larger than the local rates of change. If a growth field variable, **a**, is specified in spatial terms, then the material rate of change, symbolized $\mathbf{Da}/\mathrm{D}t$, can be evaluated as

$$\frac{\mathrm{D}\mathbf{a}}{\mathrm{D}t} = \frac{\partial \mathbf{a}}{\partial t} + \mathbf{v} \cdot \nabla \mathbf{a} \qquad (4.1)$$

Material derivative (cell specific) = Local derivative (site specific) + Convective rate of change (displacive)

where $t$ is time, and $\mathbf{v}$ is growth velocity (rate of displacement from a material origin in the plant). Equation (4.1) says that the material rate of change equals the rate of change at the position instantaneously occupied by the tissue element, plus the rate of change due to displacement to a new position. An important implication for experimental work is that material rates of change can be calculated if data are obtained on the spatial distribution of both growth velocity and the variable of interest.

In growing leaves, unlike apices, the distinction between local and material changes is less obvious and may be numerically small. Leaf development is rarely steady, so local rates of change are significant. If there is no spatial gradient in the growth variable of interest, then the convective term is zero. Inspection of equation (4.1) reveals that the material rate of change equals the local rate of change in this case. Thus, if the variable is homogeneously distributed over the leaf surface, only local changes needed be evaluated. For quantities which vary with position, evaluation of material rates of change still requires data on spatial distribution of the variable plus the growth velocity.

A useful expression for leaf growth analysis is the strain rate. The growing leaf, like an expanding fluid, consists of a continuum of material elements, each of which may experience expansion, as well as displacement due to expansions of other elements in the leaf. Leaf morphogenesis may be described quantitatively by a field of strain rate tensors, as explained in the next section. Leaves of several species have been analysed for strain rate distributions. These studies include the work of Richards & Kavanagh (1943), Erickson (1966), Archer, Lockhart & Deshmukh (1973), and R. Plant & W. K. Silk (unpublished).

The continuity equation is another useful product of fluid dynamics. This relates the local production rate (due to biosynthesis and/or import) to the observed concentration and growth velocity. In the concluding section, use of this equation is illustrated with data on growth and dry matter in grape leaves.

Historically, several botanists have perceived the analogy between plant growth and fluid flow. Schüepp (1926) described apical dome development with 'Verschiebungskurven', growth trajectories or pathlines. Within the past decade there has been widespread application of the formal aspects of fluid dynamics to growth studies. Niklas (1977) proposed a related analogy to continuum mechanics and used the continuum mechanical approach in subsequent models (particularly Niklas & Mauseth, 1980) of cell size as a result of growth and partitioning rates in the cactus apex. Archer has studied reaction wood formation in continuum mechanical terms (summarized in Wilson & Archer, 1977). Green (1976) used a graphical version of the continuity equation to model the distribution of cell lengths in the root as a result of partitioning and strain rates. Silk & Erickson (1978, 1979) described the relevance of concepts and equations of fluid dynamics to problems of plant development. Tognetti & Winley (1980) modelled the growth of a column of age- and position-dependent cells. Gandar (1980) used data of Erickson & Goddard (1951) with the continuity equation to analyse cell production rates and dry matter production rates in the primary corn root.

The strain rate distribution on the higher plant apical dome in steady growth has not been measured. Inaccessibility of the apex for marking and time lapse photography has thwarted attempts at detailed growth analysis. Nevertheless, one theoretical and one empirical study have shed light recently on apical development. Hejnowicz modelled apical growth in terms of displacement velocities (Hejnowicz & Nakielski, 1979; Hejnowicz, unpublished). He wondered, given the velocity of marks applied to the surface of the apex, what can be inferred about the velocities of the internal cellular elements. The answer was given essentially as a stream potential function and implied that lines of equal velocity are parabolas orthogonal to parabolas formed by the displacement lines. An enlightening experimental study was performed by Green and coworkers on the development of adventitious stems of *Graptopetalum*. After extensive marking of the lamina region which would produce the new apex, Green recorded photographically the origin and development of a primordial axis. Using the Mohr's circle method of mechanics to analyse strain, Green & Poethig (1982) were able to show that the plane of cell division does not coincide statistically with a plane of minimum shear. Their work refutes a hypothesis of long standing.

### Mathematical formulation of growth field kinematics

Growth can be specified by a velocity field which describes at points on the leaf surface the rates at which marks are moving away from a given origin, such as the leaf tip or the point of insertion of the petiole. The velocity field itself depends on the choice of origin and provides relatively little insight into morphogenesis. More informative is the rate of change of velocity with position, the velocity gradient, $\delta v_s/\delta s$ (where $v_s$ denotes the component of **v** in direction $s$). It describes the local stretch rate of a line segment (in length length$^{-1}$ time$^{-1}$) and does not change if the origin is changed. Some relevant properties of the velocity gradient are explained in the early growth analyses described in the following section. The quantity $\delta v_x/\delta x$ is the strain rate of the line in the $x$-direction; the quantity $\delta v_y/\delta y$ is the strain rate of the line in the $y$-direction; and $\delta v_x/\delta y + \delta v_y/\delta x$, the so-called shear strain rate, is the rate (in radians time$^{-1}$) at which the line segments originally in the $x$- and $y$-directions are rotating toward each other. The stretch rate or velocity gradient is different in different directions, but the sum of the velocity gradient in two perpendicular directions $\delta v_x/\delta x + \delta v_y/\delta y$, gives a quantity, $\nabla \cdot \mathbf{v}$, the divergence of velocity, which is independent of the choice of orientation of the axes and which measures the rate of area change locally. If the $x$- and $y$-strain rates and the shear-strain rate are known, then the rate of strain in any direction, $L_\theta$, (assuming $\theta$ is the angle between the $x$-axis and the line in question) can be calculated from

$$L_\theta = \frac{\partial v_x}{\partial x}\sin^2\theta + \left(\frac{\partial v_x}{\partial y} + \frac{\partial v_y}{\partial x}\right)\sin\theta\cos\theta + \frac{\partial v_y}{\partial y}\cos^2\theta \qquad (4.2)$$

The direction in which strain rate is greatest, $\theta_{max}$, can be found by solving

$$\tan 2(\theta_{max}) = \left(\frac{\partial v_x}{\partial y} + \frac{\partial v_y}{\partial x}\right) \Big/ \left(\frac{\partial v_x}{\partial x} - \frac{\partial v_y}{\partial y}\right) \qquad (4.3)$$

The velocity field, then, is the empirical basis for calculation of the more informative velocity gradient. A general formulation of the velocity gradient is the strain rate tensor. Probably the best way to specify the morphogenesis of the leaf at an instant in time is with a field of tensor pairs: the strain-rate tensor and the vorticity tensor characterize respectively the local expansion rate and the rate of change in orientation. Explanations of the strain rate tensor are given in many fluid dynamics and continuum mechanics texts (e.g. Crandall, Dahl & Lardner, 1972; Batchelor, 1967).

The strain rate tensor is written

$$e_{ij} = \frac{1}{2}\left(\frac{\partial v_i}{\partial x_j} + \frac{\partial v_j}{\partial x_i}\right) \tag{4.4}$$

where $v_i$ is velocity and $x_i$ is the position vector. This expression is shorthand (indicial) notation for an expression with nine components

$$M \equiv \frac{1}{2}\begin{pmatrix} \frac{\partial v_x}{\partial x} + \frac{\partial v_x}{\partial x} & \frac{\partial v_x}{\partial y} + \frac{\partial v_y}{\partial x} & \frac{\partial v_x}{\partial z} + \frac{\partial v_z}{\partial x} \\ \frac{\partial v_y}{\partial x} + \frac{\partial v_x}{\partial y} & \frac{\partial v_y}{\partial y} + \frac{\partial v_y}{\partial y} & \frac{\partial v_y}{\partial z} + \frac{\partial v_z}{\partial y} \\ \frac{\partial v_z}{\partial x} + \frac{\partial v_x}{\partial z} & \frac{\partial v_z}{\partial y} + \frac{\partial v_y}{\partial z} & \frac{\partial v_z}{\partial z} + \frac{\partial v_z}{\partial z} \end{pmatrix} \tag{4.5}$$

In leaves with negligible change in thickness during growth the strain rate tensor reduces to an expression with four components

$$N = \frac{1}{2}\begin{pmatrix} \frac{\partial v_x}{\partial x} + \frac{\partial v_x}{\partial x} & \frac{\partial v_x}{\partial y} + \frac{\partial v_y}{\partial x} \\ \frac{\partial v_y}{\partial x} + \frac{\partial v_x}{\partial y} & \frac{\partial v_y}{\partial y} + \frac{\partial v_y}{\partial y} \end{pmatrix} \tag{4.6}$$

In the expanded matrix $M$ or $N$, elements in the $i^{th}$ row and $j^{th}$ column are equal to elements in the $j^{th}$ row and $i^{th}$ column. For instance, the element in row 1, column 2 is the same as the element in row 2, column 1. The matrix is therefore said to be *symmetric*. It can be proved (cf. Boas, 1966, Chapter 10) that matrix $N$ can be *diagonalized by a similarity transformation* to another matrix $D$

$$D = \frac{1}{2}\begin{pmatrix} 2\frac{\partial V_X}{\partial X} & 0 \\ 0 & 2\frac{\partial V_Y}{\partial Y} \end{pmatrix} \tag{4.7}$$

which has all nondiagonal elements equal to zero. That is, of four elements of matrix $D$, only the elements in row 1, column 1; and row 2, column 2 are different from zero. Physically, matrix $D$ describes in a rotated coordinate system $X$, $Y$, the same deformation that matrix $N$ describes in the original $x$, $y$ system. Furthermore, the components of matrix $D$ equal the strain rates in the directions of *principal strain.* These are the strain rates in the direction of greatest strain and in the direction of least strain. The two principal strains are necessarily at right angles and do not rotate relative to each other. The methods for diagonalizing $M$ and for finding the direction of $X$ relative to $x$ have been known for more than one hundred years and can be shown to produce the equations (4.2) and (4.3).

In addition to straining, the leaf element may undergo a change in orientation. Since the leaf is a continuum, the elements do not spin independently in place. But if the bottom of the element is being displaced faster than the top of the element, there will be a net average rotation. We might average the tangential component of velocity along the circumference of a small circle and divide by the radius of the circle to obtain a measure of the orientation change. This is related to evaluating the components of the vorticity tensor, $v_{ij}$.

$$v_{ij} = \frac{1}{2}\left(\frac{\partial v_i}{\partial x_j} - \frac{\partial v_j}{\partial x_i}\right) \tag{4.8}$$

In a flat leaf which grows without change in thickness the important vorticity term is

$$w = \left(\frac{\partial v_y}{\partial x} - \frac{\partial v_x}{\partial y}\right) \tag{4.9}$$

where the vorticity $w$ is twice the average rotation rate (in radians $\text{time}^{-1}$) about an axis perpendicular to the plane of the element.

Physically, the strain rate tensor $D$ may be visualized as a circle of unit area which is deforming during growth into a larger ellipse. The area of the ellipse equals the magnitude of the local growth or strain rate; the major axis points in the direction of maximum strain; and the minor axis points in the direction of minimum strain. The ellipse axes remain at right angles to each other, but the element as a whole has a net rotation rate given by half the vorticity. These physical interpretations have been used in presenting strain rate analyses in growing leaves.

### Strain rates in growing leaves

The growth of a leaf can be characterized by the field of strain rate tensors, as described above. Leaves of several species have been analysed in these fluid dynamical terms. The first presentation of strain rate distributions was by Richards & Kavanagh (1943), who used data of Avery (1933) on the tobacco leaf. Richards & Kavanagh discussed the representation of the changing leaf geometry by velocities of material points. They introduced the important idea that the growth rate or strain rate is given by the gradient in the relative growth velocity. They recognized that growth rates can vary in different directions and pointed out the difference between isotropic growth (strains equal in different directions locally) and isogenic growth (strains equal in different directions and also at different positions). In a mathematical appendix they derived the expression relating growth in an arbitrary direction, $\theta$, relative to the $x$-axis, to growth in two orthogonal directions, $x$ and $y$ (equation 4.2). Another important derivation was the equation relating the $x$- and $y$-direction strain rates to the direction of maximum strain (equation 4.3). From standard ideas of vector analysis the authors recognized that the relative rate of volume increase equals the sum of the strains in three orthogonal directions.

In Avery's data Richards & Kavanagh did not have access to information on the chronological age of the marked leaves. They resorted to using leaf length as the time variable. Thus they computed rates of change of position with leaf length, then differentiated these morphological velocities with respect to the $x$-coordinate and the $y$-coordinate. Because they were unable to calculate velocity in temporal terms, Richards & Kavanagh did not compare the strain rates at different stages. They published the strain distribution at each of four stages in relative terms. The area of the strain ellipse depicting the largest strain was given the value 100, and the remaining strains were scaled proportionately. Results showed a different strain rate distribution at each stage (Fig. 4.1). The youngest stage, with leaf length 85 mm, has the least homogeneous strain rate distribution; strain rate at the tip is only 20% of the rate at the basal area. At later stages, the basal third of the leaf continues to have the growth maximum, but the tip grows at progressively larger fractions of the maximum. At stage II (leaf 172 mm) the tip grows 40%; at stage III (leaf 264 mm) the tip grows 50%, and at stage IV (leaf 300 mm) the tip has 70% of the maximum strain rate. Avery's contention that that tip stops growing when the leaf is at the

2 mm stage appears refuted by the strain analysis. Another interesting result was that the growth in stage I is anisotropic. Although Richards & Kavanagh did not publish the numerical results for the $\theta$ distribution, their figure shows strain ellipse axes associated with the marginal meristem at stage I are parallel, with the long axis rotated perhaps 120° from the $x$-axis. The later stages are more isotropic as well as more homogeneous in their growth rate distribution.

The analysis of Richards & Kavanagh (1943) on the tobacco leaf might have been extended to other species to provide a quantitative description of other patterns of leaf maturation. However, the ideas in the 1943 paper were not much used until 1966 when Erickson published a computer program for leaf growth analysis. This work, which has been reviewed elsewhere (Erickson, 1976; Erickson & Silk, 1980), was one of the first applications of high-speed computers to botany. The leaf analysis used the equations of Richards & Kavanagh (1943) and provided innovations in both data collection and analysis.

Photographs of a growing *Xanthium* leaf were taken on three successive days. A grid of 250 equally spaced points was marked on the photograph which was intermediate in the time series. (Distance between the grid points corresponded to 3 mm on the leaf.) With an ingenious stereo-imaging technique, Erickson identified the same

Fig. 4.1. First strain rate analysis of leaf growth. Arms of crosses are proportional in length to magnitudes of principal strains and are oriented in directions of principal strains in *Nicotiana*. (Redrawn from Richards & Kavanagh, 1943.)

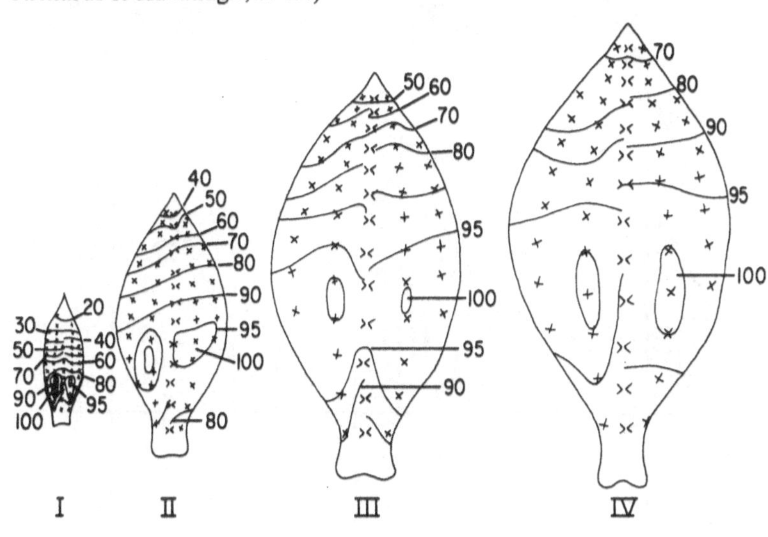

material points on the earlier and later photographs. The original grid was bowed in one direction in the earlier print and in the opposite sense in the later print (Fig. 4.2). A digitizer was used to find the $x$, $y$ coordinates of the grid points in the first and third photographs. This information was recorded on computer cards to serve as input for a program which computed growth velocities of each mark and then evaluated the growth gradient in the $x$- and $y$-directions. The program used differentiating formulae based on the fitting of polynomials to equally spaced points. This method had the advantage that the patch length and the degree of the polynomial could be varied, so that the optimum amount of smoothing could be chosen to reveal the growth pattern.

Fig. 4.2. Growth record for computerized strain rate analysis of *Xanthium*. The grid of equally spaced points was marked on the middle photograph, and corresponding points were identified on the earlier and later photographs. (Adapted from Erickson, 1966.)

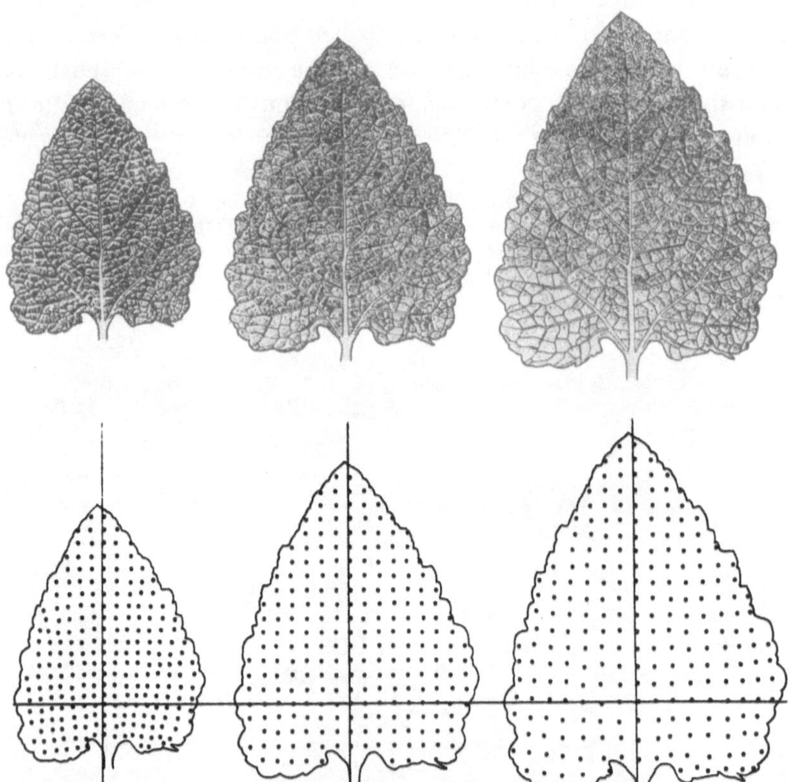

The *Xanthium* analysis, unlike the earlier study of *Nicotiana*, gave the strain rates in chronological terms (Fig. 4.3). The growth is heterogeneous (different at different positions). Like the *Nicotiana* leaf at 85 mm (stage I of Richards & Kavanagh), *Xanthium* at 65 mm (leaf plastochron index, LPI, 3.3) grows most rapidly at the base, where the strain rate is 0.60 $d^{-1}$, while strain at the tip is only 0.05 $d^{-1}$. Isopleths of equal strain rate run approximately horizontally across the leaf and dip basally as they approach the margin. Since the time rate of progression of the plastochron index is well known for *Xanthium*, it is possible to convert these values to a plastochron (p) scale. If the leaves were growing under conditions similar to those described by Erickson in 1966, at LPI 3.3, the leaf base is growing 1.27 $p^{-1}$, and the tip is growing 0.10 $p^{-1}$. While heterogeneous, the growth is rather isotropic (similar in different directions at the same point). The arms of the strain ellipse axes are equal in length, and their orientation appears almost random (Fig. 4.3*b*). This is to be expected, since $\theta$ is indeterminate when growth is isotropic. Only along the margins of the leaf is there regular orientation, with strain in the vertical direction exceeding strain in the horizontal direction.

Fig. 4.3. Strain rate analysis of *Xanthium* leaf. (*a*) Lines represent isopleths of equal strain rate: 0.6, 0.5, 0.4, 0.3, 0.2, 0.1 $d^{-1}$. (*b*) Crosses are axes of strain rate ellipses and indicate directions and magnitudes of principal strains. (Adapted from Erickson, 1966.)

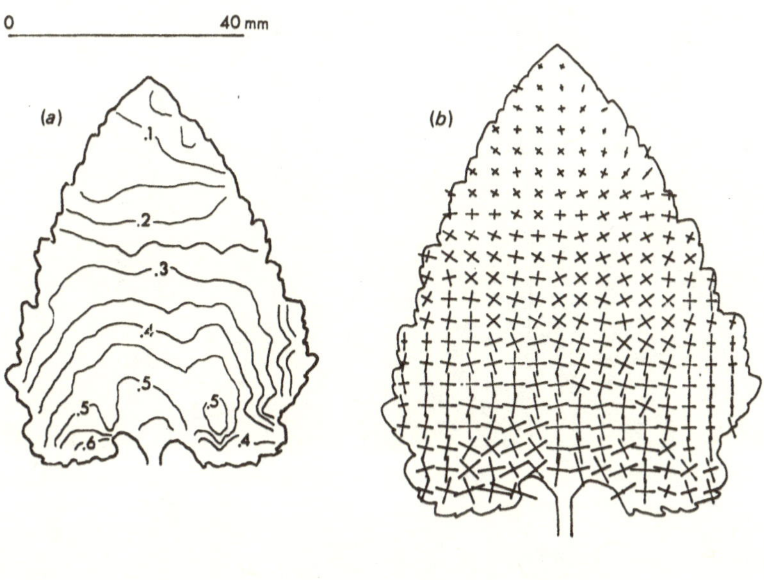

Because the base of the leaf is expanding more rapidly than the tip, leaf cells are experiencing a change in orientation during the growth. Silk & Erickson (1979) were able to quantify this aspect of the morphogenesis by computing the vorticity over the surface of the *Xanthium* leaf. The original data on growth velocities were used with equation (4.9) of the preceding section. Results were expressed as angular velocities in radians per day (Fig. 4.4). Rate of element rotation increases with distance from the midrib. Tissue elements along the left basal margin rotate as much as 0.12 rad $d^{-1}$ (0.25 rad $p^{-1}$), while elements along the right margin rotate counterclockwise at the same rate.

Fig. 4.4. Vorticity of *Xanthium* leaf. Lines represent isopleths of equal angular velocity: 0, 0.025, 0.050 radians $d^{-1}$. (From Silk & Erickson, 1979.)

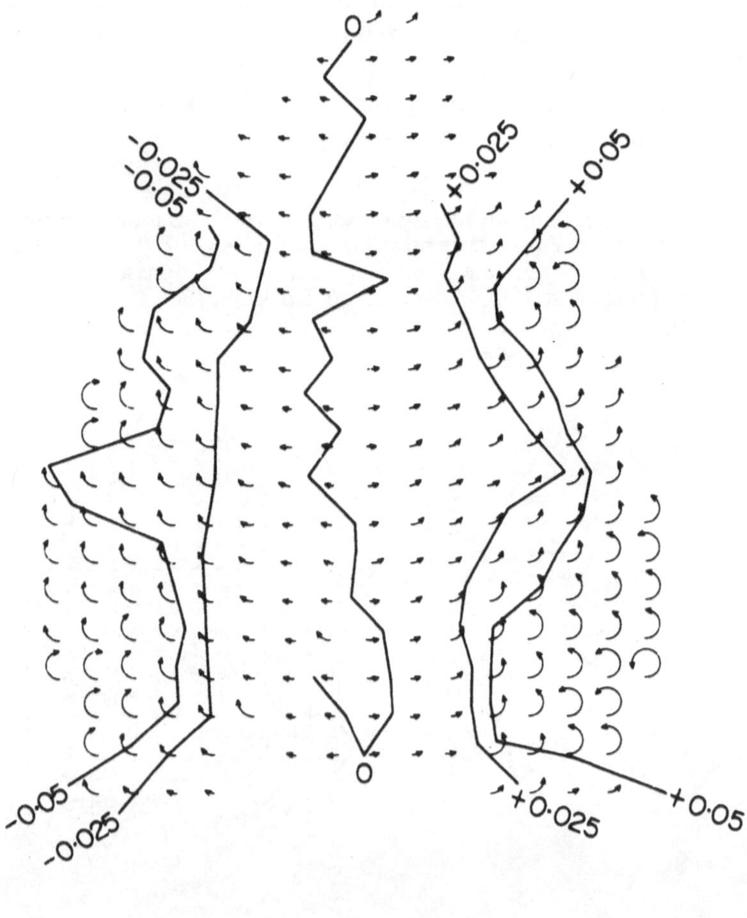

A further advance in analysis of leaf morphogenesis was made by Archer *et al.* (1973), who studied the growth of the cucumber leaf. This work extended previous studies in several ways. Analyses were performed at two, four, six and eight days after observation began. This enabled the investigators to discover the changes in growth pattern with leaf development. A finite element method was used, which permitted use of a data set of nonequally spaced points. An equation from mechanics was introduced to find the principal strains directly from the reference strains and shear strain; a test of the numerical methods was also made.

For the cucumber leaf analysis, more than 100 vein intersections were identified on each of a series of leaf photographs. Each point was located within a polygon formed by lines connecting neighbouring points, and areas of the polygons were computed. The local areas were tabulated as a function of a time for an eight-day period, and the resulting growth data were fitted to the monomolecular or 'logistic' equation. Erickson's instantaneous growth analysis was thus extended to a time series of growth rate distributions. Results showed that in cucumber, like tobacco, the tip matures earlier than the base. A tissue element near the tip reached 25%, 50% and 75% of its final size at, respectively, 1.8, 4.7 and 7.7 days after observations began. An element nearer the base reached the same stages of growth 4.7, 6.2 and 8.4 days after the beginning of the experiment. For most of the observation period the strain rate increased with distance from the tip. At $t=2$ days and $t=4$ days, strain at the tip was less than $0.25\,d^{-1}$ and strain at the base of the midrib was more than $0.50\,d^{-1}$. At $t=6$ days the leaf tip was growing $0.13\,d^{-1}$ and the base of the midrib was growing $0.37\,d^{-1}$. By day 8 the range had dropped from $0.10\,d^{-1}$ near the tip to $0.18\,d^{-1}$ near the base. Since leaf lengths were not published, it is not possible to express the results on a plastochron basis. The direction of maximum strain changed significantly from period to period and from element to element and exhibited no clear trend. When strain rate found from the plane strain equations was compared to the relative rate of area change, the two methods were found to be in reasonable agreement. This provided a good test of the finite element numerical methods.

Recently, Richard Plant and I analysed the growth of grape leaves. Although we conceived the project independently, we found that our approach bore many similarities to the earlier work of Archer *et al.* (1973), including the analysis through several plastochrons, the use of a finite element method, and the test that the relative rate of area increase be close in value to the sum of the principal strains. Results of our analysis and use of the growth data to compute production rates are discussed in the next section.

### Local dry matter production rates in growing grape leaves

Leaf dry matter accumulates from production processes (biosynthesis and import) and simultaneous dilution of substance by the cellular water uptake associated with growth. The net biomass production rate is thus related to both concentration and growth rate. The relationships are given by the continuity equation of fluid dynamics. For a two-dimensionally expanding leaf surface the local rate of deposition of substance, $D$, can be calculated from

$$
\underset{\substack{\text{Deposition rate}\\ \text{(local)}}}{D} = \underbrace{\underset{\substack{\text{Observed}\\ \text{concentration}\\ \text{change}}}{\frac{\partial \varrho}{\partial t}} + \underset{\substack{\text{Convective}\\ \text{concentration}\\ \text{change}}}{v_x \frac{\partial \varrho}{\partial x} + v_y \frac{\partial \varrho}{\partial y}}}_{\text{(material)}} + \underset{\substack{\text{stretch}\\ \text{component}\\ \text{(local)}}}{\varrho(\nabla \cdot \mathbf{v})} \tag{4.10}
$$

where $v_x$ is the $x$ component of velocity, and $v_y$ is the $y$-component of velocity. In this section the components of equation (4.10) are estimated in growing leaves of *Vitis vinifera*, cv. Ruby Cabernet.

Leaves were photographed at daily intervals, and a piece of graph paper was included in each photograph for scaling purposes. The lamina was clamped between two plexiglass plates and backlit with a portable flash to highlight the venation pattern (Fig. 4.5*a*). Lengths of blade plus petiole were also recorded to permit calculation of LPIs assuming a reference length of 30 mm. The leaf photographs data were analysed by a triangular finite element method designed in collaboration with Richard Plant. Eighty-three vein intersections were identified on each of five successive photographs. The resulting grid of nonequally spaced points was used to triangulate the surface into 119 elements (Fig. 4.5*b*). The origin of the coordinate system was chosen to be the basal point of intersection of the main veins. A length corresponding to 5 mm on the leaf was laid off from the origin along the midrib to become the $y$-axis. The $x$, $y$ coordinates of the points were then recorded with a Tektronix 4956 digitizer interfaced with a desk-top computer. The digitizing program scaled the distances to millimetres and rotated the points to a reference frame with $y$-axis along the leaf midrib. The computer program for growth analysis used two-dimensional linear interpolation to estimate point velocities, strain rates in $x$- and $y$-directions, and computational error. Interpolation formulae were taken from Zienkiewicz (1961).

The strain rate analysis revealed that at LPI 2 the growth strains are in the range 0.15–0.19 $d^{-1}$ (0.34–0.43 $p^{-1}$) over much of the interior leaf area (Fig. 4.6). Growth strains rates are smaller (0.04–0.13 $d^{-1}$, 0.10–0.30 $p^{-1}$) at the leaf tip, at the periphery of the apical regions of the lobes, and in a small region near the base of the midrib. The growth is for the most part isotropic (not shown).

To determine the spatial distribution of dry matter density, a rectangular grid was drawn on one quadrant of a leaf (Fig. 4.7). Spacing between adjacent points was 5 mm. Each grid square was represented by a two digit number, with the first digit representing the index of the row and the second digit representing the index of the column. Thus square 32 was located in the third row, second column of the grid and extended from 10 to 15 mm in the $x$-direction and from 5 to 10 mm in the

Fig. 4.5. Leaf photograph (*a*) and triangulated surface (*b*) for finite element analysis of strain rate in *Vitis*.

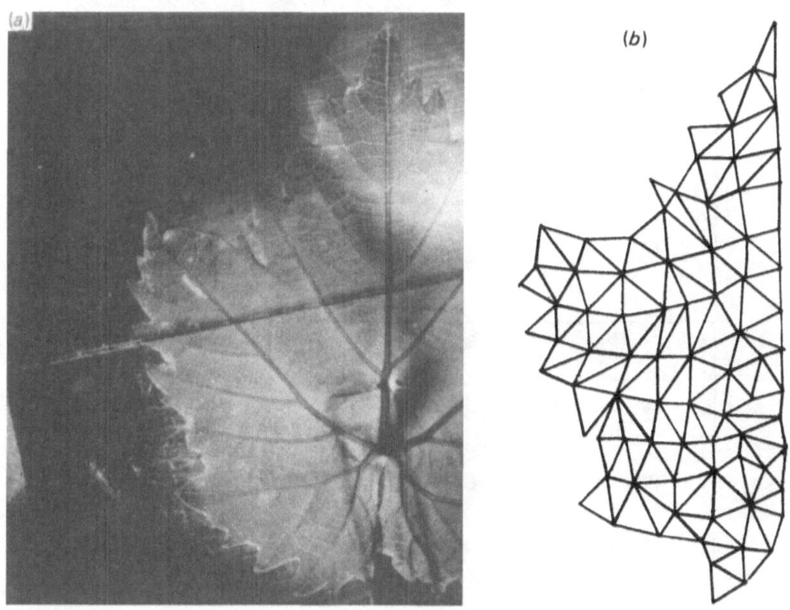

*y*-direction. A paper punch was used to cut a circle (area 7.92 mm$^2$) from the centre of each grid square. The leaf discs were dried 96 hours at 60 °C in open glass vials which were capped in the presence of desiccant and allowed to cool to room temperature just before leaf discs were

Fig. 4.6. Strain rate distribution in leaf of *Vitis* at LPI 2.

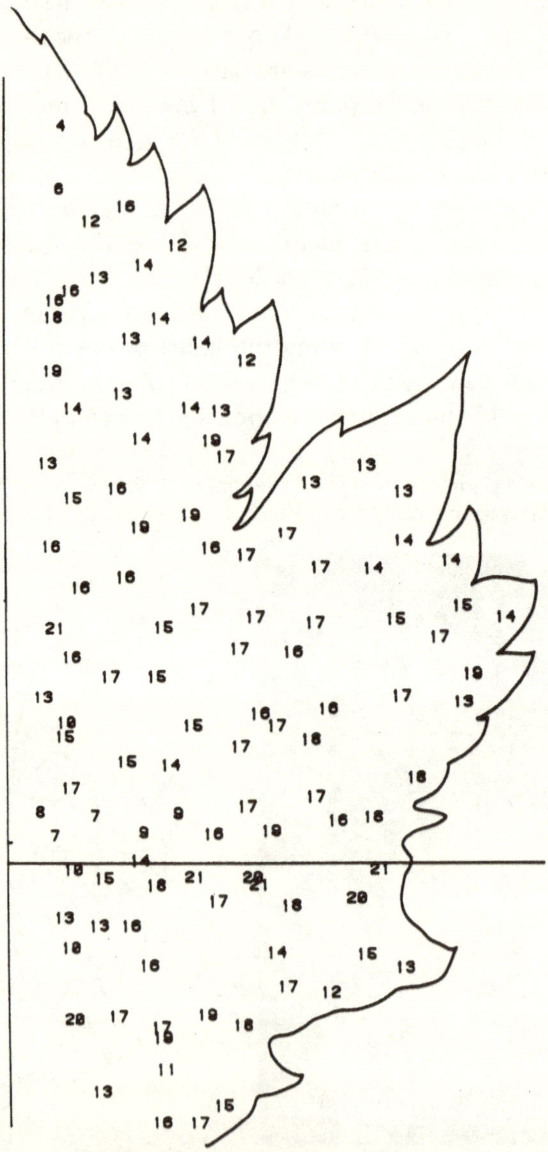

weighed. Discs were punched from three leaves, LPIs 1, 2 and 3, on each of ten plants. 'Density,' $\varrho$ of equation (4.10), was recorded for each grid square as dry weight per unit of punched area (Fig. 4.7). The information on spatial distribution of density was then used to estimate spatial gradients in density of the leaf of LPI 2. The $x$-gradient, $\delta\varrho/\delta x$, and the $y$-gradient, $\delta\varrho/\delta y$, were calculated by two-point and three-point difference formulae. The local rate of change of dry weight density, $\delta\varrho/\delta t$, was estimated on a per plastochron basis for each grid square by subtracting density in a particular leaf from density of the corresponding square in the next younger leaf. Means of differences for ten plants were used in subsequent calculations. Calculating local density changes from data on neighbouring leaves implicitly assumes dry weight density to vary with developmental stage and not with leaf serial number. This assumption is open to question. Most plants are heteroblastic to some extent, and many annuals have no two leaves alike. But preliminary evidence (Freeman, unpublished) indicates that for *V. vinifera* plots of dry matter density vs. LPI are similar for many successive leaves.

Dry weight proved to be distributed homogeneously over the leaf surface (Fig. 4.7, Table 4.1). Only the region of the largest vein (grid squares 11, 22 and 33) is significantly more dense than neighbouring squares. Thus, convective rates of density change are mostly negligible.

Local rates of change, $\delta\varrho/\delta t$, are small but probably significant, especially between LPI 2 and LPI 3, when $\delta\varrho/\delta t$ is about $3\,\mathrm{g\,m^{-2}\,p^{-1}}$ (Table 4.1). It is interesting that the region of the largest vein (grid squares 11, 22 and 33) decreases in density between LPI 2 and LPI 3.

Fig. 4.7. Distribution of dry weight density in three successive leaves of *Vitis*. Numbers are averages of density (in $\mathrm{g\,m^{-2}}$) for ten plants.

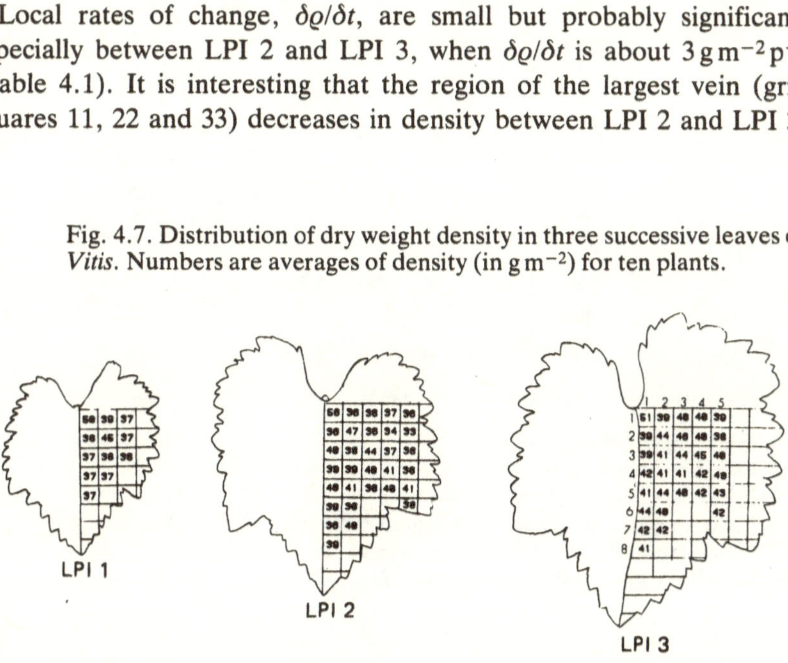

Table 4.1. *Components of local dry matter production rates in leaves of* Vitis

| (1) Spot | (2) $\varrho$ | (3) $\nabla \cdot \mathbf{v}$ | (4) $\delta\varrho/\delta t$ | (5) $\varrho((\nabla \cdot \mathbf{v})$ | (6) $D$ |
|---|---|---|---|---|---|
| 11 | 56 | 0.34 | −2.1 | 19.0 | 16.9 |
| 12 | 36 | 0.30 | 2.2 | 10.8 | 13.0 |
| 13 | 38 | 0.42 | 2.9 | 16.0 | 18.9 |
| 14 | 37 | 0.40 | 3.0 | 14.9 | 17.9 |
| 15 | 36 | | 4.5 | | |
| 21 | 36 | 0.34 | 3.6 | 12.2 | 15.8 |
| 22 | 47 | 0.38 | −1.0 | 18.0 | 17.0 |
| 23 | 36 | 0.39 | 7.5 | 14.0 | 21.5 |
| 24 | 34 | 0.41 | 4.1 | 13.9 | 18.0 |
| 25 | 33 | 0.38 | 2.7 | 12.5 | 15.2 |
| 31 | 40 | 0.41 | 4.3 | 16.4 | 20.7 |
| 32 | 38 | 0.38 | 3.8 | 14.4 | 18.2 |
| 33 | 44 | 0.40 | −2.1 | 17.6 | 15.5 |
| 34 | 37 | 0.36 | 7.8 | 13.3 | 21.1 |
| 35 | 38 | 0.36 | 2.5 | 13.7 | 16.2 |
| 41 | 39 | 0.35 | 3.6 | 13.7 | 17.3 |
| 42 | 39 | 0.41 | 2.4 | 16.0 | 18.4 |
| 43 | 40 | 0.37 | 1.9 | 14.8 | 16.7 |
| 44 | 41 | 0.32 | 1.3 | 13.1 | 14.4 |
| 45 | 38 | 0.30 | 2.6 | 11.4 | 14.0 |
| 51 | 40 | 0.40 | 1.0 | 16.0 | 17.0 |
| 52 | 41 | 0.34 | 3.6 | 13.9 | 17.5 |
| 53 | 38 | 0.35 | 1.7 | 13.3 | 15.0 |
| 54 | 40 | | 3.1 | | |
| 55 | 41 | | 3.3 | | |
| 61 | 39 | 0.40 | 5.4 | 15.6 | 21.0 |
| 62 | 38 | 0.34 | 2.7 | 12.9 | 15.6 |
| 65 | 38 | | 6.5 | | |
| 71 | 36 | 0.27 | 2.3 | 9.7 | 12.0 |
| 72 | 40 | 0.27 | 3.5 | 10.8 | 14.3 |
| 81 | 39 | 0.10 | 3.4 | 3.9 | 7.3 |

(1) Grid square designation.
(2) Disc density, $\varrho$ (g $m^{-2}$).
(3) Local strain rate, $\nabla \cdot \mathbf{v}$ ($p^{-1}$).
(4) Local derivative, $(\delta\varrho/\delta t)$, evaluated between LPI 2 and LPI 3 (g $m^{-2}$ $p^{-1}$).
(5) Strain component, $\varrho(\nabla \cdot \mathbf{v})$, (g $m^{-2}$ $p^{-1}$).
(6) Local dry matter deposition rate, $D$ (g $m^{-2}$ $p^{-1}$).

Apparently, growth (volume increase) is faster than dry matter production in the vein. Another possible interpretation is that the tissue below the vein stretches to displace the vein from the centre of the grid square. This would cause the vein to comprise a smaller fraction of the leaf disc at LPI 3 than at LPI 2. In theory, this would be accommodated by an increase in the convective term; but the present grid size is too coarse to allow detection of the increase. More work must be done to verify that vein stretch exceeds dry matter deposition in the second plastochron.

Over most of the leaf the largest component of the dry-matter production rate is the stretch component, $\varrho(\nabla \cdot \mathbf{v})$ which is approximately $14\,\mathrm{g\,m^{-2}\,p^{-1}}$ ($6\,\mathrm{g\,m^{-2}\,d^{-1}}$). The vein region, with large $\varrho$ and large $\nabla \cdot \mathbf{v}$, has the greatest stretch component; but the negative local derivative results in a production rate similar to that found in the other interior leaf regions. Near the leaf tip, in grid squares 71, 72 and 81, the stretch component is small; and in square 81 the stretch component falls to a value approaching the local rate of change.

The measurements and calculations described in this section were designed to show the utility of knowing the spatial distribution of both a physiological variable (dry weight density) and the strain rate. In a way, the grape leaf at LPI 2 proved a rather disappointing subject for this study. Its homogeneity in both strain rate and density implied that local production rates could have been estimated much more easily as the change in total dry matter per plastochron, averaged over the leaf area. However, the drop in dry matter production rates near the tip would have been obscured by the simpler experiment. Moreover, the approach I have described to evaluate the terms of equation (4.10) is quite general. From the earlier growth analyses described in the second section of this chapter, we know that many leaves do have significant gradients in growth rate, particularly in earlier stages of development. For these leaves it would be necessary to evaluate at least the terms shown in the table to know the spatial distribution of the production rates. (If the physiological variable were not homogeneously distributed it would be necessary to evaluate the convective rate of change also.) To understand the physiology of growth, we must be able to correlate the spatial and temporal distribution of production rates with other growth-related processes and presumed causal agents of growth.

The author's research is supported by Grant PCM-81-00296 from the National Science Foundation.

## References

Archer, R., Lockhart, J. & Deshmukh, R. (1973). On the distribution of leaf area growth. Technical Report No. EM73-2, Engineering Research Institute, University of Massachusetts, Amherst, MA.

Avery, G. S., Jr. (1933). Structure and development of the tobacco leaf. *American Journal of Botany*, **20,** 513–64.

Batchelor, G. (1967). *An Introduction to Fluid Dynamics.* Cambridge University Press.

Boas, M. (1966). *Mathematical Methods in the Physical Sciences.* New York: John Wiley & Sons.

Crandall, S., Dahl, N. & Lardner, T. (1972). *An Introduction to the Mechanics of Solids.* New York: McGraw-Hill.

Erickson, R. O. (1966). Relative elemental rates and anisotropy of growth in area: a computer programme. *Journal of Experimental Botany*, **17,** 390–403.

Erickson, R. O. (1976). Modeling of plant growth. *Annual Review of Plant Physiology*, **27,** 407–34.

Erickson, R. O. & Goddard, D. R. (1951). An analysis of root growth in cellular and biochemical terms. *Growth Symposium*, **10,** 89–116.

Erickson, R. O. & Silk, W. K. (1980). The kinematics of plant growth. *Scientific American*, **242,** 134–51.

Gandar, P. W. (1980). The analysis of growth and cell production in root apices. *Botanical Gazette*, **141,** 131–8.

Green, P. B. (1976). Growth and cell pattern formation on an axis: critique of concepts, terminology and modes of study. *Botanical Gazette*, **137,** 187–202.

Green, P. B. & Poethig, S. (1982). On the biophysics of plant organogenesis. *Symposium of the Society for Developmental Biology*, **40,** (in press).

Hejnowicz, Z. & Nakielski, J. (1979). Modeling of growth in shoot apical dome. *Acta Societas Botanicae Poloniae*, **48,** 423–41.

Niklas, K. J. (1977). Application of finite element analyses to problems in plant morphology. *Annals of Botany*, **41,** 133.

Niklas, K. J. & Mauseth, J. D. (1980). Simulations of cell dimensions in shoot apical meristems: Implications concerning zonate apices. *American Journal of Botany*, **67,** 715–32.

Richards, O. W. & Kavanagh, A. J. (1943). The analysis of the relative growth gradients and changing form of growing organisms: illustrated by the tobacco leaf. *American Naturalist*, **77,** 385.

Schüepp, O. (1926). Meristeme. *Handbuch der Pflanzenanatomie*, **4,** 1–186.

Silk, W. K. & Erickson, R. O. (1978). Kinematics of hypocotyl curvature. *American Journal of Botany*, **65,** 310–19.

Silk, W. K. & Erickson, R. O. (1979). Kinematics of plant growth. *Journal of Theoretical Biology*, **76,** 481–501.

Tognetti, K. & Winley, G. (1980). The growth of a column of age and position dependent cells. *Mathematical Biosciences*, **50,** 59–74.

Wilson, B. & Archer, R. (1977). Reaction wood: Induction and mechanical action. *Annual Review of Plant Physiology*, **28,** 23–43.

Zienkiewicz, O. C. (1961). *The Finite Element Method.* London: McGraw-Hill.

# 5

# The shoot apex in transition: flowers and other organs

R. W. KING

## Introduction

The vegetative shoot apex is an indeterminate meristem which continuously produces leaf primordia. However, it may form other organs including cataphylls, flowers, inflorescences, spines, tendrils and thorns, some of which may terminate its meristematic activities. In woody species especially, the apex may abort or form resting buds. Such differentiation can sometimes occur without a prior vegetative phase. For example, there is clear and early diversity in the anatomy and physiology of some originating meristems such as lateral meristems, inflorescence buds, spikelet initial sites, areoles, and short and long shoots. Often, however, there is a switch from a vegetative pattern of development and the earliest signs of this bifurcation are seen at the organ primordium stage.

One approach to understanding the processes controlling the transition between the production of leaves and of other organs is to look for the first biochemical differences at organ evocation as distinct from later processes of visible differentiation. Genetic control must be involved in the final specification of organ type. So, also, must particular chemicals including plant growth regulators. On the other hand, change in geometry involving correlative (spatial) and temporal relationships must also be important. These two controls (biochemical and geometric) are discussed later in relation to floral differentiation.

Reviews of many aspects of differentiation in plants have been published recently (e.g. Cutter, 1969; Evans, 1971, 1975; Steeves & Sussex, 1972; Zeevaart, 1976; Halperin, 1978; Bernier, 1980) and to focus on essential points, few comprehensive listings of references are necessary here. Most information is for studies with angiosperms and then mostly with herbaceous species rather than woody perennials.

## Early events of organogenesis

### *General features of biochemical specification of organs*

Little is known of the key events involved in the formation of dormant buds, thorns, spines or tendrils. Spine and leaf primordia produced by the areole meristem in *Opuntia* differ in that there are fewer divisions and a longer plastochron in the leaf mode of differentiation (Mauseth, 1976).

By contrast, dormancy-induced changes in the apical meristem of axillary buds have been examined mostly in terms of changes occuring on the release of dormancy: activation of division of a population of cells previously held in the $G_1$ phase of the cell cycle or increases in DNA and RNA content of the apex and a decrease in the DNA/histone ratio (Naylor, 1958; Dwivedi & Naylor, 1968; Usciati, Cadoccioni & Guern, 1972). It is not known if the transition to dormancy is an exact reverse of these changes occurring on loss of dormancy. Nor is it known which processes are essential in controlling the transition to the formation of one or other organ. Plant growth regulators appear to be important as 'triggers'. For instance, there have been many reports that leaves produce signals responsible for induction of bud dormancy (see Saunders, 1978). Lateral bud dormancy as studied by Naylor (1958) appears to involve indoleacetic acid (see Phillips, 1975). Also the transformation of the *Opuntia* areole is influenced by applied growth substances (Mauseth, 1976). By contrast, processes of floral organogenesis have been examined in much detail but little is known of endogenous hormonal controls.

### *Precision in the timing of floral organogenesis*

An essential factor for progress with studies of floral organogenesis has been the use of photoperiodic control of flowering as an accurate trigger to switch development rapidly from a vegetative to a floral pathway. In some of the species which respond to a single photocycle, such as the short-day plants *Pharbitis nil, Xanthium strumarium* and *Chenopodium rubrum*, flowers form rapidly (5–8 days) in response to a single dark period of 10–16 h duration. Such a response is seen whether exposure is of seedlings or adult plants (Evans, 1975). Often there is no requirement for a prolonged period of vegetative growth. By contrast, in some long-day plants such as *Lolium temulentum* and *Sinapis alba*, there is a requirement for vegetative growth prior to sensitivity to photoperiod but it is not always clear if the vegetative period is associated with changes in the leaves, the apex or both.

Perception of daylength can be ascribed to the leaves. The time of earliest response at the apex is therefore dependent on when floral stimulus is generated in the leaves and how fast it is exported to the shoot apex. The timing of stimulus transmission is known for a number of species including *Anagallis arvensis* (Taillandier, 1978a), *Chenopodium* (King, 1971, 1975), *Lolium* (Evans, 1960), *Pharbitis* (King, Evans & Wardlaw, 1968); *Sinapis* (Kinet, Bodson, Alvinia & Bernier, 1971) and *Xanthium* (see Zeevaart, 1980). The nature of stimuli to flowering, and of inhibitors, has been discussed recently by Zeevaart (1980).

In general, the stimuli to flowering reach the apex within 2–15 h of the end of photoinductive treatment of the leaf. In *Pharbitis* the processes are very rapid. As illustrated in Fig. 5.1, photoresponse of the cotyledon is completed after 12–13 h of darkness. No flowering results following removal of the cotyledon 14 h after the start of the dark period but its removal after another 1–2 h is no longer inhibitory. Thus, it may be assumed that the stimulus to flowering was exported between hour 14 and 16. This stimulus moves with assimilate (Ogawa & King, 1979) in the phloem (Kavon & Zeevaart, 1978) where it is transported at 35 cm $h^{-1}$ (King, Evans & Wardlaw, 1968). Thus, in the seedling it must arrive at the shoot apex soon after export begins.

With increases in sensitivity of the biochemical techniques for detecting early changes at the apex, questions have been asked about the precision of timing of arrival of stimuli at the apex (see Bernier, 1980; Gressel, Zilberstein, Strausbauch & Arzee, 1978). Techniques involving leaf or cotyledon removal estimate the timing of events relating to the slowest component of a floral stimulus if there is more than one. If the stimulus is a multicomponent complex then one component can arrive earlier and trigger biochemical change prior to the arrival of the last component. In *Pharbitis* there may be only a short lag (2 h) between saturation of induction (hours of darkness, Fig. 5.1) and saturation of stimulus export (hours to cotyledon removal, Fig. 5.1). If the critical photoperiod for production of an early component of the stimulus was much shorter than that shown for the last component, then differences of arrival times at the apex could be greater than 2 h. However, the intricacy of such argument far exceeds the resolving power of current experimental data. For example, in *Pharbitis* the duration of darkness required for flowering may shift from 9 to 14 h depending on where the experiments are performed, on the environment imposed during flower development (King & Evans, 1969), whether chemicals are applied (Ogawa & King, 1979), and on whether induction is of cotyledons or adult leaves. Furthermore, photoperiodic induction may be perceived weakly by the shoot apex itself (Baldev, 1962; Gressel, Zilberstein, Porath & Arzee, 1980). Nevertheless, despite these real and speculative complications it cannot be denied that the leaf performs a central role in production of the stimulus to flowering.

Fig. 5.1. Flowering response of *Pharbitis* as affected by length of dark period (○) or by removal of the induced cotyledons at various times from the start of a 16 h dark period (●). From Ogawa & King (1979).

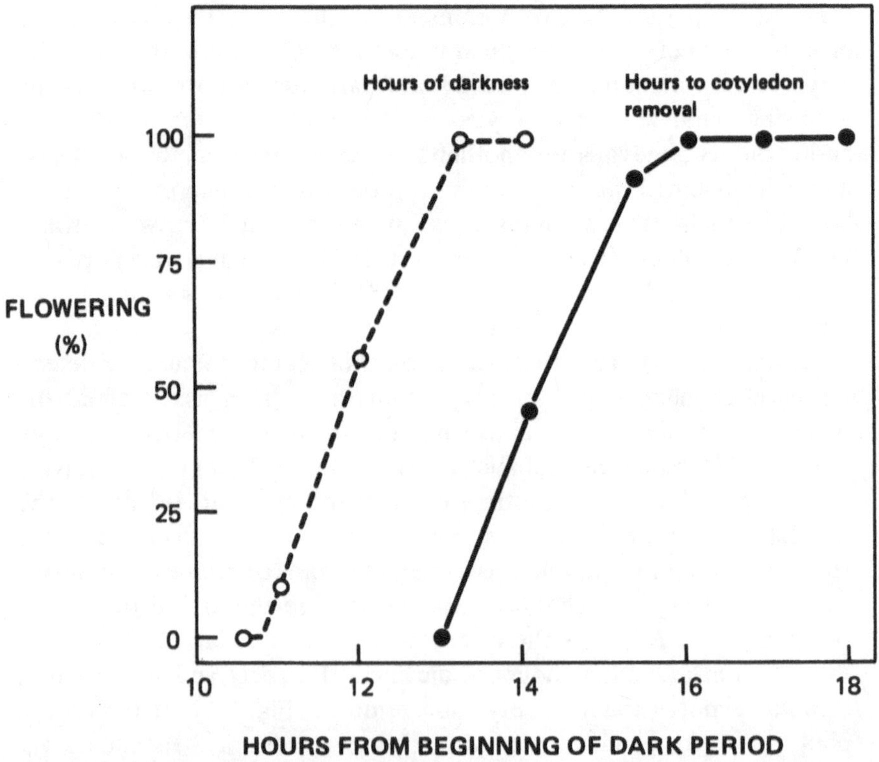

*Structural features of floral organogenesis*

Three cytologically and histologically distinct aggregates of cells can be recognized in the vegetative apical meristem of dicotyledonous species (Fig. 5.2) as summarized in Chapter 1. The central zone (CZ) of the vegetative apical meristem is relatively inactive showing infrequent cell division (Gifford & Corson, 1971). With flowering there is a general activation of the apex and much of the pattern of zonation is lost. There is an increase in RNA and DNA staining of the cells of the CZ, mitochondria numbers increase and the vacuoles become smaller. More cells divide and the duration of the cell cycle decreases dramatically especially in the CZ (Table 5.1).

Based on the observed changes in cell division and nucleic acid content, a number of plant cytologists including Buvat (1952) and Nougarède (1967) have ascribed a specific organogenic function to the CZ. They speculated that the apparent quiescence of the CZ cells in vegetative apices and their activation on flowering (e.g. Table 5.1) indicated that floral organs originated directly from this waiting meristem, '*méristème d'attente*'. It does not follow, however, that activation means that organs form from these specific CZ cells. Moreover, when more recent studies of apices are combined with earlier results, it is clear that the concept faces some problems (see Bernier, 1980).

Fig. 5.2. Generalized structure of the vegetative shoot apex. Cytohistological zones are: central zone (CZ), peripheral zone (PZ) and pith meristem (PM). Further zonation is seen in the distinction between tunica and corpus.

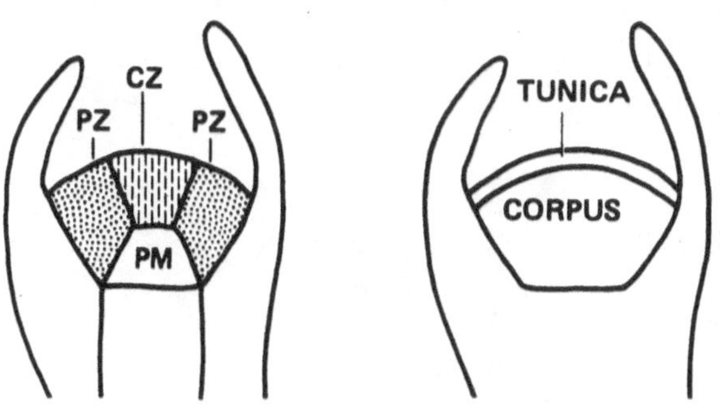

Table 5.1. *Cell cycle duration in the apical primordium in transition from vegetative to reproductive development*

| | Cell cycle duration (h) | | | | |
|---|---|---|---|---|---|
| | Vegetative | | Floral | | |
| Species | Central zone | Peripheral zone | Central zone | Peripheral zone | Reference |
| *Datura stramonium* | 76 | 36 | 46 | 28 | Corson (1969) |
| *Sinapis* | 287 | 157 | 35 | 25 | Bodson (1975) |
| *Coleus blumei* | 250 | 130 | 130 | 120 | Saint-Côme (1971) |
| *Silene* (total apex) | 20–24 | | 6–8 | | Miller & Lyndon (1976) |

## Biochemical and ultrastructural aspects of the transition to flowering

### *Early RNA and protein synthesis*

Early changes of floral evocation at the shoot apex on transition to flowering are seen in the content of metabolites, proteins and nucleic acids (Table 5.2). In the long-day plant *Lolium* an increase in RNA synthesis was found at the time estimated for arrival of floral stimulus at the apex, i.e. the morning after the single long day. This, and enhanced protein synthesis, was found from a micro-autoradiographic study of $^{35}S$ and $^{3}H$-orotic acid incorporation at presumptive sites for flower formation between adjacent leaf primordia (Fig. 5.3). Incorporation of $^{32}P$ label into nucleic acids supported these findings and established that the increase (35–45%) was very transient (Evans, Knox & Rijven, 1970). Histological techniques were only sensitive enough to detect increases in RNA staining two days later by which time the total extractable RNA in the inflorescence had increased by 46%, DNA by 16%, dry weight by 25% and length by 10%. Subsequent studies with shoot tips of *Sinapis*

Table 5.2. *Metabolic and enzyme increases associated with transition to flowering in cells of the whole apex or axillary spikelet sites of* Lolium temulentum *and in cells of the apical meristem of* Sinapis alba *and the shoot tip or apical meristem of* Pharbitis nil *(timing in hours after start of induction; floral primordia are visible after 70 h or longer)*

| Timing (h) | Cell component | *Lolium* | *Pharbitis* | *Sinapis* |
|---|---|---|---|---|
| 0–30 | Invertase activity | | | + |
| | Acid phosphatase activity | | | + |
| | Ribonuclease activity | | + | |
| | Starch content | | | + |
| | Permeability to metabolites | + | | |
| | Precursor incorporation: | | | |
| | protein synthesis | + | | |
| | RNA synthesis | + | | + |
| | DNA synthesis | | + | + |
| 30–50 | Ribonuclease activity | | | + |
| | Succinic dehydrogenase activity | | | + |
| 50 | New proteins detectable | | | + |
| | RNA and DNA content | + | | |

Adapted from Bernier (1980).

and *Pharbitis* have indicated comparable increases in RNA synthesis associated with, or even preceding, the time of arrival of the last component of the floral stimulus.

There are many problems with the use of labelling studies to measure synthesis of nucleic acids *in situ*. For example, change in precursor uptake or in the size of an endogenous precursor pool could alter incorporation, while competition occurs with other metabolic pathways and the rate of degradation of the product could also alter apparent synthesis rates; there are rapid and large (two-fold) fluctuations in the RNA-ase content of the shoot tip of *Pharbitis* at the end of an inductive dark period (Gressel, Zilberstein & Arzee, 1970). A further difficulty is

Fig. 5.3. Pattern of incorporation of (*a*) $^3$H-orotic acid and (*b*) $^{35}$S as grains per 10.9 $\mu$m$^2$ in apices of *Lolium*. The left-hand side of each apex gives average values for vegetative apices, the right-hand side values for apices 26 h after induction began (from Evans, 1969).

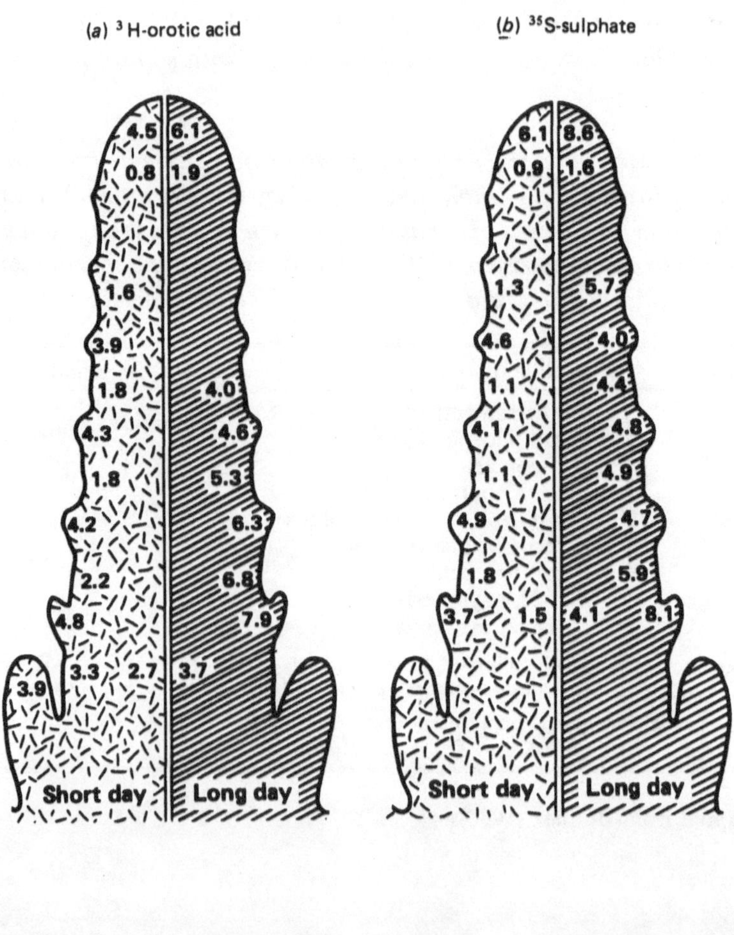

that frequently the whole shoot tip has been studied and more often than not the apical meristem is swamped by the inclusion of a 50-fold or greater amount of leaf tissue. However, despite these criticisms, it seems unlikely that similar stimulation of RNA content would have resulted from an artifact of technique with all of the three species examined.

If early stimulation of RNA and protein synthesis is a common and essential response, then, flowering should be inhibited by RNA and protein synthesis inhibitors such as 2-thiouracil, actinomycin D, cycloheximide and ethionine. Not only has this been established for *Lolium* (Fig. 5.4), *Anagallis, Pharbitis, Sinapis, Xanthium* (see summary in Bernier 1971, 1980) but these compounds are only inhibitory over the early hours of evocation (i.e. 0–24 h). Moreover, their application and

Fig. 5.4. Effect of time of injection of actinomycin D (10 $\mu$g plant$^{-1}$) near the shoot apex on floral induction in plants of *Lolium* exposed to one long day. The hatched area indicates the low-intensity light period of the long day. After Evans (1964).

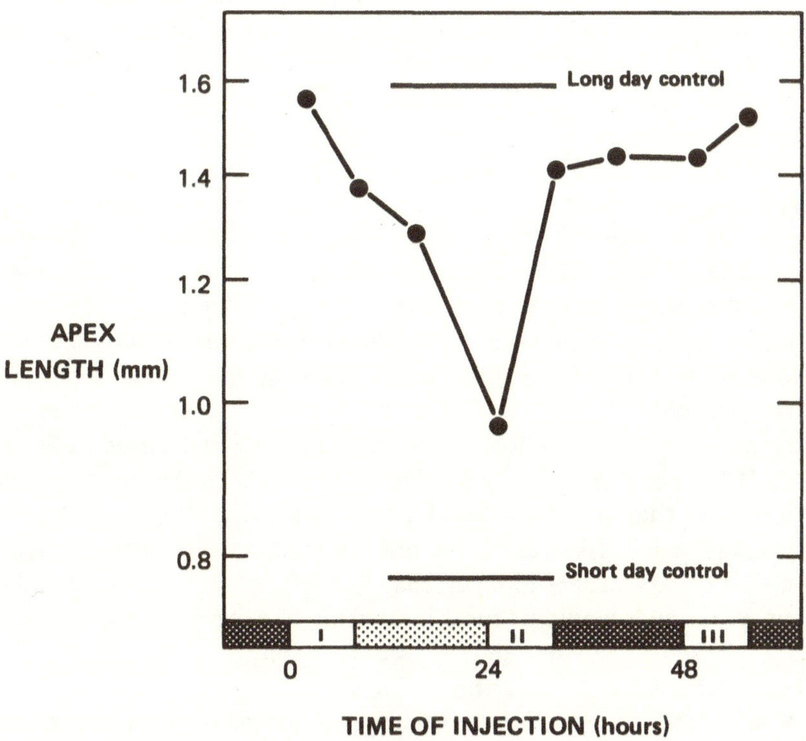

action is at the shoot tip and treatment with the appropriate complementary metabolite reverses the inhibition (e.g. methionine for ethionine). Thus, synthesis of protein and RNA must be important for early events of evocation.

It is not known yet whether these inhibitor studies indicate the synthesis of new species of RNA and protein during early evocation or merely reflect synthesis of more of existing kinds. The precursor incorporation studies favour the first suggestion but it could also be argued that perception of floral stimulus is impossible without the maintenance of normal RNA and protein synthesis. Using double-labelling techniques, Pryke & Bernier (1978a), established that ribosomal and soluble RNA does increase at the apex on flowering of *Sinapis*. No change could be detected in messenger RNA (mRNA) fractions but specific new mRNAs may constitute a very small proportion of total mRNA and a minute fraction of total RNA. For studying specific proteins sensitive histoimmunochemical techniques have been employed by Pierard, Jacqmard, Bernier & Salmon (1980), who looked for early changes at the shoot tip in proteins antigenically similar to those in fully formed flowers. This technique was adopted in preference to electrophoretic approaches which had been found to be too insensitive (Stiles & Davies, 1976). Pierard *et al.* were able to detect the appearance of two 'floral' proteins during the transition to flowering but these occurred quite late in development (40–60 h).

Although the studies with *Lolium*, *Pharbitis*, *Sinapis*, and possibly *Xanthium*, indicate the synthesis of more and perhaps unique RNA and protein at the time of flower evocation, recent studies by Miller & Lyndon (1977) with *Silene* raise some doubts about the generality of this conclusion. *Silene* requires more than six long day cycles for maximum induction and, as found with other species, there was an early increase in spectrophotometrically detectable RNA in the apical meristem. However, exposure to fewer LD or to $GA_3$ did not induce flowering but did induce an increase in RNA. Conversely, in plants induced to flower at 13 °C rather than 20 °C the change in RNA of the apex was no different to that in non-induced plants held at 20 °C, Possibly, the changes seen in RNA in *Silene* are not essential for its flowering. On the other hand, spectrophotometry may be less sensitive than labelling techniques and it would be most interesting to examine RNA synthesis using labelled precursors and to identify when, if at all, inhibitors of RNA synthesis block flowering of *Silene*.

Many of the other biochemical changes observed during the transition to flowering seem relatively unspecific. For example, the early increase in the content of acid invertase in the shoot tip of *Sinapis* (Table 5.2) can also be induced by raising the light intensity during non-inductive short days (Pryke & Bernier, 1978b). Similarly, there can be little doubt that appropriate non-inductive conditions can lead to an increase in the content of starch in the shoot tip.

### *Ultrastructural changes*

Focusing more closely on the florally evoked shoot apex, Havelange (1980) has summarized the early ultrastructural changes seen in cells of the apical meristem. Predictably, in both long day (LD) and short day (SD) plants, there are many changes on the transition to flowering (Table 5.3). Late changes in dictyosomes and size of the nucleolus and ribosome numbers might be expected in a tissue on the brink of organogenesis. The significance of the earlier changes is less certain although the larger number of mitochondria could indicate that an increased respiratory energy supply is involved in early events of flowering.

Table 5.3. *Ultrastructural changes associated with transition to flowering in cells of the apical meristem of* Xanthium strumarium, Sinapis alba *and* Spinacia oleracea *(timing in hours after start of induction; floral primordia are visible after 70 h)*

| Timing (h) | Cell character | *Xanthium* | *Sinapis* | *Spinacia* |
|---|---|---|---|---|
| 0–18 | Vacuole size | 0 | – | – |
| | Mitochondrion number | + | + | + |
| | Chloroplast number | + | 0 | + |
| | Cell size | 0 | – | |
| 18–50 | Cell size | + | + | |
| | Nucleus size | 0 | + | |
| | Dictyosome number | + | + | + |
| 50–70 | Nucleolus size | + | + | |
| | Cytoplasm size | + | + | |

0, no change; +, increase; –, decrease.
Adapted from Havelange (1980).

*Early cell division*

A common feature of response to photoperiod is a rise in the mitotic index. This is first seen 40–48 h after the inductive cycle in *Xanthium* (Thomas, 1963; Jacqmard, Raju, Kinet & Bernier, 1976) after 36 h in *Pharbitis* (Wada, 1968) after 30 h in *Chenopodium* (King, 1971) 16–20 h in *Anagallis* (Taillandier, 1978b) and 26–30 h in *Sinapis* (Bernier, Kinet & Bronchart, 1967). It is not seen until quite late in other species; after 7–8 days for *Silene* (Miller & Lyndon, 1976), which requires several cycles for induction, and after 3–4 days in *Lolium* (Knox & Evans, 1968).

At some stage after evocation more cells are required to support the growth which occurs at the time of primordium initiation. As discussed above, this late increase is achieved by a shortening of the cell cycle (Table 5.1) and also by the entry into division of the cells of the CZ. By contrast, the initial increase in the mitotic index for *Sinapis* at 26–30 h occurs well before 'visible' organogenesis at 50–60 h. 'Visible' organogenesis could, however, be a poor indicator of essential growth. It may be that the early changes in mitosis are not so distinct from the onset of organogenesis and growth. For instance, during floral evocation in *Pharbitis*, the early increase in mitosis at the shoot apex after 36–48 h is not greatly in advance of a visible increase in the rate of leaf initiation at 60 h (Wada, 1968).

Evidence for an early rise in the mitotic index has only been reported for three species, *Sinapis*, *Anagallis* and *Xanthium*. The results are clearest for *Sinapis*. Microspectrophotometric measurements of the DNA content of nuclei have established that up to 50% of the cells in the vegetative apex are present in the post-DNA-synthesis $G_2$ stage of the cell cycle (Jacqmard, Miksche & Bernier, 1972). At the time of floral evocation (26–38 h) these cells are released into cell division. The proportion of cells released from $G_2$ is much smaller in *Anagallis* (15%) (Taillandier, 1978b) and very few cells are released and release is quite slow in *Xanthium* (<5%) (Jacqmard *et al.*, 1976).

The early release of $G_2$ cells into division probably has little effect on growth. The rate of cell division (i.e. cell cycle length) does not change until later when organ formation occurs (e.g. at 50–60 h in *Sinapis*). There will, however, be partial synchronization of cell division. Bernier (1971) argues that this synchronization is essential for early events of floral evocation. He suggests that synchronization of mitosis allows an amplified synthesis of floral information which may be expressed only at one phase of the cell cycle.

A problem with this concept is that the early stimulation of mitosis can occur without flowering following non-inductive changes in daylength, or with cytokinin application (see Bernier, 1980). Also, colchicine, which should block mitosis does not inhibit flowering in *Sinapis* (Bernier, 1971) or *Pharbitis* (Zeevaart, 1962). By contrast, when applied at the time of the long day, 5-fluorodeoxyuridine (5-FDU) blocked the normal $G_2$ release in *Sinapis* (Bodson & Bernier, 1972). Nevertheless, flowering was not inhibited by 5-FDU until later (36–50 h) during floral differentiation (Kinet, Bodson, Alvinia & Bernier, 1971). Slow reestablishment of mitosis following thymidine application (see Bodson & Bernier, 1972) allowed floral differentiation to proceed without an early $G_2$ release. There are further arguments against the importance for flowering of an early $G_2$ release. Especially critical is the total absence or small early mitotic release in species such as *Lolium* (Knox & Evans, 1968), *Silene* (Francis & Lyndon, 1978), and *Xanthium* (Jacqmard *et al.*, 1976).

## Aspects of 'habituated' flowering

### *An early critical phase in the development of flowers*

The transition to flowering is not always precisely regulated by environment and many species progress autonomously to flowering. Nevertheless, the same sorts of apical responses may be involved in both photoperiodically sensitive and insensitive species. It appears that the transformation to flowering involves an early phase for flower determination and this precedes visible organogenesis.

The existence of such a critical phase in woody perennials has been demonstrated in citrus by Nir, Goren & Leshem (1972). They found that applying $GA_3$ inhibited flowering and that escape from this inhibition was very dramatic – occurring over a period of about 6 days; anatomical studies showed organogenesis to occur after this time. Similar early escape from inhibition of flowering in response to low temperature shocks has also been found in a daylength-insensitive cultivar of sunflower (Marc & Palmer, 1978).

The most recent illustration of an early sensitive period is in wheat where a 3-day exposure to short days at high temperature induced floret sterility (Frankel, Knox & Considine, 1981). The critical period was apparently less than 24 h for any one floret and occurred over the period between lemma initiation and floret formation. Histochemical and cytological analysis showed that epidermal cell proliferation was restricted and there was a block to the increase in the RNA staining which was so characteristic of fertile florets. These changes were therefore apparent well before formation of recognizable primordia on the dome of the floret, but at the same time that essential events involving RNA syntheses occur in *Lolium* (see above).

The significance of these findings, especially those with wheat, is two-fold. Firstly, whether or not the mechanism of induction involves transmission of a 'photoperiodic signal', the apical transitions at floral organogenesis include increases in RNA content and in cell division. Secondly, briefly and just at the time of organogenesis, there is an extreme change in sensitivity to environmental and chemical perturbation. Clearly, even with species, such as citrus, which flower autonomously, progression to flowering is neither gradual nor inexorable.

### *'New' cells give new organs*

In some species flowering becomes fixed or persistent. The consequence is that flower initiation can be obtained in isolated plant parts cultured *in vitro*. Such flower formation on excised plant parts has been obtained from internode segments of *Nicotiana tabacum* and various other *Nicotiana* species (Aghion-Prat, 1965), stem segments from flowering plants of *Lunaria* (Pierik, 1965), root pieces of *Cichorium intybus* (Paulet & Nitsch, 1964), excised epidermal layers of *Nicotiana* (Tran Thanh Van, 1973), tendrils of *Vitis* (Srinivasan & Mullins, 1978), stem and leaf pieces of *Passiflora* (Scorza & Janick, 1980) and leaf discs of *Perilla* (Tanimoto & Harada, 1980). Stem pieces and epidermal layers of tobacco have been examined extensively and subsequent discussion is restricted to this species although similar responses may be known for other species.

Stem pieces of certain cultivars of tobacco produce floral buds when taken from flowering plants (Aghion-Prat, 1965). However, there is a gradient of flowering potential down the stem; buds on basal segments grow out vegetatively and those from nearer to the floral apex form flower buds (see Fig. 5.5). Not all cultivars of tobacco show flowering of stem pieces; only vegetative buds form from explants from flowering plants of the photoperiodically sensitive types such as *Nicotiana tabacum* c.v. Maryland Mammoth and *Nicotiana sylvestris* (Aghion-Prat, 1965).

The permanency and nature of the organogenic change in tobacco has been examined by a number of workers. Potential to flower is transmitted through a small epidermal explant in which carry-over of a large pool of morphogen is unlikely (Tran Thanh Van, 1973). Ability to form floral buds is also transmitted to callus cells and these cells retain their capacity to flower through at least three cycles of subculture (Konstanti-

Fig. 5.5. Gradient of flowering expressed in explants from the epidermis of tobacco plants (*Nicotiana tabacum* L. Wisconsin 38). Adapted from Tran Thanh Van (1973).

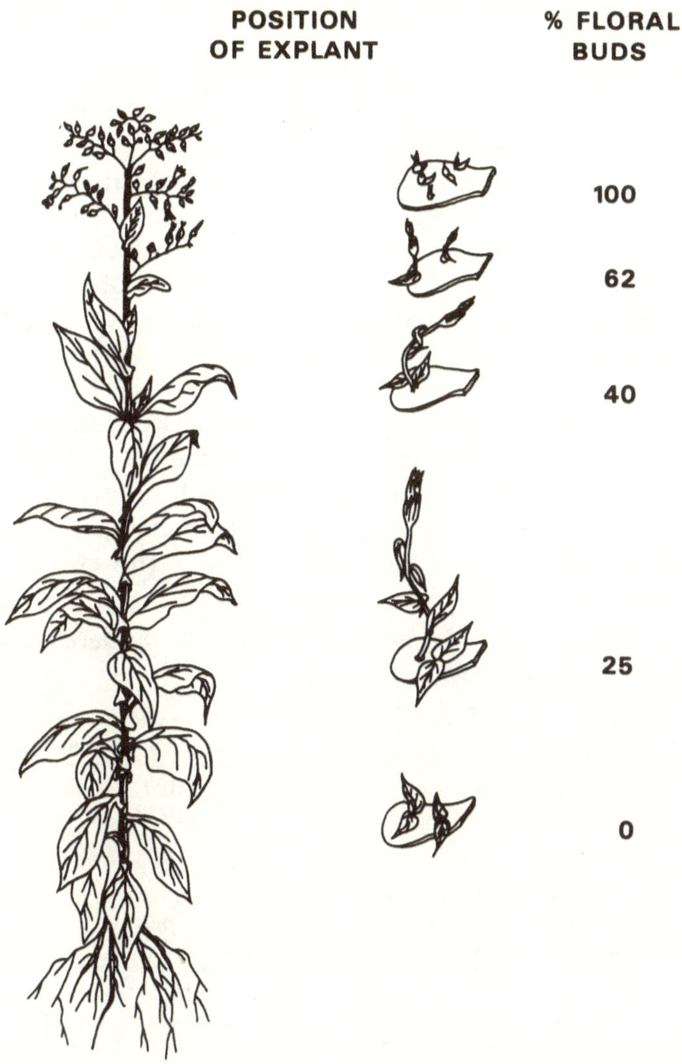

nova, Aksenova, Bavrina & Chailakhyan, 1969). Apparently the morphogenic potential of the cells has been permanently changed.

More recently, Wardell (1976, 1977) has sought evidence of this cellular change in the DNA of the tissue. He found a gradient down the stem in DNA content. The apical, 'florally committed' stem tissue contained six- to ten-fold more DNA than the basal, vegetatively committed stem tissue. This could be an inconsequential finding reflecting differences in cell numbers per unit weight of tissue. Nevertheless, combining all Wardell's data, flowering was promoted when DNA extracted from floral stem tissue was applied to marginally induced plants.

These results could explain why flowering persists in tobacco stem explants but confirmation is required in replicated trials with 'vegetative' and 'floral' DNA extracted from similar tissue by the same techniques. DNAase treatment of an extract obliterated the response but there was no valid control in this experiment. It would also be of value to examine responses to other DNA extracts perhaps of a non-plant origin. Lastly, it needs to be shown if the DNA applied to the stem actually reaches cells of the nearby shoot, and whether it is intact and is incorporated.

Recent microspectrophotometric studies of Nagl, Frisch & Frölich (1979) also raise the question of floral DNA. Their results (Table 5.4) for flower buds of *Rhoeo* shows that telophase (i.e. 2c) nuclear DNA content increased significantly (20–30%) with appearance of visible flower parts (stage 1–4, Table 5.4). Clearly, therefore, differential DNA synthesis could be part of floral differentiation. However, it would be

Table 5.4. *Differential DNA synthesis during development of floral buds of* Rhoeo discolor

| Floral bud stage | | DNA pg nucleus$^{-1}$ |
|---|---|---|
| Vegetative bud | | 16.7 |
| Generative bud Stage | 1 | 19.7[a] |
| | 2 | 21.5[a] |
| | 3 | 19.0[a] |
| Pollen meiosis | 4 | 16.2 |
| Fully formed flower | | 16.2 |
| Root tip | | 16.8 |

[a] Significantly different at 1% level from value for vegetative bud.
After Nagl, Frisch & Frölich (1979).

wrong in the absence of further studies to imply that such DNA is a trigger to flowering rather than a response to it. The apex to base gradient of flowering in tobacco argues for positional determination but it also requires prior flowering and, as shown by recent studies, the trigger may be transmissible floral factors which diffuse out of floral apices of tobacco, and induce flowering of vegetative buds (Aksenova, Konstantinova & Bavrina, 1980).

## Meristems and primordia in transition

### *Growth activation and floral differentiation*

The consequence of stimulation of cell division associated with flowering is an enhanced growth of the floral apex. In *Lolium*, for example, Evans & Rijven (1967) found that relative growth rates increased from 0.05 $d^{-1}$ to 0.45 $d^{-1}$ at early floral differentiation. Other changes are stimulation of primordia initiation, the plastochron shortens (Table 5.5) and there is a considerable enlargement of the apical dome (Fig. 5.6). Thus, relative to plants in non-inductive conditions, there are dramatic and simultaneous changes of apex volume and plastochron associated with flowering.

On the basis of such findings and knowing that there are metabolic changes in young leaf primordia as well as in the apical meristem Evans (1971, 1975) proposed that flowering involved 'general activation of growth of the shoot apex, leading to an increase in its size and elimination of its vegetative pattern of activity, thereby allowing a new floral geometry to be established'. The crucial feature of this concept, as seen by Lyndon (1978), is that change in geometry on floral organogenesis must involve a reduction or removal of the influence of the leaves on the positioning of subsequent floral primordia. The enhanced RNA and protein synthesis occurring during floral evocation need not, however, be directly related to this growth activation.

To gain further perspective on this concept we need information on three questions, namely: What is the timing of change of growth? Is there growth activation of all organs or is it related specifically to floral organs? Which aspects of apical activation – dome size, plastochron, growth rate, primordium size – are essential for floral organogenesis?

On the question of timing of events, it is clear from a number of studies that, associated with the earliest visible signs of flowering (the increase in primordium number) there is always a simultaneous increase

Table 5.5. *Stimulation of primordium initiation during early stages of the transition to flowering*

| Species | Rate of primordia initiation (hours primordium$^{-1}$) | | Reference |
|---|---|---|---|
| | Vegetative | Early flowering | |
| *Chrysanthemum* SDP | 24 | 6.5 | Horridge & Cockshull (1979) |
| *Pharbitis* SDP 21 °C | 21.8 | 7.5 | King & Evans (1969) |
| 28 °C | 13.3 | 5 | |
| *Lolium* LDP | 41 | 4 | Knox & Evans (1966) |
| *Hordeum* LDP in SD | 72 | 36 | Dale & Wilson (1979) |
| in LD | 60 | 12 | |
| *Silene* 20 °C | 117[a] | 12–24 | Lyndon (1977) |
| *Xanthium* | 46 | 30 | Erickson & Meicenheimer (1977) |

[a]Pair of primordia.

Fig. 5.6. Effect of photoperiod on the initiation of primordia (▲—△) and increase in diameter of the apical dome (●——○) of *Chrysanthemum morifolium*. Inductive SD (△, ○) 12 h photoperiod; non-inductive LD 24 h light (▲, ●). Adapted from Horridge & Cockshull (1979).

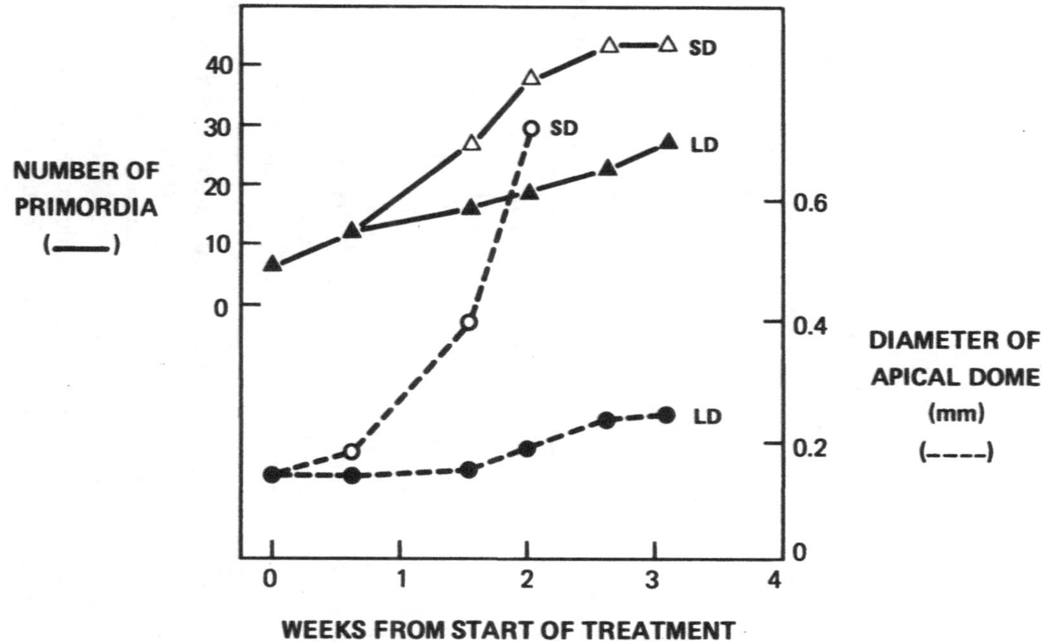

in apical dome volume (e.g. Fig. 5.6). This is so whether plants are grown with differing mineral nutrition (Dale & Wilson, 1979), photoperiod (Horridge & Cockshull, 1979), or temperature (Lyndon, 1977). However, this apical growth occurs later than the essential changes in RNA and protein synthesis which occur during floral evocation. Growth activation occurs at a time when treatment with inhibitors no longer blocks flowering (see Bernier, 1971).

On the question of general or specific growth responses of flowering, there are many reports of stimulation of lateral bud production (e.g. Popham & Chang, 1952; Evans, 1960; Gifford & Tepper, 1961; Thomas, 1962; King & Evans, 1969). Growth of primordial leaves may also be stimulated as in *Trifolium* (Thomas, 1961) and *Lolium* (Fig. 5.3). There is, however, no activation of a bud which is not destined to flower in either *Trifolium* (Thomas, 1962) or *Pharbitis* (King & Evans, 1969). In *Pharbitis*, for example, if the axillary bud at node 2 remains vegetative it produces primordia at the same rate as the identical bud on a non-induced plant. Potential flower bud sites at node 3 and higher are stimulated. Thus, there is clear target specificity in the action of floral stimulus. Stimulation of growth only occurs in those meristematic regions which contribute to the inflorescence.

To determine which aspects of apical growth changes are important for flowering it is essential to know how 'growth' is changing. As Lyndon has shown for leaf growth (Chapter 1) and also for floral development (Lyndon, 1977) there are many interacting parameters influencing apical growth. Given that apical volume increases and the plastochron becomes shorter as illustrated for chrysanthemum (Fig. 5.6), then these changes could be accommodated simply by increasing the relative volume growth rate ($R_d$). However, the changes on transition to flowering are more complex. On flowering the size of primordia is decreased at their initiation. Thus there is a reduction in $\Delta V$, the natural logarithm of the volume of apex required to restore apex volume at the end of the plastochron. In fact, by reducing only the volume replacement factor ($\Delta V$) so that the growth rate per plastochron exceeds $\Delta V$ ($R_p > \Delta V$) it is possible to accommodate an increasing volume and decreasing plastochron. Thus, growth rate need not necessarily increase for flowering to occur. Clearly, a variety of factors may change at the same time during the transition to flowering.

Lyndon (1977) has presented the only comprehensive analysis of floral organogenesis using the above relationships and some of his data for *Silene* are summarized in Table 5.6. Under his growth conditions the

Table 5.6. *Growth of the apical meristem of* Silene *during the transition to flowering; visible flowering began about 9 days after LD*

| Photoperiod sequence | Days after start of induction | Maximum apex volume each plastochron ($10^5$ $\mu m^3$) | Plastochron (days) | Volume growth per day ($R_d$) | Volume growth per plastochron ($R_p$) | ln volume increase to reform dome each plastochron ($\Delta V$) |
|---|---|---|---|---|---|---|
| SD | 0 | 4.5–8.0 | 3.9 | 0.68 | 2.64 | 2.41 |
| LD | 3 | 7.0 | — | 0.70 | 2.74 | 2.44 |
| | | | | 0.75 | 2.50 | 2.49 |
| | 6 | 7.5 | — | | | |
| | 7 | — | 3.9 | 1.06 | 2.76 | 2.26 |
| | 8 | — | 3.4 | | | |
| SD | 9 | 10.4 | 2.6 | | | |
| | 10 | — | 1.1 | 0.54 | 1.60 | 1.73 |
| | 11 | — | 0.7 | | | |
| | 12 | — | 0.5 | | | |

Adapted from Lyndon (1977).

first visible sepals are formed at 8–9 days from the start of an exposure to 7 long days. The first clear response to inductive LD is the transient increase in $R_d$ followed later by slow growth during floral organogenesis. Other changes followed, e.g. reduction in the plastochron, in $R_p$ and in $\Delta V$. Over the whole period, as also in the SD controls, apex volume increased continuously since $R_p$ was greater than $\Delta V$. With the change in $R_d$, however, and the slight drop in $\Delta V$, apical dome volume increased even more just at the time of flowering (i.e. $R_p \gg \Delta V$). Subsequently, decline in both $R_d$ and the plastochron led to a reduction in $R_p$, but since $\Delta V$ changed only slightly apical dome volume now diminished ($R_p < \Delta V$). That dome volume did decrease is interesting as there is also a reduction in primordium size on flowering in *Silene*; sepals at initiation are about one-third the size of leaf primordia and petals and stamen primordia are about one-sixth the size (Lyndon, 1978).

Lyndon's analysis highlights the diversity of responses which can contribute to changes seen on flowering. Change in dome volume or growth rates are only some of the controlling factors which it is necessary to understand. Further, the apical volume of the SD controls also increases with time but flowering does not result, while conversely the apical volume of a meristem may be much smaller and yet flowering can still occur (R. F. Lyndon, unpublished). Thus, increase in apical volume to some specific, critical size cannot be essential for flowering in *Silene*. The alternative that Lyndon prefers is that the reduced primordium size associated with more rapid initiation and dome enlargement allows transition to a higher order phyllotaxis and this change is then followed, soon, by change in gene expression associated with committal of primordia to floral organogenesis.

It is clear that there is still much to be learnt of control processes in floral organogenesis. There is considerable support in Lyndon's work for the concept suggested by Evans (1971, 1975). However, changed apical geometry may reflect quite complex changes in 'growth'. It also remains unclear how the first events of evocation relate to these later stages of floral organogenesis. Later we return to this issue in discussion of models of organogenesis.

### *Alternative developmental options for meristems*

Apical meristems can show a range of developmental pathways. When cataphylls form in dormant buds leaf production is suspended, not terminated. Sometimes, vegetative apices may abscind as part of overwintering in woody species such as *Syringa* (Garrison & Wetmore, 1961). If photoperiodic treatment is inappropriate flowering apices may become dormant, as in hop (*Humulus lupulus*) (Thomas & Schwabe, 1969), or they may abscind and degenerate as in certain grasses (e.g. *Bouteloua eriopoda*; Schwartz & Koller, 1975).

In some instances, especially in lateral buds, the meristem may never produce leaves. This is true for the anlagen meristem of grape which is committed to tendril production from inception. Similarly, in members of the Cactaceae such as *Opuntia* the areole bud forms spines directly without first forming leaves (Mauseth, 1976). However, when cytokinins are applied to the areole or anlagen meristem in culture their development is switched (Mauseth, 1976; Srinivasan & Mullins, 1978, 1979) to produce leaf initials.

Striking flexibility of developmental options is seen from studies with explants of tobacco epidermis. In culture these explants may form flowers directly without first forming leaves (Tran Thanh Van, 1973). However, by varying the balance of auxins to cytokinins, the type and concentration of sugar, and environment, it is possible to switch development between production of callus, roots, vegetative shoots or flowers (Fig. 5.7).

Clearly, committal to a particular path of organogenesis can occur early in the development of the apical meristem. It is too narrow a view to consider organogenesis as a switching or transition from leaf

Fig. 5.7. Control of morphogenesis in tobacco epidermis explants by alteration in growth regulator balance and nutrition. After Tran Thanh Van (1977).

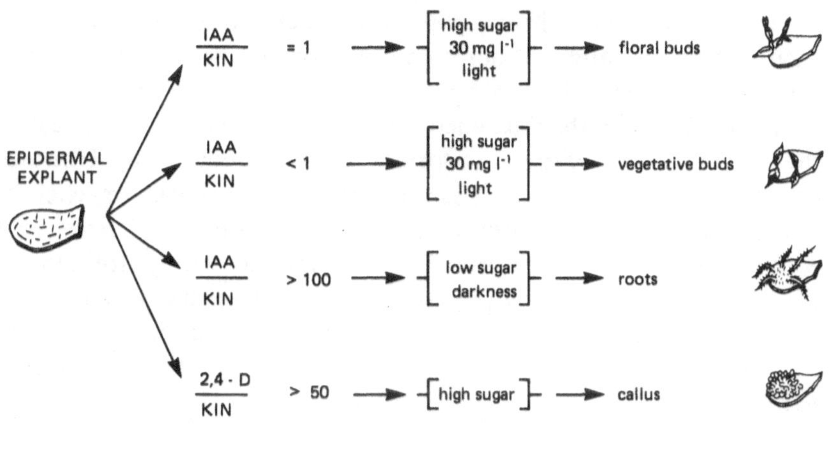

production to the formation of other organs. The question, 'Is a flower organ the outcome of a gradual transformation of developing leaves?' places undue emphasis on homology of final form. We are really concerned with much earlier events at the apical meristem and early primordium stages.

### *Interconvertibility of organs*

Not only do meristems manifest a variety of developmental options but interconvertibility has been established between many of the organs produced by the shoot apex. For instance, in *Bougainvillea* flowers may start to form but regress to thorns as a consequence of inadequate floral induction (Hackett & Sachs, 1968).

Possibly the best documented instances of interconvertibility are those between vegetative and floral buds. This is often associated with inadequate induction as seen in the vegetative outgrowth of otherwise inhibited lateral buds of *Kalanchoë* (Lang, 1965). The other extreme is seen in the vegetative reversion which occurs from the tip of the carpel of otherwise fully formed flowers of *Impatiens* (Krishnamoorthy & Nanda, 1968). Atypical, vegetative flowers are often formed in other species, too, after inadequate induction (e.g. *Cosmos*, wheat, *Anagallis*; Vince-Prue, 1975). Clearly, no matter what the pattern of organogenesis, initially all organs show a degree of developmental flexibility. The question is, for how long and to what degree does an organ retain this quality?

### *When is organogenesis irrevocable?*

It appears that irrevocable commitment of the form of an organ occurs at or just after primordium initiation on the bare apical meristem. This is reasonable for *Opuntia* where there is a clear anatomical distinction between the thickened epidermal cells and inner fibre cells of the spine and cells destined to form a leaf (Mauseth, 1976). However, experiments involving changes in either the physical or chemical environment provide better evidence of the timing of irreversible commitment of an organ. Foster (1935) examined cataphyll formation in *Carya* and found that early removal of foliage leaves blocked formation of cataphylls and allowed vegetative growth to proceed. Development was irreversible if the primordium was 90–100 $\mu$m long at the time of defoliation.

In studies with flowering, stepwise reversal of the apex from floral organogenesis back to leaf production has been reported in a number of species. The environmental switches in these instances have been photoperiod or temperature change (see also page 121). However, in general, only final form has been examined and no information has been available on the earliest changes. In *Pharbitis*, however, daily microscopic examination has been combined with precisely timed switching experiments. These experiments involved a study of both irrevocable commitment to flower formation and its reversal to vegetative growth by high temperature and the converse, irrevocable commitment with age to vegetative rather than to floral organogenesis (King & Evans, 1969).

The point at which a vegetative pattern of growth is irrevocably established in *Pharbitis* can be seen from the scanning electron microscope pictures in Fig. 5.8. The lateral axillary bud at node 3 could be induced to flower (e.g. Fig. 5.8*b*) when a single short-day inductive treatment was imposed on these seedlings 4.5 days after sowing. Floral stimulus arrived at this apex within 15–17 h of the start of this short-day

Fig. 5.8. Effect of age of axillary bud at node 3 on flowering of *Pharbitis*. (*a*) Side view of shoot tip at the time of photoperiodic induction. (*b*) Vertical view of apical meristem at lateral bud at node 3 at the end of 15 h darkness, i.e. 4.5 days after sowing. (*c*) Vertical view of similar apex at the end of 15 h darkness 5.5 days after sowing. Horizontal bar is 100 $\mu$m in length. (R. W. King unpublished.)

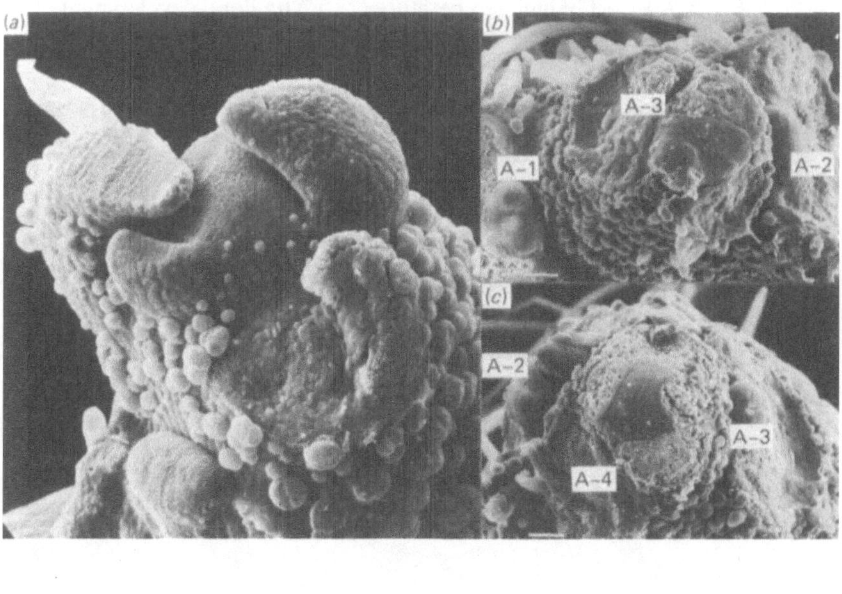

treatment. The lateral shoot meristem at this time had yet to produce a fully formed primordium. However, if photoperiodic induction was delayed by 24 h then this same lateral apex (now 24 h older) could not be induced to flower (King & Evans, 1969). It had now produced one primordium and possibly a second vestigial one (Fig. 5.8*c*). At this time it was still barely distinguishable from its potentially floral counterpart induced the day before. Clearly, quite subtle but, nevertheless, irrevocable changes occurred during the 24 h delay, either to the structure of the apex and primordium or to the spatial arrangement of primordia. Similar age-dependent changes in sensitivity of lateral buds have been described for *Anagallis arvensis* by Ballard & Grant-Lipp (1964).

The converse response in *Pharbitis*, the reversal of floral development, has been studied by applying a high temperature (28 °C vs. 21 °C) to the seedlings after a fixed inductive treatment. The timing of the temperature treatment is not so important. What is significant is that whorls of floral primordia were visible on lateral buds at node 3 130 h after the start of induction (Fig. 5.9*a*). These whorls were characteristic of sepals, petals and anthers. No such pattern or number of primordia were apparent on lateral buds of the same age but on non-induced vegetative plants, 30 h later, the floral arrangement of primordia had reverted to the spiral pattern of the vegetative plants (Fig. 5.9*b*). Temporarily, at least, primordium production ceased (King & Evans, 1969) and the previously formed floral primordia reverted to vegetative

Fig. 5.9. Effect of high temperature (28 °C) on floral development of axillary buds at node 3 of *Pharbitis*. Structure of primordium at 130 h (*a*) and 160 h (*b*) for induced plants. (R. W. King unpublished.)

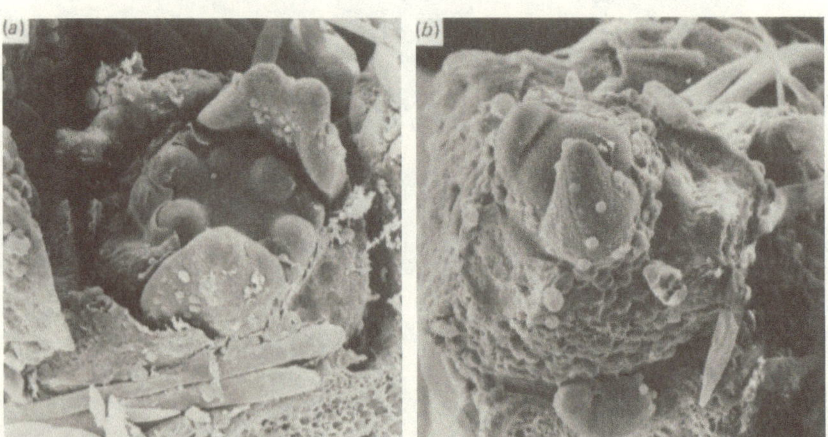

growth and to development as leaves. Therefore, until a primordium is more than an undifferentiated mass of cells its developmental future is flexible. Floral organogenesis requires more that just an increase in primordia number and a change in phyllotaxis. Apparently, irrevocable committal occurs after differentiation of the primordium.

## Models of apical organogenesis

### *General*

Models of apical organogenesis are only discussed here in the context of the transition to flowering. There may, however, be similar control mechanisms for all apices whether the product is a spine, a thorn, or a flower. For example, until the target primordia show some differentiation they can be switched from one developmental pathway to another. Spine or flower initials can both be switched to vegetative forms and thorns and flowers may be interchangeable.

For floral organogenesis there are two main models: organogenesis involves early expression of new genes; organogenesis involves geometry changes consequent on general activation of the apex.

So little is known of detailed mechanisms that neither model necessarily excludes the other. New genes may be expressed both early and late in a sequence of development. Specific new genes might be activated but general activation of genes, and even of specific genes, might also be essential for change in geometry. Furthermore, how the eventual three-dimensional structure of each floral organ is generated is only explained in vague terms by either model.

### *New organs from early expression of new genes*

Biochemical and histological evidence supporting the concept of early expression of new genes was presented above (page 115). It is principally studies with inhibitors of RNA synthesis which provide evidence for an essential requirement for new RNA in early events of floral evocation. It is apparent, however, that these inhibitors are not always specific. Actinomycin D, for example, inhibits DNA synthesis as well as RNA synthesis at the apical meristem of *Pharbitis* (Arzee, Zilberstein & Gressel, 1975). Also, isolation of new mRNA has yet to be achieved. Nevertheless, histochemical studies and experiments involving precursor incorporation (Fig. 5.3) do confirm that there is an early increase in RNA and protein synthesis. For the future it will be of interest to see further results with the immunochemical approach as pioneered by Bernier. It is also possible that cDNA techniques could be applied to the development of assays for 'floral' mRNA. The suggestion of Nagl and coworkers (page 124) of extra floral DNA per nucleus should also be examined, especially in plants where the timing of organogenesis is known.

The emphasis placed on early expression of 'floral' genes is logical in that inhibitors of RNA synthesis are only effective over the first hours of floral evocation (Fig. 5.4). However, other interpretations of these observations are possible. These inhibitors also reduce growth so that its maintenance or even enhancement might be required for perception or expression of stimulus although enhanced growth is not apparent as an early event of evocation. Another explanation ties general growth inhibition to a time-dependent action of floral stimuli. If the stimulus is labile as it seems to be in a number of species, then flowering would be depressed if the apex failed to import stimulus or if, although imported, its early action was delayed for a brief period after photoperiodic induction. Unfortunately, without knowing the nature of the stimuli, this hypothesis cannot be tested unequivocally.

Another issue as yet unresolved is whether new genes are switched on more-or-less instantaneously as suggested by the inhibitor studies or whether there is progressive expression of new blocks of genes. There is some evidence for a progression of decisive steps for floral organogenesis. Each class of floral organ may grow, initially, as an uncommitted organ which passes through a period of competence before final determination. The studies of Heslop-Harrison (1964) on the effects of temperature shocks or change in photoperiod on flower development in corn provide one illustration of progressive expression. If new gene expression is involved then it is clear that, following early gene activation, determination of the various organs involves progressive unfolding of gene action. Heslop-Harrison proposed a chain of relays successively activating new gene complexes for each organ type.

The sequential bifurcation of development seen in corn is also evident in other flowering responses. In *Impatiens*, for example, transfer to non-inductive long days even as late as carpel initiation can induce a regression to vegetative growth at the presumptive carpel site (Krishnamoorthy & Nanda, 1968). Experiments in which the developing apex is bisected at various stages of floral organogenesis also reveal that the floral meristem passes through a succession of physiological states which regulate the type of organ formed. In *Primula*, for example, bisection of the floral apex after a particular primordium is apparent (sepal, petal, etc.) leads to formation of the next organs in the sequence, both at the cut surface as well as at their normal positions (Cusick, 1956). Transforming ability at the cut surface progresses with organogenesis which indicates that progressive gene expression is part of pattern and organ formation.

Clearly studies involving environmental and other manipulations of floral organogenesis present problems for a simple, early, one-step response involving new gene expression. On the other hand, the 'relay' or 'cascade' aspect of sequential gene activation as suggested by Heslop-Harrison is not compatible with some of the physiological evidence. For example, a single inductive treatment may be adequate for the transition to flowering but continuation of development may require continued inductive cycles, as in *Impatiens* (see Evans, 1969, 1975). In other species early floral evocation may occur in non-inductive photoperiods but further floral development may show a strict photoperiod requirement as in *Caryopteris*. Alternatively, additional inductive cycles can cause more rapid floral development even if morphogenesis has already begun. Such responses imply that action of floral stimuli is not necessarily specific for first events of flowering. If gene activation is required then it may be for 'more-of-the-same'. Continued apical activation may be the sole requirement for flowering.

### *A new floral geometry*

Recently, a number of models of floral development have been suggested to explain floral organogenesis (see Thornley, 1972; Schwabe & Wimble, 1976; Charles-Edwards, Cockshull, Horridge & Thornley, 1979). The only model to approach the question of apical geometry in any complete way is that of Charles-Edwards and coworkers for *Chrysanthemum*. The main strength of this model is in highlighting interactions between apical dome size, plastochron duration and primordium size in determining phyllotactic pattern. The assumption made is that primordia are a source of 'inhibitor' present in the dome and that the 'inhibitor' pattern is established by recently formed leaf primordia. The 'new' geometry of floral organogenesis requires disruption of these gradients (patterns) of distribution of 'inhibitor'.

A central feature of this model of flowering is the achievement of a critical volume of the apex. Flowering, seen as a change in phyllotaxis, occurs when apical dome volume is between 2 and 5 × $10^{-12}$ $m^3$. This range is set from actual measurements of Horridge & Cockshull (1979) as illustrated in Fig. 5.6. Recent mathematical models of the control of the phyllotaxis of leaves highlight the importance of apical size for Fibonacci phyllotaxis (Mitchison, 1977; Veen & Lindenmayer, 1977) which follows as a mathematical necessity of apex size (see also Chapter 3). It is interesting that the values derived from Mitchison's model indicate a change in apical radius (Fig. 5.10) that is not unreasonable for generation of the higher order phyllotactic values commonly found in flowers (see Bernier, 1980). Moreover, apical enlargement and a shift in phyllotactic pattern from (2,3) to (3,5) have been shown in *Xanthium*, following application of gibberellic acid (Maksymowych & Erickson, 1977). Such treatment with gibberellic acid is known to promote flowering under marginally inductive conditions (Lang, 1965). Further, similar transition changes in phyllotaxis and apical volume were evident for plants induced to flower without gibberellin treatment (Erickson & Meicenheimer, 1977).

The problem with this model is that not all species show increased apical size and reduced plastochron on flowering – for example *Perilla* (Nougarède, Bronchart, Bernier & Rondet, 1964) and hop (Thomas & Schwabe, 1970). Further, in *Silene*, both small and large apices may be

Fig. 5.10. Computer simulated relationship between apical radius and phyllotaxis. From Mitchison (1977).

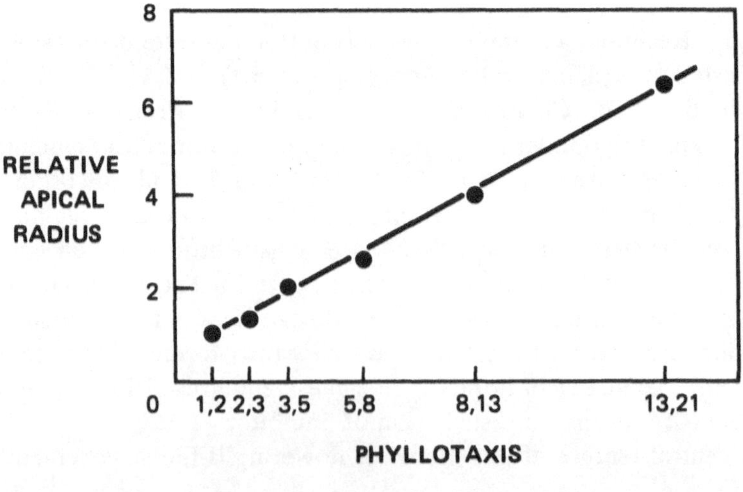

converted to flowers (R. F. Lyndon, unpublished). Lyndon (1977) considered that dome size relative to primordia size might be a crucial factor – not the absolute size of the dome. In addition his analyses indicate the complex manner in which interrelationships at the apex could be altered by change in dome size, primordium size and plastochron.

It is clear that further studies are required to establish which changes determine the pattern of phyllotaxis at flowering. It also remains to be shown whether these changes lead to specification of organ form. It is quite possible that genes for organ form are only switched on after the positions of primordia are established. This fits with the observation that organogenesis becomes irrevocable only when primordia are clearly visible and that geometry change may be important at this stage. Changes in siting of organs could also be a first step in progressive gene expression at the time of formation of various floral organs as suggested by Heslop-Harrison (1964).

On the other hand, whilst change in apex size and phyllotaxis may be sufficient to explain at least partially floral organogenesis in species which flower autonomously, it does not explain the early and essential biochemical changes occurring on photoperiodic induction of flowering. In this regard, it will be especially interesting to see what controls flowering in those situations where a few initial cells or a bare apex proceed directly to flowering without a prior period of vegetative growth as, for instance, in epidermal explants of tobacco, or in the grass apex.

The continued stimulus of Dr L. T. Evans is gratefully acknowledged. Mr Mike Moncur, CSIRO, Division of Land Use Research is thanked for the electron microscope photographs of shoot apices.

## References

Aghion-Prat, D. (1965). Néoformation de fleurs *in vitro* chez *Nicotiana tabacum* L. *Physiologie Végétale*, **3**, 229–303.

Aksenova, N. P., Konstantinova, T. N. & Bavrina, T. V. (1980). Reciprocal influence of explants with vegetative and generative morphogenesis in culture *in vitro*. *Soviet Plant Physiology*, **26**, 985–7.

Arzee, T., Zilberstein, A. & Gressel, J. (1975). Immediate intraplumular distribution of macromolecular synthesis following floral induction in *Pharbitis nil*. *Plant and Cell Physiology*, **16**, 505–11.

Baldev, B. (1962). *In vitro* studies of floral induction on stem apices of *Cuscuta reflexa* Roxb. – short-day plant. *Annals of Botany*, **26**, 173–80.

Ballard, L. A. T. & Grant-Lipp, A. E. (1964). Juvenile photoperiodic sensitivity in *Anagallis arvensis* L. subsp. foemina (Mill.) Schinz and Thell. *Australian Journal of Biological Sciences*, **17**, 323–37.

Bernier, G. (1971). Structural and metabolic changes in the shoot apex in transition to flowering. *Canadian Journal of Botany*, **49**, 803–19.

Bernier, G. (1980). The sequences of floral evocation. In *Physiologie de la Floraison. Colloques Internationaux du CNRS No. 285*, pp. 129–68.

Bernier, G., Kinet, J.-M. & Bronchart, R. (1967). Cellular events at the meristem during floral induction in *Sinapis alba* L. *Physiologie Végétale*, **5**, 311–24.

Bodson, M. (1975), Variation in the rate of cell division in the apical meristem of *Sinapis alba* during transition to flowering. *Annals of Botany*, **39**, 547–54.

Bodson, M. & Bernier, G. (1972). Effet de la 5-fluorodésoxyuridine sur l'activité mitotique du méristème caulinaire de *Sinapis alba* L. *Bulletin de la Société Royale de Botanique de Belgique*, **105**, 119–27.

Buvat, R. (1952). Structure, évolution et fonctionnement du méristème apical de quelques dicotylédones. *Annales des Sciences Naturelles (Botanique) Série 11*, **13**, 199–300.

Charles-Edwards, D. A., Cockshull, K. E., Horridge, J. S. & Thornley, J. H. M. (1979). A model of flowering in *Chrysanthemum*. *Annals of Botany*, **44**, 557–66.

Corson, G. E. (1969). Cell division studies of the shoot apex of *Datura stramonium* during transition to flowering. *American Journal of Botany*, **56**, 1127–34.

Cusick, F. (1956). Studies of floral morphogenesis. I. Median bisections of flower primordia in *Primula bulleyana* Forrest. *Transactions of the Royal Society of Edinburgh*, **13**, 153–66.

Cutter, E. G. (1969). *Plant Anatomy: Experiment and Interpretation*. pp. 343. London: Arnold.

Dale, J. E. & Wilson, R. G. (1979). The effects of photoperiod and mineral nutrient supply on growth and primordia production at the stem apex of barley seedlings. *Annals of Botany*, **44**, 537–46.

Dwivedi, R. S. & Naylor, J. M. (1968). Influence of apical dominance on the nuclear proteins in cells of the lateral bud meristem in *Tradescantia paludosa*. *Canadian Journal of Botany*, **46**, 289–98.

Erickson, R. O. & Meicenheimer, R. D. (1977). Photoperiod induced change in phyllotaxis in *Xanthium*. *American Journal of Botany*, **64**, 981–8.

Evans, L. T. (1960). Inflorescence initiation in *Lolium temulentum*. I. Plant age and leaf area on sensitivity to photoperiodic induction. *Australian Journal of Biological Sciences*, **13**, 123–31.

Evans, L. T. (1964). Inflorescence initiation in *Lolium temulentum*. VI. Effects of some inhibitors of nucleic acid, protein, and steroid biosynthesis. *Australian Journal of Biological Sciences*, **17**, 24–35.

Evans, L. T. (1969). The nature of flower induction. In *The Induction of Flowering: Some Case Histories*, pp. 457–80. South Melbourne: Macmillan and Co.

Evans, L. T. (1971). Flower induction and the florigen concept. *Annual Review of Plant Physiology*, **22**, 365–94.

Evans. L. T. (1975). *Daylength and the Flowering of Plants*. pp. 122. Menlo Park, California: Benjamin.

Evans. L. T., Knox, R. B. & Rijven, A. H. G. C. (1970). The nature and localization of early events in the shoot apex of '*Lolium temulentum*' during floral induction. In *Cellular and Molecular Aspects of Floral Induction*, ed. G. Bernier, pp. 192–206. London: Longman.

Evans, L. T. & Rijven, A. H. G. C. (1967). Inflorescence initiation in *Lolium temulentum* L. XI. Early increases in the incorporation of $^{32}P$ and $^{35}S$ by shoot apices during induction. *Australian Journal of Biological Sciences*, **20**, 1033–42.

Foster, A. S. (1935). A histogenic study of foliar determination in *Carya buckleyi* var. *arkansana*. *American Journal of Botany*, **22**, 88–147.
Francis, D. & Lyndon, R. F. (1978). Early effects of floral induction on cell division in the shoot apex of *Silene*. *Planta*, **139**, 273–80.
Frankel, O. H., Knox, R. B. & Considine, J. A. (1981). The development of the wheat flower: genetics and physiology. In *Wheat Science – Today and Tomorrow*, ed. L. T. Evans & W. J. Peacock, pp. 167–90. Cambridge University Press.
Garrison, R. & Wetmore, R. H. (1961). Studies on shoot tip abortion: *Syringa vulgaris*. *American Journal of Botany*, **48**, 789–95.
Gifford, E. M. & Corson, G. E. (1971). The shoot apex in seed plants. *Botanical Review*, **37**, 143–230.
Gifford, E. M. & Tepper, H. B. (1961). Ontogeny of the inflorescence in *Chenopodium album*. *American Journal of Botany*, **48**, 657–67.
Gressel, J., Zilberstein, A. & Arzee, T. (1970). Bursts of incorporation into RNA and ribonuclease activities associated with induction of morphogenesis in *Pharbitis*. *Developmental Biology*, **22**, 31–42.
Gressel, J., Zilberstein, A., Porath, D. & Arzee, T. (1980). Demonstration with fiber illumination that *Pharbitis* plumules also perceive flowering photoinduction. In *Photoreceptors and Plant Development*, ed. J. de Greef, pp. 525–30. Antwerpen: Antwerpen University Press.
Gressel, J., Zilberstein, A., Strausbauch, L. & Arzee, T. (1978). Photoinduction of *Pharbitis* flowering: relationship to RNA synthesis and other metabolic events. *Photochemistry and Photobiology*, **27**, 232–40.
Hackett, W. P. & Sachs, R. M. (1968). Experimental separation of inflorescence development from initiation in *Bougainvillea*. *Proceedings of the American Society of Horticultural Science*, **92**, 615–21.
Halperin, W. (1978). Organogenesis at the shoot apex. *Annual Review of Plant Physiology*, **29**, 239–62.
Havelange, A. (1980). The quantitative ultrastructure of the meristematic cells of *Xanthium strumarium* during the transition to flowering. *American Journal of Botany*, **67**, 1171–8.
Heslop-Harrison, J. (1964). Sex expression in flowering plants. In *Meristems and Differentiation*, pp. 109–12. Brookhaven Symposium in Biology No. 16.
Horridge, J. S. & Cockshull, K. E. (1979). Size of the *Chrysanthemum* shoot apex in relation to inflorescence initiation and development. *Annals of Botany*, **44**, 547–56.
Jacqmard, A., Miksche, J. P. & Bernier, G. (1972). Quantitative study of nucleic acids and protein in the shoot apex of *Sinapis alba* during transition from the vegetative to the reproductive condition. *American Journal of Botany*, **59**, 714–21.
Jacqmard, A., Raju, M. V. S., Kinet, J-M. & Bernier, G. (1976). The early action of the floral stimulus on mitotic activity and DNA synthesis in the apical meristem of *Xanthium strumarium*. *American Journal of Botany*, **63**, 166–74.
Kavon, D. L. & Zeevaart, J. A. D. (1978). Simultaneous inhibition of translocation of photosynthate and of floral stimulus by localized low-temperature treatment in the short-day plant *Pharbitis nil*. *Planta*, **144**, 201–4.
Kinet, J-M., Bodson, M., Alvinia, A. M. & Bernier, G. (1971). The inhibition of flowering in *Sinapis alba* after the arrival of the floral stimulus at the meristem. *Zeitschrift für Pflanzenphysiologie*, **66**, 49–63.
King, R. W. (1971). Timing in *Chenopodium rubrum* of export of the floral stimulus from the cotyledons and its action at the shoot apex. *Canadian Journal of Botany*, **50**, 687–702.
King, R. W. (1975). Multiple circadian rhythms regulate photoperiodic flowering responses in *Chenopodium rubrum*. *Canadian Journal of Botany*, **53**, 2631–8.

King, R. W. & Evans, L. T. (1969). Timing of evocation and development of flowers in *Pharbitis nil. Australian Journal of Biological Sciences*, **22,** 559–72.

King, R. W., Evans, L. T. & Wardlaw, I. F. (1968). Translocation of the floral stimulus in *Pharbitis nil* in relation to that of assimilates. *Zeitschrift für Pflanzenphysiologie*, **59,** 377–88.

Konstantinova, T. N., Aksenova, N. P., Bavrina, T. V. & Chailakhyan, M. K. (1969). On the ability of tobacco stem calluses to form vegetative and generative buds in culture *in vivo. Doklady Botanical Sciences*, **187,** 82–5.

Knox, R. B. & Evans, L. T. (1966). Inflorescence development in *Lolium temulentum*. VIII. Histochemical changes at the shoot apex during induction. *Australian Journal of Biological Sciences*, **19,** 233–45.

Knox, R. B. & Evans, L. T. (1968). Inflorescence initiation in *Lolium temulentum* L. XII. An autoradiographic study of evocation in the shoot apex. *Australian Journal of Biological Sciences*, **21,** 1083–94.

Krishnamoorthy, H. N. & Nanda, K. K. (1968). Floral bud reversion in *Impatiens balsamina* under non-inductive photoperiods. *Planta*, **80,** 43–51.

Lang, A. (1965). Physiology of flower initiation. In *Encyclopedia of Plant Physiology*, vol. XV/I, ed. W. Ruhland, pp. 1380–536. Berlin: Springer-Verlag.

Lyndon, R. F. (1977). Interacting processes in vegetative development and in the transition to flowering at the shoot apex. In *Integration of Activity in the Higher Plant*, ed. D. H. Jennings, pp. 221–50, *Symposia of the Society for Experimental Biology No. 31.* Cambridge University Press.

Lyndon, R. F. (1978). Phyllotaxis and the initiation of primordia during flower development in *Silene. Annals of Botany*, **42,** 1349–60.

Maksymowych, R. & Erickson, R. O. (1977). Phyllotactic change induced by gibberellic acid in *Xanthium* shoot apices. *American Journal of Botany*, **64,** 33–44.

Marc, J. & Palmer, J. H. (1978). Determination of the length of the vegetative and pre-floral stages in the day-neutral plant *Helianthus annuus* by chilling pulses. *Journal of Experimental Botany*, **29,** 367–73.

Mauseth, J. D. (1976). Cytokinin- and gibberellic acid-induced effects on the structure and metabolism of shoot apical meristems in *Opuntia polyacantha* (Cactaceae). *American Journal of Botany*, **63,** 1295–301.

Miller, M. B. & Lyndon, R. F. (1976). Rates of growth and cell division in the shoot apex of *Silene* during the transition to flowering. *Journal of Experimental Botany* **27,** 1142–53.

Miller, M. B. & Lyndon, R. F. (1977). Changes in RNA levels in the shoot apex of *Silene* during the transition to flowering. *Planta*, **136,** 167–72.

Mitchison, G. J. (1977). Phyllotaxis and the Fibonacci series. *Science*, **196,** 270–5.

Nagl, W., Frisch, B. Frölich, E. (1979). Extra-DNA during floral induction. *Plant Systematics and Evolution* (Supplement), **2,** 111–18.

Naylor, J. M. (1958). Control of nuclear processes by auxin in axillary buds of *Tradescantia paludosa. Canadian Journal of Botany*, **36,** 221–32.

Nir, I., Goren, R. & Leshem, B. (1972). Effects of water stress, gibberellic acid and 2-chloroethyltrimethyl-ammonium chloride (CCC) on flower differentiation in 'Eureka' lemon trees. *Journal of the American Society of Horticultural Science*, **97,** 774–8.

Nougarède, A. (1967). Experimental cytology of the shoot apical cells during vegetative growth and flowering. *International Review of Cytology*, **21,** 203–351.

Nougarède, A., Bronchart, R., Bernier, G. & Rondet, P. (1964). Comportement du méristème apical du *Perilla nankinensis* (Lour.) Decne en relation avec les conditions photopériodiques. *Revue Génerale de Botanique*, **71,** 205–38.

Ogawa, Y. & King, R. W. (1979). Indirect action of benzyladenine and other chemicals on flowering of *Pharbitis nil* Chois: action by interference with assimilate translocation from induced cotyledons. *Plant Physiology*, **63,** 643–9.

Paulet, P. & Nitsch, J. P. (1964). La néoformation de fleurs sur cultures *in vitro* de racines de *Cichorium intybus* L. : étude physiologique. *Annales de Physiologie Végétale*, **6,** 333–45.
Phillips, I. D. J. (1975). Apical dominance. *Annual Review of Plant Physiology*, **26,** 341–67.
Pierard, D., Jacqmard, A., Bernier, G. & Salmon, J. (1980). Appearance and disappearance of proteins in the shoot apical meristem of *Sinapis alba* in transition to flowering. *Planta*, **150,** 397–405.
Pierik, R. L. M. (1965). Regulation of morphogenesis by growth regulators and temperature treatment in isolated tissues of *Lunaria annua* L. *Proceedings of the Koninklijke Nederlandse Akademie van Wetenschappen Series C*, **68,** 324–32.
Popham, R. A. & Chan, A. P. (1952). Origin and development of the receptacle of *Chrysanthemum morifolium. American Journal of Botany*, **39,** 329–39.
Pryke, J. A. & Bernier, G. (1978a), RNA synthesis in the apex of *Sinapis alba* in transition to flowering. *Journal of Experimental Botany*, **29,** 953–61.
Pryke, J. A. & Bernier, G. (1978b). Acid invertase activity in the apex of *Sinapis alba* during transition to flowering. *Annals of Botany*, **42,** 747–9.
Saint-Côme, R. (1971). Durée du cycle mitotique chez le *Coleus blumei* Benth. durant les phases préflorale et reproductrice. *Comptes Rendus Hebdomadaires des Seánces de l'Académié des Sciences Série D*, **272,** 44–7.
Saunders, P. (1978). Phytohormones and bud dormancy. In *Phytohormones and Related Compounds – A Comprehensive Treatise*, vol. II, ed. D. S. Letham, P. B. Goodwin & T. J. V. Higgins, pp. 423–45. Amsterdam: Elsevier/North Holland Biomedical Press.
Schwabe, W. W. & Wimble, R. H. (1976). Control of flower initiation in long- and short-day plants. A common model approach. In *Perspectives in Experimental Biology*, vol. 2, Botany, ed. N. Sunderland, pp. 41–57. Oxford: Pergamon Press.
Schwartz, A. & Koller, D. (1975). Photoperiodic control of shoot-apex morphogenesis in *Bouteloua eriopoda. Botanical Gazette*, **136,** 41–9.
Scorza, R. & Janick, J. (1980). *In vitro* flowering of *Passiflora suberosa* L. *Journal of Horticultural Science*, **105,** 892–7.
Srinivasan, C. & Mullins, M. G. (1978). Control of flowering in the grapevine (*Vitis vinifera* L.). Formation of inflorescences *in vitro* by isolated tendrils. *Plant Physiology*, **61,** 127–30.
Srinivasan, C. & Mullins, M. G. (1979). Flowering in *Vitis*: conversion of tendrils into inflorescence bunches of grapes. *Planta*, **145,** 187–92.
Steeves, T. A. & Sussex, I. M. (1972). *Patterns in Plant Development*. pp. 302. Englewood Cliffs, NJ: Prentice-Hall.
Stiles, J. I. & Davies, P. J. (1976). RNA metabolism in apices of *Pharbitis nil* during floral induction. *Plant and Cell Physiology*, **17,** 825–33.
Taillandier, J. (1978a). La Floraison du Mouron rouge: Arrivée du Stimulus Florale. *Zeitschrift für Pflanzenphysiologie*, **87,** 285–96.
Taillandier, J. (1978b). L'Induction florale chez l'*Anagallis arvensis. Zeitschrift für Pflanzenphysiologie*, **87,** 395–411.
Tanimoto, S. & Harada, H. (1980). Hormonal control of morphogenesis in leaf explants of *Perilla frutescens* Britton var. crispa Decaisne f viride crispa Makino. *Annals of Botany*, **45,** 321–7.
Thomas, G. G. & Schwabe, W. W. (1969). Factors controlling flowering in the Hop (*Humulus lupulus* L.). *Annals of Botany*, **33,** 781–93.
Thomas, G. G. & Schwabe, W. W. (1970). Apical morphology in the hop (*Humulus lupulus*) during flower initiation. *Annals of Botany*, **34,** 849–59.
Thomas, R. G. (1961). The relationship between leaf growth and induction of flowering in long-day plants (LDP). *Naturwissenschaften*, **48,** 108.

Thomas, R. G. (1962). The initiation and growth of axillary bud primordia in relation to flowering in *Trifolium repens* L. *Annals of Botany*, **26**, 329–44.

Thomas, R. G. (1963). Floral induction and the stimulation of cell division in *Xanthium*. *Science*, **140**, 54–6.

Thornley, J. H. M. (1972). A model of a biochemical switch and its application to flower initiation. *Annals of Botany*, **36**, 861–71.

Tran Thanh Van, M. (1973). Direct flower neoformation from superficial tissue of small explants of *Nicotiana tabacum* L. *Planta*, **115**, 87–92.

Tran Thanh Van, M. (1977). Regulation of morphogenesis. In *Plant Tissue Culture and Its Biotechnological Application*, ed. W. Barz, E. Reinhard & M. H. Zenk, pp. 367–85. New York, Berlin, Heidelberg: Springer-Verlag.

Usciati, M., Cadoccioni, M. & Guern, J. (1972). Early cytological and biochemical events induced by a 6-benzylaminopurine application on inhibited axillary buds of *Cicer arietinum* plants. *Journal of Experimental Botany*, **23**, 1009–20.

Veen, A. H. & Lindenmayer, A. (1977). Diffusion mechanism for phyllotaxis. Theoretical physico-chemical and computer study. *Plant Physiology*, **60**, 127–39.

Vince-Prue, D. (1975). *Photoperiodism in Plants*, pp. 444. London: McGraw-Hill.

Wada, K. (1968). Studies on the flower initiation in *Pharbitis* seedlings, with special reference to the early stimulation of cell division at the flower-induced shoot apex. *Botanical Magazine, Tokyo*, **81**, 46–7.

Wardell, W. L. (1976). Floral activity in solutions of deoxyribonucleic acid extracted from tobacco stems. *Plant Physiology*, **57**, 855–61.

Wardell, W. L. (1977). Floral induction of vegetative plants supplied a purified fraction of DNA from stems of flowering plants. *Plant Physiology*, **60**, 885–91.

Zeevaart, J. A. D. (1962). DNA multiplication as a requirement for expression of floral stimulus in *Pharbitis nil*. *Plant Physiology*, **37**, 296–304.

Zeevaart, J. A. D. (1976). Physiology of flower formation. *Annual Review of Plant Physiology*, **27**, 321–48.

Zeevaart, J. A. D. (1980). Perception, nature and complexity of transmitted signals. In *Physiologie de la Floraison*, pp. 59–90. Colloques internationaux du CNRS No. 285.

# Summary and Discussion

T. A. STEEVES

It was evident in the papers presented in this session, and in the lively discussion of them, that the orderly appearance of leaf primordia at the shoot apex and their early development are the subjects of active current research. Many problems remain unresolved, but significant advances are being made.

In describing cellular phenomena which occur before and during leaf initiation, Lyndon suggested that the surface layer (referred to as epidermis) might exert a restraining effect which curtails the cellular polarity of the emerging primordium. Green briefly described experiments on *Graptopetalum* in his laboratory which emphasize the importance of biophysical factors, particularly patterns of cell wall reinforcement, in determining changes in cell polarity. However, Williams pointed out that the first events do not occur in the surface layer and questioned the determining role of stress in the apex. The validity of data on frequency and orientation of cell divisions in the apex was questioned from the point of view of statistical significance by Harte, and Woolhouse stressed the need for precise data on spindle orientation because the use of only a few categories could obscure patterns and lead to a false impression of randomness. To both of these criticisms Lyndon responded by pointing out the difficulty of obtaining adequate data and the necessity of making the best use of the available information. Sachs also questioned the physical role of the epidermis, pointing out that when wounded it sometimes gapes and sometimes does not. Bernier also noted that the polarity changes in emerging primordia are not uniform but vary in different plants. It was felt that the changes are too variable to provide a fundamental mechanism.

In Larson's paper, emphasis shifted away from the surface layer and evidence was presented which could be interpreted as indicating that the developing procambial system, perhaps through the flow of nutrients and hormones, controls the initiation of leaf primordia. There was general appreciation for the remarkably precise analysis of a complex pattern but doubts were expressed as to the determining role of this pattern upon leaf initiation. Maksymowych raised the matter of the lack of experimental evidence for such a determining role and it was pointed out that such evidence would be difficult to obtain because of the probability that establishment of important biochemical gradients precedes any visible procambial differentiation. Sachs observed that if a leaf primordium is removed surgically the procambium which would normally be connected with it does not develop. Thus the primordium appears to be necessary for the development of the procambial strand, but this does not prove that the reverse relationship is not true. Dale questioned whether there is always a one-to-one relationship between a precocious procambial strand and an incipient leaf primordium. Although many such examples have been described, it cannot be asserted that the available information confirms a causal relationship. Williams argued that there are primary events which occur in the apex which establish a pattern of organization and that this organization cannot be attributed to other causes such as procambial patterns. Poethig called attention to experiments in which surgical interventions in the surface region of the shoot apex bring about organizational changes which would be difficult to attribute to the underlying vascular tissue.

The concept emerged in discussion that two highly organized systems – the phyllotactic pattern in the apex and the acropetally developing underlying vascular system – are closely correlated and in constant communication. Possibly, neither can be said to control the other. The suggestion was made by Goodwin that the same regulating factors might control both. The need for a holistic approach was stressed. Finally it was noted that recent anatomical and palaeobotanical evidence relating to the phylogenetic origin of leaf-oriented vascular systems should be given more consideration.

Erickson's paper dealing with phyllotactic patterns in the shoot apex presented an opportunity for the continuation of this discussion. Lindenmayer asked what is the relationship between such geometric models and the actual mechanisms which account for patterns in the apex since various mechanisms such as available space, pressure or inhibitory fields can satisfy the same model. The purpose of the models, however, is not to explain the mechanism but rather to present a precise picture of the phenomenon to aid in seeking mechanisms. Williams pointed out that geometric models are static and urged that the developmental aspect be included. On the other hand, static geometry can provide a basis for modelling developmental processes. Moreover, the models are not strictly speaking static. They might better be called instantaneous and they reflect growth processes in a steady state. In this sense they are dynamic.

It was suggested that in making the functional interpretations of geometric patterns both physical and biochemical processes should be considered. Green, however, re-emphasized the over-riding importance of physical phenomena in the apex and argued that increasing knowledge of cell wall structure in the apex is providing a basis for such mechanisms. He also noted that older primordia may exert a positive effect upon primordial initiation, not necessarily an inhibiting influence as is often assumed. Jarvis suggested that geometric models, even if static, provide a useful tool for describing morphology accurately where knowledge of plant structure is necessary for other purposes. In essence, the models are seen as part of an effort to describe growth phenomena in more precise terms.

Silk's paper continued this theme and her analysis in mechanical terms of strain rates in a developing leaf was seen as a valuable contribution to the precise description of growth patterns. Poethig commented on his recent studies of differences in leaf growth patterns in different varieties of tobacco for which the genetic basis is known. Precise knowledge of the growth patterns is essential for the understanding of events at the cell level and environmental influences upon them. Harte questioned whether the assumption of a steady state is justified. Silk replied that it is only an approximation, perhaps less valid for the leaf than for the stem or root. It is normally a useful concept. Erickson pointed out that it is appropriate to select experimental material which does approximate a steady state to facilitate this kind of analysis. Jarvis objected that it is not reasonable to ignore the third dimension, i.e. thickness, in the analysis of leaf growth. Even though the number of cell layers is genetically determined and is established very early in leaf growth, leaf thickness varies considerably in response to environmental variations. To this Silk replied that growth in the third dimension is considered to be negligible and strain rate analyses in two dimensions are thought to be accurate. Nevertheless, this assumption has not been tested and it should be verified. Lindenmayer observed that it is not possible to do this without sacrificing the leaf which precludes the type of analysis which was reported.

The final paper by King offered a somewhat different approach and dwelt extensively upon the process of determination or commitment of a shoot apex to development of reproductive structures. It was noted that this aspect of the process of leaf development itself had not been dealt with in the Symposium, perhaps reflecting the fact that little recent effort has been devoted to this subject. Lyndon, noting that transformation to flowering commonly leads to a reduction in the size of leaves produced, wondered whether the transformation creates conditions in the plant which suppress or decrease leaf development. This might be associated with the activation of the shoot apex. It was pointed out that thorns also show a similar reduction in leaf size which is associated with activation of the apex. However, Kemp noted that there are cases where the transformation is accompanied by an increase in leaf size. Bernier observed that the search for phenomena which are specific to the flowering process may be misleading. Processes which are not specific may nevertheless be essential. Williams observed that studies of plants which transform abruptly after a single inductive cycle may be misleading and he pointed out that such plants may be examples in which extensive geometric re-organization of the apex is not required. He urged more comparative study of different flowering patterns. Sachs commented that even in a plant such as *Xanthium* the transition is not truly abrupt because if additional induction cycles are given more flowers are produced. There was discussion of the reality of the commitment of a shoot apex to floral development. Woolhouse called attention to certain diseases in which environmental changes can cause a reversal of the host from flowering to vegetative development. There may be no such thing as irreversible commitment.

The general impression gained from this session was that many questions concerning leaf initiation and early development which have been under discussion for a long time without resolution are still before us. The significant point, however, lies in the nature of the data which are being considered. They are very different from those of even a few years ago and they reflect the real progress which is being made. To a large extent what we now see is a more precise and meaningful description of developmental phenomena. New methods of analysis are producing empirical data which accurately record the events of leaf initiation and early development. This has not yet led to revolutionary new interpretations but it does appear to pave the way for such interpretations in the not too distant future.

# PART II

## Leaf growth and the development of function

The early growth described in the first section continues uninterrupted and in an organized and regulated way. In this section we follow the development of a leaf up to the stage of full expansion, overlapping to some degree with a part of the material presented earlier, notably in Chapter 4. Chapters 6 and 7 are complementary describing the general features of leaf development and the influence of the environment; they include some discussion on the number of leaves produced, perhaps even more important than the final size of leaves in determining the extent of the leaf surface. This aspect is taken up again in Chapter 17.

However, Chapters 8, 9 and 10 keep fairly close to the central theme of the growth and development of a leaf, dealing with specific and difficult areas – the role of growth substances, photomorphogenesis and photosynthesis. It is impossible in this framework to develop every aspect along approaches favoured by every reader; for example, those whose primary interest is in photosynthesis may wish to proceed from Chapter 10 to Chapters 12 and 13; those concerned with water relations may go from Chapter 7 to Chapter 12; or those basically concerned with transport may read Chapter 6 and then Chapter 14. But the basic theme of leaf growth is followed and we trust the reader will emerge with a reasonable picture of current understanding and of those areas where concepts are still fragmentary and uncertain, taking him to the final section where the emphasis is on functioning of the mature leaf and its eventual decline.

# 6

# General features of the production and growth of leaves

J. E. DALE AND F. L. MILTHORPE

## Leaf production

### *Rates of primordial initiation and leaf appearance*

The time interval between initiation of successive leaf primordia, the plastochron, varies appreciably between species and environments but is usually constant over short periods in any one species grown in a constant environment. In the oil palm, for example, the plastochron is measured in units of weeks whereas in many rapidly growing annual plants such as *Epilobium adenocaulon*, it may be less than two days. In this context, a primordium is recognized as 'initiated' when it reaches some arbitrarily defined size (often a recognizable bulge when viewed under a dissecting microscope at a given magnification or sometimes a measured volume or number of cells). It is already an organ in which appreciable growth has occurred and is continuing as described in Chapter 1.

Although there may be ontogenetic drifts leading to a shortening of the plastochron with increasing age – e.g. *Linum usitatissimum* (Williams, 1975) and *Picea abies* (Gregory & Romberger, 1972) – it is fairly constant during much of the vegetative phase in an environment with only small variations. Each primordium on a dicotyledonous plant increases in size (volume, weight, number of cells, etc.) at a more-or-less exponential rate up to the stage of unfolding from the terminal bud (Fig. 6.1). In wheat, however, the relative rate of growth increases at an intermediate stage and then decreases by the time of emergence from the surrounding sheaths. The exponent – that is, the relative growth rate – usually decreases with successive leaves on the plant.

The continuing growth of the apical part of the stem (of dicotyledons) and its subtended leaves leads eventually to the spatial separation of a leaf from the apical bud and its exposure to full light. This stage of unfolding of a leaf from the terminal bud is also used to assess the rate of leaf production, the time between the visual appearance of successive leaves being termed the *phyllochron*. Although also usually constant with time, it is often longer than the plastochron; i.e. the time between initiation and unfolding is greater than that between initiation of successive leaves. Hence, the number of leaves in the apical bud tends to increase during ontogeny until flower initiation. Cutter (1965) has suggested that the length of time spent in the proximity of the apex is important in determining the final form of shape of the leaf, the longer this interval the more the leaf approaches an 'adult' shape. However, this finding may simply reflect a non-causal correlation.

Both the reciprocals of the plastochron and of the phyllochron, the rates of leaf initiation and of appearance of unfolded leaves respectively, are influenced greatly by irradiance and by temperature. For example, the rate of leaf initiation, d$N$/d$t$, in *Cucumis* grown at 25 °C (Fig. 6.2) is related to irradiance, $Q_v$(MJ m$^{-2}$ d$^{-1}$) by

$$dN/dt = 0.40 + Q_v/(0.8\ Q_v + 1.95)$$

Fig. 6.1. Growth in length (mm) of the fourth leaf of subterranean clover with time. The vertical arrow indicates time of commencement of unfolding. (Data of Williams, 1975.)

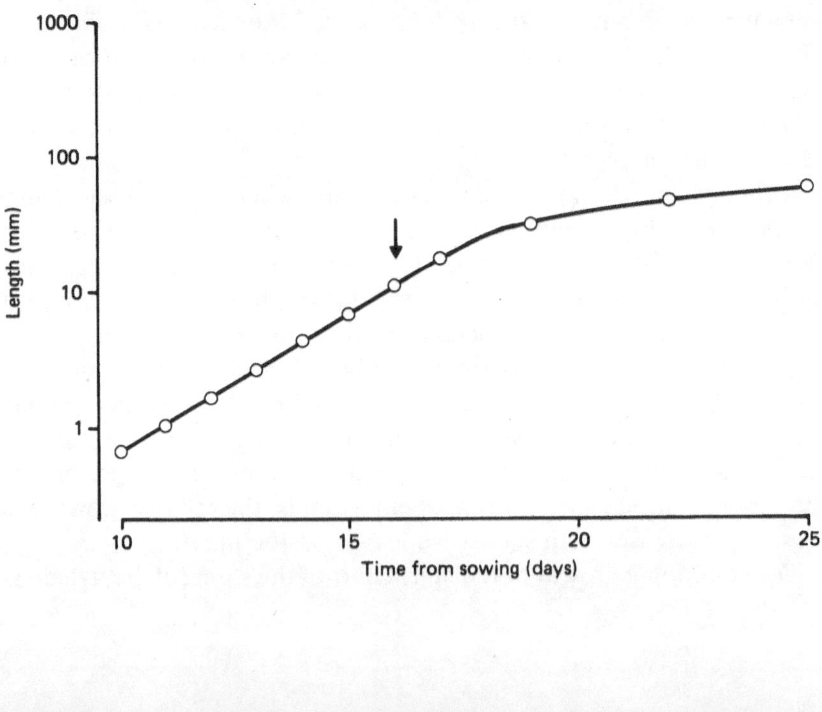

The more rapid leaf production under high irradiance implies more rapid growth at the stem apex and faster rates of cell division there. Whether this is the result of enhanced assimilate supply or other photomorphogenetic responses is not clear. In barley, the rate of initiation was less in 8-h than in 16-h photoperiods of equal total irradiance, with smaller differences at 25 °C than at 20 °C (Dale & Wilson, 1979). The relationship between rate of initiation and temperature appears to follow much the same general curve as that of other processes, i.e. an asymmetric parabola with 'minimum', 'optimum' and 'maximum' temperatures, the range between $T_{min}$ and $T_{opt}$ being greater than that between $T_{opt}$ and $T_{max}$. However, this conclusion is largely conjectural as no study of any one species is known which covers the entire range. Most studies such as that of Milthorpe (1959) have included only three or four temperatures. Unpublished data of J. E. Dale (Table 6.1) show that in barley grown at 10, 15, 20 and 25 °C the plastochron shortens less between the two highest temperatures than between the three lowest, but even at 25 °C there was no evidence for the plastochron lengthening. Findings of Raper, Thomas, Wann & York (1975) are difficult to interpret: they transferred tobacco seedlings raised at 26°C/22°C day/night temperatures to temperatures

Fig. 6.2. The effect of irradiance on the rate of production of leaves (primordia and unfolded leaves) by cucumber. Plants emerged into the light on day 4 (vertical arrow). (Redrawn from Milthorpe & Newton, 1963.)

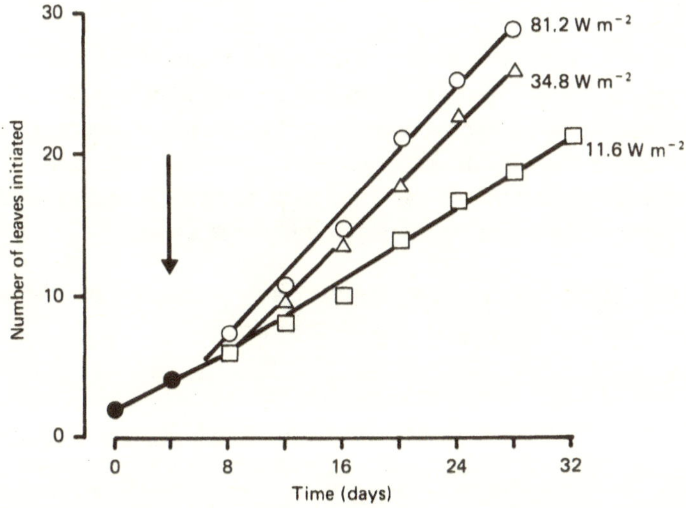

ranging from 18 °C/14 °C to 34 °C/30 °C and found that subsequent leaf initiation rates decreased slightly with temperature over this range whereas leaf appearance rates, based on the phyllochron, increased. Influences of acclimation (cf. Chapter 12) and flower initiation may well have played a large part in these responses.

### *Axillary bud development*

In assessing the effect of environmental factors on the increasing leaf area of a plant it is important to separate effects on the area of individual leaves from those on the number of unfolding leaves. Factors such as light may increase leaf number by increasing the rate of unfolding at the stem apex or by increasing the number of apices producing leaves. The initiation and subsequent development of axillary buds are therefore important determinants of the total leaf area of a plant.

Axillary buds usually arise some three or more plastochrons after the initiation of the subtending leaf primordium. They are thus formed some distance from the apical dome and well down on the flank of the apex. They develop at a relative growth rate only slightly less than that of the subtending leaf in *Trifolium subterraneum* (Williams, 1975). Tiller buds in wheat and barley grow initially at a constant relative growth rate but this begins to decline before the bud emerges. Tiller buds develop in very close contact with surrounding tissues of the leaf and the internode meristem which may exercise physical constraint upon them. Further growth leading to an emergent tiller depends upon nutrient supply, especially nitrogen (Aspinall, 1961; Halse, Greenwood, Lapins & Boundy, 1969), and light flux density (Fletcher & Dale, 1974). Hormonal relationships in axillary bud development are discussed in Chapter 8.

In small grain cereal crops, leaves on the three or more tillers normally produced quickly constitute a major part of the total leaf area, despite the fact that they may be smaller than equivalent leaves on the main stem, and produced more slowly (Fletcher & Dale, 1977).

Table 6.1. *The effect of temperature on the average plastochron (h) for leaves 5–7 of plants of Proctor barley*

| | Temperature °C | | | |
|---|---|---|---|---|
| | 10 | 15 | 20 | 25 |
| Plastochron | 148 | 118 | 67 | 59 |

*The leaf plastochron index*

The simplest measure of size of an unfolding leaf often is its length. At the beginning of the unfolding period growth in length is exponential (Fig. 6.1) and this coupled with the near constancy in rate of leaf appearance enables a developmental index to be calculated. For the plant this is expressed by the plastochron index (PI), and for the leaf by the leaf plastochron index (LPI) (Erickson & Michelini, 1957; Lamoreaux, Chaney & Brown, 1978).

$$\mathrm{PI} = n + \ln L_n - \ln \lambda / (\ln L_n - \ln L_{n+1})$$

$$\mathrm{LPI} = [n + \ln L_n - \ln \lambda / (\ln L_n - \ln L_{n+1})] - i$$

where $n$ is the serial number on the stem of the youngest leaf whose length exceeds that of an arbitrary value, $\lambda$, which is often taken as 10 mm, $L_n$ and $L_{n+1}$ are lengths of leaves $n$ and $n+1$, and $i$ is the serial number on the stem, counting from below, of the leaf for which LPI is being determined. Table 6.2 shows the relationship between LPI, chronological age, leaf length and the proportion of final size reached for a hypothetical example; nomograms enabling LPI to be readily determined have been described by Erickson (1960).

It may be noted in passing that the derivation of LPI depends more upon leaf emergence, i.e. the phyllochron, than on leaf initiation rates, the plastochron. Its use is restricted to leaves where near constancy of phyllochron and exponential increase in length occur; it cannot be used for cereal leaves where, by the time measurements can be begun, growth in length of the emerged lamina is not exponential. Nevertheless, this index has been used in important studies (Lamoreaux, *et al.*, 1978) where variation between plants, and sampling problems, make use of chronological age unsatisfactory.

Table 6.2. *Data for age, length, percentage of final size and LPI for a model leaf taking 9 days to reach final size from commencing unfolding at a length of 10 mm; the values in brackets are predicted from the LPI calculation*

| | Age, days after attaining $\lambda$ of 10 mm | | | | | | |
|---|---|---|---|---|---|---|---|
| | (−6) | (−2) | 0 | 2 | 4 | 6 | 9 |
| Length | 1 | 5 | 10 | 20 | 40 | 80 | 160 |
| % of final length | 0.6 | 3.1 | 6.3 | 12.5 | 25 | 50 | 100 |
| LPI | −2.9 | −0.95 | 0 | 0.7 | 1.15 | 3.1 | 8.2 |

*Growth after unfolding or emergence*

The exponential relationships of leaf length, volume, area, weight, etc. with time (Fig. 6.1) continue until after unfolding of the leaf (dicotyledons) or emergence from the enclosing sheaths (cereals and grasses) and then decline, giving the familiar S-shaped curves characteristic of post-primordial growth. This phase covers some 0.8 of the total growth in length, some 0.95 of that in dry weight and generally more than 0.99 of the increase in area. Possibly because of the greater ease of observation, it has received more attention than the equally important pre-unfolding phase. For example, a number of mathematical relations belonging to the asymptotic logistic family have been used to describe the changes of area, $A$, with time, $t$; these range from a modified monomolecular ($m=0$) (Constable & Rawson, 1980)

$$A = A_0 + B[1 - \exp(-ct^p)]$$

through the Gompertz ($m \rightarrow 1$) (Amer & Williams, 1957)

$$A = A_{max} \exp(-be^{-ct})$$

and the simple logistic ($m=2$) (Clough & Milthorpe, 1975)

$$A = A_{max}/[1 + b \exp(-ct)]$$

to the generalized logistic relation (Dennett, Auld & Elston, 1978; Causton & Venus, 1981)

$$A^{1-m} = A_{max}^{1-m} [1 \pm b \exp(-ct)]$$

where $A_0 + B = A_{max}$, the final area of the leaf, and $b$, $c$, $m$ and $p$ are constants for that leaf. The Gompertz relation often proves to be adequate. Analysis of leaf expansion using a model which includes cellular parameters has been presented by Horie, de Wit, Goudrian & Bensink (1979).

Growth rates are not uniform over the expanding lamina but vary both spatially and temporally; the central region adjacent to the major veins shows higher relative expansion rates than the peripheral regions (Avery, 1933; Erickson, 1966, see also Chapter 4). Expansion ceases first at the leaf tip and last in the base – both in dicotyledons (Maksymowych, 1962; Erickson, 1966) and grasses, in which all emerged portions have finished expanding (Begg & Wright, 1962). This basipetal trend for expansion is also mirrored in tissue maturation and the activity of many physiological processes, as will be seen later.

The final sizes of leaves along a stem often show a progressive increase with position; this may be related to either increased rates of expansion over much the same or even shortened durations of the expansion phase, as in *Capsicum* (Steer, 1971), or to longer durations of expansion at progressively decreasing rates as in wheat and *Trifolium subterraneum* (Williams, 1975). Those leaves formed late in the season or immediately before flowering show a decrease in size with transition to the formation of bud scales or bracts often occurring. Leaf thickness also increases as the lamina expands and often continues after full area of the blade is reached (Friend, Helson & Fisher, 1962). As thickness of the leaf increases so too does dry weight, and such changes can often be substantial (Fig. 6.3). These increases are largely accounted for by changes in cuticular waxes and polysaccharides, including cell wall materials. As a result the leaf often becomes stiffer and more resistant to mechanical damage and to disease. The increase in thickness after full

Fig. 6.3. The dry weight of 3-cm segments of the second leaf of Proctor barley at the time of full expansion on day 15 (○) and at day 20 (●). For all parts of the leaf except the terminal segment the difference is significant at $p = 0.01$. The inset shows the change in dry weight of the whole leaf with time; the arrow marks the point of full lamina expansion.

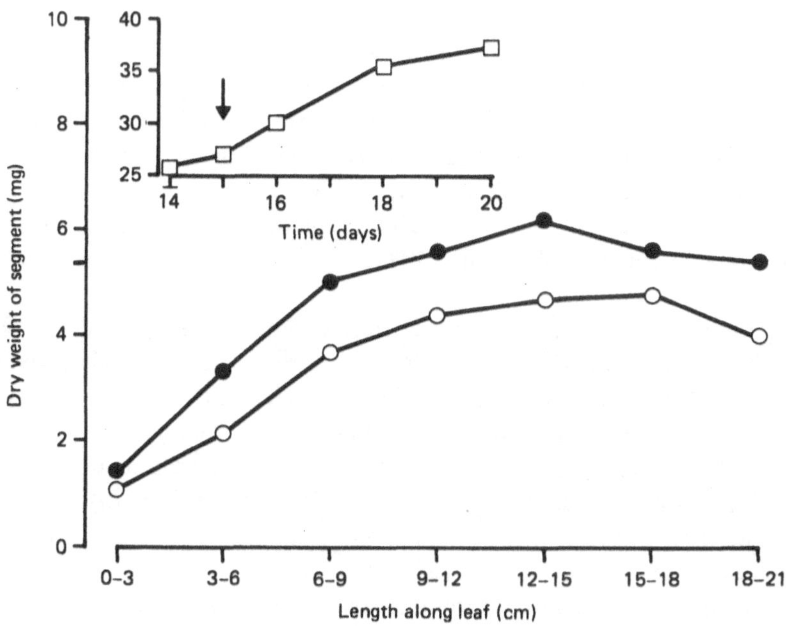

area is reached comes at a time when net photosynthesis is tending to fall rather than increase.

## The Cellular Basis of Leaf Growth

### *Growth of the primordium*

The meristematic regions in the developing primordium have been described by Esau (1965), Cutter (1971) and Dale (1982) and are summarized in Fig. 6.4. In the newly initiated primordium an apical meristem is usually established although this is not universal (Coleman & Greyson, 1976). Divisions in the intercalary regions are also found and with time these become more important as apical growth becomes less distinct. With the establishment of the primordium axis intercalary growth continues in the dicotyledons and in broad-leaved monocotyledons, and by the time that the primordium has reached a length of 80–200 $\mu$m marginal and sub-marginal meristems are established along the lateral flanks. It is the development of these meristems which marks the onset of lamina formation in dorsiventrally flattened leaves; in grasses and many other monocotyledons marginal growth is not seen and instead there is the establishment of a basal meristem at the junction of lamina and sheath (Langer, 1980). From its earliest initiation the primordium is associated with procambial tissue (Girolami, 1954; Lerstens, 1965; Larson, 1977; see Chapter 2).

Cell division kinetics in the various regions of the primordium have not been studied. Indeed, there are substantial technical difficulties in the study of young primordia, especially in deciding primordial boundaries with the adjacent stem tissues. Nevertheless, microdissection (Sunderland & Brown, 1956, 1976; Williams & Rijven, 1965, 1970) and serial reconstruction (Williams, 1975) methods have provided useful information.

By the time it is visible the primordium contains several hundred cells and has already undergone eight or more cycles of cell division (Hannam, 1968). In tobacco, a further 12 cycles of division occur prior to leaf unfolding. So growth of the primordium is mainly as a result of cell division although a 3–4 fold increase in mean cell volume does occur (Sunderland & Brown, 1956; Williams & Rijven, 1965, 1970).

From their studies on *Lupinus albus* Sunderland & Brown (1976) concluded that rates of cell division fall in successive primordia. The reasons for this are not entirely clear. It may be that, as the apex

Fig. 6.4. The changing pattern of meristematic activity at different developmental stages of a laminate, petiolate leaf. Arrows indicate main sites and directions of growth.

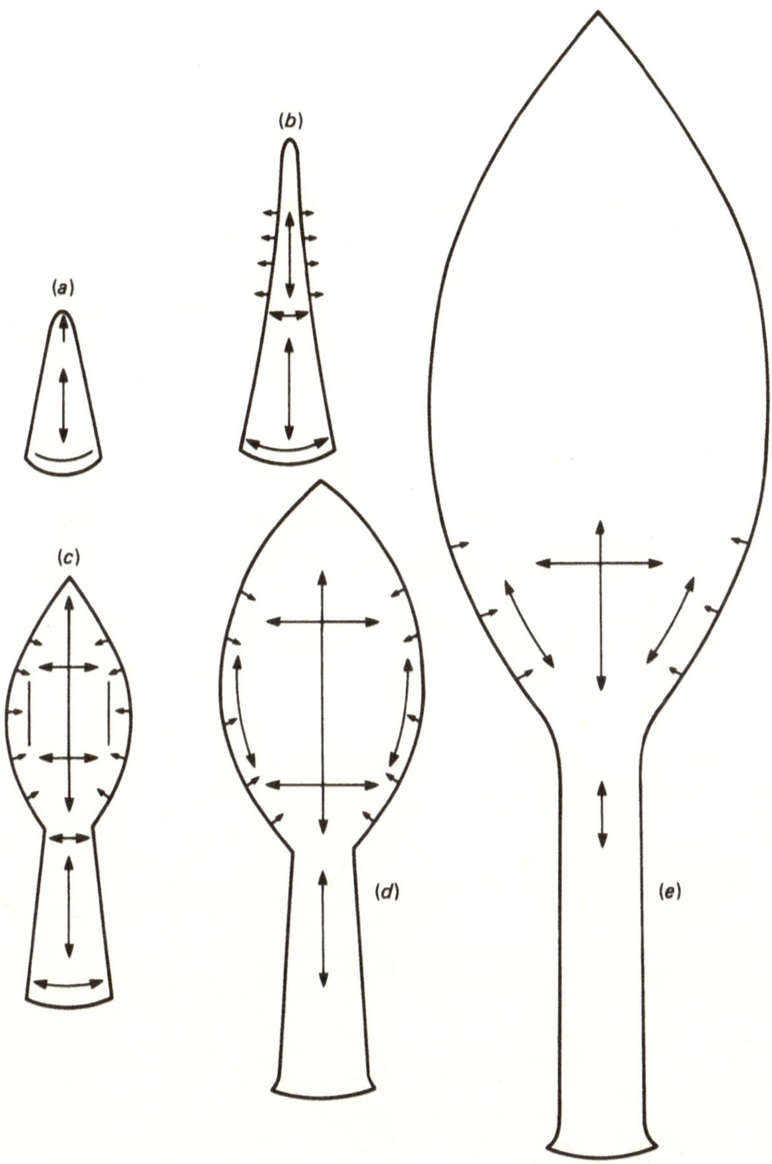

enlarges due to the accumulation of primordia waiting to unfold, a degree of nutrient stress occurs which is most marked in the youngest primordia. Since these plants were grown in boxes it is entirely possible that mineral nutrients stress could have occurred. However, other studies (Williams, 1975) support their findings. Sunderland & Brown (1976) also found that the relative rate of division fell with time in individual primordia (Fig. 6.5). This could reflect a general decrease in division rate or alternatively the emergence of a particular cell population having a longer cell cycle time than the rest. As the primordium grows the procambial tissues increase and a slower division rate in these enlarging cells is possible. It is the growth of these cells that accounts for much of the increase in average cell volume within the primordium.

In general, the slower relative rates of division in successive primordia are matched by similar falls in relative rates of increase of length, fresh weight and volume (Williams, 1975). However, these falls are compensated for by an increase in length of the primordial phase so that cell number of successive leaves is higher at the onset of unfolding. As the length of the period over which divisions occur during expansion may also increase, final cell number in successive leaves is also larger.

Fig. 6.5. The changing number of cells in successive primordia of *Lupinus albus*, with time expressed as number of plastochrons. Note that the initial cell number falls in successive primordia and that the increase in cell number for any one primordium is not exponential, while the slope of the curve declines for successive primordia. (Redrawn after Sunderland & Brown, 1976.)

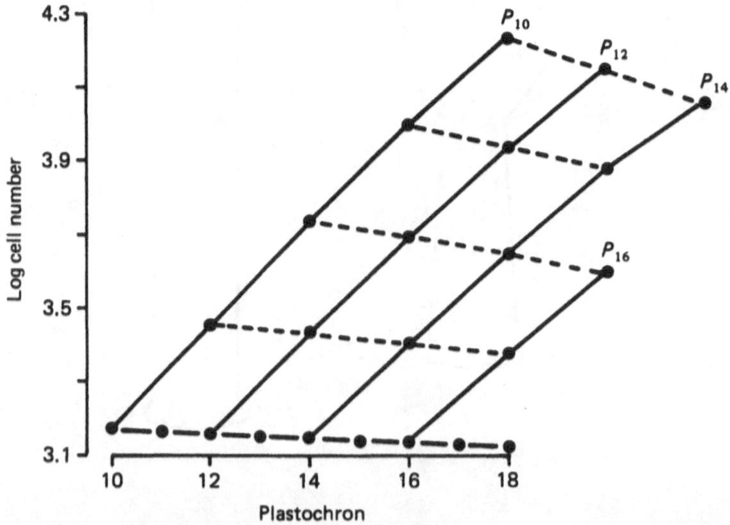

## *Changes in cell number and size during unfolding*

### *The early phase*

Marginal and sub-marginal meristems are formed on the primordial flanks long before leaf unfolding commences (Fig. 6.4). The activity of these meristems leads to the first stages of lamina formation and the establishment of a layered structure. Within these meristems the pattern of divisions is not regular and although the majority of divisions are in the anticlinal plane at least 15% are in the periclinal plane (Maksymowych & Erickson, 1960; Fuchs, 1968; Dale, 1976, 1982). However, although the marginal and sub-marginal meristems are important in the definition of the prospective lamina boundaries, expansion of the lamina results from meristematic activity of the tissues proximal to the margins which commence general division and constitute a plate meristem (Schüepp, 1966). This pattern of meristematic activity is seen in the dicotyledonous leaf, with variations occurring in some compound leaves such as tomato (Coleman & Greyson, 1976) and in broad-leafed monocotyledons such as *Hosta* (Pray, 1963). In grasses marginal growth is absent and cell divisions occur only in the basal meristem which is enclosed by the sheaths and leaf bases of older leaves. In the visible portion of an unfolding grass leaf both cell division and cell enlargement have ceased (Begg & Wright, 1962; Davidson & Milthorpe, 1966).

Development of the lamina is matched at all stages by associated differentiation of provascular tissues extending outwards from the central axis of the primordium which is destined to form the main vein (Avery, 1933). So it is that by the time that unfolding commences the main features of leaf form are established, with extensive differentiation of vascular and mesophyll tissues.

### *Cell division kinetics*

Cell division, in the plate meristem, continues throughout much of the period of lamina expansion. As a result it is not uncommon to find that 90% or more of the cells in the mature leaf have been produced after unfolding commences (Sunderland, 1960; Hannam, 1968; Maksymowych, 1973). In the eighth leaf of tobacco grown under controlled environment conditions, Hannam (1968) found 99% of palisade cells to have their origin in divisions after unfolding.

Since cell division occurs over much of the period of lamina expansion, the time over which growth of the blade is by cell expansion alone can be quite short although this depends upon species (Dale, 1976). The shape of the curve for cell-number increase in the expanding leaf is characteristically sigmoid and the early exponential phase is often concluded by the time that the leaf is between 10 and 20% of the final size. Thereafter, the mean relative rate of cell division falls steadily and the apparent doubling time for cell number increases. A number of descriptions of the changes in cell number have been published (Milthorpe & Newton, 1963; Dale, 1970; Clough & Milthorpe, 1975) and have been reviewed by Dale (1976). The available data are consistent with the idea that as leaf expansion proceeds so the proportion of cells continuing into further division decreases (Wilson, 1966) until ultimately divisions cease altogether. The difficulties of identifying the population of dividing cells, and of deciding whether concurrent changes in cell cycle times are occurring, is further compounded by the fact that meristematic behaviour varies temporally and spatially over the leaf.

Although cell divisions are found throughout the whole of the expanding leaf, they cease first in the distal regions and last in the proximal, basal parts of the lamina (Avery, 1933; Maksymowych, 1963; Saurer & Possingham, 1970; Steer, 1971). This basipetal trend is seen also in differentiation, those cells at the leaf tip becoming fully differentiated earlier than those at the base. The grass leaf represents an extreme example of this, the cells at the tip being fully mature before divisions have ceased in the basal meristem.

In grasses, cell division continues in the sheath meristem for a short while after that in the blade meristem has ceased; sheath elongation continues after full size of the lamina has been reached. Rather scanty data are available for petioles of broad-leafed species but, in *Xanthium*, cell number continues to increase there for about 0.5 LPI units after it has ceased in the blade (Maksymowych, 1963).

As well as the gradient in cell divisions along the leaf there are also differences in the behaviour of the different layers. Divisions continue longest in the palisade cells of the mesophyll and cease first in the upper epidermis, although formation of stomata may continue after divisions elsewhere in the epidermis have ceased.

### *Change in cell volume*

A four-fold increase in mean cell volume is not unusual during the primordial phase but during unfolding very much larger changes in cell size occur. For example, in the second leaf of sunflower mean cell volume increases about 22-fold during expansion (Sunderland, 1960) and in primary leaves of *Phaseolus vulgaris* volume increases of about 15-fold are found (Murray, 1968; Verbelen & de Greef, 1979). These increases in size occur in parallel with changes in cell number and reflect differentiation of the various tissues of the leaf; it is non-dividing cells that contribute much of the increase in mean cell volume.

Because cell number changes are so much greater than those in cell size it appears that leaf size is determined mainly by cell number (Humphries & Wheeler, 1963). Thus in leaves of *Cucumis* and *Helianthus* grown in low light intensities smaller leaves are associated with smaller numbers of cells (Milthorpe & Newton, 1963; Dengler, 1980). In both these species and in *Populus euamericana* (Pieters, 1974) light intensity has little effect on epidermal cell size although in *Capsicum annuum* (Schoch, 1972) and probably in *Impatiens parviflora* (Hughes, 1959) epidermal cell size is greater under shade conditions, although vertical height of the cells is reduced.

### *Cell differentiation in the expanding leaf*

The classic studies of Avery (1933) on this topic have now been supplemented by important work by Maksymowych (1963, 1973) on *Xanthium*, Dengler, Mackay & Gregory (1975) on *Fagus*, Coleman & Greyson (1976) on *Lycopersicon*, and Dengler (1980) on *Helianthus*. Two aspects of differentiation are of special interest: namely, development of the vascular tissue and development of the mesophyll.

#### *Vascular development*

Because of its importance to studies on translocation the course of phloem maturation in the leaf has been extensively studied, albeit in largely descriptive terms. The primary vein (midrib) procambium develops acropetally and appears in the base of the primordium when it is from 50–100 $\mu$m in length (Lerstens, 1965; Coleman & Greyson, 1976). By the time that the primordium is 250 $\mu$m long development of the secondary veins has begun and that of higher-order veins follows rapidly. In this early development, phloem is differentiated before xylem in all veins from the primary to minor veins of the fifth or higher orders. Lamina initiation is correlated with the appearance of immature sieve tubes which, in tobacco, differentiate more rapidly in successive leaves (Hannam, 1968). Since development and maturation of phloem is in the acropetal direction – into the primordium from the base (Esau,

1965) – there will be vascular connections to other parts of the plant from the earliest stages of leaf development; the significance of this to the nutrition of the developing leaf is obvious and is discussed in more detail below.

In the major veins, xylem development follows that of the phloem by a short interval. Initiation, like that of phloem, is acropetal but in the higher-order veins differentiation from procambial cells can be both acropetal and basipetal. Whereas phloem maturation appears to be a regular rather precisely controlled process, that of xylem vessels and tracheids 'gives the impression of being somewhat casual and imprecise. Immature tracheary elements . . . [in *Trifolium wormskioldii*] . . . were frequently noted in minor veins of almost mature leaflets which were expanded, exposed and therefore presumably transpiring and undergoing photosynthesis. In old, mature leaves it was not uncommon to see vein endings seemingly isolated from the rest of the vascular system. These anomalies are invariably the result of an intervening procambial cell failing to mature into a tracheary element to complete the xylem connection . . . [but] there is a normal phloem connection' (Lerstens, 1965).

Maturation of the leaf as a whole proceeds basipetally and this is true also of the higher-order and minor veins. In the cereal leaf vascular differentiation is complete in each part of the lamina as it appears; Patrick (1972a, b, c) has described this development in the lamina and sheath of the sixth leaf of wheat and the stem vasculature, and explored the influence of this development on the movement of assimilates. In species with bicollateral bundles adaxial phloem matures earlier than the abaxial sieve elements (Turgeon & Webb, 1976).

### *Differentiation of the mesophyll and epidermis*

In contrast to the vascular tissues, where local variations between leaves make observation difficult, there is considerable quantitative information on development of epidermal and palisade cells, notably with *Xanthium* (Maksymowych, 1963, 1973). In this species meristematic cells in palisade and epidermis have similar dimensions at LPI −1.0, being about $10 \times 8\,\mu m$ in height and diameter, giving a volume of around $500\,\mu m^3$. Rates of increase in height and cell area for the two types of cell are shown in Table 6.4. Increase in height begins earlier in the palisade than in the epidermis, and before leaf unfolding starts; as might be expected the maximum rate of increase in height for the epidermis is less than a seventh of that of the palisade mesophyll

Table 6.3. *Tissue and intercellular space volumes for leaves of sunflower grown in 100% and 25% daylight*

| | 100% daylight | 25% daylight |
|---|---|---|
| Leaf area ($cm^2$) | 21.6 | 9.0 |
| *Tissue Volume* ($\mu m^3$): | | |
| Upper epidermis | 18.7 (5.3%) | 14.9 (6.6%) |
| Palisade mesophyll | 148.8 (42.5%) | 92.2 (43.0%) |
| Spongy mesophyll | 164.1 (46.8%) | 96.8 (44.1%) |
| Lower epidermis | 19.1 (5.4%) | 14.0 (6.3%) |
| Total | 350.7 | 217.9 |
| *Intercellular Space Volume* ($\mu m^3$): | | |
| Palisade mesophyll | 62.2 | 28.0 |
| Spongy mesophyll | 85.2 | 58.6 |
| Total | 147.4 | 86.6 |
| Intercellular space as % of total volume | 29 | 35 |

After Dengler (1980).

Table 6.4. *The absolute growth rates for vertical extension ($\mu m$ $d^{-1}$) and increase in horizontal surface area ($\mu m^2$ $d^{-1}$) of cells of the upper epidermis and palisade in the seventh leaf of* Xanthium

| | Vertical extension | | Horizontal area increase | |
|---|---|---|---|---|
| LPI | Epidermis | Palisade | Epidermis | Palisade |
| −1.5 | | 0.03 | | |
| −1 | | 0.11 | 0.6 | |
| −0.5 | | 0.37 | 1.4 | |
| 0 | | 0.63 | 2.3 | −0.7 |
| 0.5 | 0.06 | 0.90 | 7.1 | −3.5 |
| 1 | 0.14 | 1.3 | 15.7 | −3.4 |
| 1.5 | 0.26 | 1.5 | 31.4 | −0.6 |
| 2 | 0.29 | 1.9 | 64.3 | 0.6 |
| 2.5 | 0.37 | 2.5 | 111.4 | 5.7 |
| 3 | 0.43 | 3.0 | 125.7 | 12.9 |
| 3.5 | 0.40 | 2.7 | 85.7 | 16.0 |
| 4 | 0.37 | 1.6 | 38.6 | 13.9 |
| 4.5 | 0.29 | 0.63 | 14.3 | 9.3 |
| 5 | 0.17 | 0.20 | 7.1 | 4.9 |
| 5.5 | 0.03 | | 2.9 | 1.2 |

Data of Maksymowych (1973).

cells. In contrast, increase in area of the upper epidermis commences earlier and spans a maximum rate ten times higher than that for the palisade; indeed, rapid anticlinal cell division in the palisade layers in the early stages of leaf expansion leads to a slight decrease in cell area in the paradermal plane from LPI 0 to LPI 1.5. Cell expansion in both layers of the leaf, in both horizontal and vertical planes, ceases at about LPI 5.5 when the leaf is about 90% of its final length. These growth trends have been found also in leaves of *Fagus grandifolia* (Dengler *et al.*, 1975) and *Helianthus* grown in full and 25% daylight (Dengler, 1980). The increase in thickness that occurs during lamina expansion is due primarily to increase in palisade cell height. This is influenced by the light regime under which the leaf expands and in high irradiances individual palisade cells are taller than under low light energies; a storied palisade is also often found in so-called sun leaves (Wylie, 1951; Dengler, 1980). The effect of light on cell expansion in *Phaseolus* leaves is currently under study by Van Volkenburgh & Cleland (1980) who have shown a direct effect of high-intensity white light on wall extensibility in tissue pieces.

Coincident with the later stages of cell enlargement is the formation of intercellular spaces in the mesophyll. These spaces are formed first in the spongy parenchyma and, some two plastochrons later, in the palisade (Isebrands & Larson, 1973; Dengler *et al.*, 1975; Dengler, 1980). The classical view of intercellular space formation advanced by Avery (1933) is that, as the cells in the leaf increase in number and expand, tensions occur between the layers which are resolved by separation of the mesophyll cells and the establishment of intercellular spaces. The driving force for separation of the palisade may well be the massive expansion of the cells of the upper epidermis, but for the spongy parenchyma in beech leaves it is suggested that differential expansion of the arms of the spongy cells themselves may result in intercellular space formation. The interesting suggestion that tissue stresses may lead to some potential procambial elements differentiating as mesophyll cells (Hara, 1962) is difficult to examine since it is not possible to follow the fate of individual cells throughout the expansion period.

The volume of intercellular space is considerably greater in the spongy compared with the palisade mesophyll and tends to be greater in plants grown under low irradiances (Table 6.3; Chabot, Jurik & Chabot, 1979; Dengler, 1980). Stomatal pores are usually continuous with intercellular spaces but whether this is due to chance, given the high incidence of intercellular space in most leaves, or to some other underlying controlling mechanism, is not known. Certainly, guard cell pairs develop asynchronously in both epidermes and are still being differentiated some time after cell divisions have ceased elsewhere in those layers (Esau, 1965). Bünning (1956) argued that developing stomata inhibit the differentiation of adjacent cells into other stomata, but that this constraint is weakened as the epidermis expands and the distance between stomata increases, thus allowing new stomata to differentiate. Although much is now known of the origin and development of the stomatal complex in many species (e.g. Stebbins & Shah, 1960) Bünning's hypothesis has never been adequately tested.

It is not known why cells in the various layers of the leaf show such dissimilar patterns of differentiation. Cells in the epidermis are continuous, in the morphological and genetical senses, with epidermal cells in the stem and tunica and it is possible that they are programmed to expand mainly in the horizontal, periclinal plane. But why adjacent cells in the mesophyll should differentiate into palisade and spongy chlorenchyma is not clear. The differences between sun and shade leaves (Dengler, 1980) suggest that the fates of developing mesophyll cells are not irrevocably determined and that it is possible for potential palisade cells to develop as spongy mesophyll. Whether the effect is a direct one of light on the individual cells, or an indirect one, perhaps through nutrient supply, is unknown. Palisade cells begin their vertical extension before unfolding (Table 6.4) and in at least some species the preliminary stages of palisade expansion can occur even in leaves kept in continuous darkness (Dale, 1964).

## The nutrition of the growing leaf

### *Mineral nutrition*

For all elements except carbon, the developing leaf is totally dependent upon supply from other parts of the plant, and ultimately on uptake from the soil. Leaf growth will therefore be affected by root growth and performance; conditions which affect root function, such as low soil temperature, soil moisture and nutrient stress also influence leaf expansion. Whether this effect is immediate or seen only over the longer term depends upon the extent to which ions already accumulated are available to buffer the effects of shortfall from the roots. Overall, remarkably little detailed information is available on the relationships between leaf and root growth, although some general results have been published (Evans, 1975; Brouwer, 1977).

Since primordia are initially small with dry weights measured in micrograms, sink size is also small. However, sink activity increases as the primordium enlarges and the requirements for phosphorus and nitrogen rise sharply as nucleic acid and protein synthesis increase. These trends were followed in the fourth leaf of wheat by Williams & Rijven (1965) who showed the relative rates of increase of both dry weight and insoluble phosphorus to rise and then fall; this decline was enhanced once the leaf commenced expansion. The relative rate of increase of protein nitrogen was constant until expansion commenced whereupon a sharp fall occurred. In the fourth leaf of *Trifolium subterraneum* (Williams & Rijven, 1970) there was no initial rise in relative rate of increase in insoluble phosphorus, indicative of total nucleic acid content, or protein nitrogen, and the decline in rate began even before the leaf unfolded. It is not clear whether these differences between species relate to the fact that growth in the wheat leaf is highly polarized so that expansion is almost complete in distal portions of the leaf before it appears.

In the very early stages of growth, the leaf depends for its supply of inorganic ions, as well as organic materials, on export from older leaves via the phloem. As the leaf grows an increasingly greater proportion comes directly from the roots via the xylem (Milthorpe & Moorby, 1969; Moorby, 1977, 1981). It eventually becomes a net exporter of mineral elements although concurrent import and export of nitrogen, phosphorus and potassium proceeds over a large part of its ontogeny. The net changes with phosphorus are particularly well seen for the second leaf of *Cucumis* (Hopkinson, 1964) where maximum import occurs before increase in dry weight and area is complete (Fig. 6.6), but at a time when cell division ceases and cell extension, as well as absolute rate of expansion in area, are maximal. Phosphorus is exported from the leaf once expansion is complete and concurrently with this there is a decline in photosynthesis and in the export of sucrose from the leaf.

The degree of recycling varies between the different mineral elements. Circulation of phosphorus is both rapid and considerable (Biddulph, 1959; Sutcliffe, 1976) and is linked to mechanisms controlling photosynthesis and respiratory activity within the leaf (see Chapters 10 and 13). In contrast, movement of calcium out of the leaf is very slow. Studies with $^{15}N$ have shown a slow accumulation into fully expanded leaves of barley, probably through the transpiration stream (Dale &

Felippe, 1977). Net loss of nitrogen from the leaf occurs from the time that the leaf is fully expanded and accelerates markedly with the onset of senescence (Carr & Pate, 1967).

### *Carbon nutrition*

Only the outermost unexpanded primordia at the stem apex are exposed to significant light intensities and, since development of mature chloroplasts, differentiation of stomata and formation of intercellular spaces are yet to come, the chemical and physical resistances to $CO_2$ fixation are extremely high. So the primordium is heterotrophic. But as the leaf unfolds, the basipetal course of anatomical maturation already mentioned is associated with a similar trend in physiological maturation as photosynthetic capacity develops (see Chapter 10).

The basic facts can be summarized as follows:

(i) Net photosynthesis increases basipetally along the leaf and reaches a maximum, on a leaf basis, when blade expansion is from 35–90% complete, depending on species (and perhaps on the methods of the investigator).

(ii) As the leaf unfolds so the rate of import of labelled carbon from other source leaves increases to reach a maximum when size is between 20–30% of the final size. Gradually import ceases, to stop when the leaf is around half final size.

Fig. 6.6. The change in area (□) of the second leaf of cucumber with time and calculated rates of imports and export of sucrose (○) and phosphorus (●). (Redrawn from Hopkinson, 1964.)

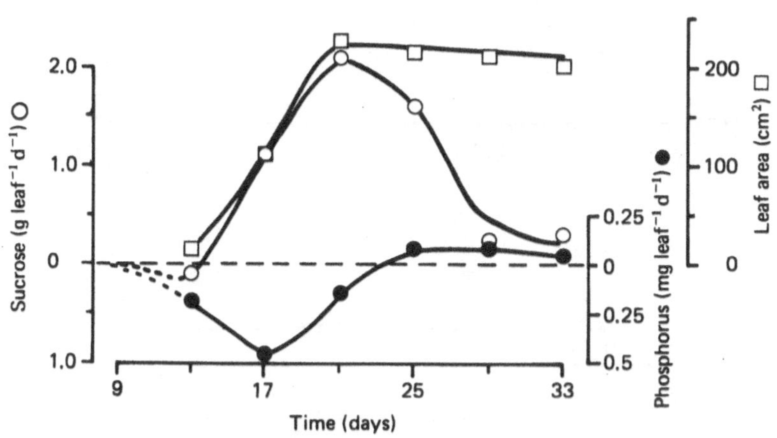

(iii) Export of locally fixed carbon begins by the time that the leaf is about 30% expanded and reaches a maximum some time after the leaf is 70% unfolded.

(iv) There is thus a period during which time import into and export out of the leaf proceed simultaneously, although not necessarily from the same part of the leaf.

These points are summarized in Fig. 6.7.

A number of questions now arise concerning the source and nature of carbon compounds supplied to the leaf. For example, where does the carbon come from? How far is locally produced carbon used in leaf growth? Is imported and locally produced assimilate utilized similarly in the growing leaf? The experimental approach using labelled carbon has provided some answers.

There is considerable evidence that the supply of carbon to the developing leaf, from early in the primordial phase, is dependent upon the vascular arrangement (Joy, 1964; Porter, 1966; Larson & Dickson, 1973; Larson, 1977; Turgeon, 1980). That is to say, primordia and

Fig. 6.7. The relationship between net photosynthesis, $F'$ and import and export of carbon for a developing leaf. The large vertical arrow indicates the point when unfolding commences, the smaller arrows mark the stage of concurrent import and export shown as a shaded area.

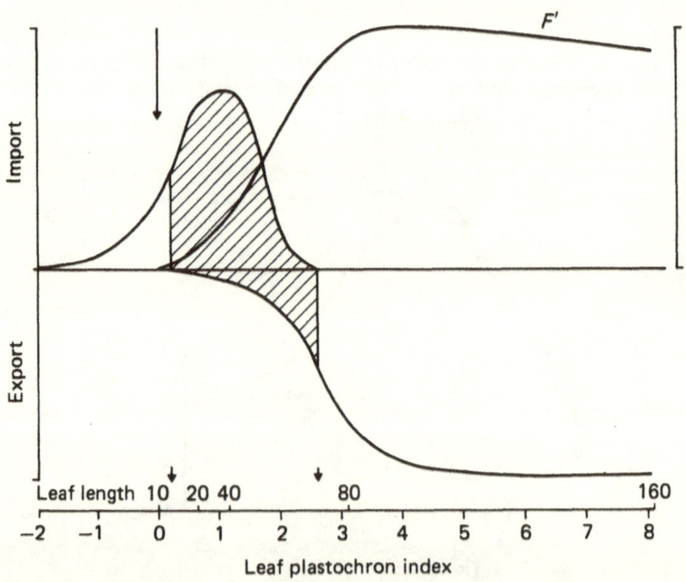

expanding leaves are supplied mainly if not exclusively from older source leaves which are in direct vascular connection with them. In some species the specificity of the vascular connection is such that feeding a source leaf results in accumulation of label in a particular sector only of the sink leaf (Barlow, 1979), with no general distribution of label. However, in many plants, including cereal grasses, the vascular network shows such extensive anastomosing that failure in supply from one source leaf can be compensated for by supply from another (Swanson, Hoddinott & Sij, 1976).

Because of the basipetal gradient in maturation the immature basal regions of the leaf may be importing while the more mature apical regions are exporting. In *Cucurbita pepo* there is evidence that assimilate from the leaf tip can be utilized for growth of the base (Turgeon & Webb, 1975). In tomato the terminal leaflet and distal pair of leaflets mature before the basal pair and export assimilate to them (Ho & Shaw, 1977). In expanding leaves of barley, feeding $^{14}CO_2$ to the mature tip region results in substantial accumulation of label in the meristematic and adjacent regions at the base of the blade (L. Anderson and J. E. Dale, unpublished); these regions are known to show high activities of acid invertase in oat (Greenland & Lewis, 1981). However, intralamina redistribution and utilization of assimilate during growth is not general (Larson, Isebrands & Dickson, 1972; Fellows & Geiger, 1974) and in those species where it does not occur, phloem characteristics (location or permeability) may preclude unloading at sinks within the leaf. Where phloem maturation in the base of the leaf is rapid, unloading may be prevented whereas slower maturation may be correlated with structure and functional features which allow assimilate to leak from the phloem and to be used elsewhere in the leaf.

It is of interest that the petiole is usually the last part of the leaf to mature. It constitutes a substantial sink in its own right and while carbon imported from other leaves is utilized in growth it will also incorporate assimilate from its own blade into chloroform-soluble and cell wall fractions (Isebrands, Dickson & Larson, 1976; Wardlaw & Mortimer, 1970).

Even where locally produced assimilate is not used elsewhere in the expanding lamina it is utilized in the maturation of tissues at the site of fixation. At least some of the increase in Fraction I protein in the developing first leaf of barley proceeds using locally produced assimilate (Blenkinsop & Dale, 1974) and incorporation into the structural polysaccharides of the cell wall is also important. Data of L. Anderson and

J. E. Dale (Table 6.5) show the importance of photosynthesis in the expanding leaf and suggest that about 60% of the carbon in the expanded third leaf of barley is derived from local fixation. Both the first and the second leaf contribute carbon to the leaf before and after unfolding commences, but these contributions become increasingly less important as lamina expansion proceeds. In the upper leaves of wheat, photosynthesis in a lamina contributes about 20% of its mass and about 50% of that of the attached sheath (Patrick, 1972c).

The expanding leaf is therefore a significant source of the carbon required for its own growth. But in one important way it appears dependent on other source leaves. The overriding importance of cell number as a determinant of leaf size has already been referred to, as has the fact that cell numbers are greater in high light conditions. There is evidence (reviewed by Dale, 1976) that older leaves supply factors necessary for cell division in the younger ones, and that supply is dependent upon the light intensity to which the source leaves are exposed. The nature of these factors remains unknown, but the fact that cell divisions cease at about the time when photosynthetic activity in the leaf is maximal suggests that the leaf itself either does not produce the necessary division factors, or alternatively cannot utilize them.

Table 6.5. *The estimated relative daily contribution of the first three leaves of Proctor barley to the carbon content of the third leaf from before emergence on day 12 to after full expansion on day 17 (values given as percentages; increase in dry weight between days 17 and 20 was about 5 mg (10%))*

| | Day 10 | 11 | 13 | 14 | 15 | 16 | 17 | 18 | 20 |
|---|---|---|---|---|---|---|---|---|---|
| Contribution (%) from | | | | | | | | | |
| 1st leaf | 77 | 83 | 37 | 23 | 7 | 7 | 3 | 2 | 1 |
| 2nd leaf | 23 | 17 | 38 | 30 | 16 | 9 | 2 | 2 | 1 |
| 3rd leaf | — | — | 25 | 46 | 77 | 84 | 95 | 96 | 98 |

By day 17 relative contributions to carbon content of the third leaf were

| | |
|---|---|
| From 1st leaf | 21% |
| From 2nd leaf | 19% |
| From 3rd leaf | 60% |

*The transition from import to export*

Two major prerequisites have to be met before an expanding leaf can become a net exporter of carbon. Firstly, its photosynthetic activity must be sufficiently high that assimilate production is in excess of that required for local use. Secondly, the transport channel through which export occurs must be competent, both structurally and functionally. Net photosynthesis reaches its maximum when the leaf is 35–39% expanded and at about the time that cell division ceases in the palisade at the base of the blade. It is at this stage or earlier that export of assimilate begins. As maturation proceeds, and requirements for locally produced carbon fall, so export increases. Export may even remain high during the early phase of decline in photosynthesis indicating that local demand is dropping faster than photosynthesis.

There is much evidence to indicate that the site of loading of assimilate for export is the sieve element/companion cell complex of the minor veins (Fellows & Geiger, 1974; Geiger, 1976; see Chapter 14). It follows that maturation of the minor veins is necessary before assimilate export can occur and this is found in sugar beet and in *Cucurbita*. In the latter, which has bicollateral bundles, although there is the usual basipetal maturation of the vascular tissues, the adaxial phloem matures earlier than the abaxial. Export is correlated with maturation of the abaxial phloem of the minor veins of the 7th order (Turgeon & Webb, 1976). This phloem is characterized by large specialized intermediary cells, which are believed to be equivalent to companion cells and to function in the loading process. Within the leaf the abaxial phloem functions in export and the adaxial in import; in tomato, a similar situation is found (Bonnemain, 1969; Ho & Shaw, 1977).

Structural maturation of the sieve elements, judged by opening of the pores in the sieve plate, does not necessarily mean that the mechanism whereby transport sugars, such as sucrose, are loaded into the phloem is fully operational; little is known about this aspect. It may be significant that accumulation of sucrose, or in *Cucurbita*, stachyose and raffinose, precedes the onset of export which appears to start only after the solute concentration in the phloem rises to a level where pressure flow is primed (Fellows & Geiger, 1974).

## References

Amer, F. A. & Williams, W. T. (1957). Leaf-area growth in *Pelargonium zonale*. *Annals of Botany*, **31**, 339–42.

Aspinall, D. (1961). The control of tillering in the barley plant. I. The pattern of tillering and its relation to nutrient supply. *Australian Journal of Biological Sciences*, **14,** 493–505.

Avery, G. S., Jr. (1933). Structure and development of the tobacco leaf. *American Journal of Botany*, **20,** 565–92.

Barlow, H. W. B. (1979). Sectorial patterns in leaves on fruit tree shoots produced by radioactive assimilates and solutions. *Annals of Botany*, **43,** 595–602.

Begg, J. E. & Wright, M. J. (1962). Growth and development of leaves from intercalary meristems in *Phalaris arundinacea* L. *Nature*, **194,** 1097–8.

Biddulph, O. (1959). Translocation of inorganic solutes. In *Plant Physiology – A Treatise*, ed. F. C. Steward, pp. 553–603. New York: Academic Press.

Blenkinsop, P. G. & Dale, J. E. (1974). The effects of nitrate supply and grain reserves on Fraction I protein level in the first leaf of barley. *Journal of Experimental Botany*, **25,** 913–26.

Bonnemain, J. L. (1969). Transport du $^{14}C$ assimilé à partir des feuilles de Tomato en voie de croissances et vers celles-ci. *Comptes Rendus Hebdomadaires des Séances de l'Académie des Sciences, Paris*, **269D,** 1660–3.

Brouwer, R. (1977). Root functioning. In *Environmental Effects in Crop Physiology*, ed. C. V. Cutting & J. J. Landsberg, pp. 229–45. London: Academic Press.

Bünning, E. (1956). General processes of differentiation. In *The Growth of Leaves*, ed. F. L. Milthorpe, pp. 18–30. London: Butterworths Scientific Publications.

Carr, D. J. & Pate, J. S. (1967). Ageing in the whole plant. In *Aspects of the Biology of Ageing: Symposium of the Society for Experimental Biology No. 21*, ed. H. W. Woolhouse, pp. 559–600. Cambridge University Press.

Causton, D. & Venus, J. (1981). *The Biometry of Plant Growth.* London: Edward Arnold.

Chabot, B. F., Jurik, T. W. & Chabot, J. R. (1979). Influence of instantaneous and integrated light flux density on leaf anatomy and photosynthesis. *American Journal of Botany*, **66,** 940–5.

Clough, B. F. & Milthorpe, F. L. (1975). Effects of water deficit on leaf development in tobacco. *Australian Journal of Plant Physiology*, **2,** 291–300.

Coleman, W. K. & Greyson, R. I. (1976). The growth and development of the leaf in tomato (*Lycopersicon esculentum*). II. Leaf ontogeny. *Canadian Journal of Botany*, **54,** 2704–7.

Constable, G. A. & Rawson, H. M. (1980). Carbon production and utilisation in cotton: inferences from a carbon budget. *Australian Journal of Plant Physiology*, **7,** 539–53.

Cutter, E. G. (1965). Recent experimental studies of the shoot apex and shoot morphogenesis. *Botanical Reviews*, **31,** 7–130.

Cutter, E. G. (1971). *Plant Anatomy: Experiment and Interpretation. Part 2. Organs.* London: Edward Arnold.

Dale, J. E. (1964). Leaf growth in *Phaseolus vulgaris*. I. Growth of the first pair of leaves under constant conditions. *Annals of Botany*, **20,** 579–89.

Dale, J. E. (1970). Models of cell number increase in developing leaves. *Annals of Botany*, **34,** 267–73.

Dale, J. E. (1976). Cell division in leaves. In *Cell Division in Higher Plants*, ed. M. M. Yeoman, pp. 315–45. London: Academic Press.

Dale, J. E. (1982). *The Growth of Leaves: Studies in Biology 137.* London: Edward Arnold.

Dale, J. E. & Felippe, G. M. (1977). Nitrogen movement into and out of the first leaf of barley. *Zeitschrift für Pflanzenphysiologie*, **84,** 77–83.

Dale, J. E. & Wilson, R. G. (1979). The effects of photoperiod and mineral nutrient supply on growth and primordia production at the stem apex of barley seedlings. *Annals of Botany*, **44,** 537–46.

Davidson, J. L. & Milthorpe, F. L. (1966). Leaf growth in *Dactylis glomerata* following defoliation. *Annals of Botany*, **30**, 173–84.

Dengler, N. G. (1980). Comparative histological basis of sun and shade leaf dimorphism in *Helianthus annuus*. *Canadian Journal of Botany*, **58**, 717–30.

Dengler, N. G., Mackay, L. B. & Gregory, L. M. (1975). Cell enlargement during leaf expansion in beech, *Fagus grandifolia*. *Canadian Journal of Botany*, **53**, 2846–65.

Dennett, M. D., Auld, B. A. & Elston, J. (1978). A description of leaf growth in *Vicia faba* L. *Annals of Botany*, **42**, 223–32.

Erickson, R. O. (1960). Nomogram for the plastochron index. *American Journal of Botany*, **47**, 350–1.

Erickson, R. O. (1966). Relative elemental rates and anisotropy of growth in area: a computer programme. *Journal of Experimental Botany*, **17**, 390–403.

Erickson, R. O. & Michelini, F. J. (1957). The plastochron index. *American Journal of Botany*, **44**, 297–305.

Esau, K. (1965). *Plant Anatomy*, 2nd edition. New York: John Wiley.

Evans, L. T. (ed.) (1975). *Crop Physiology: Some Case Histories*. Cambridge University Press.

Fellows, R. J. & Geiger, D. R. (1974). Structural and physiological changes in sugar beet leaves during sink to source conversion. *Plant Physiology*, **54**, 877–85.

Fletcher, G. M. & Dale, J. E. (1974). Growth of tiller buds in barley: effects of shade treatment and mineral nutrition. *Annals of Botany*, **38**, 63–76.

Fletcher, G. M. & Dale, J. E. (1977). A comparison of main-stem and tiller growth in barley: apical development and leaf-unfolding rates. *Annals of Botany*, **41**, 109–16.

Friend, D. J. C., Helson, V. A. & Fisher, J. E. (1962). Leaf growth in Marquis wheat, as regulated by temperature, light intensity and daylength. *Canadian Journal of Botany*, **40**, 1299–311.

Fuchs, M. C. (1968). Localisation des divisions dans le méristème marginal des feuilles des *Lupinus albus* L., *Tropaeolum peregrinum* L., *Limonium sinuatum* (L.) Miller et *Nemophila maculata* Berth. *Comptes Rendus Hebdomadaires des Séances de l'Académie des Sciences, Paris*, **267D**, 722–5.

Geiger, D. R. (1976). Phloem loading in source leaves. In *Transport and Transfer Processes in Plants*, ed. I. F. Wardlaw & J. B. Passioura, pp. 167–83. London: Academic Press.

Girolami, G. (1954). Leaf histogenesis in *Linum usitatissimum*. *American Journal of Botany*, **41**, 264–73.

Greenland, A. J. & Lewis, D. H. (1981). The acid invertases of the developing third leaf of oat. I. *New Phytologist*, **88**, 265–77.

Gregory, R. A. & Romberger, J. A. (1972). The shoot apical ontogeny of the *Picea abies* seedling. I. Anatomy, apical dome diameter, and plastochron duration. *American Journal of Botany*, **59**, 587–97.

Halse, N. J., Greenwood, E. A. N., Lapins, P. & Boundy, C. A. P. (1969). An analysis of the effects of nitrogen deficiency on the growth and yield of a Western Australian wheat crop. *Australian Journal of Agricultural Research*, **20**, 987–98.

Hannam, R. V. (1968). Leaf growth and development in the tobacco plant. *Australian Journal of Biological Sciences*, **21**, 855–70.

Hara, N. (1962). Histogenesis of foliar venation of *Daphne pseudo-mezereum*. *Botanical Magazine of Tokyo*, **75**, 107–13.

Ho, L. C. & Shaw, A. F. (1977). Carbon economy and translocation of $^{14}C$ in leaflets of the seventh leaf of tomato during leaf expansion. *Annals of Botany*, **41**, 833–48.

Hopkinson, J. M. (1964). Studies on the expansion of the leaf surface. IV. The carbon and phosphorus economy of a leaf. *Journal of Experimental Botany*, **15**, 125–37.

Horie, T., de Wit, C. T., Goudrian, J. & Bensink, J. (1979). A formal template for the development of cucumber in its vegetative stage. *Proceedings of the Koninklijke Nederlandse Akademie van Wetenschappen, Series C*, **82**, 433–79.

Hughes, A. P. (1959). Effects of environment on leaf development in *Impatiens parviflora. Journal of the Linnaean Society of London (Botany)*, **56**, 161–5.

Humphries, E. C. & Wheeler, A. W. (1963). The physiology of leaf growth. *Annual Review of Plant Physiology*, **14**, 385–410.

Isebrands, J. G., Dickson, R. E. & Larson, P. R. (1976). Translocation and incorporation of $^{14}C$ into the petiole from different regions within developing cottonwood leaves. *Planta*, **128**, 185–93.

Isebrands, J. G. & Larson, P. R. (1973). Anatomical changes during leaf ontogeny in *Populus deltoides. American Journal of Botany*, **60**, 199–208.

Joy, K. W. (1964). Translocation in sugar beet I. Assimilation of $^{14}CO_2$ and distribution of materials from leaves. *Journal of Experimental Botany*, **15**, 485–94.

Kemp, D. R. (1980). The location and size of the extension zone of emerging wheat leaves. *New Phytologist*, **84**, 729–38.

Lamoreaux, R. J., Chaney, W. R. & Brown, K. M. (1978). The plastochron index: a review after two decades of use. *American Journal of Botany*, **65**, 586–93.

Langer, R. H. M. (1980). *How Grasses Grow*, 2nd Edition. London: Edward Arnold.

Larson, P. R. (1977). Phyllotactic transitions in the vascular system of *Populus deltoides* Bartr. as determined by $^{14}C$ labelling. *Planta*, **134**, 241–9.

Larson, P. R. & Dickson, R. E. (1973). Distribution of imported $^{14}C$ in developing leaves of eastern cottonwood according to phyllotaxy. *Planta*, **11**, 95–112.

Larson, P. R., Isebrands, J. G. & Dickson, R. E. (1972). Fixation patterns of $^{14}C$ within developing leaves of eastern cottonwood. *Planta*, **107**, 301–14.

Lerstens, N. (1965). Histogenesis of leaf venation in *Trifolium wormskioldii* (Leguminosae). *American Journal of Botany*, **52**, 767–74.

Maksymowych, R. (1962). An analysis of leaf elongation in *Xanthium pennsylvanicum* presented in relative elemental rates. *American Journal of Botany*, **49**, 7–13.

Maksymowych, R. (1963). Cell division and cell elongation in leaf development of *Xanthium pennsylvanicum. American Journal of Botany*, **50**, 891–901.

Maksymowych, R. (1973). *Analysis of Leaf Development.* Cambridge University Press.

Maksymowych, R. & Erickson, R. O. (1960). Development of the lamina in *Xanthium italicum* represented by the plastochron index. *American Journal of Botany*, **47**, 451–9.

Milthorpe, F. L. (1959). Studies on the expansion of the leaf surface. I. The influence of temperature. *Journal of Experimental Botany*, **10**, 233–49.

Milthorpe, F. L. & Moorby, J. (1969). Vascular transport and its significance in plant growth. *Annual Review of Plant Physiology*, **20**, 117–38.

Milthorpe, F. L. & Newton, P. (1963). Studies on the expansion of the leaf surface. III. Influence of radiation on cell division and leaf expansion. *Journal of Experimental Botany*, **14**, 483–95.

Moorby, J. (1977). Integration and regulation of translocation within the whole plant. *Integration of Activity in the Higher Plant: Symposium of the Society for Experimental Biology No. 31*, ed. D. H. Jennings, pp. 425–54. Cambridge University Press.

Moorby, J. (1981). *Transport Systems in Plants.* London: Longman.

Murray, D. (1968). Light and leaf growth in *Phaseolus.* Ph.D. thesis, University of Edinburgh.

Patrick, J. W. (1972a). Vascular system of the stem of the wheat plant. I. Mature state. *Australian Journal of Botany*, **20**, 49–63.

Patrick, J. W. (1972b) Vascular system of the stem of the wheat plant. II. Development. *Australian Journal of Botany*, **20**, 65–78.

Patrick, J. W. (1972c). Distribution of assimilate during stem elongation in wheat. *Australian Journal of Biological Science*, **25**, 455–67.

Pieters, G. A. (1974). The growth of sun and shade leaves of *Populus euamericana* 'robusta' in relation to age, light intensity and temperature. *Mededelingen Landbouwhogeschool, Wageningen*, **74**, 1–106.

Porter, H. K. (1966). Leaves as collecting and distributing agents of carbon. *Australian Journal of Science*, **29**, 31–40.

Pray, T. R. (1963). Origin of vein endings in Angiosperm leaves. *Phytomorphology*, **13**, 60–81.

Raper, C. D., Jr., Thomas, J. F., Wann, M. & York, E. K. (1975). Temperatures in early transplant growth: influence on leaf and floral initiation in tobacco. *Crop Science*, **15**, 732–3.

Saurer, W. & Possingham, J. V. (1970). Studies on the growth of spinach leaves (*Spinacea oleracea). Journal of Experimental Botany*, **21**, 151–8.

Schoch, P. G. (1972). Effects of shading on structural characteristics of the leaf and yield of fruit in *Capsicum annuum* L. *Proceedings of American Society for Horticultural Science*, **97**, 461–4.

Schüepp, O. (1966). *Meristeme*. Basel: Birkhauser-Verlag.

Stebbins, G. L. & Shah, S. S. (1960). Developmental studies of cell differentiation in the epidermis of monocotyledons. II. Cytological features of stomatal development in the Gramineae. *Developmental Biology*, **2**, 477–500.

Steer, B. T. (1971). The dynamics of leaf growth and photosynthetic capacity in *Capsicum frutescens* L. *Annals of Botany*, **35**, 1003–15.

Sunderland, N. (1960). Cell division and expansion in the growth of the leaf. *Journal of Experimental Botany*, **11**, 68–80.

Sunderland, N. & Brown, R. (1956). Distribution of growth in the apical region of the shoot of *Lupinus albus. Journal of Experimental Botany*, **7**, 126–45.

Sunderland, N. & Brown, R. (1976). Development during vegetative growth in the apical region of the shoots of *Lupinus albus. Annals of Botany*, **40**, 199–212.

Sutcliffe, J. F. (1976). Regulation in the whole plant. In *Encyclopedia of Plant Physiology. Transport in Plants. IIIB. Tissues and Organs*, ed. U. Lüttge & M. G. Pitman, pp. 395–417. Berlin: Springer-Verlag.

Swanson, C. A., Hoddinott, J. & Sij, J. W. (1976). The effects of selected sink leaf parameters on translocation rates. In *Transport and Transfer Processes in Plants*, ed. I. F. Wardlaw & J. B. Passioura, pp. 347–56. London: Academic Press.

Turgeon, R. (1980). The import to export transition: experiments in *Coleus blumei. Berichte der Deutschen Botanischen Gesellschaft*, **93**, 91–7.

Turgeon, R. & Webb, J. A. (1975). Leaf development and phloem transport in *Cucurbita pepo*: carbon economy. *Planta*, **123**, 53–62.

Turgeon, R. & Webb, J. A. (1976). Leaf development and phloem transport in *Cucurbita pepo*: maturation of the minor veins. *Planta*, **129**, 265–9.

Van Volkenburgh, E. & Cleland, R. E. (1980) Proton excretion and cell expansion in bean leaves. *Planta*, **148**, 273–8.

Verbelen, J. P. & de Greef, J. A. (1979). Leaf development of *Phaseolus vulgaris* L. in light and darkness. *American Journal of Botany*, **66**, 970–6.

Wardlaw, I. F. & Mortimer, D. C. (1970). Carbohydrate movement in pea plants in relation to axillary bud growth and vascular development. *Canadian Journal of Botany*, **48**, 299–37.

Williams, R. F. (1975). *The Shoot Apex and Leaf Growth*. Cambridge University Press.

Williams, R. F. & Rijven, A. H. G. C. (1965). The physiology of growth in the wheat plant. 2. The dynamics of leaf growth. *Australian Journal of Biological Sciences*, **18**, 721–43.

Williams, R. F. & Rijven, A. H. G. C. (1970). The physiology of growth in subterranean clover. 2. The dynamics of leaf growth. *Australian Journal of Botany*, **18**, 149–66.

Wilson, G. L. (1966). Studies of the expansion of the leaf surface. V. Cell division and expansion in a developing leaf as influenced by light and upper leaves. *Journal of Experimental Botany*, **17**, 440–51.

Wylie, R. B. (1951). Principles of foliar organisations shown by semi-shade leaves from ten species of deciduous dicotyledon trees. *American Journal of Botany*, **38**, 355–61.

# 7

# Environmental influences on leaf expansion

N. TERRY, L. J. WALDRON AND S. E. TAYLOR

## Introduction

A growing plant experiences two distinctly different environments: the above-ground atmospheric environment and the root environment of the soil. The variables of the aerial environment which may potentially influence leaf expansion include temperature, light and carbon dioxide. Soil variables which influence leaf expansion include water and mineral nutrient availability, the salt concentration of the soil solution and soil temperature. Soil temperature affects leaf expansion through its influence on rates of water and mineral absorption by roots and through its effect on leaf temperature, particularly on the near-ground meristematic tissue of leaves of grasses. Leaf expansion, because it is very sensitive to factors in both environments, may frequently limit plant growth and productivity.

## Temperature

### *General aspects – the Arrhenius relation*

Plant growth involves numerous enzyme-catalysed biosynthetic chemical reactions, each of which is rate-controlled by temperature. According to Arrhenius, the rate of a chemical reaction is related to temperature as follows:

$$\text{Rate of reaction} = A \exp(-E_a/RT) \qquad (7.1)$$

where $A$ is referred to as the frequency function, $E_a$ is the activation energy, $R$ is the gas constant, and $T$ the Kelvin temperature. The frequency function, $A$, relates to the number of molecules colliding per unit time and is proportional to $T^{\frac{1}{2}}$. Over the range of biologically significant temperatures, $A$ varies little and is commonly taken to be constant. Thus, the temperature dependence of the 'rate of reaction' (7.1) is controlled essentially by exp $(-E_a/RT)$, the fraction of the total number of colliding molecules with energy equal to or greater than $E_a$. Where equation (7.1) holds a plot of ln rate against $1/T$ gives a straight line with a slope of $-E_a/R$. This is referred to as an Arrhenius plot.

Arrhenius plots have been found to fit certain plant growth data reasonably well. Weidner & Ziemens (1975) found that the rate of protein synthesis in wheat seedlings increased over a temperature range of 15–30 °C (Fig. 7.1). Above 30 °C, the relation was no longer linear, reaching a maximum at 33 °C and then declining with further increase in temperature. Such a decline is often attributed to protein denaturation and to increased proteolytic enzyme activity. Equation (7.1) may be applied much more generally, however, to complex growth processes such as cell enlargement. Kemp & Blacklow (1980) showed that the extension rate of wheat leaves was linearly related to $1/T$ from 5 to 26 °C.

Fig. 7.1. An Arrhenius plot of the rate of protein synthesis by wheat seedlings. The reciprocal of Kelvin temperature is shown on the abscissa with the corresponding temperatures in Celsius shown on the upper scale. (After Weidner & Ziemens, 1975).

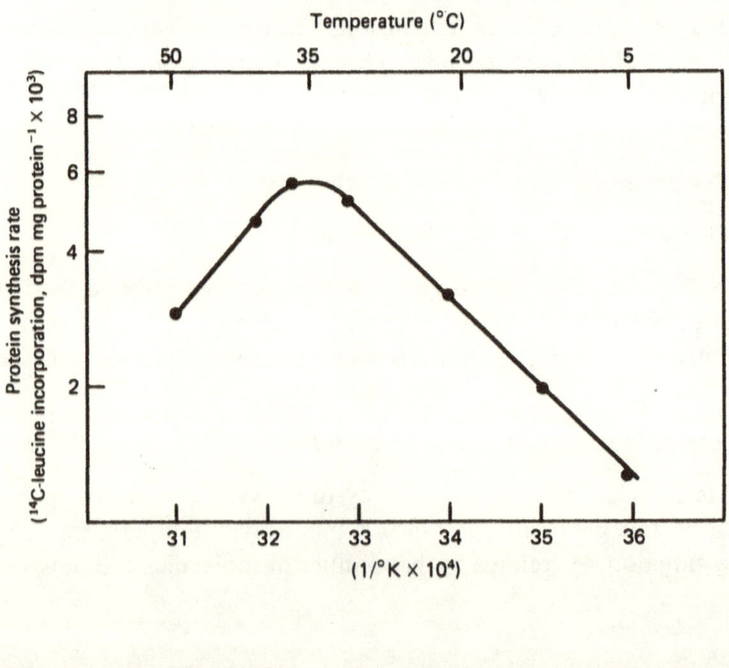

Others (Bagnell & Wolfe, 1978) have found more complex relationships.

The temperature coefficient, $Q_{10}$, which is the factor by which a given rate of a biological reaction increases for a 10 °C temperature rise, is related to activation energy by:

$$Q_{10} = \exp(10E_a/R\bar{T}^2), \text{ where } \bar{T} = \sqrt{[T(T + 10)]} \qquad (7.2)$$

$Q_{10}$'s for protein synthesis and leaf extension calculated from equation (7.2) using published $E_a$ values or Arrhenius plots give values for the two processes at 25 °C ranging from 1.8 to 3.0.

## Leaf expansion, photosynthesis and plant growth

The optimum temperature for the growth of many plants lies in the range of 20–30 °C. Since increase in dry matter is mainly the result of the accumulation of carbon products, plant growth is often analysed in terms of the influence of environmental factors on (1) the rate of photosynthesis per unit area and (2) the rate of increase in total leaf area. Temperature affects growth more by altering the expansion of the leaf surface than by altering the rate of photosynthesis. Potter & Jones (1977) showed that responses to changes in temperature were largely determined by the extent to which newly formed dry matter was partitioned into leaf surface, there being much less effect of temperature on net assimilation rate (Fig. 7.2). Forde, Whitehead & Rowley (1975) showed that photosynthesis and starch accumulation was less sensitive to temperature than leaf growth in *Paspalum dilatatum*.

The expansion of leaf surface depends on a number of factors including rates of leaf production and senescence, branching and tillering, and the rate and duration of leaf expansion. Leaves are produced more rapidly as the temperature increases to 20 or 30 °C (Terry, 1968; Fukai & Silsbury, 1976; Dennett, Elston & Milford, 1979) (Fig. 7.3). Once formed, the growth of individual leaves is also usually most rapid between 20 and 30 °C (Peet, Ozbun & Wallace, 1977; Auld, Dennett & Elston, 1978). The duration of leaf growth, however, often increases with decrease in temperature below 20–25 °C (Auld *et al.*, 1978; Dennett *et al.*, 1979).

Decrease in temperature produces significant morphological and anatomical changes. Leaves tend to become wider and shorter, and the length of the petiole and midrib may be reduced (Terry, 1968). Charles-Edwards, Charles-Edwards & Sant (1974) also showed that low temperature increased leaf thickness and the amount of leaf tissue per unit area due to a greater mesophyll cell size. Peet *et al.* (1977) found that leaf thickness, specific leaf weight, and the extractable activity of proteins, such as ribulose bisphosphate carboxylase, were increased by low temperature.

Fig. 7.2. Relative growth rate, relative leaf area expansion rate and net assimilation rate for nine species grown under three temperature regimes (21/10°C, 32/21 °C and 31/27 °C, day/night). Relative growth rate, which increased with increase in temperature, was correlated strongly with relative leaf area expansion rate, but poorly with net assimilation rate. (After Potter & Jones, 1977.)

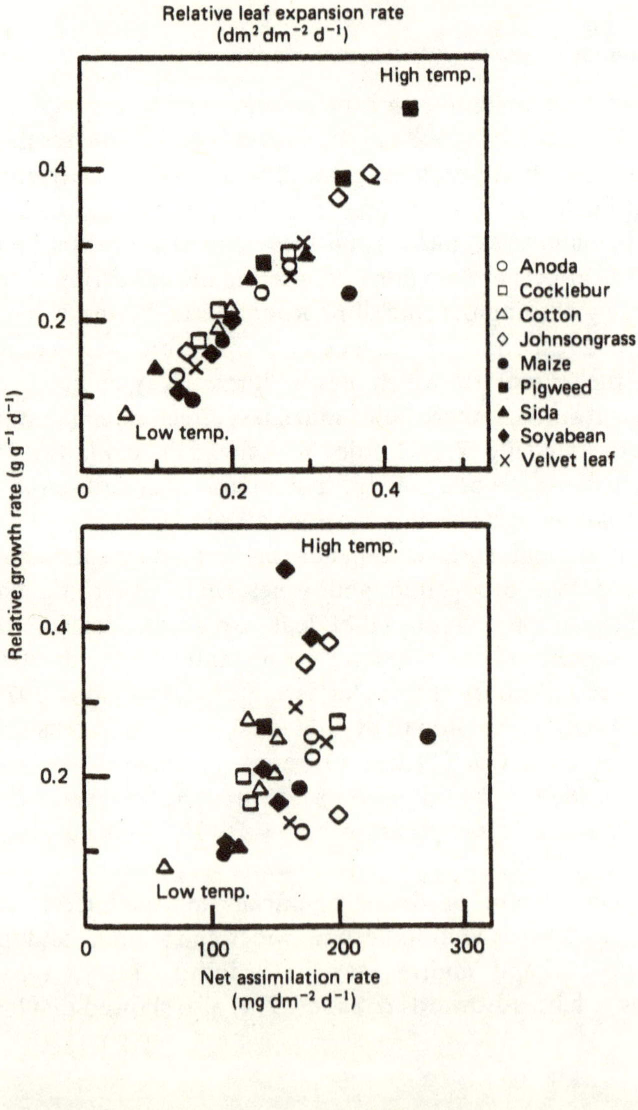

*Leaf growth in grasses*

In grasses, leaf growth proceeds by cell division and cell extension at the base of the leaf (see Chapter 6). There is evidence that the temperature of this region controls the growth rate of the leaf. Watts (1972) found that when the apical meristem and the region of cell expansion at the base of the leaf of maize were maintained at 25 °C, the rate of leaf extension was less affected by variation in temperature elsewhere in the plant. If, however, root and shoot temperatures were maintained at 25 °C and the meristem temperature was varied, there were rapid changes in leaf extension depending on the temperature of meristem. Watts (1974) also found that leaf extension rates varied widely in the field if there were large diurnal variations of air temperature, but were constant if the temperature of the shoot meristem was maintained at 30–34 °C (Fig. 7.4). The growth of maize and ryegrass is largely determined by soil temperature because the growing points are below the soil for much of the vegetative growth (Duncan, 1975; Peacock, 1975). There is some evidence however that temperature effects on the growing point may depend on the availability of water (Gallagher & Biscoe, 1979).

Fig. 7.3. Rate of leaf production by subterranean clover at four temperatures ranging from 15 to 30 °C. (After Fukai & Silsbury, 1976.)

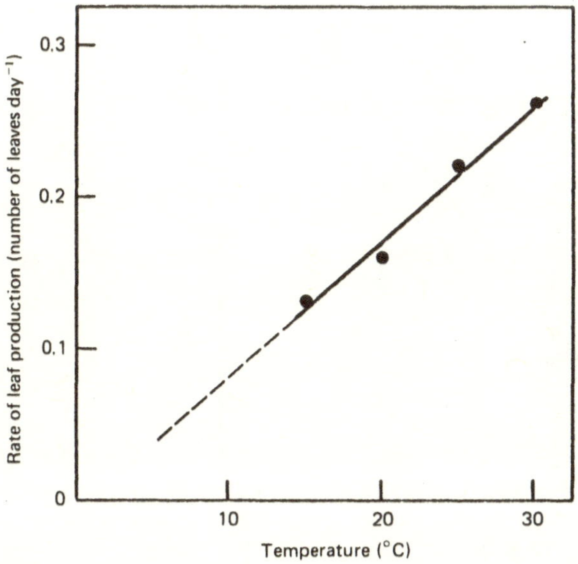

*Influence of soil temperature*

Low soil temperature reduces the absorption of water by roots, a fact which has been observed with many plants. The effect is greatest on warm climate species subjected to cool root temperatures (Kramer, 1969). Reducing the root zone temperature of cotton from 25° to 10 °C reduced transpiration to about 20% of its value at 25 °C. However, a similar reduction in root temperature of collard only reduced the transpiration rate to 80% of its value at 25 °C. Mechanisms by which low temperature may restrict the uptake of water include: (i) increased viscosity of water in plant tissues, (ii) increased resistance of the plant tissues caused by decreased cell membrane permeability, (iii) decreased active salt uptake and accumulation, and (iv) decreased root growth (Kramer, 1969).

Fig. 7.4. Changes in leaf extension rate of *Zea mays* with increase in air temperature (solid line); where the temperature of the leaf meristem and extension zone is maintained constant at 30–34 °C, leaf extension is independent of air temperature (broken line). (From Watts, 1974.)

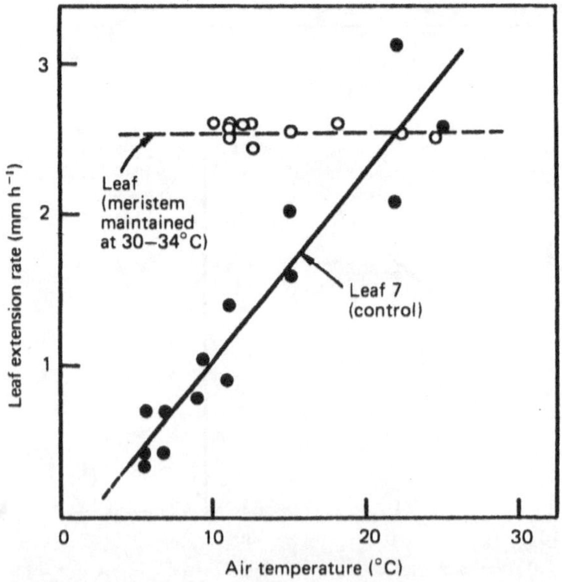

## Light

### *Light intensity effects on growth and leaf expansion*

In the early vegetative stages of plant development increases in irradiance may accelerate both plant dry matter production and expansion of the total leaf surface (Doley, 1978; Ketring, 1979). The greater leaf surface expansion is due to faster production of new leaves and to more rapid expansion of individual leaves (see also Chapter 6). Leaf cell division rate, final cell number and cell size are enhanced under high irradiance (Milthorpe & Newton, 1963; Ludlow & Wilson, 1971).

In a study of sugar beet growth, Terry (1968) concluded that the main effect of increased light was to increase the rate of production of photosynthate so that the plant proceeded through its ontogeny faster. Thus different plant growth attributes were related to one another in a manner largely independent of light intensity (Fig. 7.5) which served to change the rate but not pattern of development. Growing leaves are dependent on an imported carbohydrate supply until they reach one-third to one-half of their final size (Fellows & Geiger, 1974; see Chapter 6), and one might expect the rate of growth of young leaves to be

Fig. 7.5. Root and shoot dry weights of sugar beet harvested over seven successive intervals at four irradiances. The single relationship indicates that irradiance level did not affect the ontogenetic development. (After Terry, 1968.)

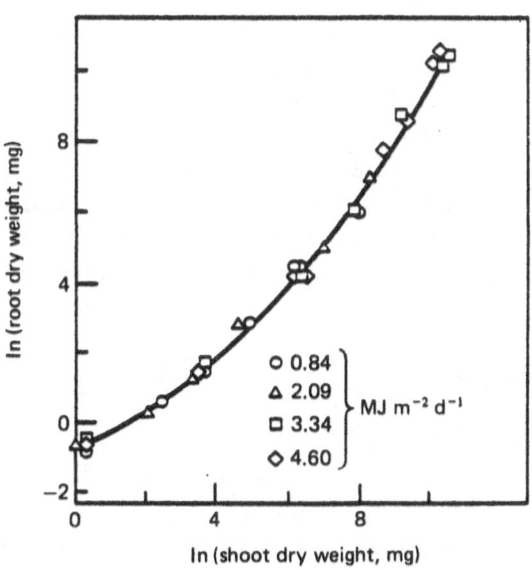

directly dependent on light intensity. Kemp (1981) compared changes in leaf extension rate of wheat with the concentration of carbohydrates in the extension zone but found leaf growth was only correlated with decreased carbohydrate concentration under conditions of intense shading.

### *Leaf morphology and anatomy*

Studies of the effect of irradiance level on growth usually show that leaf area per unit plant dry weight (leaf area ratio) increases in low light. This is partly because of an increase in the leaf area to leaf weight ratio (e.g. Doley, 1978). Leaves grown under low light intensities are thinner than those grown under high light: Nobel, Zaragoza & Smith (1975) showed that under low light leaves of *Plectranthus* were only one-third of the thickness under high light due to shorter cells in the top two layers of the palisade, fewer layers of palisade, and reduced size and frequency of spongy mesophyll cells. Similar reductions in palisade mesophyll with low light were found in *Eucalyptus* (Doley, 1978). Leaves grown in low light may also be longer and narrower than leaves grown in high light (Wilson & Cooper, 1969; Nobel, 1976).

The light level during growth has pronounced effects on chloroplast development. Boardman (1977) noted that chloroplasts of shade plants are usually larger, have a greater chlorophyll content, a greater proportion of chlorophyll b to chlorophyll a, and a smaller stromal volume than chloroplasts of high-light plants. Also, under low light, chloroplasts tend to have well-formed grana with more grana per plastid than high-light chloroplasts (Ballantine & Forde, 1970; Crookston, Treharne, Ludford & Ozbun, 1975). High-light intensities may also result in an increased number of chloroplasts per unit leaf area (Louwerse & Zweerde, 1977).

Effects of light on the development of the photosynthetic apparatus are considered in Chapter 10 and on photosynthesis in the mature leaf in Chapter 12. Photomorphogenetic effects are dealt with in Chapter 9.

## **Carbon dioxide supply**

### *Enhancement of growth by $CO_2$*

Plant growth is accelerated when concentrations of $CO_2$ are elevated above atmospheric levels in easily enclosed spaces such as greenhouses or growth chambers. Ford & Thorne (1967) observed that the total dry weight of sugar beet, barley and kale was increased by 50% when the ambient $CO_2$ concentration in growth chambers was raised from 300 to 1000 p.p.m. Carbon dioxide enrichment has also enhanced the growth of tomato (Hurd & Thornley, 1974), wheat (Gifford, 1977; Neales & Nicholls, 1978), and soybean and cotton (Mauney, Fry & Guinn, 1978). Neales & Nicholls (1978) found that most of the growth response to $CO_2$ in wheat was between 200 and 400 p.p.m. (Fig. 7.6) although growth continued to increase up to 600 p.p.m.

Carbon dioxide enhancement effects are usually accompanied by an increase in the rate of expansion of plant leaf surface (Cooper & Brun, 1967; Hurd & Thornley, 1974; Neales & Nicholls, 1978) as a result of increase in the rate of growth of individual leaves rather than through leaf production: Ho (1977) found that $CO_2$ enrichment increased individual leaf area by as much as 80% (Fig. 7.7), while Ford & Thorne (1967) and Gifford (1977) found that $CO_2$ enhancement had little effect on the rate of leaf production but increased tiller production.

Carbon dioxide enrichment leads to changes in plant development. Plants tend to branch more and there may be an increase in the root to shoot ratio (Kramer, 1981). Leaves tend to be thicker (Ho, 1977) and to have a greater fresh or dry weight per unit area (Ford & Thorne, 1967;

Fig. 7.6. Plant dry weight of wheat after 17 days of growth at different ambient $CO_2$ concentrations. (After Neales & Nicholls, 1978.)

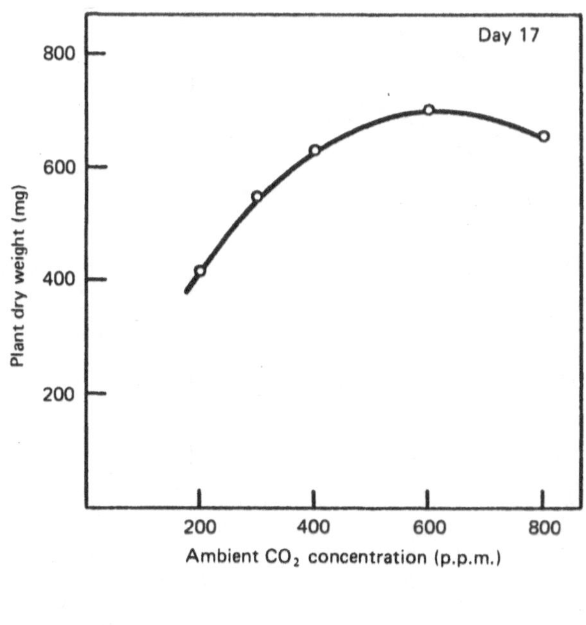

Neales & Nicholls, 1978). Madsen (1973) found that the increased leaf thickness with high $CO_2$ was due to an increase in the volume (but not number) of cells in the palisade and mesophyll tissue.

### *Mechanism of $CO_2$ enhancement*

As with increase in light intensity, the most plausible mechanism for increased growth in high $CO_2$ is that $CO_2$ enrichment leads to faster rates of photosynthesis and therefore an increased rate of supply of carbohydrate for growth. Carbon dioxide enhancement certainly leads to faster photosynthesis; Bishop & Whittingham (1968) showed that the photosynthesis of a tomato plant growing and assimilating in 1000 p.p.m. $CO_2$ was about three times that of a plant growing and assimilating in normal air (about 300 p.p.m. $CO_2$). Hurd (1968) found that high $CO_2$ increased growth and net assimilation rate in tomato and concluded 'there were no indications that beneficial effects of $CO_2$ enrichment operate other than through increased photosynthesis'.

Fig. 7.7. Influence of ambient $CO_2$ concentration on growth of a tomato leaf. (After Ho, 1977.)

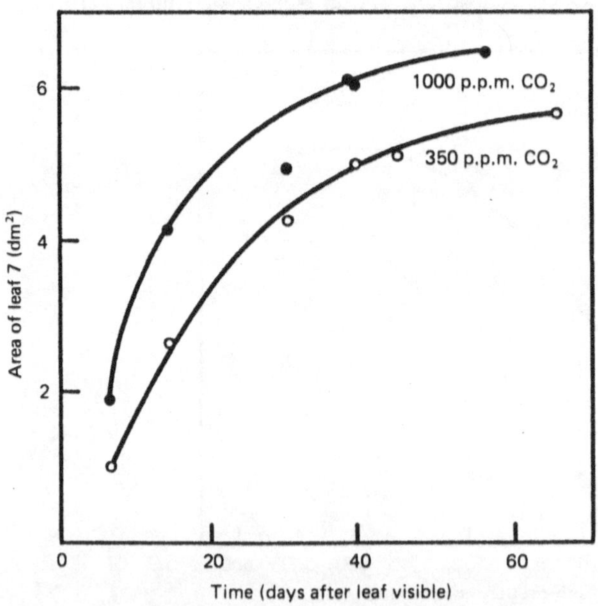

*Possible limitation to $CO_2$ enhancement*

Several plant growth studies have indicated that increases in plant growth resulting from $CO_2$ enrichment are manifested only in the early vegetative stage of growth and that the beneficial effect of $CO_2$ disappears as the plants age (see review by Kramer, 1981). Neales & Nicholls (1978), and Mauney *et al.* (1978), showed that net assimilation and relative growth rates were increased early but not late in vegetative growth. The question of whether $CO_2$ enrichment has beneficial effects on plant growth is complicated by the fact that certain growth parameters change during ontogeny. For example, in the experiments of Neales & Nicholls (1978), relative growth rates decreased substantially as plants increased in size (Fig. 7.8). High $CO_2$ accelerates development, and comparisons of relative growth rates between high and low $CO_2$ treatments at the same chronological age may not be meaningful sinee low $CO_2$ plants are smaller and have intrinsically larger relative growth rates.

Carbon dioxide enhancement may be limited or modified by other environmental factors. In some studies the effects of high $CO_2$ on plant growth were greater at low light intensities (Gifford, 1977), but in others

Fig. 7.8. Relative growth rate of wheat at three stages of plant growth as a function of ambient $CO_2$ concentration. The largest plants (1636 mg) show the lowest relative growth rates at all $CO_2$ concentrations. (After Neales & Nicholls, 1978.)

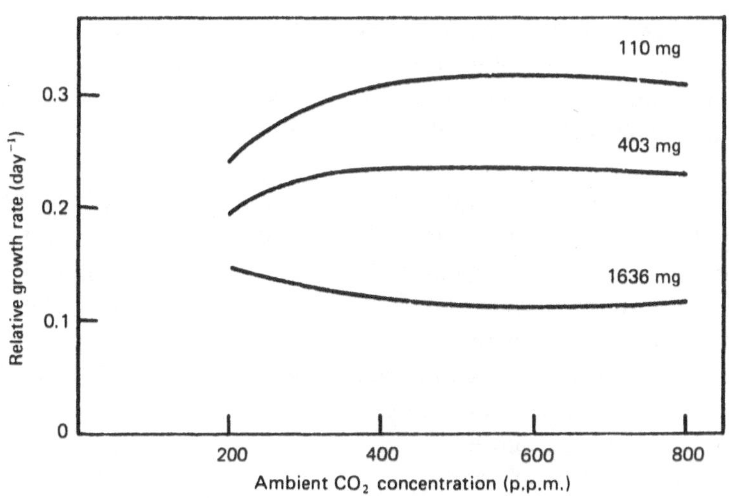

net assimilation rate and relative growth rate were increased more by $CO_2$ enrichment at high than at low light (Ford & Thorne, 1967; Bishop & Whittingham, 1968; Hurd & Thornley, 1974). The availability of water and mineral nutrients may also be important. Gifford (1977) showed that benefits of $CO_2$ enrichment were greater when water was limiting.

### Mineral nutrient supply

At least 13 different mineral elements are generally recognized as being essential for the growth of most plants. Nitrogen and potassium are usually required in the greatest amounts (Ingestad, 1972). The growth of leaves has long been known to be especially sensitive to application of nitrogen, which increases leafiness in many crops. Seedlings of birch responded rapidly and linearly to a change in the nitrogen supply by effects on the expansion of the leaf surface (Ingestad & Lund, 1979). Robson & Deacon (1978) reported that increased nitrogen levels resulted in faster leaf elongation, greater leaf length and area, and increased tillering in ryegrass. Bhat, Nye & Brereton (1979) observed an increase in the growth rate and leaf area ratio with increasing tissue nitrogen concentrations in rape. Increased nitrogen supply may increase the number of cells per leaf, the rate of cell expansion, and mean cell size (Terry, 1970).

Potassium is thought to function as an osmoticum in several physiological processes in plants, including stomatal opening (Fischer & Hsiao, 1968), water transfer into root xylem and root pressure (Mengel & Kirkby, 1980). Potassium may also have a special function as an osmoticum for leaf expansion (Mengel & Kirkby, 1980). It is of interest that Marschner & Possingham (1975) showed that increasing the potassium supply increased cell size in leaf discs of sugar beet and spinach.

The influence of mineral nutrient supply on leaf growth is complicated by the fact that many mineral elements are redistributed between different parts of the growing plant. Although a young growing leaf is a net importer of mineral nutrients, it may fairly early on in its ontogeny export certain nutrients via the phloem to other growing tissues (see also Chapter 6). In plants with low nitrogen contents, as much as 70% of the nitrogen in a leaf may be remobilized and translocated to younger leaves before the leaf dies; in high-nitrogen plants, however, nitrogen may not be redistributed until the leaves are near death (Robson & Deacon,

1978). Several studies (e.g. Graham & Ulrich, 1974) have shown that potassium may be readily exported and about 50% of the potassium taken up by rapidly developing young leaves may be derived from within the plant (particularly older leaves) rather than external sources (Greenway & Pitman, 1965). Some elements, however, are less mobile than potassium, e.g. iron, while calcium and boron appear to be immobile in the phloem.

Deficiency of some mineral nutrients may result in greater effect on the expansion of the leaf surface than on the rate of photosynthesis per area (Brix & Ebell, 1969; Bouma, Dowling & Wahjoedi, 1979). Furthermore, many mineral deficiencies, including nitrogen deficiency, increase the tissue concentration of soluble carbohydrates, suggesting that utilization of photosynthate in growth is restricted more than photosynthate production (Robson & Deacon, 1978; Bhat *et al.*, 1979).

## Water and Salt Stress

Leaf growth is rapidly reduced with the onset of water stress or with an increase in the salt concentration of the root medium. Before osmotic adjustment to salt takes place, plants exposed to a sudden increase in salinity may become temporarily water stressed with the leaves exhibiting a loss of turgor. Because of the involvement of salt stress with plant water relations, particularly with respect to osmoregulation, water and salt stress are discussed together.

### *Role of water in leaf enlargement*

Leaf enlargement is highly sensitive to water stress. It is one of the first growth processes to be affected by a decrease in leaf water potential (Hsiao, 1973). For example, leaf enlargement in corn, soybean and sunflower has been shown to cease at a leaf water potential of −0.4 MPa (Fig. 7.9; Boyer, 1970), and at −0.7 MPa in corn (Acevedo, Hsiao & Henderson, 1971). The sensitivity to water stress is because the uptake of water into leaf cells provides the mechanical force for leaf enlargement.

In sunflower, turgor pressure of 0.65 MPa was required before leaf enlargement occurred (Boyer, 1968). This is the threshold pressure required to stretch the cell wall irreversibly and is called the yield stress, $Y$. The relative rate of increase in cell volume is proportional to the difference between turgor pressure, $\psi_p$, and yield stress, $Y$ (Green,

Erickson & Buggy, 1971; Hsiao, Acevedo, Fereres & Henderson, 1976; Bunce, 1977):

$$\frac{1}{V}\left(\frac{\mathrm{d}V}{\mathrm{d}t}\right)_{\mathrm{c}} = m(\psi_{\mathrm{p}} - Y) \qquad (7.3)$$

where $1/V(\mathrm{d}V/\mathrm{d}t)_{\mathrm{c}}$ is the relative rate of change in cell volume ($s^{-1}$) and $m$ is a constant of proportionality ($s^{-1}\mathrm{Pa}^{-1}$). In the special instance when growing cells are considered to be elongating cylinders which increase in length only, $m$ is referred to as cell extensibility.

Equation (7.3) makes no allowance for the possibility that cell expansion may be limited by the rate of water uptake; for example, as the cell expands it must absorb water at a rate sufficient to maintain $\psi_{\mathrm{p}} > Y$. The rate of water uptake by leaf cells, $(\mathrm{d}V/\mathrm{d}t)_{\mathrm{w}}$, is governed by the water potential difference across the cell wall/membrane boundary, $\psi_{\mathrm{w}}^{\mathrm{in}} - \psi_{\mathrm{w}}^{\mathrm{ex}}$, and by $L_{\mathrm{p}}$, the hydraulic conductivity of that boundary:

$$\left(\frac{\mathrm{d}V}{\mathrm{d}t}\right)_{w} = -L_{\mathrm{p}}A\,(\psi_{\mathrm{w}}^{\mathrm{in}} - \psi_{\mathrm{w}}^{\mathrm{ex}}) \qquad (7.4)$$

where $A$ is the area of the cell wall/membrane barrier and $L_{\mathrm{p}}$ has units of $\mathrm{m\ s^{-1}Pa^{-1}}$ (Ray, Green & Cleland, 1972; Boyer & Wu, 1978). Since expansion growth over the short-term is largely the consequence of

Fig. 7.9. Dependence of leaf enlargement rate and photosynthetic rate of sunflower on leaf water potential. (After Boyer, 1967.)

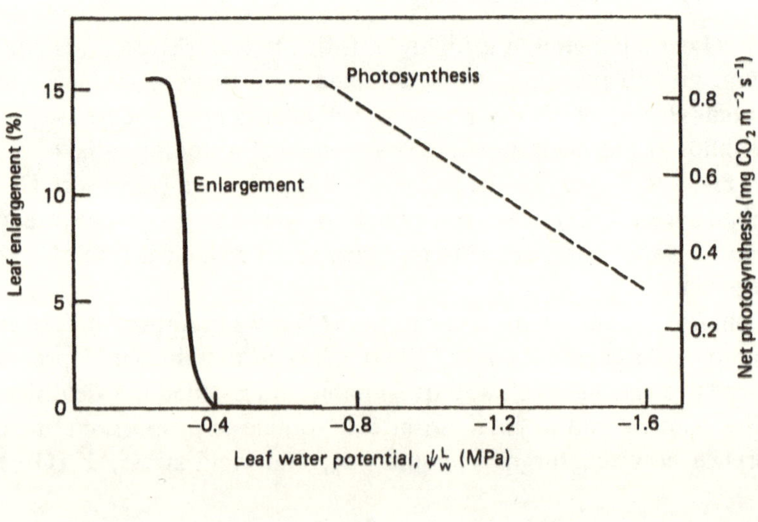

water uptake, i.e. $dV_w = dV_c$, equations (7.3) and (7.4) may be combined to form:

$$\frac{1}{V}\left(\frac{dV}{dt}\right)_c = \frac{L_p A m}{mV + L_p A}\left(\psi_w{}^{ex} - \psi_w{}^{in} - Y\right) \qquad (7.5)$$

In equation (7.5), $\psi_w{}^{ex} - \psi_w{}^{in}$ is the driving force causing water to enter cells.

In spite of its limitations, equation (7.3) has been more commonly related to plant growth phenomena than equation (7.5). Boyer (1968), Hsiao *et al.* (1976), Bunce (1977), and Sepaskhan & Boersma (1979) showed that for specific sets of environmental conditions, leaf enlargement rate was linearly related to turgor pressure. Bunce obtained a linear relationship for each of three environmental conditions (Fig. 7.10) and associated the slope and intercept of each with cell extensibility and yield stress, respectively, implying that equation 7.3 was applicable with $m$ and $Y$ being different for each of the experimental conditions.

The yield stress, $Y$, may be increased by the laying down of additional cell wall material or the strain hardening of existing cellulose microfibrils; $Y$ is decreased by large increases in turgor pressure (Grenetz & List, 1973).

Based on data obtained with coleoptiles and stem segments, it has been shown that artificial acidification of the cell wall (Rayle & Cleland, 1970) or the application of auxin (Cleland, 1967) reduces $Y$. The decrease in $Y$ with cell wall acidification is evidence for the theory that proton excretion by the cell causes the cell walls to be loosened prior to cell enlargement (Rayle & Cleland, 1970). The decrease in $Y$ with auxin is evidence for the theory that auxin operates on cell enlargement by increasing cell wall plasticity, possibly by promoting cell proton excretion (Cleland 1973; Marré, Lado, Rasi-Caldogno & Colombo, 1973; Rayle & Cleland, 1980). In leaves, elongation has been shown to occur in response to proton excretion induced artificially by fusicoccin (Marré, Colombo, Lado & Rasi-Caldogno, 1974), or by light (van Volkenburgh & Cleland, 1980) but has not been shown to occur in response to auxin.

Auxin may also affect $L_p$, which in the oat coleoptile was decreased to as little as 5% of its initial value when the tissue was deprived of exogenous auxin causing a reduction in extension growth in accordance with equation (7.5). $L_p$ and growth were restored when auxin was reapplied (Boyer & Wu, 1978).

### *Salt and leaf enlargement*

The presence of high salt concentrations in the root medium may result in growth enhancement in halophytes (Yeo & Flowers, 1980) and some crop plants (Milford, Cormack & Durrant, 1977). In most plants, however, salinity reduces both plant growth and the expansion of the leaf surface. The latter effect of salt may be associated with both a reduction in the number of leaves produced and the size of individual leaves (Meiri & Poljakoff-Mayber, 1970).

When sugar beet were treated with NaCl, the expansion of individual leaves stopped immediately; after one day of adjustment leaf expansion resumed but at a slower rate (Fig. 7.11). The reduction in leaf expansion rate with salinity resembles water stress in that it may be associated with decreased turgor pressure. In a factorial experiment involving salt and water stress Sepaskhan & Boersma (1979) found leaf elongation rates of wheat seedlings to be linearly related to turgor pressure; in terms of equation (7.3), the cell extensibility decreased and $Y$, the yield stress, increased with increasing salinity.

### *Adjustment to stress*

Leaves are able to grow despite the fact that very low water potentials may develop in response to water deficits or increased

Fig. 7.10. Leaf elongation rates as a function of leaf turgor pressure of soybeans grown in three different environments. (After Bunce, 1977.)

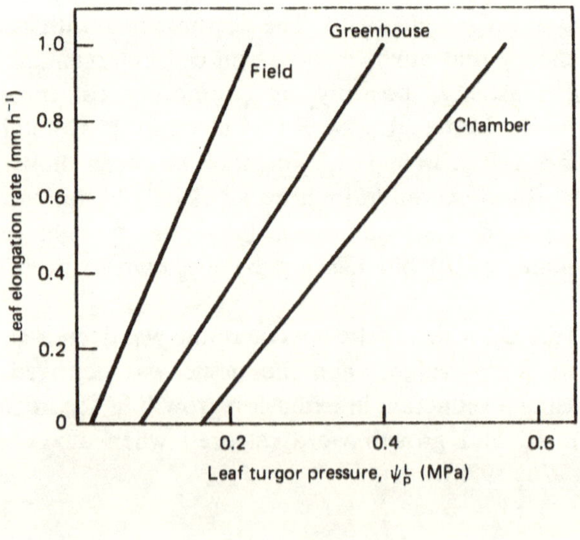

salinity. To understand how this occurs, it is necessary to consider the different components which together determine leaf water potential, $\psi_w{}^L$, i.e.

$$\psi_w{}^L = \psi_p{}^L + \psi_s{}^L + \psi_m{}^L \tag{7.6}$$

where $\psi_p{}^L$, $\psi_s{}^L$ and $\psi_m{}^L$ are the turgor, solute and matric potentials, respectively. $\psi_w{}^L$ is taken as zero unless the leaf is appreciably desiccated (Boyer, 1967). Thus, if $\psi_w{}^L$ declines as a result of water or salt stress, $\psi_p{}^L$ may be maintained if $\psi_s{}^L$ decreases by an equal or greater amount. Such an adjustment permits leaves to develop sufficient mechanical pressure for cell enlargement and is referred to as osmoregulation. It has been observed in soybean hypocotyls (Meyer & Boyer, 1972) and in leaves (Biscoe, 1972). Osmoregulation may result from internal generalization of organic solutes, or by the absorption of salts (Ramati, Lipschitz & Waisel, 1979). Halophytic plants adjust $\psi_s{}^L$ mainly by salt absorption (Greenway & Munns, 1980).

Fig. 7.11. The time-course for the extension of a sugar beet leaf over a six-day period with a two-day episode of salinization with 100 mol m$^{-3}$ NaCl. (From L. J. Waldron & N. Terry, unpublished.)

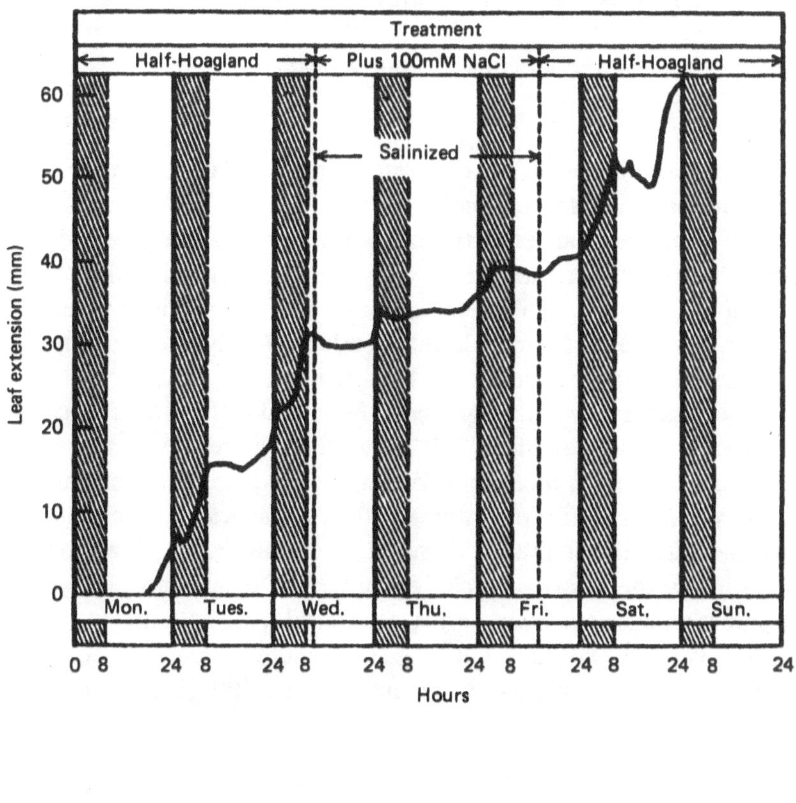

When water is resupplied to a stressed plant, recovery may be rapid and complete provided the stress is mild (Acevedo *et al.*, 1971), but not if the desiccation is prolonged (Boyer, 1970). Plants may also recover rapidly from salt stress as shown by the immediate increase in leaf extension rate when sugar beets were transferred from a saline medium to half-Hoagland's culture solution (Fig. 7.11). The rapidity of this recovery suggests that the response was almost certainly physical, i.e. the rapid, sudden increase in leaf growth rate was due directly to the large increase in water potential at the roots which was immediately transmitted to the leaf.

### *Relation of leaf water potentials to environmental factors*

The water supply in the soil and the evaporative demand of the atmosphere are environmentally determined factors which strongly influence leaf water potentials. The plant must regulate its water relations in response to changes in these factors. Movement of water through the plant to the atmosphere occurs in response to differences in water potentials along the flow path. The flow is often compared to an electrical circuit in which the water flux, $J$(m s$^{-1}$), is governed by a transport equation analogous to Ohm's law:

$$J = \Delta\psi_w / r \qquad (7.7)$$

where $\Delta\psi_w$ is the potential difference and $r$ is the resistance to flow between points in the flow path.

The various resistances in the transpiration stream in this soil–plant–atmosphere circuit are not simple constants, but may vary with changing conditions (Rawlins, 1963). For example, stomata partially regulate the resistance at the leaf/atmosphere boundary. As the soil dries not only does the soil water potential fall, but the hydraulic resistance in the soil increases markedly. Although these facts make the situation complex for exact analysis, equation (7.7) shows that where flow resistances are high, potential differences must be large to maintain a given flux. As a result, when transpiration rates are large, leaf water potentials ($\psi_w{}^L$) must be much lower than soil water potentials ($\psi_w{}^S$). Therefore, $\psi_w{}^L$ varies over the diurnal cycle of potential evapotranspiration with minimum values usually occurring near midday. If potential evapotranspiration is high, $\psi_w{}^L$ may fall to values where stomatal closure occurs even when $\psi_w{}^S$ is high, as for example, in a culture solution.

*Cell division*

Cell division, as well as cell enlargement in leaves, is also sensitive to water stress. Terry, Waldron & Ulrich (1971) showed that cell division in leaves of young sugar beet plants was sensitive to relatively small decreases in $\psi_s$ of the root medium (Fig. 7.12). Subsequently, Meyer & Boyer (1972) showed that cell division in etiolated, intact soybean hypocotyls was as sensitive as cell elongation to low water potentials, while Kirkham, Gardner & Gerloff (1972) concluded that cell division in isolated radish cotyledons was more sensitive to turgor pressure than cell enlargement. On the other hand, Clough & Milthorpe (1975), found that while leaf cell expansion was rapidly reduced by a small water deficit, cell division was less sensitive, continuing at a reduced rate even after leaf expansion had ceased.

Salt stress may also affect cell division. Hayward & Long (1941) showed that cambial activity was inhibited by high concentrations of salts while Wignarajah, Jennings & Handley (1975) showed that cell division in bean leaves was curtailed earlier in salt-treated plants than in control plants.

Fig. 7.12. The relative rate of cell division of sugar beet as the solute potential of the culture solution was lowered ($\Delta\psi$) by additions of polyethylene glycol to half-strength Hoagland solution. (After Terry *et al.*, 1971.)

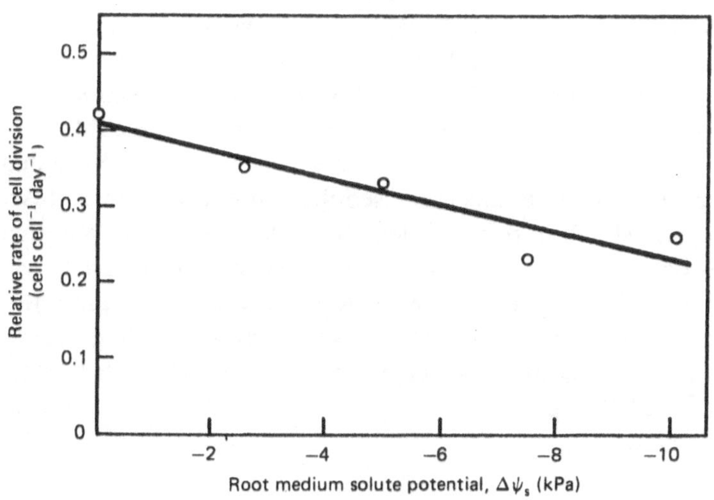

*Photosynthesis*

Both water and salt stress appear to have less effect on photosynthesis than on leaf enlargement, especially after osmotic adjustment occurs. Boyer (1970) found that in water-stressed corn, soybean and sunflower, enlargement was much more sensitive to water stress than photosynthesis or respiration (Fig. 7.9). Bunce (1978) found that although water stress from low humidity resulted in reduction in leaf expansion in soybean and cotton, it did not reduce photosynthesis even though leaf diffusive resistance was increased 1.5-fold; furthermore, soil drought reduced leaf expansion before photosynthesis was affected. Leach (1980) observed that the main effect of water stress on spring barley under field conditions was a 40% reduction in leaf area with no reduction in the rate of photosynthesis; leaf area was reduced due to the survival of fewer tillers and to premature senescence of leaves.

That salinity has relatively small effects on photosynthesis was shown by Nieman (1962) who concluded that the rate of photosynthesis was not a limiting factor in twelve species of salt-stunted crop plants (Table 7.1). Research in our laboratory also indicates that leaf expansion in sugar beet is very sensitive to salt stress, while photosynthesis is much less sensitive. Photosynthesis was relatively little affected by salt stress in the salt-tolerant Bermuda grass (Ackerson & Youngner, 1975) and *Spartina alterniflora* (Longstreth & Strain, 1977).

There is indirect evidence that photosynthesis does not limit growth in either water- or salt-stressed plants since there may be a build-up in the tissue levels of carbohydrates during either type of stress. Barlow, Boersma & Young (1976) found that when the water potential of the soil was decreased from −0.035 to −0.25 MPa, leaf elongation rates decreased 44% while soluble carbohydrate content increased 42% in leaves of corn (Fig. 7.13). Similarly, Gauch & Eaton (1942) showed that excess chloride or sulphate salts increased the concentration of sucrose, total sugars, and starch in barley. Nieman & Clark (1976) showed that salinity caused a build-up of sugar phosphates in mature, photosynthesizing leaves of corn. These investigators speculated that the transport of carbohydrates to the growing tissues may be limiting growth rather than photosynthesis; however, Nassery, Ogata, Nieman & Maas (1978) concluded that phloem transport did not limit growth of salt-stressed sesame or pepper.

Fig. 7.13. The time-course of soluble carbohydrate concentration of *Zea mays* shoots growing under water-stressed conditions. (After Barlow *et al.*, 1976.)

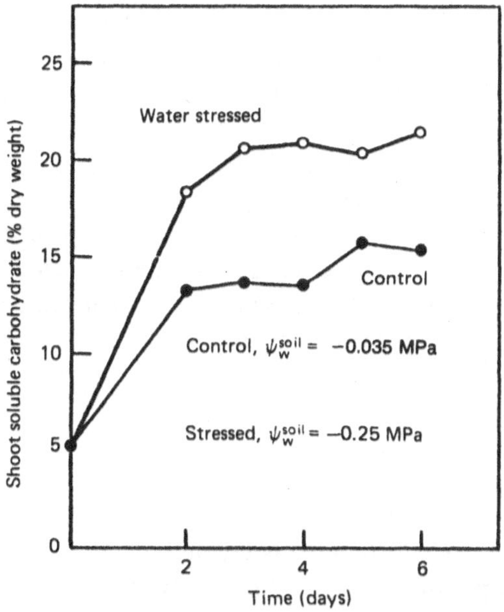


Table 7.1. *Influence of salinity on the photosynthetic rate of eleven species of crop plants*

| Species | Leaf disc photosynthetic $O_2$ production ($mg\ m^{-2}\ s^{-1}$) Control | $\Delta\psi_s$ medium[a] −0.4 MPa |
|---|---|---|
| Bean | 0.09 | 0.16 |
| Beet | 0.17 | 0.17 |
| Cabbage | 0.21 | 0.23 |
| Lettuce | 0.09 | 0.13 |
| Mustard | 0.18 | 0.16 |
| Onion | 0.13 | 0.12 |
| Pepper | 0.18 | 0.20 |
| Radish | 0.21 | 0.21 |
| Spinach | 0.24 | 0.23 |
| Tomato | 0.20 | 0.21 |
| Turnip | 0.22 | 0.22 |

[a]Medium solute potential was lowered by addition of NaCl. After Nieman (1962).

*Changes in leaf morphology and anatomy*

Salt stress may have pronounced effects on leaf structure. Hayward & Long (1941) showed that NaCl salinity increased leaf succulence and that NaCl and $Na_2SO_4$ salinity increased leaf thickness; this is due to an increased thickness of the spongy mesophyll, the palisade layer being thinner under salt treatment (Wignarajah *et al.*, 1975). On the other hand, increased mesophyll thickness in salt-stressed bean, cotton and *Atriplex* was due to an increase in length of palisade cells and to an increased number of spongy cell layers (Longstreth & Nobel, 1979). Water stress can also induce other morphological changes in leaves; O'Toole & Cruz (1980) found that the degree of leaf rolling in rice was linearly related to the water potential of the leaf.

*Does salinity reduce growth by inducing a water stress?*

Although much of the recent literature discounts the earlier idea that salinity retards growth by inducing a water stress, there is some evidence in support of this view (Maas & Nieman, 1978). Although osmoregulation appears to permit a sufficient gradient of water potential to be developed to transport water from the root medium into leaf cells, plant growth is still suppressed roughly in proportion to the decrease in water potential ($\psi_s$) of the growth medium (Bernstein & Hayward, 1958; Hoffman, Shalhevet & Meiri, 1980). Also salinity results in less growth suppression when the humidity around leaves is increased (Nieman & Paulsen, 1967; Hoffman & Rawlins, 1971). Furthermore, osmoregulation, while attenuating salt and water stress effects on growth, may not entirely offset them because growth may be retarded during the time required for adjustment. This may be particularly true over the daily cycle when there may be brief periods of high transpiration.

## Conclusions

The inescapable conclusion from this review is that environmental variables influence plant growth to a very large degree via their effects on leaf expansion. With the exception of light and $CO_2$ supply, environmental influences on photosynthesis appear in general to be less pronounced than those on leaf expansion. In the leaf expansion process, cell enlargement seems to be particularly sensitive to temperature and to water and salt stress. Because of the complex nature of the interaction of environmental variables, and because of the particular sensitivity of leaf extension to water stress, we believe that future studies of leaf growth should provide for continuous and simultaneous monitoring of leaf extension, leaf temperature and leaf water status over short as well as long time periods. Also, the role of osmoregulation in maintaining turgor pressures for leaf expansion is of crucial importance, especially with regard to changes during the diurnal cycle and in rapidly growing rather than mature leaf tissue.

## References

Acevedo, E., Hsiao, T. C. & Henderson, D. W. (1971). Immediate and subsequent growth response of maize leaves to changes in water status. *Plant Physiology*, **48**, 631–6.

Ackerson, R. C. & Youngner, U. B. (1975). Responses of bermudagrass to salinity. *Agronomy Journal*, **67**, 678–81.

Auld, B. A., Dennett, M. D. & Elston, J. (1978). The effect of temperature changes in the expansion of individual leaves of *Vicia faba* L. *Annals of Botany*, **42**, 877–88.

Bagnell, D. J. & Wolfe, J. A. (1978). Chilling sensitivity in plants: do the activation energies of growth processes show an abrupt change at a critical temperature. *Journal of Experimental Botany*, **29**, 1231–42.

Ballantine, J. E. & Forde, B. J. (1970). The effect of light intensity and temperature on plant growth and chloroplast ultrastructure in soybean. *American Journal of Botany*, **57**, 1150–9.

Barlow, E. W. T., Boersma, L. & Young, J. L. (1976). Root temperature and soil water potential effects on growth and soluble carbohydrate concentration of corn seedlings. *Crop Science*, **16**, 59–62.

Bernstein, L. & Hayward, H. E. (1958). Physiology of salt tolerance. *Annual Review of Plant Physiology*, **9**, 25–46.

Bhat, K. K. S., Nye, P. H. & Brereton, A. J. (1979). The possibility of predicting solute uptake and plant growth response from independently measured soil and plant characteristics. VI. The growth and uptake of rape in solutions of constant nitrate concentration. *Plant and Soil*, **53**, 137–67.

Biscoe, P. V. (1972). The diffusion resistance and water status of leaves of *Beta vulgaris*. *Journal of Experimental Botany*, **23**, 930–40.

Bishop, P. M. & Whittingham, C. P. (1968). The photosynthesis of tomato plants in a carbon dioxide enriched atmosphere. *Photosynthetica*, **2**, 31–8.

Boardman, N. K. (1977). Comparative photosynthesis of sun and shade plants. *Annual Review of Plant Physiology*, **28**, 355–77.

Bouma, D., Dowling, E. J. & Wahjoedi, H. (1979). Some effects of potassium and magnesium on the growth of subterranean clover (*Trifolium subterraneum*). *Annals of Botany*, **43**, 529–38.

Boyer, J. S. (1967). Matric potential of leaves. *Plant Physiology*, **42**, 213–17.

Boyer, J. S. (1968). Relationship of water potential to growth of leaves. *Plant Physiology*, **43**, 1056–62.

Boyer, J. S. (1970). Leaf enlargement and metabolic rates in corn, soybean, and sunflower at various leaf water potentials. *Plant Physiology*, **46**, 233–5.

Boyer, J. S. & Wu, G. (1978). Auxin increases the hydraulic conductivity of auxin-sensitive hypocotyl tissue. *Planta*, **139**, 277–37.

Brix, H. & Ebell, L. F. (1969). Effects of nitrogen fertilization on growth, leaf area, and photosynthesis rate in douglas-fir. *Forest Science*, **15**, 189–96.

Bunce, J. A. (1977). Leaf elongation in relation to leaf water potential in soybean. *Journal of Experimental Botany*, **28**, 156–61.

Bunce, J. A. (1978). Effects of water stress on leaf expansion, net photosynthesis, and vegetative growth of soybeans and cotton. *Canadian Journal of Botany*, **56**, 1492–8.

Charles-Edwards, D. A., Charles-Edwards J. & Sant, F. I. (1974). Leaf photosynthetic activity in six temperate grass varieties grown in contrasting light and temperature environments, *Journal of Experimental Botany*, **25**, 715–24.

Cleland, R. (1967). A dual role of turgor pressure in auxin-induced cell elongation in *Avena* coleoptiles. *Planta*, **77**, 182–91.

Cleland, R. (1973). Auxin-induced hydrogen ion excretion from *Avena* coleoptiles. *Proceedings of the National Academy of Sciences, USA*, **70**, 3092–3.

Clough, B. F. & Milthorpe, F. L. (1975). Effect of water deficit on leaf development in tobacco. *Australian Journal of Plant Physiology*, **2**, 291–300.

Cooper, R. L. & Brun, W. A. (1967). Response of soybean to a carbon dioxide-enriched atmosphere. *Crop Science*, **7**, 455–7.

Crookston, R. K., Treharne, K. J., Ludford, P. & Ozbun, J. L. (1975). Response of beans to shading. *Crop Science*, **15**, 412–16.

Dennett, M. D., Elston, J. & Milford, J. R. (1979). The effect of temperature on the growth of individual leaves of *Vicia faba* in the field. *Annals of Botany*, **43**, 197–208.

Doley, D. (1978). Effects of shade on gas exchange and growth in seedlings of *Eucalyptus grandis* Hill ex Maiden. *Australian Journal of Plant Physiology*, **5**, 723–38.

Duncan, W. G. (1975). Maize. In *Crop Physiology: Some Case Histories*, 1st edn, ed. L. T. Evans. Cambridge University Press.

Fellows, R. J. & Geiger, D. R. (1974). Structural and physiological changes in sugar beet leaves during sink to source conversion. *Plant Physiology*, **54**, 877–85.

Fischer, R. A. & Hsiao, T. C. (1968). Stomatal opening in isolated epidermal strips of *Vicia faba*. II. Responses to KCl concentrations and the role of potassium absorption. *Plant Physiology*, **43**, 1953–8.

Ford, M. A. & Thorne, G. N. (1967). Effect of $CO_2$ concentration on growth of sugar-beet, barley, kale, and maize. *Annals of Botany*, **31**, 629–44.

Forde, B. J., Whitehead, H. C. M. & Rowley, J. A. (1975). Effect of light intensity and temperature on photosynthetic rate, leaf starch content and ultrastructure of *Paspalum dilatatum*. *Australian Journal of Plant Physiology*, **2**, 185–95.

Fukai, S. & Silsbury, J. H. (1976). Response of subterranean clover communities to temperature. I. Dry matter production and plant morphogenesis. *Australian Journal of Plant Physiology*, **3**, 527–43.

Gallagher, J. N. & Biscoe, P. V. (1979). Field studies of cereal leaf growth. III. Barley leaf extension in relation to temperature, irradiance, and water potential. *Journal of Experimental Botany*, **30**, 645–55.

Gauch, H. G. & Eaton, F. M. (1942). Effect of saline substrate on hourly levels of carbohydrates and inorganic constituents of barley plants. *Plant Physiology*, **17**, 347–65.

Gifford, R. M. (1977). Growth pattern, carbon dioxide exchange and dry weight distribution on wheat growing under differing photosynthetic environments. *Australian Journal of Plant Physiology*, **4**, 99–110.

Graham, R. D. & Ulrich, A. (1974). Retranslocation of potassium in *Beta vulgaris* L. under conditions of low sodium supply. *Australian Journal of Plant Physiology*, **1**, 387–96.

Green, P. B., Erickson, R. O. & Buggy, J. (1971). Metabolic and physical control of cell elongation rate. *In vivo* studies in *Nitella*. *Plant Physiology*, **47**, 423–30.

Greenway, H. & Munns, R. (1980). Mechanism of salt tolerance in nonhalophytes. *Annual Review of Plant Physiology*, **31**, 149–90.

Greenway, H. & Pitman, M. G. (1965). Potassium retranslocation in seedlings of *Hordeum vulgare*. *Australian Journal of Biological Science*, **18**, 235–47.

Grenetz, P. S. & List, A. (1973). A model for predicting growth responses in plants to changes in external water potential: *Zea mays* primary roots. *Journal of Theoretical Biology*, **39**, 29–45.

Hayward, H. E. & Long, E. M. (1941). Anatomical and physiological responses of the tomato to varying concentrations of sodium chloride, sodium sulphate, and nutrient solutions. *Botanical Gazette*, **102**, 437–62.

Ho, L. C. (1977). Effects of $CO_2$ enrichment on the rates of photosynthesis and translocation of tomato leaves. *Annals of Applied Biology*, **87**, 191–200.

Hoffman, G. J. & Rawlins, S. L. (1971). Growth and water potential of root crops as influenced by salinity and relative humidity. *Agronomy Journal*, **63**, 877–80.

Hoffman, G. J., Shalhevet, J. & Meiri, A. (1980). Leaf age and salinity influence water relations of pepper leaves. *Physiologia Plantarum*, **48**, 463–9.

Hsiao, T. C. (1973). Plant response to water stress. *Annual Review of Plant Physiology*, **24**, 519–70.

Hsiao, T. C., Acevedo, E., Fereres, E. & Henderson, D. W. (1976). Stress metabolism. Water stress, growth, and osmotic adjustment. *Philosophical Transactions of the Royal Society of London*, **B273**, 479–500.

Hurd, R. G. (1968). Effects of $CO_2$-enrichment on the growth of young tomato plants in low light. *Annals of Botany*, **32**, 531–42.

Hurd, R. G. & Thornley, J. H. M. (1974). An analysis of the growth of young tomato plants in water culture at different light integrals and $CO_2$ concentrations. I. Physiological aspects. *Annals of Botany*, **38**, 375–88.

Ingestad, T. (1972). Mineral nutrient requirements of cucumber seedlings. *Plant Physiology*, **52**, 332–8.

Ingestad, T. & Lund, A. B. (1979). Nitrogen stress in birch seedlings. I. Growth technique and growth. *Physiologia Plantarum*, **45**, 137–48.

Kemp, D. R. (1981). The growth rate of wheat leaves in relation to the extension zone sugar concentration manipulated by shading. *Journal of Experimental Botany*, **32**, 141–50.

Kemp, D. R. & Blacklow, W. M. (1980). Diurnal extension rates of wheat leaves in relation to temperature and carbohydrate concentrations of the extension zone. *Journal of Experimental Botany*, **31**, 821–8.

Ketring, D. L. (1979). Light effects on development of an indeterminate plant. *Plant Physiology*, **64**, 665–7.

Kirkham, M. B., Gardner, W. R. & Gerloff, G. C. (1972). Regulation of cell division and cell enlargement by turgor pressure. *Plant Physiology*, **49**, 961–2.

Kramer, P. J. (1969). *Plant and Soil Water Relationships: A Modern Synthesis*, 1st ed, pp. 195–200. New York: McGraw-Hill.

Kramer, P. J. (1981). Carbon dioxide concentration, photosynthesis and dry matter production. *Bioscience*, **31**, 29–33.

Leach, J. E. (1980). Photosynthesis and growth of spring barley: some effects of drought. *Journal of Agricultural Science, Cambridge*, **94**, 623–35.

Longstreth, D. J. & Nobel, P. S. (1979). Salinity effects on leaf anatomy. Consequences for photosynthesis. *Plant Physiology*, **63**, 700–3.

Longstreth, D. J. & Strain, B. R. (1977). Effects of salinity and illumination on photosynthesis and water balance of *Spartina alterniflora* Loisel. *Oecologia*, **31**, 191–9.

Louwerse, W. & Zweerde, W. V. D. (1977). Photosynthesis transport and leaf morphology of *Phaseolus vulgaris* and *Zea mays* grown at different irradiances in artificial and sun light. *Photosynthetica*, **11**, 11–21.

Ludlow, M. M. &. Wilson, G. L. (1971). Photosynthesis of tropical pasture plants. II. Temperature and illuminance history. *Australian Journal of Biological Science*, **24**, 1065–75.

Maas, E. V. & Nieman, R. H. (1978). Physiology of plant tolerance to salinity. In *Crop Tolerance to Suboptimal Land Conditions*, vol. 32, ed. G. A. Jung, pp. 277–99. *American Society of Agronomy. Special Publication.*

Madsen, E. (1973). Effect of $CO_2$-concentration on the morphological, histological and cytological changes in tomato plants. *Acta Agriculturae Scandinavica*, **23**, 241–6.

Marré, E., Colombo, R., Lado, P. & Rasi-Caldogno, F. (1974). Correlation between proton extrusion and stimulation of cell enlargement. Effect of fusicoccin and of cytokinins on leaf fragments and isolated cotyledons. *Plant Science Letters*, **3**, 139–50.

Marré, E., Lado, P., Rasi-Caldogno, F. & Colombo, R. (1973). Correlation between cell enlargement in pea internode segments and decrease in the pH of the medium of incubation. I. Effects of fusicoccin, natural and synthetic auxin, and mannitol. *Plant Science Letters*, **1**, 179–84.

Marschner, H. & Possingham, J. V. (1975). Effect of $K^+$ and $Na^+$ on growth of leaf discs of sugar beet and spinach. *Zeitschrift für Pflanzenphysiologie*, **75**, 6–16.

Mauney, J. R., Fry, K. E. & Guinn, G. (1978). Relationship of photosynthetic rate to growth and fruiting of cotton, soybean, sorghum, and sunflower. *Crop Science*, **18**, 259–63.

Meiri, A. & Poljakoff-Mayber, A. (1970). Effects of various salinity regimes on growth, leaf expansion, and transpiration rate of bean plants. *Soil Science*, **190**, 26–34.

Mengel, K. & Kirkby, E. A. (1980). Potassium in crop production. *Advances in Agronomy*, **33**, 59–110.

Meyer, R. F. & Boyer, J. S. (1972). Sensitivity of cell division and cell elongation to low water potentials in soybean hypocotyls. *Planta*, **198**, 77–87.

Milford, G. F. J., Cormack, W. F. & Durrant, M. J. (1977). Effects of sodium chloride on water status and growth of sugar beet. *Journal of Experimental Botany*, **28**, 1380–8.

Milthorpe, F. L. & Newton, P. (1963). Studies of the expansion of the leaf surface. II. The influence of radiation on cell division and leaf expansion. *Journal of Experimental Botany*, **14**, 483–95.

Nassery, H., Ogata, G., Nieman, R. H. & Maas, E. V. (1978). Growth, phosphate pools, and phosphate mobilization of salt-stressed sesame and pepper. *Plant Physiology*, **62**, 299–31.

Neales, T. F. & Nicholls, A. O. (1978). Growth responses of young wheat plants to a range of ambient $CO_2$ levels. *Australian Journal of Plant Physiology*, **5**, 45–59.

Nieman, R. H. (1962). Some effects of sodium chloride on growth, photosynthesis and respiration of twelve crop plants. *Botanical Gazette*, **123**, 279–85.

Nieman, R. H. & Clark, R. A. (1976). Interactive effects of salinity and phosphorus nutrition on the concentration of phosphate and phosphate esters in mature photosynthesizing corn leaves. *Plant Physiology*, **57**, 157–61.

Nieman, R. H. & Paulsen, L. L. (1967). Interactive effects of salinity and atmospheric humidity on the growth of bean and cotton plants. *Botanical Gazette*, **128**, 69–73.

Nobel, P. S. (1976). Photosynthetic rates of sun versus shade leaves of *Hyptis emoryi* Torr. *Plant Physiology*, **58**, 218–23.

Nobel, P. S., Zaragoza, L. J. & Smith, W. K. (1975). Relation between mesophyll surface area, photosynthetic rate and illumination level during development for leaves of *Plectranthus parviflorus*. *Plant Physiology*, **55**, 1067–70.

O'Toole, J. C. & Cruz, R. T. (1980). Response of leaf water potential, stomatal resistance, and leaf rolling to water stress. *Plant Physiology*, **65**, 428–32.

Peacock, J. M. (1975). Temperature and leaf growth in *Lolium perenne*. II. The site of temperature perception. *Journal of Applied Ecology*, **12**, 115–23.

Peet, M. M., Ozbun, J. L., & Wallace, D. H. (1977). Physiological and anatomical effects of growth temperature on *Phaseolus vulgaris* L. cultivars. *Journal of Experimental Botany*, **28**, 57–69.

Potter, J. R. & Jones, J. W. (1977). Leaf area partitioning as an important factor in growth. *Plant Physiology*, **59**, 10–14.

Ramati, A., Lipschitz, N. & Waisel, Y. (1979). Osmotic adaptation in *Panicum repens*. Differences between organ, cellular, and subcellular levels. *Physiologia Plantarum*, **45**, 325–31.

Rawlins, S. L. (1963). Resistance to water flow in the transpiration stream. *Connecticut Agricultural Experimental Station Bulletin*, **664**, 69–85.

Ray, P. M., Green, P. B. & Cleland, R. (1972). Role of turgor in plant cell growth. *Nature*, **239**, 163–4.

Rayle, D. L. & Cleland, R. (1970). Enhancement of wall loosening and elongation by acid solution. *Plant Physiology*, **46**, 250–3.

Rayle, D. L. & Cleland, R. E. (1980). Evidence that auxin-induced growth of soybean hypocotyls involved proton excretion. *Plant Physiology*, **66**, 433–7.

Robson, J. J. & Deacon, M. J. (1978). Nitrogen deficiency in small closed communities of S24 ryegrass. II. Changes in the weight and chemical composition of single leaves during their growth and death. *Annals of Botany*, **42**, 1199–213.

Sepaskhan, A. R. & Boersma, L. (1979). Elongation of wheat leaves exposed to several levels of matric potential and NaCl-induced osmotic potential of soil water. *Agronomy Journal*, **71**, 848–52.

Terry, N. (1968). Developmental physiology of sugar beet. I. The influence of light and temperature on growth. *Journal of Experimental Botany*, **19**, 795–811.

Terry, N. (1970). Developmental physiology of sugar beet. II. Effects of temperature and nitrogen supply on the growth, soluble carbohydrate content and nitrogen content of leaves and roots. *Journal of Experimental Botany*, **21**, 477–96.

Terry, N., Waldron, L. J. & Ulrich, A. (1971). Effects of moisture stress on the multiplication and expansion of cells in leaves of sugar beet. *Planta*, **97**, 281–9.

van Volkenburgh, E. & Cleland, R. E. (1980). Proton excretion and cell expansion in bean leaves. *Planta*, **148**, 273–8.

Watts, W. R. (1972). Leaf extension in *Zea mays*. II. Leaf extension in response to independent variation of the temperature of the apical meristem, of the air around the leaves, and of the root-zone. *Journal of Experimental Botany*, **23**, 713–21.

Watts, W. R. (1974). Leaf extension in *Zea mays*. III. Field measurements of leaf extension in response to temperature and leaf water potential. *Journal of Experimental Botany*, **25**, 1085–96.

Weidner, M. & Ziemens, C. (1975). Preadaptation of protein synthesis in wheat seedlings to high temperature. *Plant Physiology*, **56**, 590–4.

Wignarajah, K., Jennings, D. H. & Handley, J. F. (1975). The effect of salinity on growth of *Phaseolus vulgaris* L. I. Anatomical changes in the first trifoliate leaf. *Annals of Botany*, **39**, 1029–38.

Wilson, D. & Cooper, J. P. (1969). Effect of light intensity during growth on leaf anatomy and subsequent light-saturated photosynthesis among contrasting *Lolium* genotypes. *New Phytologist*, **68**, 1125–35.

Yeo, A. R. & Flowers, T. J. (1980). Salt tolerance in the halophyte *Suaeda maritima* L. Dum: Evaluation of the effect of salinity upon growth. *Journal of Experimental Botany*, **31**, 1171–83.

# 8

## Hormonal influences on leaf growth

P. B. GOODWIN AND M. G. ERWEE

### Introduction

Animal hormones are chemical messengers, produced at low concentrations in one region of the organism (either a separate organ in the body, or a separate organelle in the cell) and moving to and producing an effect in another region. Insulin and cyclic adenosine monophosphate are examples of inter-organ and inter-organelle hormones respectively. Plant hormones may be defined as compounds which show analogous behaviour in plants, such as auxins from leaves, inducing vascular differentiation in the stem (Steeves & Sussex, 1972) and lateral root production (McDavid, Sagar & Marshall, 1972). It has not always been possible to demonstrate plant hormone movement; for example, in the ethylene induction of ripening of climacteric fruit. However, even here short distance intra- or inter-cellular movement may occur. At present, demonstration of hormone movement cannot be regarded as a prerequisite for the demonstration of hormonal control.

How then do we decide whether a particular hormone influences a process such as leaf growth? There are three avenues of approach. In increasing order of technical difficulty, but of scientific rigour, they are the pharmacological, the correlative and the molecular approach.

The pharmacological approach depends on the availability of pure hormones and of specific antagonists. High concentrations of the hormone, or antagonist, are applied to the tissue on the general assumption that if a clear cut response is seen, then the hormone may have something to do with the response in the normal situation. The response to applied hormones may also have practical utility, as for example with the auxin-promoted rooting of cuttings, or the ethylene-promoted flowering of bromeliads. However, gibberellic acid, the most commonly applied gibberellin, is known to be native to only a handful of plant species (Graebe & Ropers, 1978). Growth retardants, such as succinic acid-2,2-dimethylhydrazide (SADH), have a broad spectrum of effects, and are not specific gibberellin antagonists (see Goodwin, 1978). On the other hand, useful information has been obtained by the use of auxin transport inhibitors, and of ethylene antagonists such as silver ions, and the ethylene synthesis inhibitor aminoethoxy vinylglycine (Lieberman, 1979). It must be borne in mind that responses seen may be abnormal effects due to high dosage, and that lack of response may be due to lack of penetration, or rapid inactivation of the applied hormone, or genetically determined insensitivity (Gale, 1979).

In the correlative approach, because attempts are made to measure the concentration of hormone in the responding organ, some of these uncertainties are resolved. The correlation between the response and the endogenous hormone level is followed during normal development, or in response to a variety of external stimuli (light, daylength, temperature, nutrition, applied hormones and antagonists), or in response to variation in genotype. There are a number of restrictions to this approach. Firstly, specific, sensitive and accurate methods for physical assay are commonly available only for ethylene, abscisic acid, and possibly indoleacetic acid. Advanced physical methods using preparative high performance liquid chromatography leading to gas-chromatography combined with mass spectrometry, with internal deuterated standards, suitable for all the major phytohormones, are in use in only a few laboratories (see Brenner, 1979). With gibberellins and cytokinins most laboratories must still rely on bioassays, or less specific physical assays. There are over 50 identified gibberellins, so that bioassay results on impure samples are of limited value. The statistical methods used for estimating significance in hormone bioassays are very often faulty, relying on replications at the assay rather than at the original plant level. Differences of less than three to four fold are unlikely to be meaningful.

A further complication is the possibility that much of the hormone in the organ is inactive – for example, in unresponsive tissues, in vacuoles, or simply in molecular forms which are inactive. Thus abscisic acid in unstressed leaves is essentially confined to the chloroplasts (Loveys, 1977). Other problems are that the amounts of not just one, but of a number of hormones are likely to change in most processes (Letham, Higgins, Goodwin & Jacobsen, 1978). Yet another possibility is that it is the rate of turnover of the specific hormone in the tissue, and not its concentration, which limits the rate of tissue response.

The premise behind the molecular approach is that the visible growth response is well down the reaction path from the initial hypothetical hormone/receptor event, and that if we could define the steps in this path we would know not only that the hormone influenced the process, but exactly how it did so. Good progress is being made in this direction with a number of animal hormones (for example Le Meur *et al.*, 1981) but in plants research is limited by a less-well developed technology for the study of phytohormones. This approach has as yet made little contribution to our understanding of leaf growth.

It should also be noted that undoubtedly there exist as yet undiscovered major classes of hormones. The recently described brassinosteroids may represent such a group (Maugh, 1981). Bearing these problems in mind, what do we know of the influence of hormones on leaf growth?

Table 8.1 summarises some of the unequivocal identifications of phytohormones in leaves based on physical techniques, where possible using gas-liquid chromatography combined with mass spectrometry (GC-MS).

## The influence of auxins on leaf growth

Auxins were the first phytohormones to be extensively studied in relation to leaf development. Avery (1935) showed that the diffusible auxin yield of tobacco leaves is at a peak in young expanding leaves, that the auxin appears to be produced over the whole leaf, and that it is exported in a polar fashion through the main veins. Highest activities were found at the leaf base. The general finding that the young expanding leaves are the richest source of auxin has been supported by a number of such studies (see Humphries & Wheeler, 1963). However, these early studies used unreliable techniques for auxin extraction, identification and quantification. In a more rigorous study of tobacco leaf development, Wightman (1977, Fig. 8.1) studied three leaf auxins, indole-3-acetic acid (IAA), phenylacetic acid and indole-3-propionic acid. The concentration of each auxin is at its peak 48 hours before the time of most rapid leaf elongation. Allen & Baker (1980), using GC-MS, found a pattern of IAA levels in leaves of *Ricinus communis* in agreement with the same concept.

The evidence for IAA synthesis by young expanding leaves is as

Table 8.1. *Phytohormones isolated from various leaf sources and identified by GC-MS*

| Species | Tissue | Phytohormone | Concentration ($ng\ g^{-1}$ fresh weight) | Reference |
|---|---|---|---|---|
| *Populus × robusta* | Mature leaves | 6-(*O*-hydroxybenzylamino)-9-β-D-ribofuronosylpurine | 100 | Horgan, Hewett, Horgan, Purse & Wareing (1975) |
| *Populus alba* | Mature leaves | *O*-β-D-glycopyranosyl-9-β-*O*-ribofuronosyl-dihydrozeatin<br>*O*-β-D-glucopyronosylzeatin | — | Letham, Parker, Duke, Summons & MacLeod (1976) |
| *Phaseolus vulgaris* | Mature leaves | dihydrozeatin-*O*-β-D-glucoside | — | Wang, Thompson & Horgan (1977) |
| *Picea sitchensis* | Young needles | zeatin riboside | 9 | Lorenzi, Horgan & Wareing (1975) |
| *Phaseolus vulgaris* | Primary leaf<br>First trifoliate leaf | indole-3-acetic acid<br>indole-3-acetic acid | 2.0<br>8.3 | McDougall & Hillman (1980) |
| *Ricinus communis* | Apical leaf | indole-3-acetic acid | 38.0 | Allen & Baker (1980) |
| *Picea sitchensis* | Needles | $GA_9$-glucosyl ester | 16 | Lorenzi, Horgan & Heald (1976) |
| *Picea sitchensis* | Young needles | iso-$GA_9$ | — | Lorenzi, Saunders, Heald & Horgan (1977) |
| *Ricinus communis* | Young leaf<br>Mature leaf | abscisic acid<br>abscisic acid | 400<br>60 | Zeevaart (1977) |
| *Lolium temulentum* | Mature leaves | abscisic acid | 1.2[a] | King, Evans & Firn (1977) |
| *Nicotiana tabacum* | Expanding leaf<br>Mature leaf | ethylene<br>ethylene | 470[b]<br>170[b] | Aharoni, Lieberman & Sisler (1979) |
| *Euphorbia lathyrus* | Leaves | ethylene | 190[b] | Sivakumaran & Hall (1978a) |

[a] Expressed on a dry weight basis. [b] Internal concentration expressed as $nl\ l^{-1}$.

follows: they are rich sources of extractable and diffusible IAA, they can be removed and replaced by IAA in a number of experimental systems, and they readily convert the auxin precursor tryptophan to IAA (see Schneider & Wightman, 1978 for a detailed discussion). Despite the fact that young expanding leaves are among the major sources of auxin in the plant (the others being young flowers, young fruit, and possibly the cambium), relatively little is known about factors influencing leaf auxin levels or export rate.

However, information is available for shoots. In species in which growth is promoted by long days the amount of extractable auxin may or may not be raised. Shoot tips from light-grown plants are more effective auxin exporters (into agar blocks) than tips from dark-grown plants. Gibberellin treatment increases the amount of diffusible and extractable auxin in stem apices. Ethylene treatment reduces the rate of auxin

Fig. 8.1. Changes of endogenous levels of indole-3-acetic acid (IAA), phenylacetic acid (PAA) and indole-3-propionic acid (IPA) in tobacco leaves at different stages of leaf development (after Wightman, 1977).

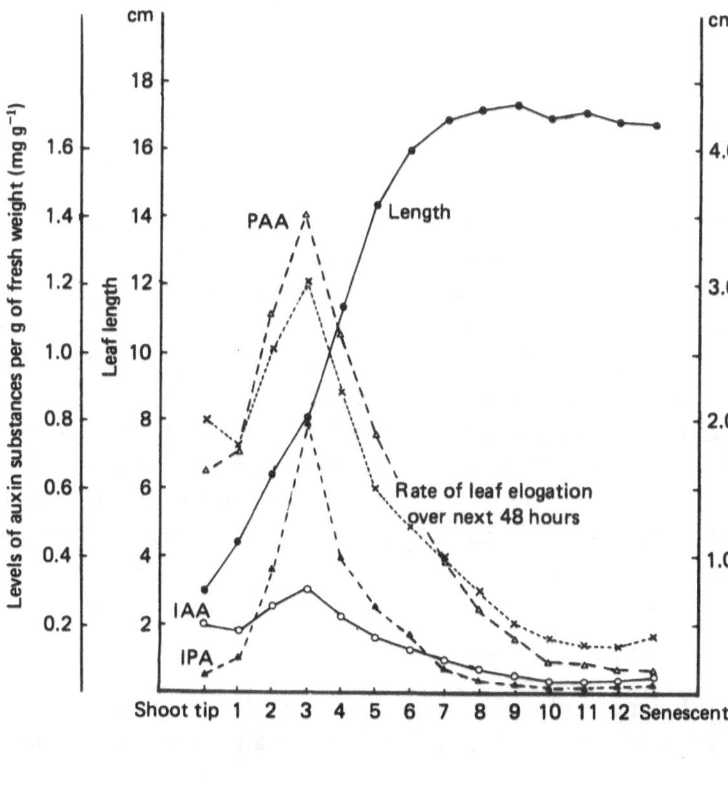

synthesis, reduces the amount of auxin diffused or extracted from apices, reduces the rate of auxin transport and increases the rate of IAA conjugation (see Goodwin, 1978). Diffusible and extractable IAA has been rigorously identified in *Phaseolus vulgaris* shoot tips which included terminal leaflets up to 2 cm long (White, Medlow, Hillman & Wilkins, 1975). McDougall & Hillman (1980), using GC-MS, found more IAA in the apex than in the unfolding leaves of *P. vulgaris*. Very likely these various measurements on shoot tips reflect changes in the auxin in the young leaves of the apex.

If auxin is a limiting factor for leaf expansion one might expect a strong response (either promotion or inhibition) to applied auxins. In at least some species application of 2,4-D promotes overall leaf expansion (Miller, Mikkelsen & Huffaker, 1962). A more common response is leaf blade hyponasty, due to expansion of cells on the abaxial side of the leaf, particularly in the veins. This is possibly due to the action of geotropic adaxial-abaxial auxin transport (Hayes, 1978). Went & Thimann (1937) summarise earlier work suggesting that auxins are vein growth factors, but there are numerous reports that auxins at high concentrations inhibit the development of the mesophyll without affecting vein extension (Gifford, 1953).

Growth of lateral buds is inhibited by applied auxins (for a review see Phillips, 1975). However, the IAA content in lateral buds increases within 24 hours of their release from apical dominance (Hillman, Math & Medlow, 1977). The increase in auxin level is probably a consequence, rather than a cause of bud outgrowth, since bud growth begins within 30 minutes of decapitation. The most common interpretation of this paradox is that the inhibition of lateral buds is an indirect effect of auxin from the apical leaves, but direct effects cannot be ruled out.

If we accept that auxin is a vein-growth factor produced by the mesophyll – a hypothesis for which there is little solid evidence – then growth of the leaf as a whole will be self sufficient for auxin, as long as mesophyll function is unimpaired. If its function is impaired, growth of the leaf as a whole will stop. Since in many chimeras with variegated leaves leaf form and size appear unaltered (Stewart, 1978), it would appear that local photosynthesis is not essential for mesophyll function in leaf development.

## The influence of ethylene

Ethylene and auxin are physiologically coupled plant hormones. Plant tissues which are rich in auxin are generally also strong producers of ethylene. Treatment with auxin at a high enough concentration induces ethylene production. Thus part of the response to auxin, especially at high doses, is due to auxin-induced ethylene production, as with auxin-induced flowering in bromeliads, and the inhibition of stem segment elongation at high auxin concentrations. In contrast, ethylene does not stimulate auxin production. Rather, it inhibits auxin synthesis and auxin transport, and promotes auxin metabolism (Osborne, 1978). Thus part of the response to ethylene is due to ethylene inhibition of auxin activity.

Plant responses to ethylene were first reported by Neljubow in 1901 (see Osborne, 1978). A common response is leaf epinasty – the opposite of the response of leaves to auxin. Ethylene is also a very effective inhibitor of leaf expansion, concentrations as low as $0.025\,\mathrm{nl\,l^{-1}}$ cause inhibition of leaf expansion in *Cucumis* (Abeles, 1973). The inhibition involves a general reduction in leaf dimensions, rather than just a reduction in mesophyll tissue. That is to say, it is distinct from the auxin inhibition of leaf growth. The inhibition of blade expansion in *Poa pratensis* and *Avena sativa* is due to a reduction in cell number (van Andel & Verkerke, 1978). Exposure of peanut or sunflower to ethylene inhibits photosynthesis, but the inhibition of leaf expansion in pea and *P. vulgaris* by ethylene is not associated with reduced photosynthesis (Kays & Pallas, 1980).

There is a diurnal cycle in ethylene production by *Vicia faba* leaves with two peaks, one at the beginning of the light period, and the other at the beginning of the dark period (El-Beltagy & Hall, 1974). The rate of ethylene production and the internal ethylene concentration is highest in rapidly expanding leaves of cotton (McAfee & Morgan, 1971), tobacco (Aharoni, Lieberman & Sisler, 1979, Fig. 8.2), olive (Lavee & Martin, 1981), and in expanding apple buds (Blanpied, 1972). However, changes with leaf expansion are small. The petiole produces ethylene more rapidly, and has a higher endogenous ethylene concentration than the leaf blade (McAfee & Morgan, 1971). Growing buds and unfolding leaves of light-grown pea seedlings are rich sources of ethylene, as are the buds and the plumular hook of etiolated peas (Burg, 1968; Goeschl, Pratt & Bonner, 1967). A red light pulse to etiolated peas causes a transient fall in ethylene production, and a transient increase in leaf expansion. The application of ethylene inhibits leaf expansion by inhibiting cell division (Apelbaum & Burg, 1972), the response being mediated by phytochrome (Goeschl *et al.*, 1967). Thus ethylene appears to control leaf expansion in etiolated peas. It may also be a major factor mediating the effect of light on leaf growth in other situations.

Drought, flooding or osmotic stress cause an increase in ethylene production, particularly by younger leaves. In tomato flooding results in leaf epinasty, due to ethylene production, as shown by measurement of internal ethylene, response to applied ethylene, and prevention by benzathiazole and high carbon dioxide levels which are antagonistic to ethylene action. The effect of flooding appears to be mediated by the roots, in that if they are removed leaves do not respond. Maintaining the plant roots in anaerobic solution culture also leads to ethylene production by the leaves (Jackson & Campbell, 1976).

The rapid loss of ethylene through the plant surface restricts the movement of ethylene in above ground parts (Zeroni, Jerie & Hall, 1977), and suggests that the factor produced by the roots is probably not ethylene. The rate of ethylene production in plant tissues appears to be limited by the level of endogenous 1-aminocyclopropane-1-carboxylic acid (ACC), the immediate precursor of ethylene (Apelbaum *et al.*, 1981), and this appears to be the factor supplied by flooded roots

Fig. 8.2. Internal concentration of ethylene in intact tobacco leaves harvested at different ages. Leaves are numbered according to their position on the stem: 1, apical expanding leaves; 2, older expanding leaves; 3, fully expanded mature leaves; 4, slight yellowing; 5, moderate yellowing; 6, advanced yellowing; 7, complete yellowing (after Aharoni *et al.*, 1979).

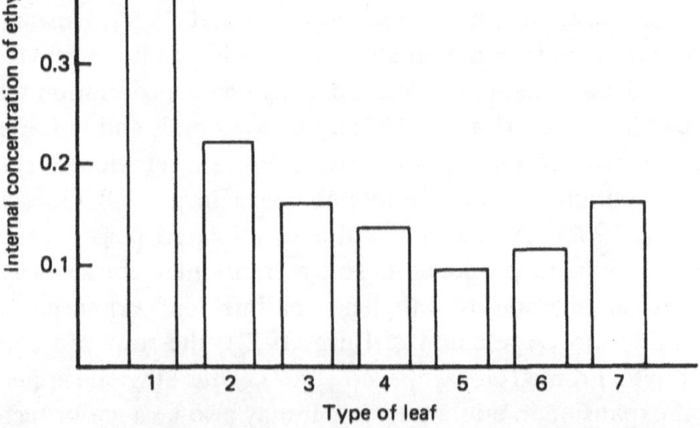

(Bradford & Yang, 1980). On the other hand, many of the adverse effects of flooding, including leaf epinasty, can be at least partially reversed by spraying the shoot with a mixture of cytokinin, gibberellin and ammonium nitrate. It is only the gibberellin component of this mixture which promotes leaf growth (Selman & Sandanam, 1972). Furthermore, isolated leaves can produce ethylene in response to osmotic stress (Hall, Kapuya, Sivakumaran & John, 1977). Although there is an increase in leaf auxin the increase in ethylene production precedes this. In fact, it is likely that the increase in auxin is caused by ethylene blocking its export from the leaf.

Flooding not only increases ethylene levels, but also causes a sharp fall in the leaf water potential, and a rise in abscisic acid. It is not known which of these factors causes the check to leaf growth, although it should be possible to find out.

Ethylene produced under stress could be involved in checking leaf expansion, or in inducing senescence and leaf abscission. Wright (1979) found that stress ethylene production was enhanced by benzyladenine and by light – factors which delay senescence. Aharoni (1978) found that detached leaves from species where no abscission occurs (cabbage, lettuce, sugarbeet, squash, tobacco) showed only a transient increase in ethylene production (probably due to a wound response). In contrast, in leaves from species showing leaf abscission (bean, castor bean, kohlrabi, orange, pepper) ethylene production continued up to high levels.

There is no clear relationship between ethylene and the outgrowth of lateral buds following decapitation. Applied ethylene may either reduce apical dominance (Skytt-Anderson, 1976), presumably by blocking auxin transport, or it may stop all bud growth, including that of the apical bud (Burg, 1973), presumably by its more direct effect on leaf growth.

Considering the ready availability of suitable instrumentation for rigorous quantification of ethylene and its possibly important role in the control of leaf expansion, it is surprising that the relationship between ethylene levels and leaf growth is not more clearly defined.

## Abscisic acid and leaf growth

Abscisic acid, like ethylene, is intimately involved in the response of plants to stress. The rapid accumulation of abscisic acid in water-stressed leaves was first reported by Wright & Hiron (1969) and since then much work has gone into studying this system (Wright, 1978).

In turgid leaves there appears to be a low rate of synthesis and degradation of abscisic acid, maintaining a constant, low level of abscisic acid. Water stress, perhaps operating through the fall in cell turgor (Pierce & Raschke, 1980), causes a rapid increase in the rate of abscisic acid synthesis. Once the abscisic acid content reaches a certain level, often ten or more times that in the unstressed leaf, the rate of synthesis falls, and this level of abscisic acid is maintained by a new balance between synthesis and metabolism. If the stress is removed the rate of abscisic acid synthesis falls, metabolism continues, and the abscisic acid content rapidly falls to that normal in turgid leaves (Zeevaart, 1980). The principal metabolites of abscisic acid under these conditions are phaseic acid and dihydrophaseic acid. At the same time as stress increases the abscisic acid level, it causes a fall in gibberellin and cytokinin content, and these recover on removal of the stress (Aharoni, Blumenfeld & Richmond, 1977). Water stresses inducing these phenomena include drought, flooding (that is, an anaerobic root system), and osmotic or salinity stress. Similar responses occur to mineral stress (Goldbach, Goldbach, Wagner & Michael, 1975; Boussiba & Richmond, 1976; Krauss, 1978) and to temperature stress (Daie & Campbell, 1981).

Water stress can cause a reduction in leaf size (Turner, 1979). In wheat leaves this is due largely to a reduction in cell number. Applied abscisic acid causes similar morphological changes (Quarrie & Jones, 1977; Fig. 8.3). Repeated foliar applications of abscisic acid reduce the final leaf size, but tiller number and leaf number are increased (Hall & McWha, 1981). The shoot apex and very young leaves are less prone to wilt than older leaves, but under stress they generally contain more abscisic acid than the rest of the plant, presumably due to translocation from the older leaves (Zeevaart, 1977; Sivakumaran & Hall, 1978b; Wright, 1978). Rather surprsingly, there appears to be no work in which measurement of changes in plant hormone levels in response to stress have been related to associated changes in leaf growth rate.

Even in plants growing in non-stress conditions, the young expanding leaves have higher abscisic acid levels than mature leaves (Goldbach *et al.*, 1975; Raschke & Zeevaart, 1976; Zeevaart, 1977; Fig. 8.4). Very high abscisic acid concentrations have been found in the vegetative shoot apex (King, Evans & Firn, 1977). This is the opposite of what might be expected if abscisic acid levels limit leaf growth. So is the fact that the *flacca* mutant of tomato (which has a low abscisic acid content), has normal leaf development except for a slower growth rate and a tendency to epinasty (Tal, 1966). The mutant has a high rate of ethylene production, which is restored to normal levels by the application of abscisic acid (Tal, Imber, Erez & Epstein, 1979).

A major line of work has evolved from the suggestion of Wareing and co-workers that abscisic acid, originally called dormin, produced by the leaves of woody plants is responsible for inducing vegetative dormancy, and that its removal during the chilling of winter allows bud burst in the spring. However, although application of high rates of abscisic acid will suppress bud growth (Perry & Hellmers, 1973; Saunders *et al.*, 1974; Robitaille & Carlson, 1976), no group with the exception of El-Antably, Wareing & Hillman (1967) has been able to produce true resting buds by this treatment. Nor has a satisfactory correlation been established

Fig. 8.3. Change of abscisic acid (ABA) content in *Xanthium* leaves at different ages expressed per unit fresh weight, and the change in dry weight of ten leaves of different ages (after Raschke & Zeevaart, 1976).

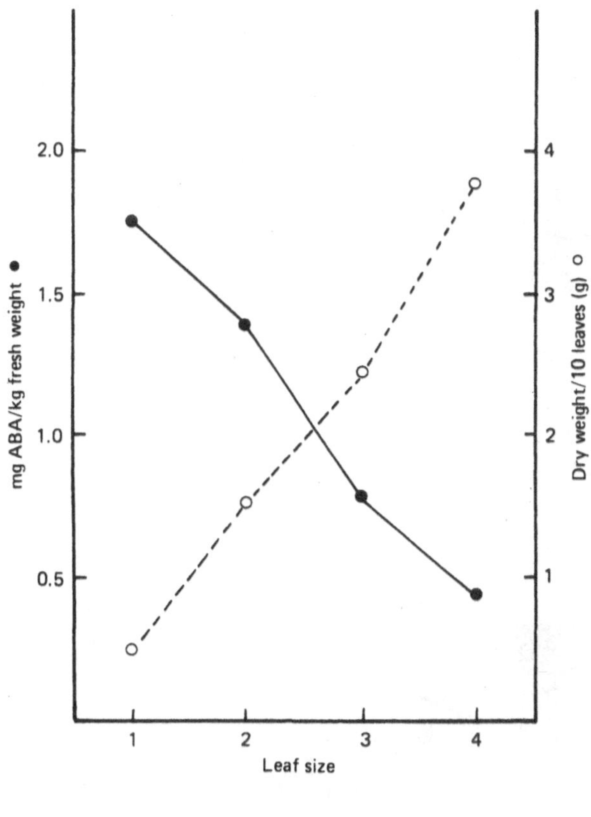

between abscisic acid level in the plant and bud dormancy. For example, Dumbroff, Cohen & Webb (1979) working on buds of *Acer saccharum*, found abscisic acid levels were high in the summer, low in the autumn (when the buds were most dormant), then rose to a peak in November–December associated with leaf fall, declining again before bud break.

Fig. 8.4. Comparison of leaf area responses to leaf area of sampled leaves in three wheat cultivars (MSL5, MSL7, T2L2). (*a*) Leaf area of leaves after injection of abscisic acid (50 μl containing 5 μg of abscisic acid) twice weekly for three weeks into the leaf sheaths at the base of the shoots. (*b*) Leaf area of leaves after exposure to three weeks of water stress. Shaded blocks represent the treated plants. Vertical bars represent LSD (P = 0.05) between controls and treated plants (after Quarrie & Jones, 1977).

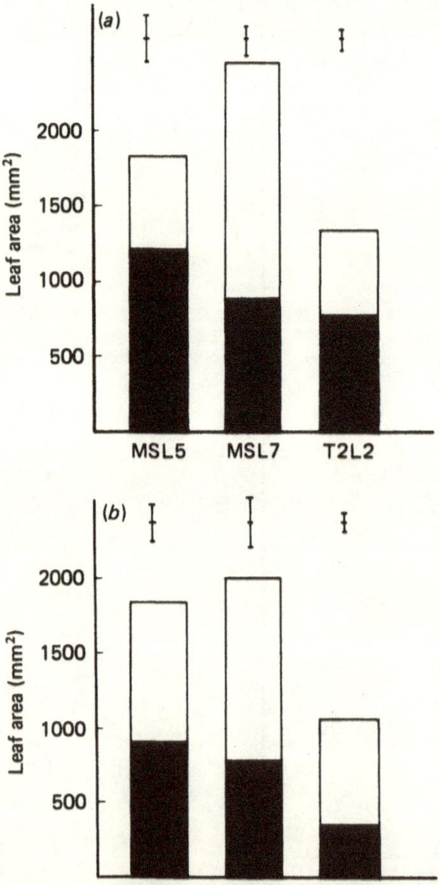

Phillips, Miners & Roddick (1980), working on *Acer pseudoplatanus* in which dormancy is highly sensitive to daylength, found no effect of daylength on abscisic acid levels. The topic is thoroughly reviewed in Saunders (1978).

Rather less work has been carried out attempting to define the role of abscisic acid in apical dominance. However, no clear correlation has been established between the repression of bud growth in apical dominance and the level of abscisic acid in the bud (Walton, 1980). Abscisic acid does not seem to play a major role in the control of bud growth.

Whether abscisic acid is involved in the hormonal balance in normal leaf expansion, or in the control of leaf growth responses to stress is not known. However, it is probable that a balance between the various hormones is important in the young expanding leaves. Abscisic acid (at low concentrations) could promote leaf expansion by inhibiting ethylene production, the latter associated with the auxin production which appears to be an integral part of the functioning of the developing leaf.

## Cytokinins and leaf growth

The root system is a major source of cytokinins for the plant (van Staden & Davey, 1979) and there is evidence that leaves depend on the roots for cytokinins. Thus detached leaves show a rapid fall in endogenous cytokinin levels. Detached bean leaves have very low cytokinin activity, but this increases sharply once adventitious roots form (Engelbrecht, 1972). The buds on defoliated plants show a rapid rise in cytokinin content, reaching four to five times that in control plants within three days (Henson & Wareing, 1977c).

Leaf cytokinin levels are also sensitive to plant nutrition. Nitrogen deficiency reduces leaf cytokinin activity in sunflower (Salama & Wareing, 1979), pumpkin (Göring & Mardanov, 1976) and *Betula pendula*, but not in *Acer pseudoplatanus* (Horgan & Wareing, 1980). Nitrate is more effective than ammonium fertilisation in enhancing plant growth and in raising leaf cytokinin content (Yoshida & Oritani, 1974; Salama & Wareing, 1979). This may reflect a better pH balance in the system. Phosphate and potassium deficiencies also reduce leaf cytokinin content, but not to the same extent as nitrogen deficiency.

On the other hand, three lines of enquiry suggest that leaves show some independence in cytokinin production. The first is that leaves often contain cytokinin types not found in the xylem sap (Henson, 1978a). Secondly, rapid transient increases in cytokinin contents are found in leaves exposed to light (Thompson, Horgan & Heald, 1975; Uheda & Kuraishi, 1977). A third line of evidence is that isolated leaves or shoots appear to produce cytokinins. Cytokinin levels in isolated leaves of *Helianthus annuus* and *Xanthium strumarium* decline when they are placed in water for three days; however, there is recovery to normal levels if the leaves are then transferred to nutrient solution (Salama & Wareing, 1979). In addition, de-rooted cuttings of *Solanum andigena* show extensive growth, and can maintain their cytokinin content for up to 40 days (Wang & Wareing, 1979).

The question of the sites of cytokinin synthesis will only be resolved by the demonstration of the synthesis of specific cytokinins in particular tissues, and from a knowledge of which forms of cytokinin are active.

A consistent relationship between endogenous cytokinin levels and leaf development is not always found. Higher cytokinin contents are found in immature leaves and apical buds than in mature leaves of *Populus* × *robusta*, *X. strumarium* and lettuce (Hewett & Wareing, 1973; Henson & Wareing, 1976; Kemp, Knavel & Hamilton, 1979). In contrast Davey & van Staden (1978) found no cytokinin activity in apices and unexpanded leaves of *Lupinus alba*. Interpretation is complicated by the fact that young leaves tend to have high levels of the free bases, whereas mature leaves have high levels of the glucoside derivatives, and a high capacity for glucosylation (Hewett & Wareing, 1973; Henson, 1978b). It is probable that the glucoside derivatives are inactive within the plant, although metabolic studies have yet to establish this.

A serious complication in studies of responses of leaf growth to applied cytokinins is the enhancement of auxin contents, and the substantial ethylene production that cytokinins induce (Osborne, 1978). This has generally been ignored, but probably explains the enhancement of cytokinin-promoted leaf expansion by cobalt ions, since these inhibit ethylene production (Lau & Yang, 1976).

Application of cytokinins can promote leaf unfolding and expansion in intact plants (Hayes, 1978; Hammond, 1979), and in green and etiolated leaf discs and detached cotyledons in a range of species (see Goodwin, 1978). The promotion of expansion in leaf discs is commonly associated with increased cell size rather than cell number in the epidermis (Grierson, Chambers & Penniket, 1977). However, cytokinins applied directly to intact plants frequently fail to increase leaf size (e.g. Herzog & Geisler, 1977) and may reduce leaf size (Richards,

1980). There is also a lack of striking responses – cytokinin treatment does not replace light in the promotion of leaf expansion. Removal of the root system does not necessarily reduce the leaf growth rate; removal of older leaves is far more effective (Went, 1938). Attempts to replace the leaf factor with cytokinins or to compensate for the effects of mineral deficiency on leaf expansion with cytokinins have not been located.

On the other hand, the situation seems clearer for bud expansion. The promotion of lateral bud outgrowth is a very common response to applied cytokinins (see Phillips, 1975; Richards, 1980). Bud break in the spring is associated with high endogenous cytokinin levels (Hewett & Wareing, 1974; Taylor & Dumbroff, 1975; Smith & Schwabe, 1980). In plants of *Salix babylonica* treated by stem girdling the level of cytokinins in buds below the girdle increases sharply before they grow out (van Staden & Brown, 1978). Applied cytokinins overcome the inhibition of lateral bud growth associated with nitrogen deficiency (Horgan & Wareing, 1980; Sharif & Dale, 1980). Long days induce bud development and also high cytokinin levels in *Bryophyllum diagremontianum* (Henson & Wareing, 1977b). Applied cytokinins are effective inducers of adventitious bud formation in tissue culture (for example, Negrutiu, Jacobs & Cachita, 1978; Barlass & Skene, 1980). Thus cytokinins appear to be essential growth factors for buds. The question is, are they essential for the leaves in the bud, for the stem tissues, for the apical meristem, or for all three?

### Gibberellins

Work on gibberellins is complicated by the very large number, over 50, found in plants, by the fact that the endogenous gibberellins are known for only a handful of plant species, and are often not readily available for use in experiments, and by the problem that few laboratories are appropriately equipped to identify and estimate gibberellins. For these reasons much of the literature on leaf gibberellins is of doubtful value. Papers in which endogenous gibberellins are assessed on fractions so crude as to contain appreciable amounts of inhibitory material, papers which base their conclusions on small differences in activity in extracts, and papers which presume that plant growth retardants are specific gibberellin antagonists have largely been ignored.

There are numerous reports of high gibberellin-like activity in young leaves, and a decline with leaf age (see Goodwin, 1978). More recently there have been a few studies on identified gibberellins. In *Picea sitchensis* the principal gibberellin in swelling buds and actively growing needles is iso$GA_9$ (Lorenzi *et al.*, 1977). During dormancy the major component is $GA_9$ glucosyl ester (Lorenzi, Horgan & Heald, 1975, 1976). In rice plants the principal gibberellin is $GA_{19}$, although trace amounts of $GA_1$ may be present. In the shoots two peaks in $GA_{19}$ content are found, the first at the third leaf stage and the second at panicle initiation (Kurogochi, Murofushi, Oto & Takahashi, 1979). Kurogochi *et al.*, suggest that $GA_{19}$ may act as a reserve pool of gibberellin, being activated by conversion to $GA_1$. However, there seems to be no necessity to invoke conversion to $GA_1$ since in normal rice cultivars $GA_1$ and $GA_{19}$ differ in the concentration required to bring about equal response only by about threefold.

Zeevaart and co-workers have studied the role of gibberellins in species in which long days induce a major increase in the rate of stem elongation. In spinach an increase in stem length is apparent after five to seven long days. Within this period there is a sharp decline in the level of $GA_{19}$ in the shoots (essentially all leaves), and a rise in $GA_{20}$, and slightly later, in $GA_{29}$. There is no change in $GA_{17}$ and $GA_{44}$ levels (Metzger & Zeevaart, 1980). $GA_{29}$ is a $2\beta$-hydroxylated gibberellin, and all such gibberellins are biologically inactive (Graebe & Ropers, 1978). It is suggested that, as proposed in rice, $GA_{19}$ is an inactive 'pool' precursor, and that long days induce the conversion of $GA_{19}$ to $GA_{20}$. In *Agrostemma githago*, stem elongation occurs within seven to eight long days. The levels of $GA_{17}$, $GA_{19}$, $GA_{20}$ and $GA_{44}$ in extracts of shoots and leaves increase to a transient peak after eight long days. $GA_1$ and epi$GA_1$ show a peak at twelve long days. $GA_{53}$ begins to increase after twelve long days (Jones & Zeevaart, 1980). In neither species are the leaf growth responses reported. However the studies do give an idea of the complexity of the gibberellins, and of their changes in response to environmental stimuli. Unravelling these phenomena represents a major challenge.

A number of workers have shown a correlation between leaf expansion and the level of gibberellin-like activity. Wheeler (1960, 1961) and Humphries & Wheeler (1964) showed such correlations in the response of *Phaseolus vulgaris* leaves to light, and to removal of cotyledons or shoot tips. Similar results have been obtained with *Pharbitis nil* cotyledons (Ogawa, 1964) and wheat leaves (Jurekova & Repka, 1973).

A number of lines of evidence suggest that leaves are a source of gibberellins. Red light causes a rapid transient increase (within 5 to 15 minutes) in extractable gibberellin-like activity in *P. nil* cotyledons, and in etiolated barley, wheat and *Pinus sylvestris* leaves (Graebe & Ropers, 1978). In etiolated pea shoot tips this treatment causes an increase in ethylene production (Goeschl *et al.*, 1967) and it would be of interest to know if both ethylene and gibberellin changes occur in this tissue. Even more striking are studies on chloroplasts. High levels of gibberellin-like activity have been found in crude preparations of chloroplasts from barley, kale, potato and pea. Red light leads to an increase in the gibberellin-like activity extractable from etioplasts. Browning & Saunders (1977) found the detergent Triton X-100 to be much more effective (1000 fold!) than the standard methanol solvent for gibberellin extraction from wheat chloroplasts, where they identified $GA_4$ and $GA_9$. This detergent, however, did not give higher yields from pea chloroplasts (Railton & Rechav, 1979).

Jones & Phillips (1966) applied the van Overbeek technique in an attempt to demonstrate gibberellin production by the shoot tip and young leaves. The gibberellin activity of extracts from leaves or shoot tips before and after diffusion into agar, together with the activity diffused into the agar, were measured. The activity in the agar plus that in the organ after diffusion may have been greater than that in the organ before diffusion, so that net synthesis may have occurred. However, the bioassayed extracts were only crudely purified, so that the presence of inhibitors complicates interpretation of the bioassays. (2-chloroethyl)-trimethylammonium chloride (CCC) did not cause a decrease in diffusible gibberellin-like activity (Jones & Phillips, 1967). There is also evidence that gibberellins move from the leaves in the phloem, with negligible movement by diffusion (Graebe & Ropers, 1978). Demonstration of synthesis of defined gibberellins has so far been achieved only in developing seeds (Hedden, MacMillan & Phinney, 1978), but this is the only way it can be shown conclusively that a particular organ has the capacity for gibberellin synthesis.

Gibberellin-like activity in apices is reduced by a range of environmental treatments which reduce shoot growth, for example by short days or long days (Proebsting, Davies & Marx, 1978), with nitrogen or potassium deficiency, or with flooding, drought or the application of abscisic acid (see Goodwin, 1978). Harada and co-workers found a sharp increase in the levels of $GA_1$, $GA_3$, $GA_9$ and $GA_8$ glucoside extractable from apices of *Althea rosea* after chilling (Harada, 1962; Harada & Nitsch, 1967; Harada & Yokota, 1970). There is some evidence that increases in gibberellins in chilled apices reflect an early stage in shoot growth, rather than inducing that growth (see Avigdori-Avidov, Goldschmidt & Kedar, 1977). These changes in gibberellin-like activity in apices may be due to changes in the content of gibberellins in the leaves.

Application of gibberellins may stimulate, have no effect (Dicks & Abdel-Kawi, 1979), or inhibit (Maksymowych, Cordero & Erickson, 1976; Tonecki, 1979) leaf area growth on intact plants, isolated cotyledons, leaf discs and leaf segments. Promotion of leaf length is very common (Gray, 1957; Aloni & Pressman, 1980) and generally petiole length is promoted more strongly than blade length. Perhaps the best evidence for a role for gibberellins in leaf growth is the conversion of the reduced leaves of certain genetic dwarfs which have a low endogenous gibberellin-like activity, to the normal phenotype by gibberellin application. This has been shown in maize, rice and *Lolium perenne* (see Goodwin, 1978). Israelstam & Davis (1979) caused dwarf pea leaves to develop towards the normal leaf type by application of gibberellin, but the endogenous gibberellin status of the plants is not known. Treatment with gibberellins alters the phyllotaxis and increases the rate of production of primordia in *Xanthium* (Maksymowych *et al.*, 1976). By and large, leaf area responses to gibberellin treatment are relatively small, a doubling of leaf size being rare, and increases are often followed by compensating reductions in the size of later-formed leaves.

There is no strong evidence for gibberellins being required as factors in bud growth. Application of gibberellins rarely terminates bud dormancy (Saunders, 1978). Applications often, but not always, increase apical dominance in intact plants, presumably as a consequence of promotion of stem elongation and auxin production. On the other hand, where apical dominance has been broken, or severely weakened (for example by cytokinin application), gibberellins enhance lateral bud outgrowth (see Phillips, 1975).

Even where growth retardants cause a sharp decline in endogenous gibberellin-like activity, there is only a limited reduction in leaf area (e.g. Dicks & Abdel-Kawi, 1979). The general impression is that gibberellins play only a modifying role in leaf growth.

## Studies on leaf development *in vitro*

Fern leaf primordia, when isolated at an appropriate stage of development, develop into shoots (Steeves, 1961). However, leaf primordia of sunflower and tobacco, when excised and cultured *in vitro* in a complete medium containing sucrose, mineral nutrients and vitamins, do not have this potential and only very limited growth occurs. Additives such as casein hydrolysate and coconut milk (a cytokinin source) stimulate growth to a limited extent (Steeves, Gabriel & Steeves, 1957). The longest sunflower leaf length reported was 14.5 mm, obtained from a primodium initially 3.4 mm long, indicating the severity of leaf reduction. These excised leaves, from normal shoot apices, lack some stimulus which usually comes from the mother plant. Leaves are therefore not autonomous morphological units and require some unknown factor if their development is to proceed normally.

The influence of hormones on leaf area has been studied in potato, using leafy single node cuttings, cultured *in vitro* on a complete nutrient medium in the light (M. G. Erwee, unpublished data; Fig. 8.5). Under these conditions leaf growth was not dependent on the addition of sucrose to the medium. The influence of hormones has been compared with that of the subtending axillary leaf whose removal reduces the area but not the number of subsequent leaves. The effect of the subtending leaf could not be substituted by any of the hormones tested, and the results suggest that it does not act as a source of auxins, gibberellins or cytokinins for leaf expansion. However, the 'mother' leaf, in the

Fig. 8.5. Histogram of (*a*) average leaf number, (*b*) total leaf area and (*c*) average leaf area produced from single node cuttings cultured *in vitro* after 21 days on a half strength Murashige and Skoog medium with 3% sucrose at 22 °C with a light intensity of 120 $\mu$mol m$^{-2}$s$^{-1}$. $C_L$ represents a control plant with a 'mother' leaf and C a control plant with no 'mother' leaf. Hormonal treatments at a 10 $\mu$M level are indicated by G, gibberellic acid; I, indole-3-butyric acid and B, benzyladenine. Explants for hormonal treatments had no 'mother' leaf. LSD are at the 5% level (M. G. Erwee, unpublished data).

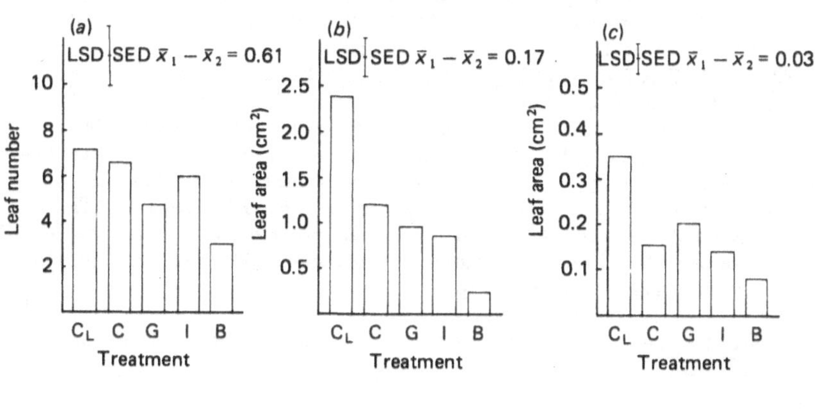

well-illuminated conditions used, is a major source of growth factors for the developing leaves. The nature of these factors, perhaps equivalent to the leaf-forming substances or 'phyllocalines' of Went (1938), remains unknown.

## Conclusion

The challenge of explaining the mechanism by which older leaves strongly promote the expansion of young developing leaves remains. They could provide something as simple as carbohydrate or as complex as a new type of phytohormone. Of the known hormones, gibberellins, auxins and abscisic acid influence leaf growth, but only to a limited degree. They appear to modify patterns of leaf development determined by other influences. Cytokinins strongly influence bud outgrowth, perhaps due to an effect on the initial expansion of leaves. However, they have little effect on leaf expansion on intact plants. The hormone which at the moment appears most likely to play a major role in leaf expansion is ethylene. It is clearly involved in the light-induced expansion of plumule leaves in pea, and has striking effects on the expansion of leaves on intact light-grown plants.

M. G. Erwee acknowledges the support of the Thomas Lawrance Pawlett Scholarship of the University of Sydney.

## References

Abeles, F. B. (1973). *Ethylene in Plant Biology*. New York: Academic Press.

Aharoni, N. (1978). Relationship between leaf water status and endogenous ethylene in detached leaves. *Plant Physiology*, **61**, 658–62.

Aharoni, N., Blumenfeld, A. & Richmond, A. E. (1977). Hormonal activity in detached lettuce leaves as affected by leaf water content. *Plant Physiology*, **59**, 1169–73.

Aharoni, N., Lieberman, M. & Sisler, H. D. (1979). Patterns of ethylene production in senescing leaves. *Plant Physiology*, **64**, 796–800.

Allen, J. R. F. & Baker, D. A. (1980). Free tryptophan and indole-3-acetic acid levels in the leaves and vascular pathways of *Ricinus communis* L. *Planta*, **148**, 69–74.

Aloni, B. & Pressman, E. (1980). Interaction with salinity of $GA_3$ induced leaf elongation, petiole pithiness and bolting in celery. *Scientia Horticulturae*, **13**, 135–42.

Apelbaum, A. & Burg, S. P. (1972). Effects of ethylene on cell division and deoxyribonucleic acid synthesis in *Pisum sativum*. *Plant Physiology*, **50**, 117–24.

Apelbaum, A., Burgoon, A. C., Anderson, J. D., Solomon, T. & Lieberman, M. (1981). Some characteristics of the system converting 1-aminocyclopropane-1-carboxylic acid to ethylene. *Plant Physiology*, **67**, 80–4.

Avery, G. S. (1935). Differential distribution of a phytohormone in the developing leaf of *Nicotiana*, and its relation to polarized growth. *Bulletin of the Torrey Botanical Club*, **62**, 313–30.

Avigdori-Avidov, H., Goldschmidt, E. E. & Kedar, N. (1977). Involvement of endogenous gibberellins in the chilling requirements of strawberry (*Fragaria × ananassa* Duch.). *Annals of Botany*, **41**, 927–36.
Barlass, M. & Skene, K. G. M. (1980). Studies on the fragmented shoot apex of grapevine. II. Factors affecting growth and differentiation *in vitro*. *Journal of Experimental Botany*, **31**, 489–95.
Blanpied, G. D. (1972). A study of ethylene in apple, red raspberry, and cherry. *Plant Physiology*, **49**, 627–30.
Boussiba, S. & Richmond, A. E. (1976). Abscisic acid and the after effect of stress in tobacco plants. *Planta*, **129**, 217–19.
Bradford, K. J. & Yang, S. F. (1980). Xylem transport of 1-aminocyclopropane-1-carboxylic acid, an ethylene precursor, in waterlogged tomato plants. *Plant Physiology*, **65**, 322–6.
Brenner, M. L. (1979). Advances in analytical methods for plant growth substance analysis. In *Plant Growth Substances*, ed. N. B. Mandava, pp. 215–44. Washington: American Chemical Society.
Browning, G. & Saunders, P. F. (1977). Membrane localised gibberellins $A_9$ and $A_4$ in wheat chloroplasts. *Nature*, **265**, 375–7.
Burg, S. P. (1968). Ethylene, plant senescence and abscission. *Plant Physiology*, **43**, 1503–11.
Burg, S. P. (1973). Ethylene in plant growth. *Pest Articles and News Summaries*, **70**, 591-7.
Daie, J. & Campbell, W. C. (1981). Response of tomato plants to stressful temperatures. *Plant Physiology*, **67**, 26–9.
Davey, J. E. & van Staden, J. (1978). Cytokinin activity in *Lupinus albus*. I. Distribution in vegetative and flowering plants. *Plant Physiology*, **43**, 77–81.
Dicks, J. W. & Abdel-Kawi, A. A. (1979). Antagonists and synergistic interaction between ancymidol and gibberellins in shoot growth of cucumber (*Cucumis sativus* L.). *Journal of Experimental Botany*, **30**, 779–93.
Dumbroff, E. B., Cohen, D. B. & Webb, D. P. (1979). Seasonal levels of abscisic acid in buds and stems of *Acer saccharum*. *Physiologia Plantarum*, **45**, 211–14.
El-Antably, H. M., Wareing, P. F. & Hillman, J. (1967). Some physiological responses to d, 1-abscisin (dormin). *Planta*, **73**, 74–90.
El-Beltagy, A. S. & Hall, M. A. (1974). Effects of water stress upon endogenous ethylene levels in *Vicia faba*. *New Phytologist*, **73**, 47–60.
Engelbrecht, L. (1972). Cytokinins in leaf-cuttings of *Phaseolus vulgaris* L. during their development. *Biochemie und Physiologie der Pflanzen*, **163**, 335–43.
Gale, M. D. (1979). Plant hormones and plant breeding. In *Genetic Variation in Hormone Systems*, vol. 2, ed. J. G. M. Shire, pp. 123–49. Florida: CRC Press, Inc.
Gifford, E. M. (1953). Effects of 2,4-D upon the development of the cotton leaf. *Hilgardia*, **21**, 607–44.
Goeschl, J. D., Pratt, H. K. & Bonner, B. A. (1967). An effect of light on the production of ethylene and the growth of the plumular portion of etiolated pea seedlings. *Plant Physiology*, **42**, 1077–80.
Goldbach, E., Goldbach, H., Wagner, H. & Michael, G. (1975). Influence of nitrogen-deficiency on the abscisic acid content of sunflower plants. *Physiologia Plantarum*, **34**, 138–40.
Goodwin, P. B. (1978). Phytohormones and growth and development of organs of the vegetative plant. In *Phytohormones and Related Compounds – A Comprehensive Treatise*, vol. 2, ed. D. S. Letham, P. B. Goodwin & T. J. V. Higgins, pp. 31–173. Amsterdam: Elsevier/North Holland Biomedical Press.
Göring, H. & Mardanov, A. A. (1976). Influence of nitrogen deficiency on K/Ca ratio and

cytokinin content of pumpkin seedlings. *Biochemie und Physiologie der Pflanzen*, **170**, 261–4.

Graebe, J. E. & Ropers, H. J. (1978). Gibberellins. In *Phytohormones and Related Compounds – A Comprehensive Treatise*, vol. 1, ed. D. S. Letham, P. B. Goodwin & T. J. V. Higgins, pp. 107–204. Amsterdam: Elsevier/North Holland Biomedical Press.

Gray, R. A. (1957). Alteration of leaf size and shape and other changes caused by gibberellins in plants. *American Journal of Botany*, **44**, 674–82.

Grierson, D., Chambers, S. E. & Penniket, L. P. (1977). Nucleic acid and protein synthesis in discs cut from mature leaves of *Nicotiana tabacum* L. and cultured on nutrient agar with and without kinetin. *Planta*, **134**, 29–34.

Hall, M. A., Kapuya, J. A., Sivakumaran, S. & John, A. (1977). The role of ethylene in the response of plants to stress. *Pesticide Science*, **8**, 217–23.

Hall, H. K. & McWha, J. A. (1981). Effects of abscisic acid on growth of wheat (*Triticum aestivum* L.). *Annals of Botany*, **47**, 427–33.

Hammond, H. D. (1979). Growth regulator interactions on morphogenesis in *Solanum* species. In *The Biology and Taxonomy of the Solanaceae*, ed. J. G. Hawkes, R. N. Lester & A. D. Shelding, pp. 357–69. London: Academic Press.

Harada, H. (1962). Growth substances and flowering. *Révue Général de Botanique*, **69**, 201–97.

Harada, H. & Nitsch, J. P. (1967). Isolation of gibberellins $A_1$, $A_3$, $A_9$ and a fourth growth substance from *Althaea rosea* Cav. *Phytochemistry*, **6**, 1695–703.

Harada, H. & Yokota, T. (1970). Isolation of gibberellin $A_8$-glucoside from shoot apices of *Althaea rosea. Planta*, **92**, 100–4.

Hayes, A. B. (1978). Auxin-cytokinin effects in leaf blade hyponasty. *Botanical Gazette*, **139**, 385–9.

Hedden, P., MacMillan, J. & Phinney, B. D. (1978). The metabolism of the gibberellins. *Annual Review of Plant Physiology*, **29**, 149–92.

Henson, I. E. (1978a). Types, formation, and metabolism of cytokinins in leaves of *Alnus glutinosa* (L.) Gaertn. *Journal of Experimental Botany*, **29**, 935–51.

Henson, I. E. (1978b). Cytokinins and their metabolism in leaves of *Alnus glutinosa* L. Gaertn: effects of leaf development. *Zeitschrift für Pflanzenphysiologie*, **86**, 363–70.

Henson, I. E. & Wareing, P. F. (1976). Cytokinins in *Xanthium strumarium*. Distribution in the plant and production in the root system. *Journal of Experimental Botany*, **27**, 1268–78.

Henson, I. E. & Wareing, P. F. (1977a). Cytokinins in *Xanthium strumarium* L.: some aspects of the photoperiodic control of endogenous levels. *New Phytologist*, **78**, 35–45.

Henson, I. E. & Wareing, P. F. (1977b). Changes in the level of endogenous cytokinins and indole-3-acetic acid during epiphyllous bud formation in *Bryophyllum daigremontianum. New Phytologist*, **79**, 225–32.

Henson, I. E. & Wareing, P. F. (1977c). The effect of defoliation on the cytokinin content of buds of *Xanthium strumarium. Plant Science Letters*, **9**, 27–31.

Herzog, H. & Geisler, G. (1977). The effects of cytokinin application on the leaf and general shoot development in spring wheat. *Zeitschrift für Acker-und Planzenbau*, **144**, 8–17.

Hewett, E. W. & Wareing, P. F. (1973). Cytokinins in *Populus × robusta:* qualitative changes during development. *Physiologia Plantarum*, **29**, 386–9.

Hewett, E. W. & Wareing, P. F. (1974). Cytokinin changes during chilling and bud burst in woody plants. In *Mechanisms of Regulation of Plant Growth*, ed. R. L. Bieleski, A. R. Ferguson & M. M. Cresswell, pp. 693–701. Wellington: Royal Society of New Zealand.

Hillman, J. R., Math, V. B. & Medlow, G. C. (1977). Apical dominance and the level of indole acetic acid in *Phaseolus* lateral buds. *Planta*, **134**, 191–3.

Horgan, R., Hewett, E. W., Horgan, J. M., Purse, J. & Wareing, P. F. (1975). A new cytokinin from *Populus × robusta*. *Phytochemistry*, **14,** 1005–8.
Horgan, J. M. & Wareing, P. F. (1980). Cytokinins and the growth responses of seedlings of *Betula pendula* Roth. and *Acer pseudoplatanus* L. to nitrogen and phosphate deficiency. *Journal of Experimental Botany*, **31,** 525–32.
Humphries, E. C. & Wheeler, A. W. (1963). The physiology of leaf growth. *Annual Review of Plant Physiology*, **14,** 385–410.
Humphries, E. C. & Wheeler, A. W. (1964). Cell division and growth substances in leaves. In *Régulateurs Naturels de la Croissance Végétale*, pp. 505–15. CNRS: Paris.
Israelstam, G. F. & Davis, E. (1979). An analysis of the effects of gibberellic acid on the leaflet structure of dwarf cultivars of pea (*Pisum sativum*). *Canadian Journal of Botany*, **57,** 1089–92.
Jackson, M. B. & Campbell, D. J. (1976). Waterlogging and petiole epinasty in tomato: the role of ethylene and low oxygen. *New Phytologist*, **76,** 21–9.
Jones, R. L. & Phillips, I. D. J. (1966). Organs of gibberellin synthesis in light grown sunflower plants. *Plant Physiology*, **41,** 1381–6.
Jones, R. L. & Phillips, I. D. J. (1967). Effect of CCC on the gibberellin content of excised sunflower organs. *Planta*, **72,** 53–9.
Jones, M. G. & Zeevaart, J. A. D. (1980). The effect of photoperiod on the levels of seven endogenous gibberellins in the long day plant *Agrostemma githago* L. *Planta*, **149,** 274–9.
Jurekova, Z. & Repka, J. (1973). Heterogeneity of the content of endogenous gibberellins in the leaves of winter wheat in relation to their insertion and ontogeny. *Biologia Plantarum*, **15,** 305–11.
Kays, S. J. & Pallas, J. E. (1980). Inhibition of photosynthesis by ethylene. *Nature*, **285,** 51–2.
Kemp, T. R., Knavel, D. E. & Hamilton, J. L. (1979). Isolation of natural cytokinins from lettuce leaves. *HortScience*, **14,** 635–6.
King, R. W., Evans, L. T. & Firn, R. D. (1977). Abscisic acid and xanthoxin content in the long day plant *Lolium temulentum* L. in relation to daylength. *Australian Journal of Plant Physiology*, **4,** 217–23.
Krauss, A. (1978). Tuberization and abscisic acid content in *Solanum tuberosum* as affected by nitrogen nutrition. *Potato Research*, **21,** 183–93.
Kurogochi, S., Murofushi, N., Oto, Y. & Takahashi, N. (1979). Identification of gibberellins in the rice plant and quantitative changes of gibberellin $A_{19}$ throughout its life cycle. *Planta*, **146,** 185–91.
Lau, O. L. & Yang, S. F. (1976). Inhibition of ethylene production by cobaltous ion. *Plant Physiology*, **58,** 114–17.
Lavee, S. & Martin, G. C. (1981). Ethylene evolution from various developing organs of olive (*Olea europaea*) after excision. *Physiologia Plantarum*, **51,** 33–8.
Le Meur, M., Glanville, N., Mandel, J. L., Gerlinger, P., Palmiter, R. & Chambon, P. (1981). The ovalbumin gene family: hormonal control of X and Y gene transcription and m-RNA accumulation. *Cell*, **23,** 561–71.
Letham, D. S., Higgins, T. J. V., Goodwin, P. B. & Jacobsen, J. V. (1978). Phytohormones in retrospect. In *Phytohormones and Related Compounds – A Comprehensive Treatise*, vol. 1, pp. 1–27. Amsterdam: Elsevier/North Holland Biomedical Press.
Letham, D. S., Parker, C. W., Duke, C. C., Summons, R. E. & MacLeod, J. K. (1977). O-glucosylzeatin and related compounds – a new group of cytokinin metabolites. *Annals of Botany*, **41,** 261–3.
Lieberman, M. (1979). Role of ethylene in plant growth, development and senescence. In *Plant Growth Substances*, ed. N. B. Mandava, pp. 115–34. Washington: American Chemical Society.

Lorenzi, R., Horgan, R. & Heald, J. K. (1975). Gibberellins in *Picea sitchensis* Carriere: seasonal variation and partial characterisation. *Planta*, **126,** 75–82.
Lorenzi, R., Horgan, R. & Heald, J. K. (1976). Gibberellin $A_9$ glucosyl ester in needles of *Picea sitchensis. Phytochemistry*, **15,** 789–90.
Lorenzi, R., Horgan, R. & Wareing, P. F. (1975). Cytokinins in *Picea sitchensis* Carriere: identification and relation to growth. *Biochemie und Physiologie der Pflanzen*, **168,** 333–9.
Lorenzi, R., Saunders, P. F., Heald, J. K. & Horgan, R. (1977). A novel gibberellin from needles of *Picea sitchensis. Plant Science Letters*, **8,** 179–82.
Loveys, B. R. (1977). The intracellular location of abscisic acid in stressed and non-stressed leaf tissue. *Physiologia Plantarum*, **40,** 6–10.
McAfee, J. A. & Morgan, P. W. (1971). Rates of production and internal levels of ethylene in the vegetative cotton plant. *Plant and Cell Physiology*, **12,** 839–47.
McDavid, C. R., Sagar, G. R. & Marshall, C. (1972). The effect of auxin from the shoot on root development in *Pisum sativum* L. *New Phytologist*, **71,** 1027–32.
McDougall, J. & Hillman, J. R. (1980). Distribution of indole-3-acetic acid in shoots of *Phaseolus vulgaris* L. *Zeitschrift für Pflanzenphysiologie*, **97,** 367–71.
Maksymowych, R., Cordero, R. E. & Erickson, D. E. (1976). Long-term developmental changes in *Xanthium* induced by gibberellic acid. *American Journal of Botany*, **63,** 1047–53.
Maugh, T. H. (1981). New chemicals promise larger crops. *Science*, **212,** 33–4.
Metzger, J. D. & Zeevaart, J. A. D. (1980). Effect of photoperiod on the levels of endogenous gibberellins in spinach as measured by combined gas chromatography-selected ion current monitoring. *Plant Physiology*, **66,** 844–6.
Miller, M. D., Mikkelsen, D. S. & Huffaker, R. C. (1962). Effects of stimulating and inhibitory levels of 2,4-D, iron, and chelate supplements on juvenile growth of field beans. *Crop Science*, **2,** 111–14.
Negrutiu, I., Jacobs, M. & Cachita, D. (1978). Some factors controlling *in vitro* morphogenesis of *Arabidopsis thaliana. Zeitschrift für Pflanzenphysiologie*, **86,** 113–24.
Ogawa, Y. (1964). Changes in amount of gibberellin-like substances in the seedlings of *Pharbitis nil* with special reference to expansion of cotyledons. *Plant and Cell Physiology*, **5,** 11–20.
Osborne, D. J. (1978). Ethylene. In *Phytohormones and Related Compounds – A Comprehensive Treatise*, vol. 1, ed. D. S. Letham, P. B. Goodwin, & T. J. V. Higgins, pp. 265–94. Amsterdam: Elsevier/North Holland Biomedical Press.
Perry, T. O. & Hellmers, H. (1973). Effects of abscisic acid on growth and dormancy of two races of red maples. *Botanical Gazette*, **134,** 283–9.
Phillips, I. D. J. (1975). Apical dominance. *Annual Review of Plant Physiology*, **26,** 341–67.
Phillips, I. D. J., Miners, J. & Roddick, J. G. (1980). Effects of light and photoperiodic conditions on abscisic acid in leaves and roots of *Acer pseudoplatanus* L. *Planta*, **149,** 118–22.
Pierce, M. & Raschke, M. (1980). Correlation between loss of turgor and accumulation of abscisic acid in detached leaves. *Planta*, **148,** 174–82.
Proebsting, W. M., Davies, P. J. & Marx, G. A. (1978). Photoperiod-induced changes in gibberellin metabolism in relation to apical growth and senescence in genetic lines of peas (*Pisum sativum* L.). *Planta*, **141,** 231–8.
Quarrie, S. A. & Jones, H. G. (1977). Effects of abscisic acid and water stress on development and morphology of wheat. *Journal of Experimental Botany*, **28,** 192–203.
Railton, I. D. & Rechav, M. (1979). Efficiency of extraction of gibberellin-like substances from chloroplasts of *Pisum sativum. Plant Science Letters*, **14,** 75–8.
Raschke, K. & Zeevaart, J. A. D. (1976). Abscisic acid content, transpiration, and

stomatal conductance as related to leaf age in plants of *Xanthium strumarium*. *Plant Physiology*, **58**, 169–74.

Richards, D. (1980). Root-shoot interactions: effects of cytokinin applied to the root and/or shoot of apple seedlings. *Scientia Horticulturae*, **12**, 143–52.

Robitaille, H. A. & Carlson, R. F. (1976). Gibberellin and abscisic acid like substances and the regulation of apple shoot extension. *Journal of the American Society for Horticultural Science*, **101**, 382–92.

Salama, A. M. S. & Wareing, P. F. (1979). Effects of mineral nutrition on endogenous cytokinins in plants of sunflower (*Helianthus annuus* L.). *Journal of Experimental Botany*, **30**, 971–81.

Saunders, P. F. (1978). Phytohormones and bud dormancy. In *Phytohormones and Related Compounds – A Comprehensive Treatise*, vol. 2, ed. D. S. Letham, P. B. Goodwin & T. J. V. Higgins, pp. 423–44. Amsterdam: Elsevier/North Holland Biomedical Press.

Saunders, P. F., Harrison, M. A. & Alvim, R. (1974). Abscisic acid and tree growth. In *Plant Growth Substances*, pp. 871–81. Tokyo: Hirokawa Publishing Company, Inc.

Schneider, E. A. & Wightman, F. (1978). Auxins. In *Phytohormones and Related Compounds – A Comprehensive Treatise*, vol. 1, ed. D. S. Letham, P. B. Goodwin & T. J. V. Higgins, pp. 29–105. Amsterdam: Elsevier/North Holland Biomedical Press.

Selman, I. W. & Sandanam, S. (1972). Growth responses of tomato plants in non-aerated water culture to foliar sprays of gibberellic acid and benzyladenine. *Annals of Botany*, **36**, 837–48.

Sharif, R. & Dale, J. E. (1980). Growth-regulatory substances and the growth of tiller buds in barley; effects of cytokinin. *Journal of Experimental Botany*, **31**, 921–30.

Sivakumaran, S. & Hall, M. A. (1978a). Effects of osmotic stress upon endogenous hormone levels in *Euphorbia lathyrus* L. and *Vicia faba* L. *Annals of Botany*, **42**, 1403–11.

Sivakumaran, S. & Hall, M. A. (1978b). Effects of age and water stress on endogenous levels of plant growth regulators in *Euphorbia lathyrus* L. *Journal of Experimental Botany*, **29**, 195–205.

Skytt-Anderson, A. (1976). Regulation of apical dominance by ethephon, irradiance and $CO_2$. *Physiologia Plantarum*, **37**, 303–8.

Smith, P. J. & Schwabe, W. W. (1980). Cytokinin activity in oak (*Quercus robur*) with particular reference to transplanting. *Physiologia Plantarum*, **48**, 27–32.

Steeves, T. A. (1961). A study of the developmental potentialities of excised leaf primordia in sterile culture. *Phytomorphology*, **11**, 346–59.

Steeves, T. A., Gabriel, H. P. & Steeves, M. W. (1957). Growth in sterile culture of excised leaves of flowering plants. *Science*, **126**, 350–1.

Steeves, T. A. & Sussex, I. M. (1972). *Patterns in Plant Development*. Englewood Cliffs, New Jersey: Prentice-Hall.

Stewart, R. N. (1978). Ontogeny of the primary body in chimeral forms of higher plants. *Symposium of the Society for Developmental Biology*, **36**, 131–60.

Tal, M. (1966). Abnormal stomatal behaviour in wilty mutants of tomato. *Plant Physiology*, **41**, 1387–91.

Tal, M., Imber, D., Erez, A. & Epstein, E. (1979). Abnormal stomatal behaviour and hormonal inbalance in flacca, a wilty mutant of tomato. *Plant Physiology*, **63**, 1044–8.

Taylor, J. G. & Dumbroff, E. B. (1975). Bud, root, and growth-regulator activity in *Acer saccharum* during the dormant season. *Canadian Journal of Botany*, **53**, 321–31.

Thompson, A. G., Horgan, R. & Heald, J. K. (1975). A quantitative analysis of cytokinin using single-ion-current-monitoring. *Planta*, **124**, 207–10.

Tonecki, J. (1979). Effect of the growth substances on plant growth and shoot apex differentiation in gladiolus (*Gladiolus hortorum* cv. Acca Laurentia). *Acta Horticulturae*, **91**, 201–4.

Turner, N. C. (1979). Drought resistance and adaptation to water deficits in crop plants. In *Stress Physiology in Crops*, ed. H. Mussel & R. C. Staples, pp. 343–72. New York: John Wiley and Sons.

Uheda, E. & Kuraishi, S. (1977). Increase of cytokinin activity in detached etiolated cotyledons of squash after illumination. *Plant and Cell Physiology*, **18**, 481–3.

van Andel, O. M. & Verkerke, D. R. (1978). Stimulation and inhibition by ethylene of stem and leaf growth of some Gramineae at different stages of development. *Journal of Experimental Botany*, **29**, 639–51.

van Staden, J. & Brown, N. A. C. (1978). Changes in the endogenous cytokinins of bark and buds of *Salix babylonica* as a result of stem girdling. *Physiologia Plantarum*, **43**, 148–53.

van Staden, J. & Davey, J. E. (1979). The synthesis, transport and metabolism of endogenous cytokinins. *Plant, Cell and Environment*, **2**, 93–106.

Walton, D. C. (1980). Biochemistry and physiology of abscisic acid. *Annual Review of Plant Physiology*, **31**, 453–9.

Wang, T. L., Thompson, A. G. & Horgan, R. (1977). A cytokinin glucoside from the leaves of *Phaseolus vulgaris* L. *Planta*, **135**, 285–8.

Wang, T. L. & Wareing, P. F. (1979). Cytokinins and apical dominance in *Solanum andigena*: lateral shoot growth and endogenous cytokinin levels in the absence of roots. *New Phytologist*, **82**, 19–28.

Went, F. W. (1938). Specific factors other than auxin affecting growth and root formation. *Plant Physiology*, **13**, 55–80.

Went, F. W. & Thimann, K. (1937). *Phytohormones*. New York: MacMillan.

Wheeler, A. W. (1960). Changes in a leaf growth substance in cotyledons and primary leaves during the growth of dwarf bean seedlings. *Journal of Experimental Botany*, **11**, 217–26.

Wheeler, A. W. (1961). Leaf-growth substances from dwarf French bean leaves. *Annual Report of Rothamsted Experimental Station, 1960*, 100.

White, J. C., Medlow, G. C., Hillman, J. R. & Wilkins, M. B. (1975). Correlative inhibition of lateral bud growth in *Phaseolus vulgaris* L. Isolation of indoleacetic acid from the inhibitory region. *Journal of Experimental Botany*, **26**, 419–29.

Wightman, F. (1977). Gas chromatographic identification and quantitative estimation of natural auxins in developing plant organs. In *Plant Growth Regulation*, ed. P. E. Pilet, pp. 77–90. Berlin, Heidelberg: Springer-Verlag.

Wright, S. T. C. (1978). Phytohormones and stress phenomena. In *Phytohormones and Related Compounds – A Comprehensive Treatise*, vol. 2, ed. D. S. Letham, P. B. Goodwin & T. J. V. Higgins, pp. 495–536. Amsterdam: Elsevier/North Holland Biomedical Press.

Wright, S. T. C. (1979). The effect of 6-benzyladenine and leaf aging treatments on the level of stress-induced ethylene emanating from wilted wheat leaves. *Planta*, **144**, 179–88.

Wright, S. T. C. & Hiron, R. W. P. (1969). (+)-Abscisic acid, the growth inhibitor induced in detached wheat leaves by a period of wilting. *Nature*, **224**, 719–20.

Yoshida, R. & Oritani, T. (1974). Studies on nitrogen metabolism in crop plants. XIII. Effects of nitrogen top-dressing on cytokinin content in the root exudate of rice plant. *Proceedings of the Crop Science Society of Japan*, **43**, 47–51.

Zeevaart, J. A. D. (1977). Sites of abscisic acid synthesis and metabolism in *Ricinus communis* L. *Plant Physiology*, **59**, 788–91.

Zeevaart, J. A. D. (1980). Changes in the levels of abscisic acid and its metabolites in excised leaf blades of *Xanthium strumarium* during and after water stress. *Plant Physiology*, **66**, 672–8.

Zeroni, M., Jerie, P. H. & Hall, M. A. (1977). Studies on the movement and distribution of ethylene in *Vicia faba* L. *Planta*, **134**, 119–25.

# 9

# Photomorphogenesis in leaves

D. VINCE-PRUE AND D. J. TUCKER

## Introduction

Photomorphogenesis concerns the regulation of plant growth and development by light, other than through photosynthesis. That light has a marked regulatory influence on growth, apart from its photosynthetic effect, has been recognized for a long time. Originally the characteristic growth patterns associated with development in darkness and collectively termed 'etiolation' were thought to derive from nutritional effects but as early as 1873, Godlewsky began to develop the idea that defective nutrition was not an adequate explanation for etiolation phenomena. MacDougal (1903) described differences between plants growing on their food reserves in the dark and those growing in the light and concluded that light had a morphogenetic influence; when etiolated plants receive light, they begin to develop the characteristics of normal green plants, internode elongation is inhibited and the leaves rapidly expand and develop chlorophyll. However, MacDougal also demonstrated that there is a considerable variation in the extent to which leaves can develop in darkness, ranging from types such as *Vicia*, in which leaves are inhibited at a very early stage, through *Calla*, in which expansion is more or less normal, to *Narcissus*, where leaves are actually larger than normal when grown in the dark. A useful generalization is that net-veined dicotyledonous leaves are unable to expand in darkness, while parallel-veined leaves of monocotyledons elongate in darkness, sometimes to a greater extent than in the light (Williams, 1956). Because of such differences, the processes under photocontrol might also be expected to show variation but only a few species have so far been studied in any detail.

Thompson & Miller (1962) carried out a series of experiments with *Pisum sativum* in order to determine at what stage in their development light first affects the form and anatomy of leaves, and to determine what developmental processes bring about the differences that appear between normal and etiolated leaves. They concluded that light had no effect on the manner of initiation or early development of leaf primordia but that it does markedly affect the rate and extent of later leaf development. In the pea plant, as also in *Vicia faba* (Butler & Lane, 1959), there was a small but consistent acceleration of the rate of production of leaf primordia in the light. In white mustard, *Sinapis alba*, on the other hand, the production of new leaf primordia almost entirely ceased in the dark (Mohr & Pinnig, 1962). These experiments were carried out with seedlings growing on their food reserves, so that photosynthesis appeared not to be directly involved. However, these reported differences between mustard on the one hand, and pea and bean seedlings on the other, may be related to the size of the food reserve available in the seed, or perhaps to the amount of differentiation of the embryo at maturity (Vince, 1964). Few studies appear to have been made on the influence of light on the rate of primordial inception in monocotyledons. Light was, however, shown to increase the mitotic rate at the shoot apex of young rice seedlings (Rolinson & Vince-Prue, 1976) especially in the summit region and in the leading flank destined to give rise to the next leaf primordium; these changes were observed after only 16 h in red light.

The effects of light on leaf expansion have been examined in more detail, but again in only a few species. Thompson & Miller (1962) could detect no differences between young leaf primordia of pea seedlings grown in light and darkness but found that light accelerated the later stages of growth, with differentiation of the leaf going to completion only in intense light. They concluded that, in pea leaves, light did not have separate effects on division, enlargement and maturation but merely accelerated the sequence of development. In the primary leaf of *Phaseolus vulgaris*, the large increase of leaf surface in the light has also been shown to result from effects on both the cell division and cell expansion phases of development (Verbelen & de Greef, 1979), but these two phases were influenced differentially. Cell division began and ended slightly earlier in the light, but it proceeded more rapidly and so resulted in the formation of twice as many palisade cells as in dark-grown leaves (Table 9.1). The most striking *qualitative* effect of light was the induction of a phase of cell expansion at the end of the period of intensive cell division. In dark-grown plants this developmental phase was completely lacking in the palisade parenchyma and there was only minimal expansion of the epidermal cells (Table 9.1). Thus, even with closely related species such as *Pisum* and *Phaseolus*, there appear to be some differences in detail in the way in which light controls leaf surface expansion. In part, these may result from the use of different experimental conditions since, at low irradiances (~1.0 lux) the major effect of light on *Vicia faba* leaves was to increase the number of cells, whereas at higher irradiances there were large increases in cell size (Butler, 1963).

The most intensively studied photoresponse in the leaves of monocotyledons has been the unrolling of the primary leaf in members of the Gramineae. The first leaf of a grass seedling is normally folded laterally in the form of a scroll inside the coleoptile. In the dark, the leaf emerges from the coleoptile by an increase in length but it remains rolled. In the light, the leaf unrolls as it emerges from the coleoptile. As with some other photomorphogenetic response such as the straightening of the plumular hook in many dicotyledons (Gee & Vince-Prue, 1976), leaf unrolling involves a differential response of cells to light and results from a greater expansion of cells of the upper mesophyll as compared with the remaining mesophyll tissues (Burström, 1942; Virgin, 1962).

Table 9.1. *Effect of light on the growth of the primary leaf of* Phaseolus vulgaris *cv. Limburg; seedlings were grown at 21 °C in darkness or in continuous white fluorescent light at 5500 lux*

| Final value | Light grown | Dark grown | Approximate increase in light |
|---|---|---|---|
| Leaf surface ($mm^2$) | 2168 | 128 | ×17.0 |
| Leaf thickness ($\mu m$) | 157 | 105 | × 1.5 |
| Upper epidermis | | | |
| cell area ($\mu m^2$) | 1555 | 104 | ×15.0 |
| cell height ($\mu m$) | 17 | 10 | × 1.7 |
| cell volume ($\mu m^3$) | 25 820 | 1016 | ×25.0 |
| Palisade parenchyma | | | |
| cell area ($\mu m^2$) | 221 | 31 | × 7.0 |
| cell height ($\mu m$) | 47 | 31 | × 1.5 |
| cell volume ($\mu m^3$) | 6949 | 962 | × 7.0 |
| cells per leaf | $8454 \times 10^3$ | $4142 \times 10^3$ | × 2.0 |

After Verbelen & De Greef (1979).

Light also has a remarkable influence on leaf shape in several species (Njoku, 1956; Allsopp, 1965; Bensink, 1971).

## Photomorphogenetic systems and photoreceptors

### *The low-energy phytochrome system*

The best known photomorphogenetic system is the red/far-red reversible reaction of the pigment, phytochrome. Following the initial report that red-light-induced germination of light-sensitive lettuce seeds could be prevented by giving a subsequent irradiation with far-red light between 700 and 750 nm (Borthwick *et al.*, 1952), action spectrum determinations for other light-controlled responses indicated that these photoreversible effects of red and far-red light occur in an extremely wide range of plant species covering an equally wide range of developmental processes (see Smith, 1975). These include effects on all aspects of leaf development from the formation of leaf primordia (Mohr & Pinnig, 1962), through leaf expansion (Downs, 1955) to their eventual senescence (Biswal & Sharma, 1976; Tucker, 1982). The photomorphogenetic system involved in these responses has become known as the low-energy (or red/far-red reversible) reaction of phytochrome so as to distinguish it from other responses to light, which require higher energies and may act through phytochrome or through some other photoreceptor.

Detailed accounts of the properties of phytochrome and the characteristics of phytochrome-mediated responses can be found in many review articles and books (e.g. Kendrick & Frankland, 1976; Marmé, 1977; Mohr, 1972; Pratt, 1978, 1979; Satter & Galston, 1976; Smith, 1975). Phytochrome is a soluble protein with a bilitriene type chromophore. Native phytochrome is a multimer of 120 000 dalton subunits, and molecular sieve and sedimentation data are consistent with a non-globular protein of about 240 000 daltons; it is, therefore, probably a dimer (Pratt, 1978). Phytochrome exists in two forms that are interconvertible by radiation (Fig. 9.1). It is synthesized in the form, Pr, which absorbs strongly in red light (with a maximum near 660 nm) and very little in far-red light: on illumination with red light, Pr is photoconverted to the Pfr form which absorbs strongly in far-red (with a maximum near 730 nm) and less strongly in red light. Both forms have weak absorption bands in blue (Pr near 380 nm and Pfr near 400 nm) and both forms have an absorption band at 280 nm due to the protein moiety of the pigment. On illumination with red or white light, a photochemical equilibrium is established almost at once, being reached in about a minute at irradiances as low as 0.1% of sunlight. The position of the equilibrium depends on the wavelength of the radiation. Using un-degraded preparations of phytochrome, the Pfr/P ratio established at 660 nm is 0.75 (Pratt, 1978); this differs slightly from the earlier estimates of 0.81 based on degraded phytochrome (Fig. 9.2). In far-red light at 730 nm, Pfr/P = 0.03; in blue, values range from about 0.35 to 0.45, and with respect to Pfr/P ratio, green light acts like red and

Fig. 9.1. Schematic representation of the major pathways of phytochrome interconversions: Pp, phytochrome precursor; dr, dark reversion; de, destruction reaction; Pd, products of the destruction reaction; x, reaction partner(s) of Pfr; solid lines, thermal (dark) reactions; dashed lines, photochemical reactions (after Mancinelli & Rabino, 1978).

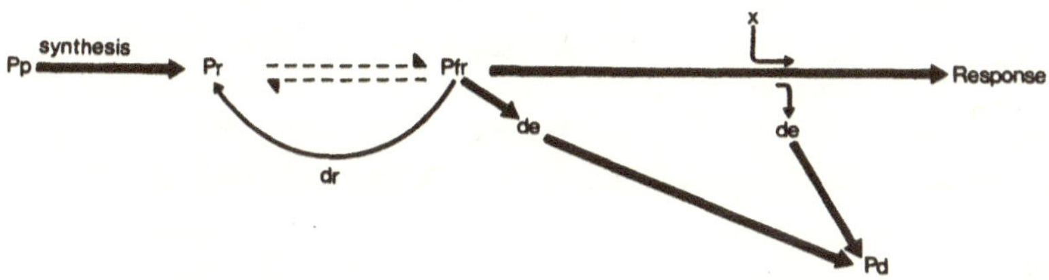


Fig. 9.2. The ratio of Pfr/P established by light of different wavelengths (calculated from photoconversion kinetics and absorption data) (after Hanke, Hartmann & Mohr, 1969).

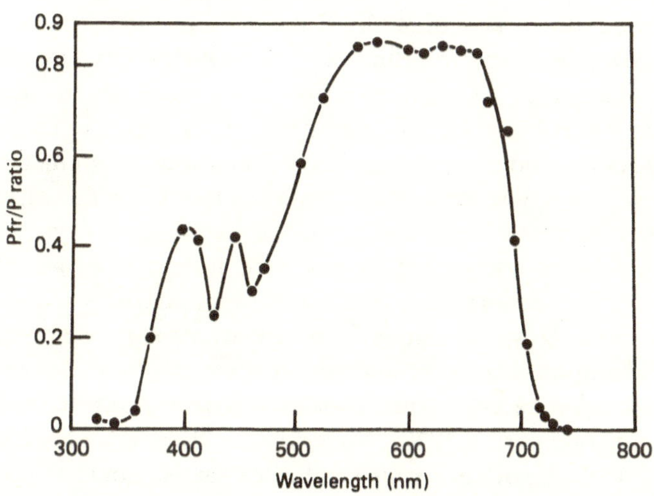

establishes a high value. While illuminated, a steady-state ratio of Pfr/P is maintained but because of the overlapping absorption spectra phytochrome cycles between the two forms at a rate which depends on both wavelength and irradiance (Kendrick & Spruit, 1973; Johnson & Tasker, 1979).

Phytochrome also undergoes a number of non-photochemical reactions (Fig. 9.1) the precise nature and function of which are poorly understood but which have been included in various models put forward to explain its mode of action. These are: (i) *de novo* synthesis in the Pr form; (ii) destruction mainly of Pfr, which involves both loss of the protein moiety (Hunt & Pratt, 1980) and loss of photoreversibility (Jabben & Deitzer, 1978); and (iii) dark reversion of Pfr to Pr, which maintains the photoreversible system. Studies of the phytochrome system in dark-grown plants have shown that only part of the total phytochrome undergoes reversion and that this process occurs more rapidly than destruction (Hopkins & Hillman, 1965). Reversion has not been observed *in vivo* in dark-grown grass seedlings (Butler & Lane, 1965) nor in the Centrospermae (Kendrick & Hillman, 1971). When dark-grown seedlings are transferred to the light, Pfr is formed and destruction occurs so that the content of phytochrome is substantially lower in light-grown than in etiolated plants (Heim, Jabben & Schäfer, 1981).

The scheme presented in Fig. 9.1. assumes that Pfr is the effector molecule for the low-energy reaction of phytochrome. The low-energy system is characterized by the following: the response is inductive (i.e. it is elicited by only brief exposures to light); the necessary exposures are small; maximum action occurs in red light near 660 nm and the response is red/far-red reversible; and the Bunsen–Roscoe reciprocity law holds (Table 9.2). Induction by a single, brief, low-irradiance light treatment and red/far-red reversibility are the operational criteria used to demonstrate phytochrome action in light-mediated responses. The enhanced association of phytochrome with a particulate cell fraction which rapidly follows a brief exposure of etiolated seedlings to red light, (see Pratt, 1978, 1979) has been interpreted as evidence for an association of phytochrome Pfr with a membrane-bound receptor and for its action on membrane function. However, although phytochrome may behave as if it were a peripheral membrane protein under some conditions, such an association has yet to be established conclusively and shown to have a physiological function (Pratt, 1978). Nevertheless, there is a substantial body of physiological evidence which indicates that in some systems,

Table 9.2. *General properties of responses mediated by the reversible reaction of phytochrome and by a high-irradiance reaction*

| Reversible reaction | High-irradiance reaction |
|---|---|
| 1. Inductive type responses | 1. Steady-state type responses |
| 2. Low-exposure requirement ($<0.2 \times 10^4$ J m$^{-2}$) Saturation induced by a single brief (few minutes) exposure at a low irradiance | 2. High-exposure requirement ($>10^4$ J m$^{-2}$) Full expression of the response requires long irradiations (hours or days) at high irradiances Responses are irradiance dependent |
| 3. Red/far-red reversibility | 3. No red/far-red reversibility |
| 4. Peak of action for red effect near 660 nm; peak of action for far-red reversal near 730 nm | 4. Peaks of action in near UV (~370 nm), blue (420–480 nm), red (~650 nm), and far-red (710–730 nm) |
| 5. Response obeys Bunsen–Roscoe reciprocity law | 5. Response does not obey Bunsen–Roscoe reciprocity law |
| 6. Photoreceptor: Phytochrome | 6. Photoreceptor(s): Phytochrome, UV-blue absorbing pigment |

After Mancinelli & Rabino (1978).

light may act by modifying membrane properties (Marmé, 1977). The intracellular distribution of phytochrome is reversibly affected by red and far-red radiation (Epel, Butler, Pratt & Tokuyasu, 1980) and, after red light, phytochrome appears to be in a more sequestered state in the cell.

A single saturating exposure of a few minutes to red light converts about 75% of the total phytochrome present to Pfr and such treatments are sufficient to bring about a considerable degree of response. For example, just 2–3 min of light is sufficient to cause some leaf expansion in dark-grown *Phaseolus* seedlings (Downs, 1955). Plants growing in their natural environments are, however, exposed daily to prolonged periods of high-intensity radiation and, in many cases, illumination for a longer duration is more effective and induces a larger response than can be obtained by a brief exposure to red light. In part, this may be due to the need repeatedly to re-establish Pfr which, after a brief irradiation, is lost from the tissue by the processes of destruction and/or reversion (Fig. 9.1). The characteristics of photomorphogenetic responses under prolonged exposures indicate, however, that this is not the only factor and that one or more 'high irradiance reactions' (HIR) also operate to regulate plant development.

### *High-irradiance reactions*

This class of light-dependent reaction responses has several general properties which are distinct from those characteristic of low-energy red/far-red reaction of phytochrome (Table 9.2). The full expression of the photoresponse requires prolonged exposures to high irradiances of visible or near-visible radiation, the magnitude is a function of both the irradiance and duration of light but does not follow the Bunsen–Roscoe reciprocity law, and the responses do not show red/far-red reversibility. When short exposures to light are given, the maximum action usually occurs near 660 nm (the absorption maximum of Pr), but with longer exposures peaks of action are found in blue, in the near ultra-violet, and in a wide range of red and far-red wavelengths. These properties of the high-irradiance reaction (HIR) cover a wide range of growth and developmental responses of higher plants to prolonged exposures, including responses with rather large differences in the relative effectiveness of the active spectral regions; for example, anthocyanin synthesis in *Sorghum* seedlings occurs only in blue light (Downs & Siegelman, 1963; Drumm & Mohr, 1978), hypocotyl elongation in *Raphanus* seedlings is inhibited by continuous red (but not by a brief pulse) as well as by blue and far-red (Jose, 1977; Jose & Vince-Prue, 1977), while cotyledon expansion in *Sinapis* seedlings is most strongly promoted in far-red (Mohr, 1964).

The HIR has attracted a great deal of interest and controversy, particularly with respect to its role in the natural environment and the identity of the photoreceptor(s). There is a good deal of evidence that both red and far-red light act through phytochrome and various models have been put forward to account for the irradiance dependency and for the action maximum which often occurs at or near 720 nm under prolonged exposures (Hartmann, 1966; Mancinelli & Rabino, 1975; Schäfer, 1975, 1976). These models have recently been reviewed in some detail (Mancinelli & Rabino, 1978; Mancinelli, 1980). Most hypotheses for the HIR involve phytochrome destruction and are based on an interaction of phytochrome with a reaction partner; a basic common concept is that maximum action occurs under conditions (i.e. in far-red light) that result in the presence of Pfr over a long period of time (Mancinelli & Rabino, 1978; Mancinelli, 1980). However, other models have emphasized that the action spectra for most HIR do not bear any obvious relationship to the concentration of Pfr (Johnson & Tasker, 1979), and it has been suggested that the HIR in the red and far-red part of the spectrum depends in some way on the rate of cycling between the two forms of the pigment in continuous light (Jose & Vince-Prue, 1978; Johnson & Tasker, 1979). Thus two actions of phytochrome have been envisaged, a reaction dependent on Pfr and another dependent on pigment cycling.

In the control of leaf growth under natural conditions it is important to note that during de-etiolation with red light, acting through the low-energy reversible reaction of phytochrome, the total phytochrome content falls and the high-irradiance response to far-red light is lost or severely reduced. Hence, the far-red HIR may be of little importance under natural conditions with prolonged daily exposures, even in canopy shade where light is enriched for far-red (Jose & Vince-Prue, 1978).

### *Specific effects of blue light*

Action spectra for HIR commonly show considerable action in the blue part of the spectrum and there have been several attempts to explain these blue-light effects in terms of phytochrome, both forms of which absorb weakly in blue light to establish a Pfr/P ratio intermediate between that in red and far-red (Hendricks & Borthwick, 1959; Hartmann, 1966; Schäfer, 1975). However, there are a number of differences between the effects of blue and red or far-red light which indicate that a specific blue-absorbing photoreceptor, which is distinct from phytochrome, operates in the control of photomorphogenesis (Thomas, 1981). For example, both blue and far-red light inhibit elongation of the hypocotyl in dark-grown lettuce seedlings. The far-red response is seen only in young seedlings, lasts for about 24 h and is eliminated by a red pre-treatment. In contrast, blue light is highly inhibitory at any age, its effect persists throughout the illumination period and is not abolished by a red pre-treatment (Turner & Vince, 1969). Distinctions between effects of blue and far-red have also been made in cucumber seedlings (Black & Shuttleworth, 1974; Thomas & Dickinson, 1979) and in cress (Hart & MacDonald, 1980). At present the identity of such a blue-absorbing photoreceptor is unknown, although the most likely candidate is thought to be a flavoprotein.

The ability of blue light to photoconvert phytochrome is only about one-hundredth of that of the red and far-red part of the spectrum (Hartmann & Unser, 1972), and phytochrome responses in natural conditions would not be greatly affected by the blue component of the light. For this reason, the demonstration of a specific effect of blue light on development in higher plants is significant. An irradiance-dependent blue-sensitive system may be involved in the perception of and response to light intensity whereas phytochrome may be primarily important in the perception of light quality as, for example, under plant canopies (Holmes & Smith, 1977a, b) or in water (Spence, 1981).

## Photomorphogenetic processes in leaves

Many investigations of the photocontrol of leaf development have been concerned solely with plants grown in the dark. Etiolated, dark-grown seedlings are useful tools for investigating photomorphogenesis since transferring such seedlings to light causes a transition to the developmental pattern of fully light-grown plants. Photomorphogenetic phenomena are, however, by no means restricted to etiolated plants and it is important also to consider the photocontrol of leaf growth under the more natural conditions experienced by plants growing in the light. In this chapter these aspects are considered separately.

*Dark-grown plants*

Went (1941) demonstrated that different wavelengths of light had different effects on the rate of growth of leaves of dark-grown pea plants, with the maximum response being in the red region of the spectrum. He further showed that this effect of red light was more pronounced when leaves were illuminated during the period of most rapid growth, but that there was no further effect when growth had stopped. In 1949, Parker, Hendricks, Borthwick & Went showed that the action spectrum for the light-stimulated increase of leaf area in etiolated *Pisum* seedlings had a peak in the red region and was similar to the action spectra for photoperiodic effects on flower initiation and for seed germination (Fig. 9.3) suggesting that these processes are

Fig. 9.3. Action spectra for the control of leaf elongation in etiolated pea seedlings and for control of floral initiation in the long-day plant barley and the short-day plants cocklebur and soybean. The ordinate shows exposure needed to achieve 100% response. (After Parker *et al.*, 1949.)

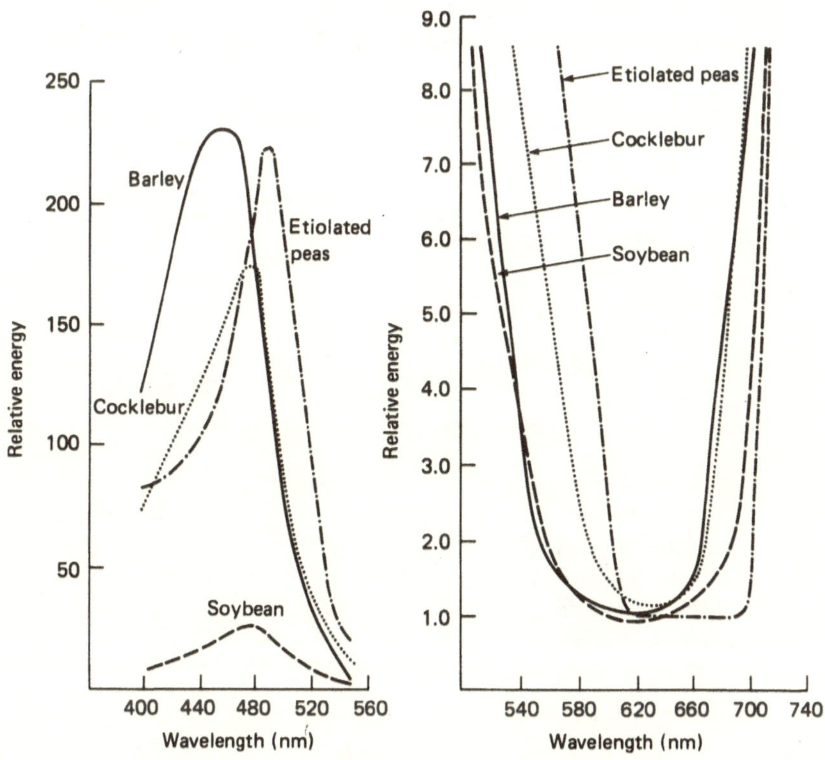

controlled by the same photoreceptor. Later, Liverman, Johnson & Starr (1955) showed that expansion in the leaf disks cut from dark-grown dwarf bean seedlings was red/far-red reversible confirming phytochrome as the photoreceptor (Table 9.3) and demonstrating that the leaves themselves were capable of perceiving and responding to light. Reversibility was independently demonstrated in intact plants by Downs (1955; Table 9.4). Downs found that far-red reversibility of the red light effect was lost very slowly. Only about 20% of the reversibility was lost after 8 h in darkness indicating an extremely stable 'active' photoreceptor. This was later confirmed by Klein (1969) who concluded that complete disappearance of Pfr could not have taken place for at least 48 h, as shown by the persistence of the sensitivity to far-red (Fig. 9.4). During this period the far-red treated seedlings were still capable of normal growth responses if a second promoting red irradiation was given to a group of seedlings that had begun to show the effects of

Table 9.3. *Promotion of expansion of disks cut from dark-grown bean leaves by red light and its reversal by far-red light; results expressed as increase in diameter of 5 mm disks after 48 h growth in darkness*

| Treatment | Duration of light treatment (h) | Increase in diameter (mm) |
|---|---|---|
| (1) Red (fluorescent source and filter) | 0.5 | 1.7 ± 0.04[a] |
| (2) Same as (1), followed by far-red | 1 | 1.1 ± 0.02 |
| (3) Red (incandescent and filter) | 0.5 | 1.6 ± 0.03 |
| (4) Same as (3), followed by far-red | 1 | 1.0 ± 0.03 |
| (5) Red (photographic safe-light, 60 W) | 0.5 | 1.5 ± 0.04 |
| (6) Same as (5), given continuously | 48 | 2.0 ± 0.02 |
| (7) Far-red control | 1 | 0.9 ± 0.02 |
| (8) Dark control | 48 | 1.0 ± 0.02 |

After Liverman *et al.* (1955).
[a] Standard error.

far-red reversal. Fig. 9.4 shows that red light re-starts the growth of leaves indicating their continued competence to respond to illumination. In some processes, e.g. regulation of flowering and seed germination, reversibility may be lost with the completion of a crucial irreversible step in the responding system other than the decay of the phytochrome pigment. In leaf growth, however, Klein found no evidence for such a step because red light-induced growth was essentially linear over at least 96 h and the slope of the response could be altered at any point (Fig. 9.4). In view of the long period that the leaves remained sensitive to far-red reversal of the red light effect, it appears that at least some fraction of the Pfr does not decay with time but remains associated with the site of its regulatory functions. Recent *in vivo* studies have confirmed that there is a stable component of the total Pfr which decays only very slowly in darkness (Heim, Jabben & Schäfer, 1981).

The red/far-red reversible effect on leaf expansion has also been demonstrated in other species. Sale & Vince (1963) found a promotive effect of red light on leaf expansion in *Tropaeolum majus* and *Pisum sativum* which was largely reversed by far-red light and concluded that the low-energy red/far-red reversible reaction is one of the systems controlling leaf development in these plants. Light also affects cotyledon expansion and this is illustrated in Fig. 9.5. Mohr (1959) found that the cotyledons of *Sinapis alba* almost completely failed to expand in complete darkness, but they enlarged following a 5-min exposure to low-intensity red light. If this irradiation was followed immediately by far-red light for up to 64 min the red effect was completely nullified: longer exposures to far-red promoted cotyledon expansion through a high-irradiance reaction (see below).

Table 9.4. *Leaf lengths of dark-grown red kidney bean seedlings exposed alternately to 2 min of red and 5 min of far-red*

| Treatment | Leaf length (mm) ± s.e. |
|---|---|
| Dark control | 17 ± 0.8 |
| Red | 44 ± 0.8 |
| Red, FR | 21 ± 0.4 |
| Red, FR, Red | 45 ± 0.2 |
| Red, FR, Red, FR | 21 ± 0.4 |
| Red, FR, Red, FR, Red | 46 ± 0.3 |

After Downs (1955).

There is evidence that short periods of low-irradiance red light promote leaf expansion through effects on cell numbers, with cell volume being increased only at much higher irradiances (Powell & Griffith, 1960; Butler, 1963). Using dark-grown plants of *Phaseolus vulgaris*, Murray (1968) showed that irradiation with white light for 10 min increased cell number of the primary leaves by more than 50%. She concluded that the effects of short irradiations probably involve

Fig. 9.4. Reversal of far-red inhibition of leaf growth in *Phaseolus* (*a*) for fresh weight (*b*) for protein, per primary leaf pair. Seven-day-old etiolated seedlings were illuminated for 24 h with white light and then either returned to the dark (L + D) or irradiated for 7 min with far-red light and then returned to the dark (L + FR + D). Six hours later a group of far-red treated plants were exposed to red light for 10 min and then returned to the dark (L + FR + R + D). One group received no illumination (D). (After Klein, 1969.)

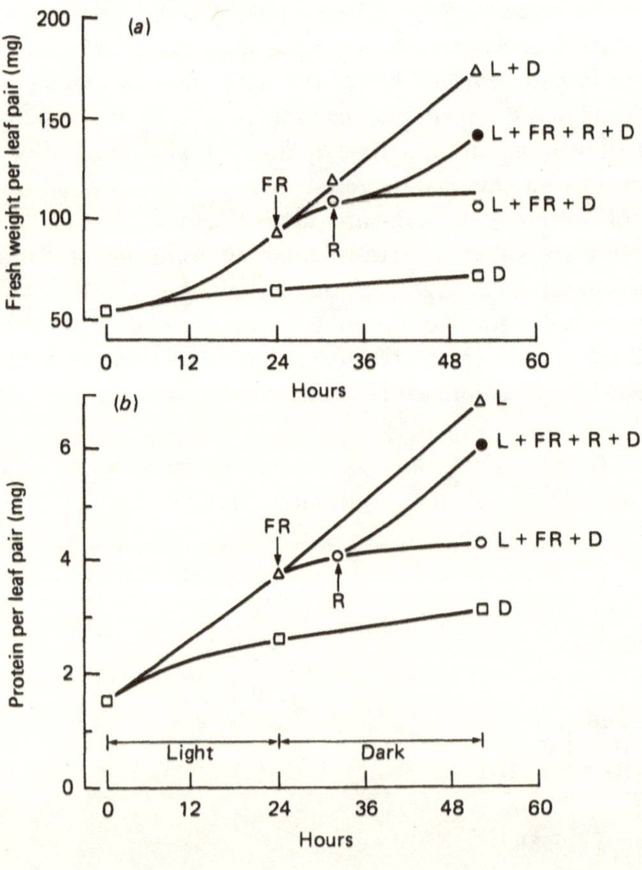

only the phytochrome system, and that the expansion of leaves of dark-grown plants or disks cut from such plants is due principally to effects on cell division. Further evidence for this theory comes from the work of Dale & Murray (1969) who found that leaves of *Phaseolus* plants grown in the dark showed an increase in fresh weight and cell number when they were irradiated for 10 min with red light; no responses to short exposures of blue or far-red were observed. If the red light exposure was followed immediately by 10-min far-red, there was no detectable increase in cell number (Table 9.5). Cotyledon expansion in *Sinapis*, however, involves only cell enlargement.

The main photomorphogenetic effect of brief periods of light on dark-grown monocotyledonous plants so far studied is the influence on leaf unrolling in grasses. Virgin (1962) determined the action spectrum for light-induced unrolling of the first leaf of wheat seedlings and found it to be promoted by red light, the promoting effect being nullified by far-red. Red/far-red reversible control of leaf unrolling has also been demonstrated in other cereals (Steiner, Price, Mitrakos & Klein, 1968;

Fig. 9.5. The influence of red and far-red light on cotyledon expansion in mustard (*Sinapis alba*). The seedlings were irradiated with 5 min red or 5 min red followed immediately by far-red for various durations as shown on the abscissa; cotyledon areas were measured 96 h after sowing. (After Mohr, 1959.)

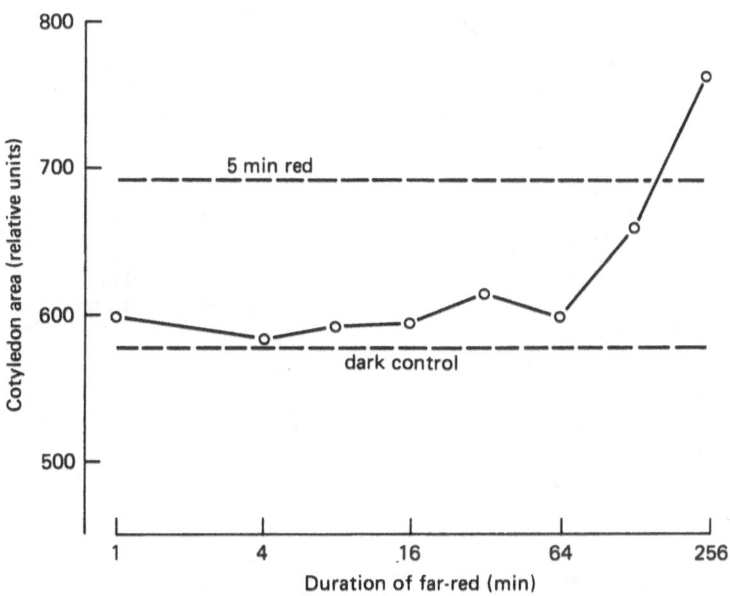

Wagné, 1964; Carr, Clements & Menhennet, 1972) and identifies phytochrome as the photoreceptor. A possible exception is rice where leaf unrolling occurred only in blue light (Katsura & Inada, 1979) and increased with increasing exposure to be maximal after 1 h at ≈0.8 W $m^{-2}$ (Sasakawa & Yamamoto, 1980); the response to blue light was counteracted by red but not by far-red and it was concluded that phytochrome and a blue-absorbing photoreceptor interact to control leaf expansion in rice. A blue light effect on leaf unrolling in barley was, however, reversed by far-red, and probably operates through the low-energy reversible reaction of phytochrome (Deutch & Deutch, 1975).

An interesting recent observation (Atkinson, Bradbeer & Frankland, 1980) is that mutant lines of barley with pigment-deficient plastids unroll

Table 9.5. *The effect of irradiation with red and far-red light on cell number of primary leaves of* Phaseolus vulgaris

| Treatment | Cell number per leaf (millions) ± s.e. |
|---|---|
| Dark control | 20.7 ± 1.31 |
| Red light | 27.9 ± 1.07 |
| Red followed by far-red | 19.2 ± 0.90 |

After Dale & Murray (1969).

Table 9.6. *Effects of red and far-red treatments on unrolling of mutant (white) and normally pigmented barley leaves*

| Light treatment | Increase in leaf width after 24 h (mm) | |
|---|---|---|
| | Saskatoon (white) | Mazurka (normally pigmented) |
| none | 0.5 | 0.6 |
| 2 min red | 0.8 | 1.8 |
| 2 min red & 8 min far-red | 0.6 | 1.3 |
| 8 min far-red | 0.8 | 1.3 |
| 8 min far-red & 2 min red | 1.0 | 1.8 |
| 24 h far-red | 1.7 | 1.8 |

After Atkinson *et al* (1980).

more slowly in response to brief red or continuous white light than do normally pigmented leaves of sister lines and also fail to show red/far-red reversibility (Table 9.6). The mutant leaves appeared to contain only about 40% of the phytochrome found in normal leaves and so Pfr levels were lower after red light (Table 9.7). It is not clear that this, in itself, is sufficient to account for the slow response but it is noteworthy that the rate of leaf unrolling in continuous far-red (which would be expected to maintain low levels of Pfr and so reduce phytochrome destruction) is similar to that in normally pigmented leaves.

As with dicotyledonous plants, the insertion of a dark period between the red and far-red irradiation treatments causes a partial failure of the reversibility of the response, the degree of reversibility remaining being dependent on the length of the dark period inserted and the species used. With barley, for example, the ability of far-red light to reverse the effects of red is lost completely after about 3 h (Fig. 9.6) but with wheat, far-red reversibility is lost after only 10 min (Cooke & Saunders, 1975).

Leaf unrolling in cereals results from an effect of light on cell expansion and this response affords perhaps the best evidence that the photocontrol of leaf expansion involves light-dependent changes in hormone levels. In the investigations of Loveys & Wareing (1971a, b), an attempt was made to evaluate the role played by gibberellins (GAs) in wheat leaf unrolling. $GA_3$ appears to substitute for red light in the stimulation of unrolling, and a rapid rise in endogenous GA-like

Table 9.7 *Changes in phytochrome content and form of pigmented and unpigmented barley leaves during 1 h of darkness after 2 min red light*

| | | Measured content (relative units) | |
|---|---|---|---|
| Leaf pigmentation | Phytochrome constituent | after 2 min red | after 1 h dark at 20 °C |
| Normal | Total phytochrome (P) | 100 | 84 |
| | Pfr | 75 | 49 |
| | Pfr/P | 0.75 | 0.56 |
| White | Total phytochrome | 37 | 30 |
| | Pfr | 28 | 10 |
| | Pfr/P | 0.75 | 0.33 |

After Atkinson *et al.* (1980).

substances occurred in etiolated barley leaf sections following a brief exposure to red light. Irradiation of homogenates of etiolated barley leaves also increased the level of extractable GA-like activity (Reid, Tuing, Durley & Railton, 1972) as did irradiation of suspensions of intact etioplasts (Cooke, Saunders & Kendrick, 1975); the GA-like substances produced in etioplasts were capable of passing into the surrounding medium within 20 min and so may be directly involved in promoting leaf unrolling. More recent data confirm that purified barley etioplasts have associated with them small but consistently detectable amounts of photoreversible phytochrome and that they respond *in vitro* to red light by an increase in the level of extractable GA-like substances (Hilton & Smith, 1980). Since the plastid envelope is generally believed to be impermeable to terpenoids and related com-

Fig. 9.6. The escape from photoreversibility of the barley leaf unrolling process. Seedlings were given red light (600 J $m^{-2}$) followed by far-red light (1800 J $m^{-2}$) after various time intervals. Projected leaf widths were measured 24 h after the red light treatment. (After Smith, 1975.)

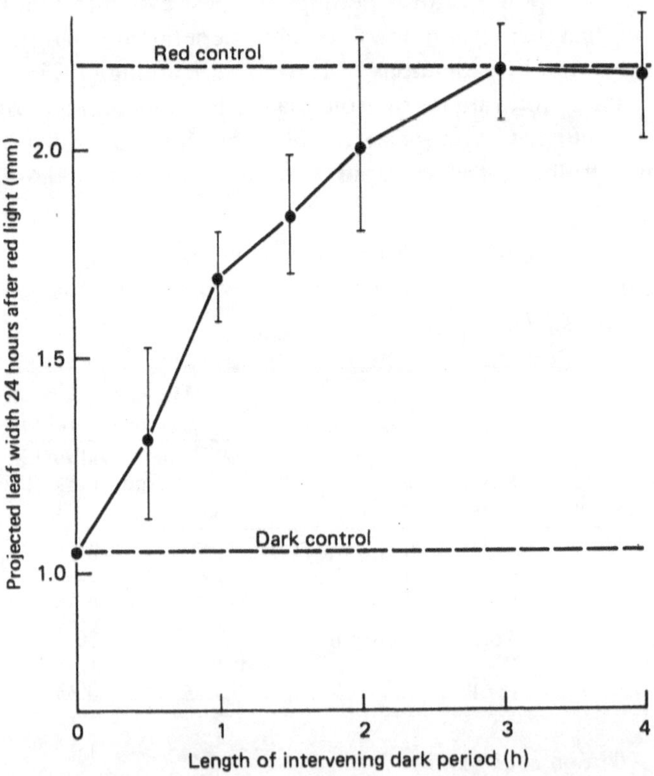

pounds (Rogers, Shah & Goodwin, 1965) it is tempting to speculate that phytochrome is associated with the etioplast envelope and induces changes in its permeability to GAs; however, the GA production in etioplasts appears to be more complex than this and inhibitor experiments indicate that there may be release from a 'bound' form or that *de novo* synthesis may be involved (Cooke *et al.*, 1975).

Further evidence for the role of gibberellins in leaf unrolling has come from studies on the effects of far-red reversal of the red light effect. Cooke & Saunders (1975) found that the rise in levels of GA-like substances induced by a short period of red light was reversed by an immediate subsequent irradiation with far-red light, but like the loss of the leaf unrolling response, this reversibility was lost if a dark period exceeding 10 min was inserted between red and far-red treatments. They also demonstrated that the stimulus for leaf unrolling can pass from irradiated to non-irradiated sections of partly irradiated leaves within 15 min and that increased levels of GA-like activity were detectable in non-irradiated sections of similarly treated leaves within the same time.

All the available evidence then indicates that leaf unrolling in monocotyledonous plants is probably mediated by certain GA-like substances, the production of which is stimulated by red light. There is little information to suggest that phytochrome-mediated changes in other hormones may be involved in the leaf unrolling process, though in barley it has been demonstrated that the levels of an extractable unidentified inhibitor of leaf unrolling decreased following red irradiation of dark-grown leaves (Carr *et al.*, 1972). The role of cytokinins in leaf expansion seems equally obscure. In developing rye leaves total leaf growth was stimulated by cytokinins in the dark but their effect did not mimic that of white light. In the dark, protein synthesis seemed to be determined by the available substrate pools irrespective of the cytokinin supply, whereas in light the incorporation of amino nitrogen into protein was clearly increased relative to that in dark-grown leaves (de Boer & Feierabend, 1978). It was concluded that the cytokinin effect on leaf growth may be mediated indirectly by correlative phenomena between different organs of the seedling.

In dicotyledonous plants the relationship between light and the hormonal control of leaf expansion is even less well understood and though both gibberellins (Liverman, 1959) and cytokinins (Miller, 1956) have been implicated in the photocontrol of leaf expansion the evidence is by no means conclusive.

*High-irradiance reactions in dark-grown plants*

The effect of the far-red HIR on 'leaf' growth has been studied in detail in cotyledons of *Sinapis alba* where light-dependent expansion appears to be governed by two reactions (Mohr, 1959). When brief irradiations were given, maximum expansion was induced by red light and the response was red/far-red reversible (Fig. 9.5). As the duration of exposure to far-red increased beyond 1 h, it failed to reverse the red effect and, after about 4 h in far-red light, cotyledon expansion was greater in far-red than in red (Fig. 9.5). The wavelength response to equal photon irradiances given for 6 h and corrected for the low-energy phytochrome reaction shows a main peak in far-red with a secondary peak in blue (Fig. 9.7). This typical HIR has the characteristics given in Table 9.2 (Schäfer & Mohr, 1974). Cotyledons of other species also

Fig. 9.7. Wavelength response curve for cotyledon expansion in *Sinapis alba* under high irradiance conditions. Light was given for 6 h and the response curve corrected for the effect of the low-energy red/far-red reversible reaction of phytochrome. (After Mohr, 1959.)

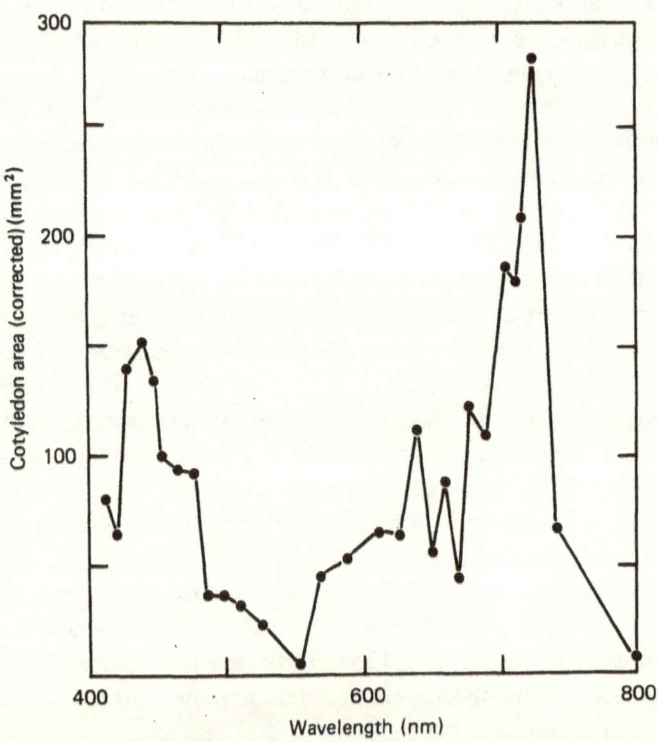

show enlargement in far-red; for example, in *Pharbitis nil* the cotyledon areas ($cm^2$) after various light treatments were 1.9 (dark), 2.7 (10 min red) and 3.6 (24 h far-red); after 24 h in white light the cotyledon area was 4.0 $cm^2$ (King & Vince-Prue, 1978). However, the response to far-red was not shown to involve a HIR. As far as the authors are aware, no critical studies have been made on true leaves. Intact etiolated leaves of *Phaseolus* were reported to have expanded markedly after 48 h in far-red light (>700 nm) and cell enlargement appeared to be predominant since expanded cells with large central vacuoles were present (de Greef, Butler & Roth, 1971). Leaves of *Pisum* and *Tropaeolum* also showed some expansion in far-red light (Table 9.8), but much less than in blue or red.

Irradiance-dependent responses to blue and red light were observed in leaves of *Pisum sativum* and *Tropaeolum majus* (Sale & Vince, 1963). Seedlings were exposed to three levels of photo-irradiance in blue and red for 0.25–24 h each day from emergence, and in some cases sequential treatments were given with different light qualities. The results indicated that leaf expansion in red light was dependent not only on the red/far-red reversible reaction of phytochrome but also on a HIR, since leaf expansion was increasingly promoted as the irradiance and duration of red light were increased. There was also some evidence that a specific

Table 9.8. *Effects of sequential irradiation treatments with different wavebands on leaf area in* Pisum sativum *(one leaflet of second leaf) and* Tropaeolum majus *(one of first leaf pair)*

| Daily irradiation treatment | | Final leaf area ($mm^2$) | |
|---|---|---|---|
| 4 h | 4 h | *Pisum* | *Tropaeolum* |
| blue | dark | 176 | 329 |
| blue | far-red | 160 | 305 |
| red | dark | 216 | 321 |
| red | far-red | 127 | 171 |
| blue | blue | 171 | 599 |
| red | red | 199 | 501 |
| blue | red | 335 | 753 |
| red | blue | 197 | 472 |
| far-red | far-red | 118 | 82 |
| dark | dark | 19 | 13 |
| Significant difference at $P = 0.05$ | | 36 | 116 |

After Sale & Vince (1963).

blue-absorbing photoreceptor may interact with phytochrome in the control of leaf expansion in both plants. Although leaf areas were similar when plants were exposed to 4 h of blue or red light at equal photon irradiance, the effect of blue, unlike that of red, was not significantly reduced by subsequent far-red. Moreover, the sequence 4 h blue followed by 4 h red resulted in much larger leaves than either waveband alone, or when the light treatments were given in the reverse order (Table 9.8). A similar interaction between blue and red light has been demonstrated in the control of light-dependent anthocyanin synthesis in *Sorghum*, where blue light was obligatory (Downs & Siegelman, 1963; Drumm & Mohr, 1978) but the nature of any interaction between phytochrome and a blue-absorbing photoreceptor is not understood at present (Mohr, 1980; Thomas, 1981).

### *Photomorphogenesis in light-grown plants*

Only recently have critical attempts been made to determine the importance of the various photomorphogenetic reaction systems for the development of plants growing in natural conditions. While de-etiolation phenomena under phytochrome control (and perhaps also dependent on blue light) are clearly important in the early stages of seedling development, the reactions which modulate leaf expansion in light-grown plants are less well understood. The experiments involve methodological difficulties because of their long-term nature and the possibility of confounding photomorphogenetic effects with those due to photosynthesis. The morphogenetic responses to light in green plants are not necessarily coupled to the presence of chlorophyll. For example Trumpf (1924) showed that normal light-adapted development can occur without the formation of chlorophyll or photosynthesis provided sufficient substrate reserves are available. However, in natural conditions, leaf growth and greening are linked and the expression of photomorphogenetic responses in the natural environment is ultimately limited by photosynthesis.

The natural radiation environment varies markedly with respect both to the irradiance and daily duration of light. In contrast, the spectral photo distribution of sunlight does not change much throughout the hours of daylight, although there are significant changes during sunset and sunrise when the red/far-red ratio decreases from the usual value of about 1.1 to a value of about 0.7 (Holmes & McCartney, 1976). Selective attenuation by vegetation, however, produces shade light much depleted in both red (and, therefore, with a low red/far-red ratio) and blue wavelengths, as well as in total irradiance (Holmes & Smith, 1977a). Selective attenuation by water, on the other hand, leads to red/far-red ratios much higher than those in terrestrial habitats (Spence, 1981). It has been suggested that photomorphogenesis in natural environments may be continually modulated through perception of the red/far-red ratio of light by phytochrome and perception of its irradiance by a blue-absorbing photoreceptor (Smith, 1981).

There have been several approaches to the study of photomorphogenesis in light-grown plants. These include growing plants under standard conditions during the main light period to ensure comparable photosynthesis and manipulating light quality at the end of the day, or during a day extension given at low irradiance, so that the energy of the supplementary light is negligible in comparison with that of the main light period. Other methods have involved growing plants for long periods in simulated shade light (i.e. with different ratios of red/far-red) but with constant levels of photosynthetically active radiation, or under lamp combinations giving different spectral photon distributions in the visible spectrum.

When plants are grown in white light but are exposed daily to brief end-of-day treatments with far-red light, leaf expansion is generally reduced (Frankland & Letendre, 1978) and leaf shape may also be modified (Kasperbauer & Hiatt, 1966; Kasperbauer & Peaslee, 1973; Sanchez, 1971). These effects are reversible by red light demonstrating participation of the low-energy phytochrome reaction in the control of leaf expansion and the determination of leaf shape in plants growing in the light. Red/far-red reversible effects on total leaf area have been reported for the woodland species, *Circaea lutetiana* (Table 9.9) and in the gloxinia, *Sinningia speciosa* (Satter & Wetherell, 1968). In the latter, far-red pulses continued to be effective after many hours in darkness, supporting the conclusion from studies with etiolated plants (Klein, 1969) that physiologically active Pfr is relatively stable in the dark. Effects on leaf shape have been reported in tobacco, where leaves of plants receiving 5 min far-red at the end of each 8 h day were narrower than those receiving red light, though the specific leaf area remained unchanged (Table 9.10). Light quality was also shown to have a striking effect on heteroblastic development in *Taraxacum officinale*, which displays a wide range of leaf shapes from a more round to a deeply incised runcinate form (Sanchez, 1971). The depth of the leaf incision was quantified by a shape index which decreased as the depth of

Table 9.9. *Effects of far-red light at the end of the day on leaf growth in* Circaea lutetiana; *measurements were made after 6 weeks treatment*

| | 12 h fluorescent light per day | 15 min far-red given at the end of the light period | Far-red followed by 15 min red |
|---|---|---|---|
| Total leaf area ($cm^2$) | 105 | 64 | 119 |
| Leaf area ratio ($cm^2$/mg dry weight) | 0.20 | 0.16 | 0.18 |
| Specific leaf area ($cm^2$/mg dry weight) | 0.44 | 0.56 | 0.51 |

After Frankland & Letendre (1978).

the leaf incisions increased. Daily exposures to brief periods of far-red light at the end of a 10 h photoperiod prevented the formation of runcinate leaves so that only rounded or slightly incised leaves developed (Fig. 9.8). These effects of far-red were reversible by red light. The amount of assimilate was not altered by the different light quality treatments but the effect of far-red light was influenced by the irradiance of the photoperiod; increasing the white light irradiance significantly decreased the effect of a given far-red exposure (Sanchez & Cogliatti, 1975). It was concluded that the photocontrol of leaf shape may be the result of an interaction between phytochrome and an irradiance-dependent reaction, assumed to be photosynthesis.

It is clear that Pfr modulates leaf expansion in light-grown plants since its absence for substantial periods during each day (following exposure to far-red light before transfer to darkness) alters the pattern of leaf development. It is less certain, however, that the modifications to leaf growth which occur in shade result directly from the change in light quality. In canopy shade, the red/far-red ratio is substantially reduced throughout the daylight hours, and in simulated canopy shade experiments such changes in light quality have been shown to modify developmental patterns in the absence of any change in light quantity (Holmes & Smith, 1977b). With respect to leaf development, however, there is some evidence that the reduced irradiance of shade light conditions may be more important than the red/far-red ratio. Plants growing in natural shade adapt to make maximum use of available light by increase in leaf area ratio and specific leaf area, as well as showing other changes such as in stem elongation and specific leaf water content.

Table 9.10. *Shape of leaves of two lines of* Nicotiana tabacum *plants that were irradiated at the end of each day with 5 min red (R), 5 min far-red (FR), or 5 min far-red followed immediately by 5 min red light (FR/R)*

| | Radiation treatment | | |
|---|---|---|---|
| Line | R | FR | FR/R |
| | Ratio of leaf length to width | | |
| Iso 1 | 1.83 | 2.13 | 1.85 |
| Iso 2 | 2.59 | 2.98 | 2.49 |

After Kasperbauer & Hiatt (1966).

In the woodland species, *Circaea lutetiana*, the pattern of leaf growth responses to natural shade was most closely simulated by varying the irradiance while maintaining a constant but high ratio of red/far-red (Frankland & Letendre, 1978). Effects of end-of-day far-red (Table 9.9) and of far-red enrichment during the photoperiod demonstrated that there are Pfr-controlled responses in *Circaea*, but there was no evidence that changes in spectral quality are necessary for the detection of shade in control of leaf growth in this plant. Similarly, in aquatic species of *Potamogeton* it appears that differences in specific leaf area must be attributed to differences in daily light level rather than to any change in light quality (Spence, 1981). However, the red/far-red ratio may be extremely important in determining leaf shape in some aquatic plants. In *Hippuris vulgaris*, for example, aerial-type leaves are produced near the surface of the water where the red/far-red ratios are lower than at

Fig. 9.8. Effect of end-of-day light quality on leaf shape in *Taraxacum*: 10 W, 10 h fluorescent light $d^{-1}$; 10 W/FR, 10 h fluorescent light $d^{-1}$ followed by 30 min far-red light; 10 W/FR/R, 10 h fluorescent light $d^{-1}$ followed by 30 min far-red light and 20 min red light. (After Sanchez, 1971.)

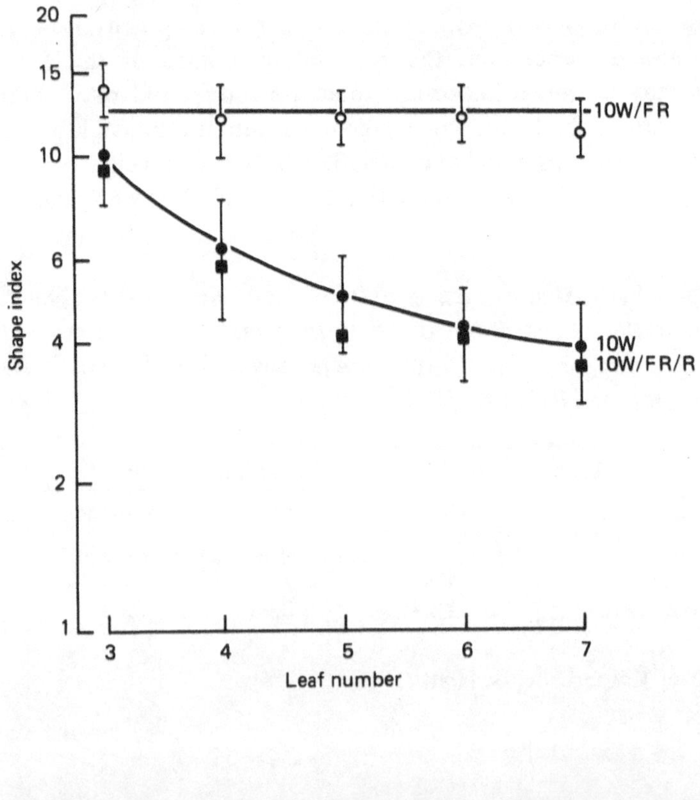

greater depths, and aerial and submerged leaf forms have been produced experimentally by changing the red/far-red ratio in controlled conditions (Spence, 1981).

Morphogenetic responses to changes in irradiance under natural conditions could be effected through phytochrome acting in a high-irradiance mode (Frankland & Letendre, 1978) or through a blue-absorbing photoreceptor; photosynthesis may also be involved. Results from growing plants under different spectral photon distributions are difficult to interpret because of the possible confounding effects of photosynthesis but, in general, increasing the relative amount of blue light in white light resulted in the production of smaller, thicker leaves than in plants grown under lamp combinations enriched in red (Warrington & Mitchell, 1976). There were, however, substantial differences between species in the degree of response. The relative importance of phytochrome and a blue-absorbing photoreceptor and their possible interactions in determining the pattern of leaf growth under natural conditions remain to be determined. Blue light appears to be required for the formation of sun-type chloroplasts, which develop few granal stacks, a higher chlorophyll a/b ratio and lower carotenoid content. The effect of high irradiance white light can be mimicked by giving low photon irradiances in blue ($5.5\,\mu mol\,m^{-2}\,s^{-1}$), whereas in red light at the same photon irradiance only shade-type chloroplasts are produced (Lichtenthaler, Buschmann & Rahmsdorf, 1980).

The recent interest in the possible effects of enhanced radiation in the ultraviolet between 280–320 nm (UV-B) such as would occur with a substantial reduction in ozone in the upper atmosphere, has led to several studies of plant responses. Any increase in UV-B levels inhibited both photosynthesis and leaf growth and even UV-B levels corresponding to present-day solar radiation caused a reduction in leaf expansion (Sisson & Caldwell, 1977). Although the protein moiety of phytochrome absorbs at 280 nm and could transfer energy to the chromophore, the phytochrome control of leaf shape that can be demonstrated in *Rumex* appeared to operate independently of UV-B (Lindoo & Caldwell, 1978).

### *Leaf senescence*

There is some evidence that phytochrome affects not only the expansion and form of leaves but is also important with respect to their senescence. Sugiura (1963) has reported that senescence of tobacco leaf disks taken from mature leaves and floated on water in darkness is influenced by the phytochrome system, with red light delaying senescence. A repetition of his experiments by Goldthwaite & Laetsch (1967) failed, however, to confirm this finding. The only published data which clearly implicate phytochrome and exclude any possible involvement of photosynthesis are those of Biswal & Sharma (1976). They found that a 5 min irradiation with red light was as effective as continuous white light in arresting the senescence of detached leaves of barley and that this red effect was reversible by far-red (Table 9.11). Recently Tucker (1982) has shown that a 5 min irradiation with red light every 24 h was as effective as continuous white light in delaying senescence of detached whole leaves and leaf disks of tomato and cucumber plants; this red effect was also far-red reversible (Table 9.12). The state of phytochrome in leaves may, therefore, be one of the factors which regulates their senescence; a high level of Pfr appears to delay senescence and the senescence of old leaves which become shaded by other leaves may be determined in part by the fact that they are exposed to low red/far-red ratios. Species may, however, differ since there appears to be no effect of phytochrome on senescence in leaves of *Phaseolus* (Goldthwaite & Laetsch, 1967).

Table 9.11. *Phytochrome regulation of senescence in detached barley leaves; results are amounts present after 120 h expressed as percentages of the initial protein and chlorophyll levels*

| Treatments | Protein | Chlorophyll |
|---|---|---|
| (1) 120 h continuous darkness | 35 | 36 |
| (2) 120 h continuous white light | 71 | 70 |
| (3) 5 min red light exposure given every 12 h | 60 | 63 |
| (4) 5 min far-red light exposure given every 12 h | 35 | 35 |
| (5) 5 min red followed immediately by 5 min far-red light exposure every 12 h | 34 | 35 |

After Biswal & Sharma (1976).

**Effects of daylength**

In addition to its direct effects on morphogenesis, phytochrome interacts with a time-measuring process to effect daylength-dependent responses. The photoperiodic mechanism has largely been studied with respect to the control of floral initiation, but daylength also affects many other aspects of the plant's development, including the growth and morphology of leaves (Vince-Prue, 1975). The underlying mechanism is apparently similar in all cases and has been shown to involve the low-energy red/far-red reversible reaction of phytochrome and, in some cases, possibly also a high-irradiance response (Vince-Prue, 1980).

Responses to daylength have been observed in the leaves of many species but only in a relatively few cases have these been demonstrated to be true photoperiodic effects, i.e. to depend on the timing of the light in a 24 h cycle and not on some other facet such as the total light energy received.

Effects of daylength on leaf growth may be complicated by the transition to flowering since bracts are usually smaller and simpler than leaves. However, when compared in the vegetative phase, the leaves of most plants seem to be larger in long days than in short days; larger leaves in short days have been recorded in a few species, e.g. in *Plantago* and *Oryzopsis*, and in some cases there is no effect, e.g. in *Salvia* and *Ageratum* (Vince-Prue, 1975). The effect of photoperiod on leaf expansion has been studied in some detail in *Callistephus chinensis* (Cockshull, 1966), where long-day treatment was found to accelerate

Table 9.12. *Effect of red and far-red light on senescence of detached cucumber leaves. Red and far-red treatments were given every 24 h. Results are amounts present after 48 h expressed as percentages of the initial protein and chlorophyll levels*

| Treatments | Protein | Chlorophyll |
|---|---|---|
| Continuous white light | 84.5 | 91.0 |
| Darkness | 60.2 | 68.5 |
| 10 min red | 83.1 | 91.8 |
| 10 min red/10 min far-red | 61.0 | 71.1 |
| 10 min far-red | 61.6. | 68.0 |

After D. J. Tucker (unpublished data).

the expansion of leaf surface and to increase the specific leaf area. As in the photoperiodic control of flowering in many plants, the effect of long days was simulated by short days together with a brief night-break interrupting the long dark period. Cessation of the night-break treatment reduced expansion even when the leaves were already three-quarters grown and so daylength continued to exert an effect throughout the entire expansion phase of the leaf.

In the control of flowering in some species, night-break treatments are relatively ineffective in bringing about long-day responses (Vince-Prue, 1975) and prolonged daily exposures to light may also be necessary in order to promote leaf expansion. This type of response is illustrated in strawberry, where short days plus night-breaks do not promote leaf growth nor inhibit flowering; long day extensions using tungsten filament lamps containing both red and far-red wavelengths both inhibited flowering and increased leaf area (Tafazoli & Vince-Prue, 1978; Vince-Prue & Guttridge, 1973).

In general, the effects of daylength on leaf area seem to result from a change in water content and distribution of dry matter, as in *Callistephus*, although the increased leaf area in strawberry resulted from an increase in cell number based on counts of epidermal cells (Arney, 1956). The usual effect of long days is to increase surface expansion, and lead to the development of thinner, less-succulent leaves with an increased specific leaf area. Such changes are not, however, invariably associated with long days and in the woodland species, *Circaea lutetiana*, the specific leaf area was higher when plants were growing in short days (Frankland & Letendre, 1978).

Changes in leaf shape in response to daylength have been observed in several terrestrial and aquatic plants. Deeply dissected submerged leaves and simple aerial leaves are characteristic of many aquatic species of *Ranunculus*. Both types are initiated under water and photoperiodic treatments in terrestrial environments indicated that the final leaf form is largely controlled by daylength. In long days, *R. aquatilis* produced land-type leaves whereas in short days typical submerged leaves developed. The semi-aquatic plant, *Proserpinaca palustris*, produced divided juvenile leaves in 8 h photoperiods and lanceolate-serrate adult leaves in 12 h photoperiods (Davis, 1967). The transition to the production of adult leaves was not correlated with flowering for floral initiation occurred only with daylengths longer than 12 h.

The alteration of leaf morphology by photoperiod can take many forms. The induction of winter resting buds by short days modifies the developmental pathway in such a way that the leaf primordia develop into scale leaves instead of foliage leaves. The formation of bulbs in onion in response to long days also involves profound modifications of growth to form scale leaves. A change in length:breadth ratio of leaves is a fairly common response and in both *Chrysanthemum* (Vince, 1955) and *Chenopodium* (Thomas, 1961) the ratio was greater in short days. In succulent plants both the degree of succulence as well as leaf morphology can be influenced by photoperiod. *Bryophyllum crenatum* and *Kalanchoe blossfeldiana* have different photoperiodic requirements for flowering but show similar leaf growth responses with a marked increase in succulence and leaf thickness in short days (Zeevaart, 1969; Schwabe, 1969).

The photoperiodic stimulus affecting leaf growth has been shown to be transmissible in both strawberry (Guttridge, 1959) and *Kalanchoe* (Harder, 1948). Thus, as with other photoperiodic events, exposing leaves to particular daylengths brings about the production of transmissible substances which can alter developmental patterns in other parts of the plant.

## Conclusions

Light affects leaf development by increasing cell division and cell expansion, and by inducing differential patterns of division and expansion of cells to bring about unrolling of leaves as in cereals or to cause changes in leaf shape.

The low-energy red/far-red reversible reaction of phytochrome has been identified in the control of these effects in leaves. It has been shown to operate in dark-grown seedlings where leaf expansion or unrolling are promoted by the formation of Pfr, and also in light-grown plants where leaf size and shape may be influenced. Phytochrome also affects leaf growth through the photoperiodic mechanism.

One or more 'high-irradiance' photomorphogenetic responses may also contribute to the regulation of leaf growth. There is some evidence for a specific blue-absorbing photoreceptor which may be important in the detection of changes in irradiance and consequent modifications of leaf development. However, irradiance-dependent effects on leaf and cotyledon growth have also been observed in red and far-red light and these have been attributed to phytochrome operating in a high irradiance mode.

## References

Allsopp, A. (1965). Heteroblastic development in cormophytes. In *Encyclopedia of Plant Physiology*, vol. 15, ed. W. Ruhland, pp. 1172–221.

Arney, S. E. (1956). Studies of growth and development in the genus *Fragaria*. VI. The effect of photoperiod and temperature on leaf size. *Journal of Experimental Botany*, **7**, 65–79.

Atkinson, Y. E., Bradbeer, J. W. & Frankland, B. (1980). Leaf unrolling in two barley mutants. In *Photoreceptors and Plant Development*, ed. J. de Greef, pp. 543–50. Antwerp: Antwerpen University Press.

Bensink, J. (1971). On morphogenesis of lettuce leaves in relation to light and temperature. *Mededelingen van de Landbouwhoogeschool te Wageningen*, **71**,(15), 1–93.

Biswal, U. C. & Sharma, R. (1976). Phytochrome regulation of senescence in detached barley leaves. *Zeitschrift für Pflanzenphysiologie*, **80**, 71–3.

Black, M. & Shuttleworth, J. E. (1974). The role of the cotyledons in the photocontrol of hypocotyl extension in *Cucumis sativus* L. *Planta*, **117**, 57–66.

de Boer, J. & Feierabend, J. (1978). Comparative analysis of the action of cytokinin and light on the growth of rye leaves. *Planta*, **142**, 67–73.

Borthwick, H. A., Hendricks, S. B., Parker, M. W., Toole, E. H. & Toole, V. K. (1952). A reversible photoreaction controlling seed germination. *Proceedings of the National Academy of Sciences, USA*, **38**, 662–6.

Burström, H. (1942). Über Entfaltung und Einrollen eines mesophilen Grasblattes. *Botaniska Notiser*, **7**, 351–62.

Butler, R. D. (1963). The effect of light intensity on stem and leaf growth in broad bean seedlings. *Journal of Experimental Botany*, **14**, 142–52.

Butler, R. D. & Lane, G. R. (1959). The study of apical development in relation to etiolation. *Journal of the Linnean Society of London, Botany*, **56**, 170–6.

Butler, W. L. & Lane, H. C. (1965). Dark transformations of phytochrome *in vivo*. *Plant Physiology*, **40**, 13–7.

Carr, D. J., Clements, J. B. & Menhenett, R. (1972). Studies on leaf unrolling in barley. In *Plant Growth Substances 1970*, ed. D. J. Carr, pp. 633–45. Berlin: Springer-Verlag.

Cockshull, K. E. (1966). Effects of night-break treatment on leaf area and leaf dry weight in *Callistephus chinensis*. *Annals of Botany*, **30**, 791–806.

Cooke, R. J. & Saunders, P. F. (1975). Photocontrol of gibberellin levels as related to the unrolling of etiolated wheat leaves. *Planta*, **126**, 151–60.

Cooke, R. J., Saunders, P. F. & Kendrick, R. E. (1975). Red light induced production of gibberellin-like substance in homogenates of etiolated wheat leaves and in suspensions of intact etioplasts. *Planta*, **124**, 319–28.

Dale, J. E. & Murray, D. (1969). Light and cell division in primary leaves of *Phaseolus*. *Proceedings of the Royal Society of London*, **B173**, 541–55.

Davis, G. J. (1967). *Proserpinaca:* Photoperiodic and chemical differentiation of leaf development and flowering. *Plant Physiology*, **42**, 667–8.

Deutch, B. & Deutch, B. I. (1975). Blue light induction of barley leaf unfolding. A phytochrome reaction? *Physiologia Plantarum*, **35**, 322–7.

Downs, R. J. (1955). Photoreversibility of leaf and hypocotyl elongation of dark grown red kidney bean seedlings. *Plant Physiology*, **30**, 468–73.

Downs, R. J. & Siegelman, H. W. (1963). Photocontrol of anthocyanin synthesis in Milo seedlings. *Plant Physiology*, **38**, 25–30.

Drumm, H. & Mohr, H. (1978). The mode of interaction between blue (UV) light photoreceptor and phytochrome in anthocyanin formation of the sorghum seedling. *Photochemistry and Photobiology*, **27**, 241–8.

Epel, B. L., Butler, W. L., Pratt, L. H. & Tokuyasu, K. T. (1980). Immunofluorescence localization studies of the $P_R$ and $P_{FR}$ forms of phytochrome in the coleoptile tips of oats, corn and wheat. In *Photoreceptors and Plant Development*, ed. J. de Greef, pp. 121–33. Antwerp: Antwerpen University Press.

Frankland, B. & Letendre, R. J. (1978). Phytochrome and effects of shading on growth of woodland plants. *Photochemistry and Photobiology*, **27**, 223–30.

Gee, H. & Vince-Prue, D. (1976). Control of the hypocotyl hook angle in *Phaseolus mungo* L.: the role of parts of the seedling. *Journal of Experimental Botany*, **27**, 314–23.

Goldthwaite, J. J. & Laetsch, W. M. (1967). Regulation of senescence in bean leaf discs by light and chemical growth regulators. *Plant Physiology*, **42**, 1757–62.

de Greef, J. A., Butler, W. L. & Roth, T. F. (1971). Greening of etiolated bean leaves in far-red light. *Plant Physiology*, **47**, 457–64.

Guttridge, C. G. (1959). Evidence for a flower inhibitor and vegetative growth promoter in the strawberry. *Annals of Botany*, **23**, 351–60.

Hanke, J., Hartmann, K. M. & Mohr, H. (1969). Die Wirkung von 'Störlicht' auf die Blütenbildung von *Sinapis alba* L. *Planta*, **86**, 235–49.

Harder, R. (1948). Vegetative and reproductive development of *Kalanchoe blossfeldiana* as influenced by photoperiodism. *Symposium of the Society for Experimental Biology*, **2**, 117–38.

Hart, J. W. & MacDonald, I. R. (1980). Photoregulation of hypocotyl growth: geotropic evidence for the operation of two photosystems. *Plant, Cell and Environment*, **3**, 189–93.

Hartmann, K. M. (1966). A general hypothesis to interpret 'high energy phenomena' of photomophogenesis on the basis of phytochrome. *Photochemistry and Photobiology*, **5**, 349–66.

Hartmann, K. M. & Unser, I. C. (1972). Analytical action spectroscopy with living systems; photochemical aspects and attenuance. *Berichte der Deutschen botanischen Gesellschaft*, **85**, 481–551.

Heim, B., Jabben, M. & Schäfer, E. (1981). Phytochrome destruction in dark- and light-grown *Amaranthus caudatus* seedlings. *Photochemistry and Photobiology*, **34**, 89–93.

Hendricks, S. B. & Borthwick, H. A. (1959). Photocontrol of plant development by the simultaneous excitation of two interconvertible pigments. *Proceedings of the National Academy of Sciences, USA*, **45**, 344–9.

Hilton, J. R. & Smith, H. (1980). The presence of phytochrome in purified barley etioplasts and its *in vitro* regulation of biologically active gibberellin levels in etioplasts. *Planta*, **148**, 312–18.

Holmes, M. G. & McCartney, H. A. (1976). Spectral energy distribution in the natural environment and its implications for phytochrome function. In *Light and Plant Development*, ed. H. Smith, pp. 467–76. London: Butterworths.

Holmes, M. G. & Smith, H. (1977a). The function of phytochrome in the natural environment. II. The influence of vegetation canopies on the spectral energy distribution of natural daylight. *Photochemistry and Photobiology*, **25**, 539–45.

Holmes, M. G. & Smith, H. (1977b). The function of phytochrome in the natural environment. IV. Light quality and plant development. *Photochemistry and Photobiology*, **25**, 551–7.

Hopkins, W. G. & Hillman, W. S. (1965). Phytochrome changes in tissues of dark-grown seedlings representing various photoperiodic classes. *American Journal of Botany*, **52**, 427–32.

Hunt, R. E. & Pratt, L. H. (1980). Radio-immunoassay of phytochrome content in green, light-grown oats. *Plant, Cell and Environment*, **3**, 91–5.

Jabben, M. & Deitzer, G. E. (1978). Spectrophotometric phytochrome measurements in light-grown *Avena sativa* L. *Planta*, **143**, 309–13.

Johnson, C. B. & Tasker, R. (1979). A scheme to account quantitatively for the action of phytochrome in etiolated and light-grown plants. *Plant, Cell and Environment*, **2**, 259–65.

Jose, A. M. (1977). Photoreception and photoresponses in the radish hypocotyl. *Planta*, **136**, 125–9.

Jose, A. M. & Vince-Prue, D. (1977). Action spectra for the inhibition of growth in radish hypocotyls. *Planta*, **136**, 131–4.

Jose, A. M. & Vince-Prue, D. (1978). Phytochrome action: a reappraisal. *Photochemistry and Photobiology*, **27**, 209–16.

Kasperbauer, M. J. & Hiatt, A. J. (1966). Photoreversible control of leaf shape and chlorophyll content in *Nicotiana tabacum* L. *Tobacco Science*, **10**, 29–32.

Kasperbauer, M. J. & Peaslee, D. E. (1973). Morphology and photosynthetic efficiency of tobacco leaves that received end-of-day red or far-red light during development. *Plant Physiology*, **52**, 440–2.

Katsura, N. & Inada, K. (1979). Blue light-induced unrolling in rice plant leaves. *Plant and Cell Physiology*, **20**, 1071–7.

Kendrick, R. E. & Frankland, B. F. (1976). *Phytochrome and Plant Growth. Institute of Biology, Studies in Biology, No. 68*, London: Edward Arnold.

Kendrick, R. E. & Hillman, W. S. (1971). Absence of phytochrome dark reversion in seedlings of the Centrospermae. *American Journal of Botany*, **58**, 424–8.

Kendrick, R. E. & Spruit, C. J. P. (1973). Phytochrome intermediates *in vivo*. I. Effects of temperature, light intensity, wavelength and oxygen on intermediate accumulation. *Photochemistry and Photobiology*, **18**, 139–44.

King, R. W. & Vince-Prue, D. (1978). Light requirement, phytochrome and photoperiodic induction of flowering of *Pharbitis nil* Chois. I. No correlation between photomorphogenetic and photoperiodic effects of light pre-treatment. *Planta*, **141**, 1–7.

Klein, A. O. (1969). Persistent photoreversibility of leaf development. *Plant Physiology*, **44**, 897–902.

Lichtenthaler, H. K., Buschmann, C. & Rahmsdorf, U. (1980). The importance of blue light for the development of sun-type chloroplasts. In *The Blue Light Syndrome*, ed. H. Senger, pp. 485–94. Berlin: Springer-Verlag.

Lindoo, S. J. & Caldwell, M. M. (1978). Ultraviolet-B radiation-induced inhibition of leaf expansion and promotion of anthocyanin production. Lack of involvement of the low irradiance phytochrome system. *Plant Physiology*, **61**, 278–82.

Liverman, J. L. (1959). Control of leaf growth by an interaction of chemicals and light. In *Photoperiodic and Related Phenomena of Plants and Animals*, ed. R. B. Withrow, pp. 161–80. Washington: American Association for the Advancement of Science.

Liverman, J. L., Johnson, M. P. & Starr, L. (1955). Reversible photoreaction controlling expansion of etiolated bean-leaf disks. *Science*, **121**, 440–1.

Loveys, B. R. & Wareing, P. F. (1971a). The red light controlled production of gibberellin in etiolated wheat leaves. *Planta*, **98**, 109–16.

Loveys, B. R. & Wareing, P. F. (1971b). The hormonal control of wheat leaf unrolling. *Planta*, **98**, 117–27.

MacDougal, D. T. (1903). Influence of light and darkness upon growth and development in plants. *Memoirs of the New York Botanical Garden*, **2**, 319.

Mancinelli, A. L. (1980). The photoreceptors of the high irradiance responses of plant photomorphogenesis. *Photochemistry and Photobiology*, **32**, 853–7.

Mancinelli, A. L. & Rabino, I. (1975). Photocontrol of anthocyanin synthesis. IV. Dose dependence and reciprocity relationships in anthocyanin synthesis. *Plant Physiology*, **56**, 351–5.

Mancinelli, A. L. & Rabino, I. (1978). The 'high irradiance responses' of plant photomorphogenesis. *Botanical Review*, **44,** 129–80.

Marmé, D. (1977). Phytochrome: membranes as possible sites of primary action. *Annual Review of Plant Physiology*, **28,** 173–98.

Miller, C. O. (1956). Similarity of some kinetin and red light effects. *Plant Physiology*, **31,** 318–19.

Mohr, H. (1959). Der Lichteinfluss auf das Wachstum der Keimblätter bei *Sinapis alba* L. *Planta*, **53,** 219–45.

Mohr, H. (1964). The control of plant growth and development by light. *Biological Review of the Cambridge Philosophical Society*, **39,** 87–112.

Mohr, H. (1972). *Lectures on Photomorphogenesis*. Berlin: Springer-Verlag.

Mohr, H. (1980). Interaction between blue light and phytochrome in photomorphogenesis. In *The Blue Light Syndrome*, ed. H. Senger, pp. 97–109. Berlin: Springer-Verlag.

Mohr, H. & Pinnig, E. (1962). Der Einfluss des Lichtes auf die Bildung von Blatt-primordien am Vegetationskerel der Keimlinge von *Sinapis alba* L. *Planta*, **58,** 569–79.

Murray, D. (1968). Light and leaf growth in *Phaseolus*. Ph.D. Thesis, University of Edinburgh.

Njoku, E. (1956). The effect of light intensity on leaf shape in *Ipomoea caerulea*. *New Phytologist*, **55,** 91–110.

Parker, M. W., Hendricks, S. B., Borthwick, H. A. & Went, F. W. (1949). Spectral sensitivities for leaf and stem growth of etiolated pea seedlings and their similarity to action spectra for photoperiodism. *American Journal of Botany*, **36,** 194–204.

Powell, R. D. & Griffith, M. M. (1960). Some anatomical effects of kinetin and red light on disks of bean leaves. *Plant Physiology*, **35,** 273–5.

Pratt, L. H. (1978). Molecular properties of phytochrome. *Photochemistry and Photobiology*, **27,** 81–105.

Pratt, L. H. (1979). Phytochrome: function and properties. In *Photochemical and Photobiological Reviews*, vol. 4, ed. K. C. Smith, pp. 59–124. New York: Plenum Press.

Reid, D. M., Tuing, M. S., Durley, R. C. & Railton, I. D. (1972). Red light-enhanced conversion of tritiated gibberellin $A_9$ into other gibberellin-like substances in homogenates of etiolated barley leaves. *Planta*, **108,** 67–75.

Rogers, L. J., Shah, S. P. J. & Goodwin, T. W. (1965). Intracellular localisation of mevalonic kinase in germinating seedlings; its importance in the regulation of terpenoid biosynthesis. *Biochemical Journal*, **96,** 7–8.

Rolinson, A. E. & Vince-Prue, D. (1976). Responses of the rice shoot apex to irradiation with red and far-red light. *Planta*, **132,** 215–20.

Sale, P. J. M. & Vince, D. (1963). Some effects of light on leaf growth in *Pisum sativum* and *Tropaeolum majus*. *Photochemistry and Photobiology*, **2,** 401–5.

Sanchez, R. A. (1971). Phytochrome involvement in the control of leaf shape of *Taraxacum officinale*. *Experentia*, **27,** 1234–7.

Sanchez, R. A. & Cogliatti, D. (1975). The interaction between phytochrome and white light irradiance in the control of leaf shape in *Taraxacum officinale*. *Botanical Gazette*, **136,** 281–5.

Sasakawa, H. & Yamamoto, Y. (1980). Effects of blue and red light on unrolling of rice leaves. *Planta*, **147,** 418–21.

Satter, R. L. & Galston, A. W. (1976). The physiological functions of phytochrome. In *Chemistry and Biochemistry of Plant Pigments*, vol. 1, ed. T. W. Goodwin, pp. 680–735. New York: Academic Press.

Satter, R. L. & Wetherell, D. F. (1968). Photomorphogenesis in *Sinningia speciosa* cv. Queen Victoria. I. Characterization of phytochrome control. *Plant Physiology*, **43,** 953–60.

Schäfer, E. (1975). A new approach to explain the 'High Irradiance Responses' of photomorphogenesis on the basis of phytochrome. *Journal of Mathematical Biology*, **2**, 41–56.

Schäfer, E. (1976). The 'High Irradiance Reaction'. In *Light and Plant Development*, ed. H. Smith. London: Butterworths.

Schäfer, E. & Mohr, H. (1974). Irradiance dependency of the phytochrome system in cotyledons of mustard (*Sinapis alba* L.). *Journal of Mathematical Biology*, **1**, 9–15.

Schwabe, W. W. (1969). *Kalanchoe blossfeldiana* Poellniz. In *The Induction of Flowering*, ed. L. T. Evans, pp. 227–46. Melbourne: Macmillan of Australia.

Sisson, W. B. & Caldwell, M. M. (1977). Atmospheric ozone depletion: reduction of photosynthesis and growth of a sensitive higher plant exposed to enhanced UV-B radiation. *Journal of Experimental Botany*, **28**, 691–705.

Smith, H. (1975). *Phytochrome and Photomorphogenesis*. London: McGraw-Hill.

Smith, H. (1981). Function, evolution and action of plant photoreceptors. In *Plants and the Daylight Spectrum*, ed. H. Smith, pp. 499–508. London: Academic Press.

Spence, D. H. N. (1981). Light quality and plant responses underwater. In *Plants and the Daylight Spectrum*, ed. H. Smith, pp. 245–75. London: Academic Press.

Steiner, A., Price, L., Mitrakos, K. & Klein, W. H. (1968). Red light effects on uptake of $^{14}C$ and $^{32}P$ into etiolated corn leaf tissue during photomorphogenic leaf opening. *Physiologia Plantarum*, **21**, 895–901.

Sugiura, M. K. (1963). Effect of red and far-red light on protein and phosphate metabolism in tobacco leaf discs. *Botanical Magazine of Tokyo*, **76**, 174–80.

Tafazoli, E. & Vince-Prue, D. (1978). A comparison of the effects of long days and exogenous growth regulators on growth and flowering in strawberry, *Fragaria* × *ananassa* Duch. *Journal of Horticultural Science*, **53**, 255–9.

Thomas, B. (1981). Specific effects of blue light on plant growth and development. In *Plants and the Daylight Spectrum*, ed. H. Smith, pp. 443–59. London: Academic Press.

Thomas, B. & Dickinson, H. G. (1979). Evidence for two photoreceptors controlling growth in de-etiolated seedlings. *Planta*, **146**, 454–50.

Thomas, R. G. (1961). Correlations between growth and flowering in *Chenopodium amaranticolor*. II. Leaf and stem growth. *Annals of Botany*, **25**, 255–69.

Thompson, B. F. & Miller, P. M. (1962). The role of light in histogenesis and differentiation in the shoot of *Pisum sativum*. II. The leaf. *American Journal of Botany*, **49**, 383–7.

Trumpf, C. (1924). Über den Einfluss intermittierender Belichtung auf das Etiolement der Pflanzen. *Botanisches Archiv*, **5**, 381–410.

Tucker, D. J. (1982). Phytochrome regulation of leaf senescence in cucumber and tomato. *Plant Science Letters* (in press).

Turner, M. R. & Vince, D. (1969). Photosensory mechanisms in the lettuce seedling hypocotyl. *Planta*, **84**, 368–82.

Verbelen, J. P. & de Greef, J. A. (1979). Leaf development of *Phaseolus vulgaris* L. in light and in darkness. *American Journal of Botany*, **66**, 970–6.

Vince, D. (1955). Some effects of temperature and daylength on flowering in the chrysanthemum. *Journal of Horticultural Science*, **30**, 34–42.

Vince, D. (1955). Some effects of temperature and daylength on flowering in the chrysanthemum. *Journal of Horticultural Science*, **30**, 34–42.

Vince, D. (1964). Photomorphogenesis in plant stems. *Biological Reviews of the Cambridge Philosphical Society*, **39**, 506–36.

Vince-Prue, D. (1975). *Photoperiodism in Plants*. London: McGraw-Hill.

Vince-Prue, D. (1980). Effect of photoperiod and phytochrome in flowering: time measurement. In *Physiologie de la Floraison*, ed. P. Champagnat & R. Jacques, pp. 91–127. Paris: CNRS.

Vince-Prue, D. & Guttridge, C. G. (1973). Floral initiation in strawberry: spectral evidence for the regulation of flowering by long-day inhibition. *Planta*, **110**, 165–72.
Virgin, H. I. (1962). Light-induced unfolding of the grass leaf. *Physiologia Plantarum*, **15**, 380–9.
Wagné, C. (1964). The distribution of the light-effect in partly irradiated grass leaves. *Physiologia Plantarum*, **17**, 751–6.
Warrington, I. J. & Mitchell, K. J. (1976). The influence of blue- and red-biased light spectra on the growth and development of plants. *Agricultural Meteorology*, **16**, 247–62.
Went, F. W. (1941). Effects of light on stem and leaf growth. *American Journal of Botany*, **28**, 83–95.
Williams, W. T. (1956). Etiolation phenomena. In *The Growth of Leaves*, ed. F. L. Milthorpe, pp. 127–35. London: Butterworths.
Zeevaart, J. A. D. (1969). *Bryophyllum*. In *The Induction of Flowering*, ed. L. T. Evans, pp. 435–56. Melbourne: Macmillan of Australia.

# 10

# The development of photosynthetic capacity in leaves

RACHEL M. LEECH AND N. R. BAKER

## Introduction

Every seedling of an autotropic higher plant becomes dependent early in its life on photosynthesis, here defined as the light-induced synthesis of carbohydrates from $CO_2$. In the majority of species photosynthesis occurs in leaves, each of which has a unique ontogeny determined to a considerable degree by the developmental stage of the plant when the leaf is produced. During its own early development each leaf acts as a sink for assimilates produced by older leaves and only when a leaf attains a photosynthetic capacity greater than its own respiration rate does it become potentially independent of the energy supply from older leaves; it then becomes capable of exporting its own assimilates to other organs and thus positively contributing to the net productivity of the plant.

Photosynthesis is a complex series of physico-chemical reactions which facilitate the transduction of light energy into chemical energy (ATP and NADPH) and the utilization of this for the reduction of $CO_2$ to carbohydrates (Fig. 10.1). The successful operation of this reaction series is dependent upon the presence of the reaction components, their specific organization within the leaf cells, and the efficient supply of substrates to the reaction sites. The assembly of the components of the photosynthetic apparatus into a physiologically functional unit is extremely complex and to understand fully the processes involved, many aspects of leaf biology need to be studied, ranging from changes in leaf anatomy and plastid structure occurring during leaf biogenesis to the genetic control of chloroplast development. The anatomical development of the leaf tissue is required to allow an adequate supply of $CO_2$

Fig. 10.1. Diagram illustrating the inter-relationships between the light harvesting, energy transduction and carbon reduction phases of photosynthesis.

Photosystem I (PSI) and Photosystem II (PSII) are two functionally distinct pigment assemblies. PSI absorbs light at longer wavelengths than PSII. The majority of the chlorophyll molecules in both photosystems function as light harvesting 'antennae' molecules but the heart of each photosystem is a reaction centre consisting of a chlorophyll dimer P700 (PSI) and P690 (PSII). Photons absorbed by antennae molecules are rapidly transferred from one chlorophyll molecule to another and eventually trapped in the reaction centre. At the reaction centres, redox carriers of lower electrochemical potential than chlorophyll become reduced initiating the biochemical events of photosynthesis and the photolysis of water restores the chlorophyll molecules to the ground state. The flow of electrons through the series of carriers (Q, X, PQ, cyt f, PC, Fd, FP, NADP) in the chloroplast thylakoid membrane creates an electrochemical potential difference of protons across the membrane. The synthesis of ATP is driven by a reverse flow of protons and involves the operation of an ATP synthase complex (Fig. 10.11). The final electron acceptor is $NADP^+$ and NADPH and ATP are energetically linked to the reduction of carbon dioxide which is catalysed by the enzyme ribulose bisphosphate carboxylase/oxygenase (Fraction I protein) to give 3-phosphoglycerate (PGA). Phosphorylation and reduction of the PGA yields triosephosphate. Some of the triosephosphate molecules immediately leave the chloroplast while others are recycled to regenerate ribulose bisphosphate. Further details of the thylakoid located processes are shown in Fig. 10.11. Key to the symbols: Q, quencher; PQ, plastoquinone; cyt f, cytochrome f; PC, plastocyanin; X, 1° acceptor of PSI; Fd, ferredoxin; FP, flavoprotein; PGA, 3-phosphoglycerate; R-5-P, ribose-5-phosphate, RuBP, ribulose bisphosphate; Mn, Manganese.

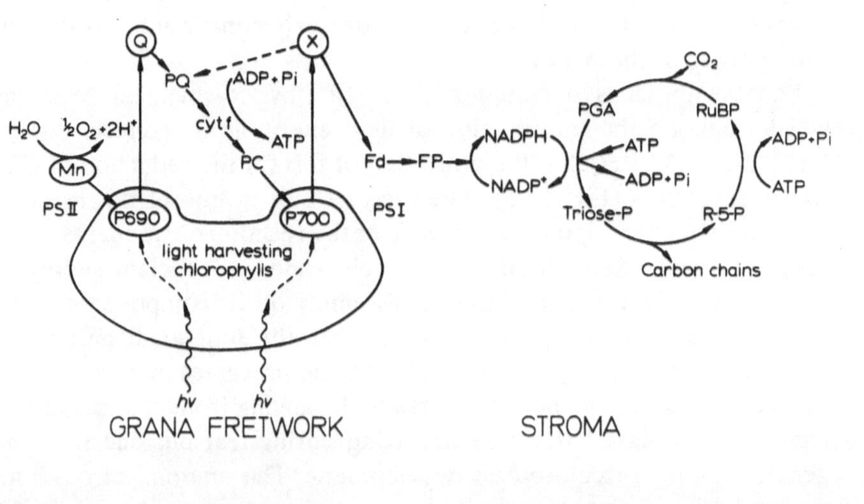

from the atmosphere to diffuse to the biochemical sites of carboxylation within the cells. The development of fully functional chloroplasts involves the biosynthesis of the components of the photosynthetic apparatus and their assembly into units with the capacity to transduce light energy into ATP and NADPH and to utilize these molecules for the enzymic reduction of $CO_2$.

It is not yet possible to provide a unified description of the development of leaf photosynthetic capacity but this is clearly under both genetic and environmental control. Different species show different patterns of biogenesis of the photosynthetic apparatus and differences are also often observed when a given species is grown under a range of environmental conditions. It is, however, possible to identify some of the major factors involved in the development of photosynthetic competence and our aim is to describe the present state of understanding of the nature and operation of these factors. We first identify the major anatomical and structural features of the leaf which determine the efficiency of transfer of $CO_2$ from the atmosphere to the carboxylating sites in the chloroplast and discuss the relevance of tissue and cellular changes to the development of photosynthetic capacity. We next describe the development of the population of photosynthetically functional chloroplasts. Current knowledge is inadequate to provide a comprehensive description of all aspects of plastid development but the major changes occurring during the synthesis and assembly of the chloroplast photosynthetic apparatus and the control of the onset of photochemical function will be described. Certain aspects of the biology of chloroplast development have been investigated more fully than others and we give greater weight to those aspects of chloroplast development which are currently receiving most attention. Finally, we attempt to integrate the different types of information and to identify those areas about which further knowledge is needed; for example, of how the physiology and cellular biochemistry changes during the leaf development.

Chloroplast development occurs in very young, growing leaves but, because of their very small size and rapid development these have proved very difficult to investigate biochemically. An exception is the young developing leaf of grasses in which cells are produced from a basal intercalary meristem and the leaf itself provides a gradient of cellular and plastid development with the youngest tissue nearest the base of the leaf and the oldest near the tip. These may be sampled at one point in time, or alternatively the natural development of photosynthetic capacity may be followed in a single identifiable leaf at it matures. However, until rather recently knowledge of the biogenesis of photochemical activity has been based almost entirely on observations of the effects of illumination on plastids from seedlings which have been previously held in continuous darkness and become etiolated. Only a short period of intense illumination is needed to initiate the conversion of the plastids in etiolated leaves – etioplasts – to chloroplasts. Because of the ease with which greening etiolated plants can be manipulated they have been extensively investigated as model systems for the study of chloroplast development (see Treffry, 1978, for a recent review). Certain aspects of the development of photosynthetic capacity in etioplasts may resemble normal development. This seems to be so with the pattern of development of primary photochemical activities but in other aspects of their development greening etioplasts differ in important respects from normal chloroplasts, for example in the rate of chlorophyll biosynthesis and the synchrony of photosynthetic unit synthesis.

We will emphasise here the development of photosynthetic activity in light-grown plants using observations from etiolated greening systems when these are the only ones yet available or when a specific aspect of etioplast development during greening appears to resemble normal plastid development.

## Leaf anatomical and morphological considerations

Large changes in morphology and anatomy occur during leaf biogenesis and are illustrated in the three sections of flag leaves of wheat in Fig. 10.2*a*, *b*, *c*. It is well established that a close relationship exists between leaf structure and photosynthetic activity in mature leaves. The internal architecture of the leaf is complex and varies considerably from one species to another, as shown by the transverse sections taken from leaves of three species frequently used for photosynthesis research. Features such as stomatal size and frequency and the fraction of leaf volume occupied by air spaces between the mesophyll cells are markedly modified during leaf development and are important in determining the rate of transfer of atmospheric $CO_2$ to the carboxylation sites within the leaf. Chloroplast number per cell is also important in determining leaf photosynthetic capacity. During the early stages of chloroplast development proplastid replication maintains the number of plastids in meristematic cells and may also lead to increase in chloroplast number at a later stage (Butterfass, 1979). Recently, considerable evidence has accumulated showing that young chloroplasts with functional grana also divide. Since chloroplast division occurs both in dicotyledons (Possingham, 1980) and in monocotyledons (Boffey, Ellis, Sellden & Leech,

Fig. 10.2. Transverse sections of young leaves of wheat (*a–c*), maize (*d*) and spinach (*e*).

The drawings are from photomicrographs of 2.5 $\mu$m thick sections of resin-embedded material. (*a–c*) *Triticum aestivum* var. Maris Dove young flag leaves (wheat): (*a*) unexpanded flag leaf within the leaf sheath; (*b*) unexpanded but exposed flag leaf; (*c*) fully expanded flag leaf; (*d*) *Zea mays* var. Kelvedon Glory (maize); (*e*) *Spinacia oleracea*, Yates Hybrid 102, fully expanded leaf.

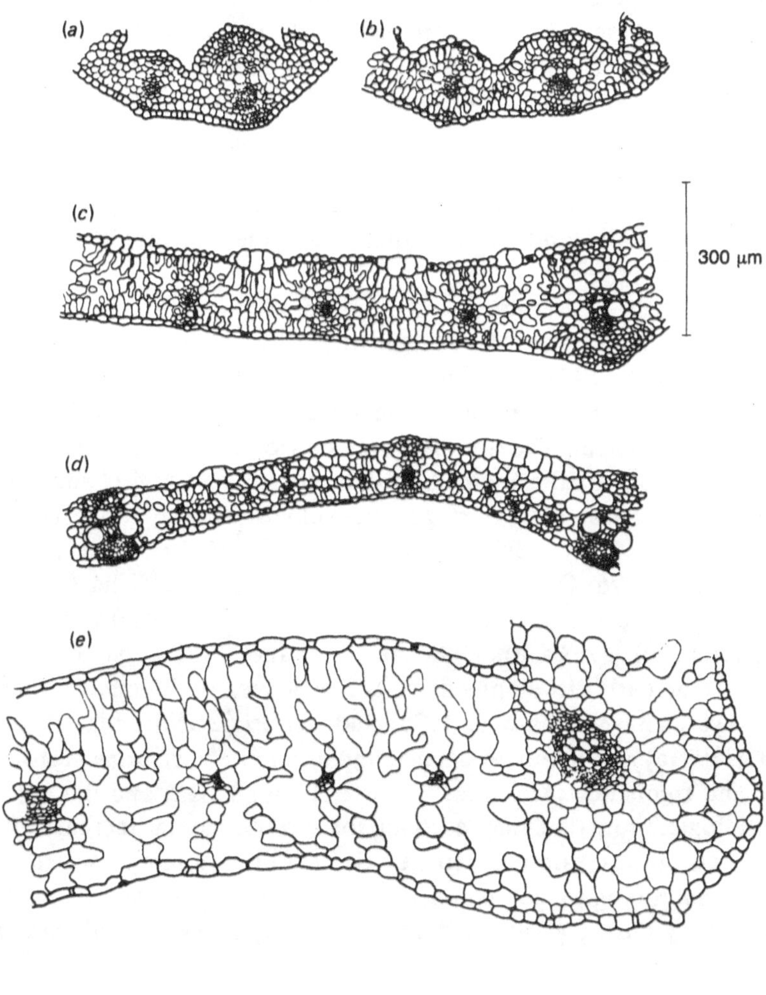

1979) after cell division has ceased, it contributes to increased chloroplast number, and therefore increased photosynthetic capacity, in the mature leaf (Leech, 1976). In spinach 80% of chloroplasts in the mature leaf are calculated to have been derived by this division process and in modern varieties of hexaploid wheat the chloroplasts divide three or four times during development (Boffey *et al.*, 1979). The chloroplast-containing cells responsible for the major photosynthetic function of the leaves of higher plants are the mesophyll cells and also in C4 plants, the cells of the bundle-sheath. The number of chloroplasts per mesophyll cell varies widely between species e.g. in cocoa 2–3, *Zea mays* 30, while some varieties of hexaploid wheat have 160 chloroplasts per cell.

It is useful when examining possible limitations to $CO_2$ assimilation during development to consider the diffusion pathway in terms of a series of resistances from the atmosphere to the sites of carboxylation within the cells. Theoretically, any one of the resistances shown in Fig. 10.3 may be the major limiting factor to $CO_2$ assimilation at a given developmental stage and this may change during leaf development. In practice the boundary layer, $r_a$, and intercellular, $r_i$, resistances represent only a small fraction of the total leaf resistance to $CO_2$ assimilation ($\Sigma r$) (Nobel 1974) and can be ignored as potential major limitations during development.

Since cuticle resistance, $r_c$, is considerably greater than stomatal resistance, $r_s$, in mature leaves and these two resistances function in parallel (Nobel, 1974), the cuticle is unlikely to play a major role in regulating $CO_2$ assimilation during development unless stomata are absent from the leaf. The two major resistances to $CO_2$ assimilation are $r_s$ and the resistance of the liquid phase of the leaf, $r_l$, and large changes occur in both these parameters during leaf development (Rawson & Hackett, 1974; Šesták, Čatský, Solárová, Strnadová & Tichá, 1975; Čatský, Tichá & Solárová, 1976; O'Toole, Ludford & Ozbun, 1977; Tichá & Čatský, 1977). In mature leaves of C3 and C4 plants $r_s$ constitutes a large proportion of $\Sigma r$ (Körner, Schael & Bauer, 1979; Nobel, 1974), however at early development stages of leaves of *Phaseolus vulgaris* (Čatský, Tichá & Solárová, 1976) and *Zea mays* (Miranda, Baker & Long, 1981a) $r_s$ is relatively small compared to $r_l$. Variations in stomatal size and frequency (Miranda, Baker & Long, 1981a) and the onset of physiological function may account for developmental increases in $r_s$. Etiolated leaves contain structurally mature stomata, which only achieve physiological competence upon the development of photosynthetic activity during greening (Lurie, 1977a, b; Ogawa, 1979). Similarly in light-grown *Zea mays* leaves, stomatal functioning appears after the onset of photosynthetic activity of the tissue (Miranda, Baker & Long, 1981a, b).

It has been widely shown that $r_l$ constitutes a major fraction of $\Sigma r$ in young leaf tissue and provides a major limitation to $CO_2$ assimilation (Ludlow & Wilson, 1971; Rawson & Hackett, 1974; Rawson &

Fig. 10.3. Resistances involved in $CO_2$ photoassimilation by a leaf. Symbols represent specific resistances to $CO_2$ diffusion or assimilation associated with leaf anatomical or physiological features. Key to resistances: $r_a$, leaf boundary layer resistance; $r_c$, cuticular resistance; $r_s$, stomatal resistance; $r_i$, intercellular space resistance; $r_m$, diffusive resistance from surface of mesophyll cells to carboxylation sites; $r_e$, resistance to $CO_2$ assimilation imposed by the rate of photochemical energy (ATP and NADPH) production; $r_x$, resistance to $CO_2$ assimilation imposed by the rate of enzymatic carboxylation; $r_1$, liquid phase resistance, a composite of $r_m$, $r_e$ and $r_x$.

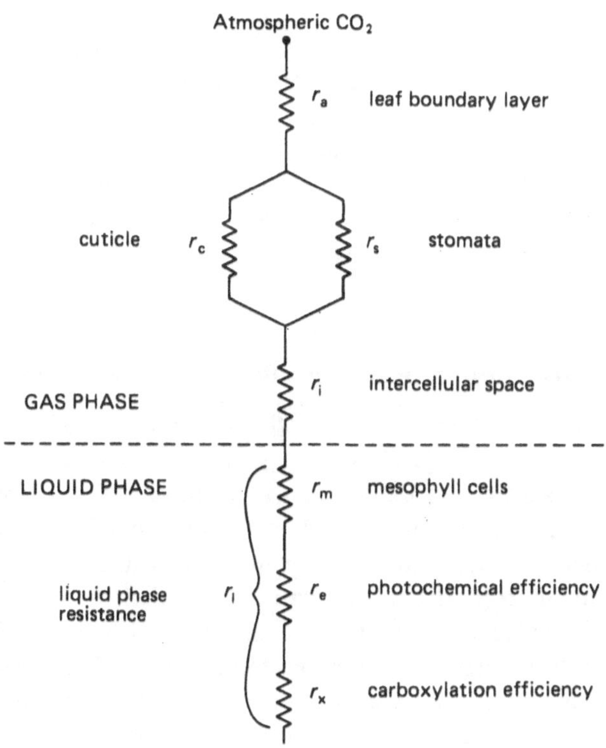

Woodward, 1976; Woodward & Rawson, 1976; Čatský, Tichá, & Solárová, 1976; Samsuddin & Impens, 1979; Miranda, Baker & Long, 1981a). The large decreases in $r_l$ during leaf development may be due to decreases in $r_m$, $r_e$ or $r_x$ or any combination of these parameters. One of the major determinants of $r_m$ is the total area of the mesophyll cell walls exposed to intercellular air spaces (Nobel, 1974), and changes in this parameter have been shown to be responsible for changes in the rate of photosynthetic $CO_2$ assimilation during leaf development (Nobel, Zaragoza & Smith, 1975). At early stages of leaf development the intercellular air spaces within the mesophyll are poorly developed (Miranda, Baker & Long, 1981a) and that $r_m$ has high values which may limit carbon assimilation.

Increases in carboxylation capacity and sites for transduction of light energy to ATP and NADPH, which occur during leaf development, will result in decreases in the carboxylation, $r_x$, and excitation, $r_e$, resistances respectively. The conversion of light energy into ATP and NADPH is dependent upon the development of functional thylakoid membranes within the chloroplasts, whilst the enzymic reduction of $CO_2$ in C3 plants is dependent upon the synthesis of the Calvin cycle enzyme systems also within the chloroplasts. In C4 plants the development of photosynthetic function requires in addition the development of enzyme systems responsible for $CO_2$ fixation within the cytoplasm of mesophyll cells and the translocation of assimilated carbon to the chloroplasts of bundle sheath cells (see Edwards & Walker, 1982, for details of C3 and C4 photosynthetic enzyme systems).

## The development of photosynthetically competent chloroplasts

The development and functional organization, including structural, molecular biological and energetic aspects, of its chloroplast population is essential before a leaf can achieve full photosynthetic competence. The structure of the photosynthetic apparatus is indivisible from its physico-chemical functions because the visible structural components of the chloroplast and the features revealed by ultrastructural examination are also the functional components which, in the fully mature chloroplast, carry out the integrated function of photosynthesis. However, for clarity, the structural changes associated with the maturation of proplastids into chloroplasts will be described first and development of photosynthetic function discussed subsequently.

### *The development of structurally complete chloroplasts*

The main features of a fully photosynthetic leaf chloroplast are illustrated in the model shown in Fig. 10.4*a* and the thin sections shown in Fig. 10.4*b*. In living cells the shape of the chloroplast is constantly changing and its surface regions are continually mobile but these dynamic aspects are lost during the fixation procedures necessary to prepare leaf material for the examination of chloroplast fine structure. All higher plant chloroplasts exhibit four structural features: the double bounding envelope encloses a proteinaceous stroma where the carbon cycle enzymes, the ribosomes (70 S), DNA and plastoglobuli are located. The chlorophyll-containing thylakoid membrane systems are fenestrated and located centrally in the chloroplast.

How does the complex architecture of the chloroplast develop? In the young meristematic leaf cells of normally grown plants, the first recognizable progenitors of chloroplasts are tiny (*c.* 1 $\mu$m in diameter) generally spherical, often pleiomorphic organelles with double bounding envelopes and few internal membranes. These organelles, which often contain small amounts of chlorophyll, are known as proplastids (Kirk & Tilney-Bassett, 1978) or more recently as eoplasts (Thomson & Whatley, 1980). The ultrastructural changes which occur during chloroplast development have been followed in expanding green leaves of *Phaseolus* (Whatley, 1977) and also in developing cells in several species of monocotyledons. All the major changes in chloroplast development occur after cell division has ceased. A diagram illustrating the sequence of structural changes which occur in plastid development is shown in Fig. 10.5. The developing proplastids accumulate several round starch grains which they lose as they pass into an amoeboid stage. The mature thylakoid system develops from the few perforated membranes of the proplastid by extensive formation of new membranes and by the folding back and stacking of these membranes to form the characteristic granal fretwork system. The later stages of chloroplast development involve a large increase in plastid size and, in granal chloroplasts, in the size and number of the grana. In the first leaf of 7-day-old maize the mesophyll chloroplasts develop from proplastids in 3 h, increase ten times in volume during development, the ribosome numbers increase fourfold and the chlorophyll and membrane lipids per plastid increase tenfold. The total internal membranes increase fiftyfold in area during chloroplast development in this plant.

The details of thylakoid membrane development and the three-

Fig. 10.4. (*a*) Diagram illustrating the major structural features of a fully photosynthetic higher plant chloroplast. (Reproduced with permission from R. A. Reid & R. M. Leech, 1980.)

(*b*) Electron micrographs of (on the left) a chloroplast profile of a chloroplast of *Zea mays* (maize) (× 10 000). On the right, higher magnification of the thylakoid fretwork (×20 000). g, grana; p, plastoglobuli; s, stroma; e, envelope; t, thylakoid. (Electron-micrographs reproduced with permission of W. W. Thomson.)

(*a*)

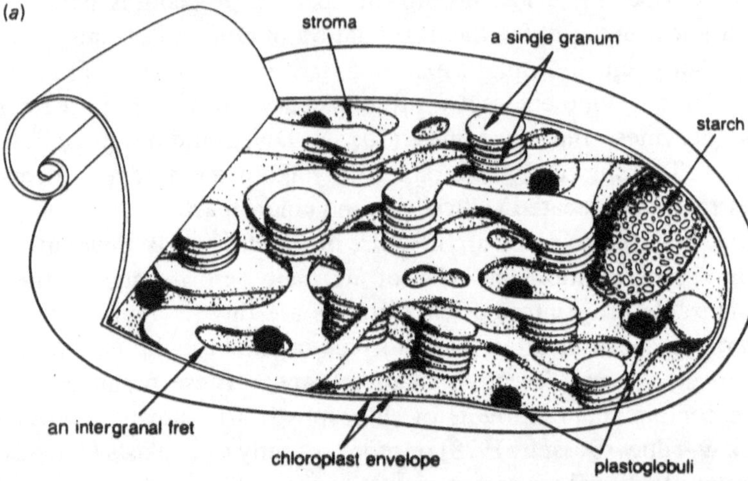


(*b*)

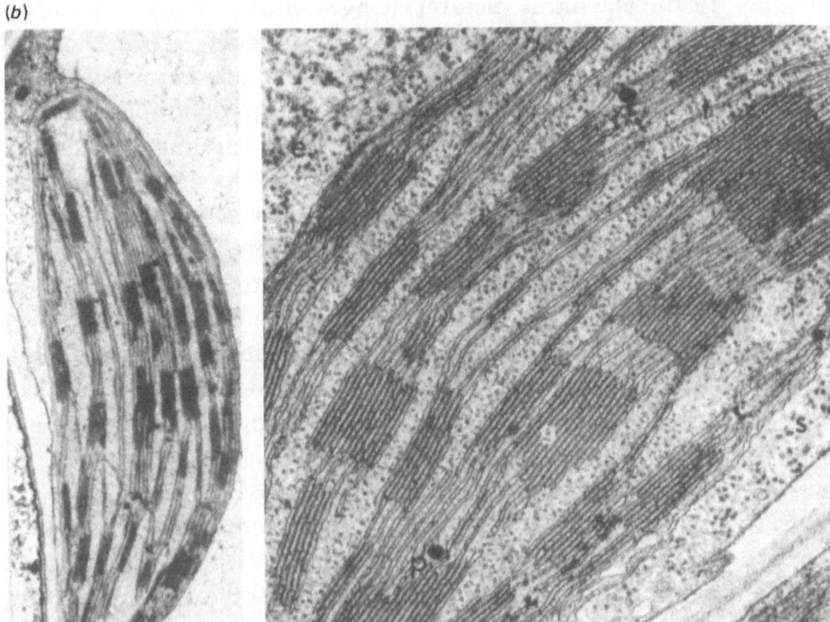

dimensional foldings and invaginations involved have been carefully followed by Brangeon and co-workers (Mustardy & Brangeon, 1978; Brangeon & Mustardy, 1979) and a model they have built from the interpretation of thousands of thin sections of developing rye-grass (*Lolium*) chloroplasts is shown in Fig. 10.6*b*. The structure of the

Fig. 10.5. Diagram illustrating the development of a proplastid into a chloroplast.

Proplastids (A) are 1 μm in diameter and repeatedly divide in the meristematic cell and then increase in size, passing through a starch-containing phase (B) and an amoeboid stage (C) before the perforated membrane plates begin to extend (D, E, G) and overlap at the young chloroplast stage (H). The young chloroplasts (H) also divide and increase in size and in the complexity of their granal stacks (I) until the fully photosynthetic chloroplast is formed. In wheat the development from stage A to stage J takes about 6 h. If leaves are held in the dark, development from stage E is arrested and an etioplast with an internal prolamellar body is formed: on illumination etioplasts (F) develop into chloroplasts (J). Hollow arrows, dark; solid arrows, light.

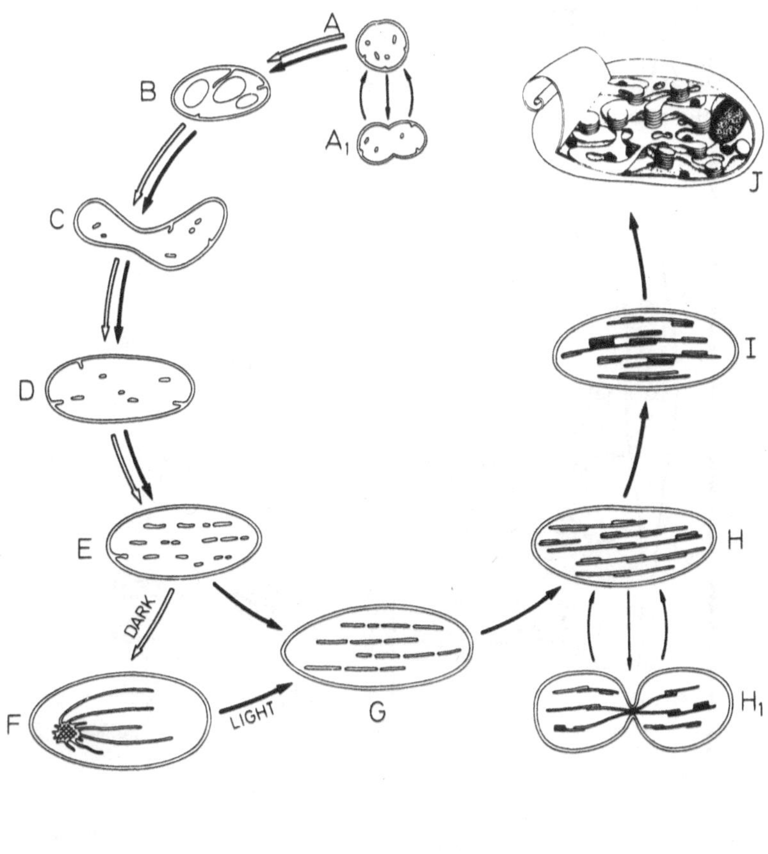

granum achieved by this developmental mechanism reflects and explains the multiple spiral structure of the granal structure in mature spinach chloroplasts revealed by Paolillo (1970).

Some mature higher plant chloroplasts never possess highly developed grana and in these plastids the distinguishing feature is numerous unstacked thylakoids distributed parallel to each other along the long axis of the chloroplast. Such chloroplasts are particularly characteristic of plants exhibiting the C4 pathway of photosynthesis and are known as agranal or lamellate chloroplasts.

If leaves are held in continuous darkness for extended periods (3–14 days), the proplastids develop into etioplasts (Weier & Brown, 1970) (see Fig. 10.5). In etiolated leaves of graminaceous monocotyledons, several types of developing proplastids and developing etioplasts are found (Robertson & Laetsch, 1974). Etioplasts are spherical or elongated plastids 1–5 $\mu$m in diameter; they lack chlorophyll and are

Fig. 10.6. Diagram illustrating the gradient of chloroplast development from the base to the tip of a young graminaceous leaf. (*a*) Cross-sectional profiles of chloroplasts as seen in electron micrographs. (*b*) Three-dimensional reconstructed views of membrane development proposed by Mustardy & Brangeon (1978) from their studies of plastid structure in rye-grass (*Lolium multiflorum*).

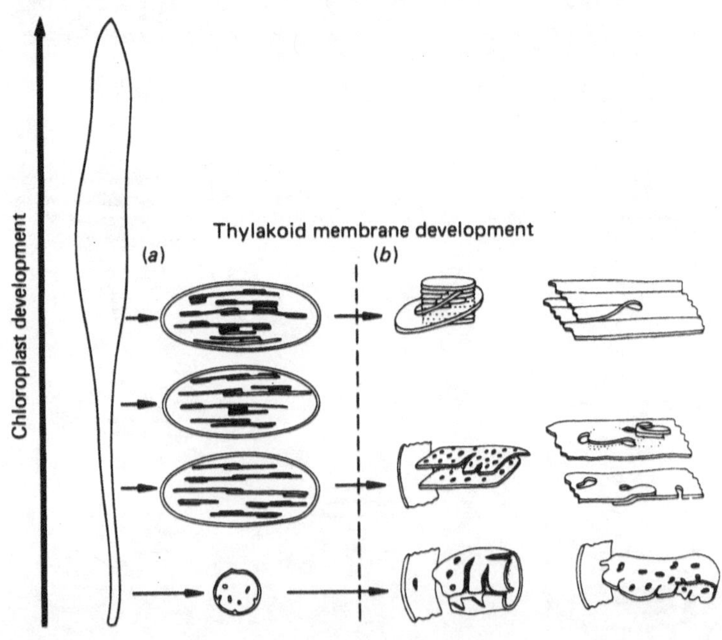

characterized by the complexity of their internal membrane system which consists of a paracrystalline lattice in the form of a strutted dodecahedron. Etioplasts are unknown in normally grown plants and should not be confused with the plastids containing small regular membrane assemblies (tubular complexes) which are an order of magnitude smaller than the central paracrystalline structure (prolamellar body) present in an etioplast (Platt-Aloia & Thomson, 1977). Brief flashes of intense illumination initiate chlorophyll synthesis and also membrane dispersion, synthesis and reassembly in etioplasts which, after a few days of continuous illumination, resemble normal chloroplasts both structurally and functionally (Treffry, 1978).

### *Thylakoid assembly and stacking*

The stacking of internal chloroplast thylakoid membranes into grana (Figs. 10.4, 10.7) which occurs during chloroplast development is a distinctive feature of many higher plant chloroplasts. Grana are

Fig. 10.7. A model showing the distribution of the major particles within the thylakoid membrane. The diagram demonstrates how the surface charge density can determine whether the membranes are stacked or unstacked. Note that in the unstacked regions there is a high negative surface charge density, due to a high proportion of PSI chlorophyll-protein complex being present, whereas this is considerably reduced in stacked regions.

particularly well developed when plants are grown under shade conditions. The molecular architecture of the thylakoids is a major determinant in the stacking process which requires continuous illumination for completion (Strasser & Butler, 1976). Generally the fluid-mosaic model of membrane structure is used to describe the thylakoid membrane as a fluid matrix of galacto- and phospho-lipids in which many different types of integral and intrinsic proteins, chlorophyll and electron transport components are embedded. The lipids, of which the majority are galactose-containing and have an extremely high content of polyunsaturated fatty acids, are considered to be arranged in a bimolecular layer with their polar groups exposed to the hydrophilic environment of the stroma and the hydrophobic tails towards the interior of the membrane. Freeze-fracture electron microscopy of the thylakoid membrane has revealed a particulate substructure, with the size and distribution of the particles differing in the stacked and unstacked regions of the chloroplast (see Fig. 10.7). It is thought that these particles contain the thylakoid proteins, chlorophylls and electron transport components of the two photosystems. Although the particles exhibit a continuous size range from *ca* 10–20 nm, generally two size categories can be recognized: small particles *ca* 9–11 nm in diameter and larger particles with diameters between 15 and 20 nm. These intrinsic membrane particles can be distinguished from the readily recognizable ATP synthase ($CF_1$–$CF_0$) complex (Hiller & Goodchild, 1981) and from Fraction I protein. On the basis of the freeze-fracture studies and specific antibody tests, it appears that the ATP synthase is exclusively located in unstacked regions of the membranes (Miller & Staehelin, 1976). Similarly the light-harvesting chlorophyll a/b protein complex is considered to be preferentially located in stacked thylakoids (Staehelin & Arntzen, 1979). A current model suggesting a particular distribution of the major membrane particles within stacked and unstacked thylakoids is shown in Fig. 10.7. In greening etiolated leaves the appearance of thylakoid stacking has been correlated with increased synthesis of the most prevalent fatty acid in photosynthetic tissue, $\alpha$-linolenic acid. However, this now appears to be a purely coincidental correlation since grana formation can be induced in rapidly greening etiolated tissue without a change in the proportion of $\alpha$-linolenic acid in the major chloroplast lipid, monogalactosyl-diglyceride (Selldén & Selstam, 1976). The application of electrostatic theory to thylakoid stacking has suggested that stacking may be induced by changes in the electrical properties of the membrane surface (Barber, 1980). Stacking may result from the efficient electrostatic screening of the surface negative charges on the membranes, which could also produce the uneven distribution of particles and surface-negative charges between stacked and unstacked thylakoid membranes as illustrated in Fig. 10.7. Thus thylakoid stacking in the developing chloroplast may be the result of changes in electrical properties of the membrane surfaces. Thylakoid stacking has been correlated with the appearance of photo-induced water splitting in the greening etiolated leaf (Strasser & Butler, 1976); it is possible that the onset of electron flow from $H_2O$ to NADP, (Figs. 10.1, 10.10) which will produce a pumping of protons into the thylakoid space and presumably a counter efflux of $Mg^{2+}$ into the stroma, will effect an electrostatic screening of the surface negative charges on the thylakoid sufficient to induce thylakoid aggregation. Staehelin, Armond & Miller, (1977) have shown that during granal stacking in etiochloroplasts (i.e. etioplasts commencing to green) developing in continuous illumination after a flash regime, increases in the size and aggregation of membrane particles occur in parallel with increasing oxygen evolution. The biological implications of thylakoid stacking are not yet fully understood. The specific aggregation of the light-harvesting chlorophyll a/b protein complexes within granal regions may be required to allow the efficient control of excitation energy distribution through this chlorophyll protein complex to photosystem I and photosystem II reaction centres.

## Development of photosynthetic function

Concomitantly with the structural development of the leaf and its chloroplasts, photosynthetic function begins. This depends on the successful completion of a series of complex correlated developmental processes which include:

(*a*) The synthesis of the constituent molecules of the photosynthetic apparatus.

(*b*) The post-translational modification of these molecules into soluble and membrane-bound macromolecular complexes.

(*c*) Organizational modifications which allow the development of chloroplast function and regulation of the integrated process of photosynthesis.

The detailed examination of individual processes and specific macromolecular syntheses, has unequivocally shown that throughout its development, the chloroplast is dependent on innumerable interactions with the other functional cellular components. The syntheses of chlorophyll, chloroplast proteins and chloroplast lipids all depend on a supply of small molecules from the cytosol. Many aspects of the transport and the assembly of the component molecules into macromolecular complexes are also under extra-chloroplastic control. The co-operative interaction of the nuclear and chloroplast genomes in the control of the synthesis of Fraction I protein is well documented (see Ellis, 1981) and many similar interactions certainly occur. Until the chloroplast achieves full photosynthetic function, the energy for its biosynthetic activities must be supplied from outside the chloroplast. The ontogeny of the chloroplast envelope and the development of its specific transport properties are associated changes of major significance. Neither the energetic changes nor changes in transport properties of the developing plastid have yet been investigated systematically but there is some evidence for a co-ordinated relationship between mitochondria and chloroplasts for energy supply in the early stages of greening (Wellburn, Hampp & Wellburn, 1981).

### *The synthesis of the constituent molecules of the photosynthetic apparatus*

Isolated intact mature chloroplasts contain three times as much lipid as protein and the thylakoid fretworks derived from them have approximately equal amounts of protein and lipid by weight, a third of the lipid fraction consisting of chlorophyll molecules. Other functional and quantitatively important components of the chloroplast include numerous carriers involved in the bioenergetic reactions of photosynthesis, biochemical components of the extra-nuclear genetic system including DNA, mRNA and ribosomes and, based on current models of chloroplast/cytoplasmic interactions, at least a thousand enzymes. All these components are required in sufficient concentration for the activity of a fully functional chloroplast. We have reliable biochemical information on some aspects of the biosynthesis of only a few of these chloroplast components and, although at maturity the fully photosynthetic chloroplast is capable of synthesizing a large variety of small molecules, it is clear that most of the macromolecular syntheses require biochemical collaboration between the chloroplast and a variety of other cellular components. The characteristics of biosynthesis during plastid development are very unclear, because little is known of the enzymology of developing leaves.

*Chlorophyll biosynthesis*

In the mature chloroplast all the chlorophyll appears to be complexed to proteins (Thornber & Markwell, 1981). In greened etiolated leaves the biosynthesis of chlorophyll appears to be the trigger which initiates the assembly of the fully functional photosynthetic apparatus (Kirk & Tilney-Bassett, 1978). Etioplasts completely lack chlorophyll but brief flashes (of the order of 1 ms) of intense illumination (5–10 mJ) are sufficient to initiate the biosynthesis of chlorophyll. Endogenous pools of protochlorophyllide are photo-converted during the first few seconds of illumination of etiolated leaves to chlorophyllide and then to chlorophyll. The enzyme catalysing its conversion is NADPH-protochlorophyllide reductase (Griffiths, 1978) but the surprising 80% inhibition of this enzyme observed on illumination remains unexplained (Mapleston & Griffiths, 1980). There is a short lag of up to an hour before the major phase of rapid chlorophyll biosynthesis begins (Treffry, 1978). Initially, the chlorophyll a/chlorophyll b ratio is high but ratios characteristic of the mature tissue are established after a few hours. In etiolated greening wheat, the ratio of 4.5 is established after 8 h (Boffey, Selldén & Leech, 1980) but in the normal tissue the progress to a ratio of 4.5 in the developing proplastids takes 24 h.

The first committed intermediate of chlorophyll biosynthesis is $\delta$-amino laevulinic acid which is derived from glutamate or glycine, or both, in the cytosol. The early stages of chlorophyll synthesis to protoporphyrin IX resemble haem biosynthesis and later the insertion of the Mg atom, addition of the cyclopentanone ring and modification of the heterocyclic rings of the porphyrin head group, lead to the formation of protochlorophyllide. The details of this pathway are given in Reid & Leech (1980). Protochlorophyllide is associated with a protein, holochrome, the apoprotein of which is similar to Fraction I protein but antigenically distinct from it. Photo-conversion of protochlorophyllide can only occur when the pigment is associated with holochrome in a group of 5–25 chromophres (MW $6 \times 10^5$ daltons). The photo-reduction of protochlorophyllide and subsequent esterification with geranyl geraniol yields free chlorophyll a which becomes associated with the thylakoid proteins and inserted into the membrane. Chlorophyll b is probably synthesized directly and not from chlorophyll a.

The assembly of the pigment-protein complexes and their apparently simultaneous insertion into the developing thylakoid membranes is of critical importance in the development of photofunction since the light-harvesting and energy-transducing reactions of photosynthesis depend on the correctness of this assembly. The details of the process are discussed later.

### *Lipid biosynthesis*

An account of photofunction development in chloroplasts would be incomplete without reference to the synthesis of the lipids which make up 50% by weight of the thylakoid membrane and are also essential components of the envelope membrane and stroma. Chloroplast lipids are mainly galactose-containing molecules characterized by their high content of polyunsaturated fatty acids. The trienoic acid, α-linolenic acid, represents 90% of the lipid fatty acid content in the mature thylakoid fretwork. The unusual lipid sulphoquinovosyl diglyceride is a diagnostic component of photosynthetic membranes but of unknown function; phosphatidyl glycerol is the single major phospholipid of thylakoids and phosphatidyl choline is the major phospholipid of the chloroplast envelope. During biogenesis of the thylakoid membranes and plastid development massive lipid biosynthesis occurs, polyunsaturation accelerates and the lipid content per plastid increases at least tenfold (Leech, Rumsby & Thomson, 1972). It is significant that these changes are quantitative not qualitative. The only new lipid molecule which appears during plastid ontogeny, either during the greening of etiolated leaf tissue or during normal chloroplast development is phosphatidyl glycerol containing the trienoic acid $\Delta^{3trans}$ hexadecenoic acid. This fatty acid becomes prominent when grana are increasing in size and at this time can represent up to 30% of the fatty acid of phosphatidyl glycerol (Selldén & Selstam, 1976). The functional significance of this thylakoid component is unknown. During early plastid ontogeny the intermediates for lipid biosynthesis are provided from the cytosol. When chloroplasts begin actively to photo-reduce $CO_2$, a comparison of chloroplast and leaf lipid biosynthesis suggests that while the chloroplasts may be able to synthesize all of the fatty acids required by the leaf (Ohlrogge, Kuhn & Stumpf, 1979) the biosynthesis of the fatty acid precursors and the desaturation of these fatty acids also involves other organelles in the leaf cell (Murphy & Leech, 1981). It seems likely that fatty acid desaturation and lipid assembly occurs at the chloroplast envelope and phosphatidyl choline from outside the chloroplast is implicated as an acyl carrier (Roughan, Holland & Slack, 1980). As in the biosynthesis of chloroplast proteins, to be considered later, the interdependence of the plastid and the rest of the cell in lipid biosynthesis is a distinctive and significant feature of plastid development.

### *Synthesis of chloroplast proteins*

The development of chloroplast photofunction is also dependent on the synthesis of numerous chloroplast proteins. These include the thylakoid membrane proteins which complex with the photosynthetic pigments, electron carriers such as iron-sulphur proteins, cytochromes, plastocyanin and soluble enzymes involved in biosynthesis and photosynthetic carbon reduction. So far most work has been on the development of those components of the mature leaf cell which are the easiest to analyse or which change most dramatically during development. For these reasons the soluble Fraction I protein has attracted the most attention (Ellis, 1979). Fraction I protein is the chloroplast enzyme ribulose bisphosphate carboxylase-oxygenase responsible for photosynthetic carbon reduction and catalyses the carboxylation of ribulose bisphosphate to yield the first product of photosynthesis, 3-phosphoglycerate. It is the most abundant protein in the natural world and represents between 70% and 90% of the soluble protein fraction of mature leaves. Quantitative analyses show that on a cellular basis it increases 20-fold during 3 days of development in wheat leaves before declining. Each chloroplast during its development from a proplastid, increases in Fraction I protein content by 6-fold.

The patterns of quantitative and qualitative changes in the leaf protein components during leaf differentiation, including the changes in Fraction I protein and its large and small subunits, can be followed by polyacrylamide gel (PAGE) electrophoresis of the sodium dodecyl sulphate (SDS) polypeptides derived from the extracted leaf protein. An example of the changing pattern of polypeptides found in the cells of wheat leaf tissue as photosynthetic capacity increases can be seen in Fig. 10.8 which shows that very young cells containing small undifferentiated proplastids possess both subunits of Fraction I protein and that the rates of synthesis of the two subunits appear to be synchronized as leaves develop. The $\alpha$ and $\beta$ subunits of the chloroplast coupling factor can be distinguished from each other after one day of development. The major insoluble protein, the light-harvesting chlorophyll a/b protein complex, becomes visible on the gels at a later stage of leaf development than Fraction I protein and then increases substantially. Many of the other protein components also increase during leaf development and, even

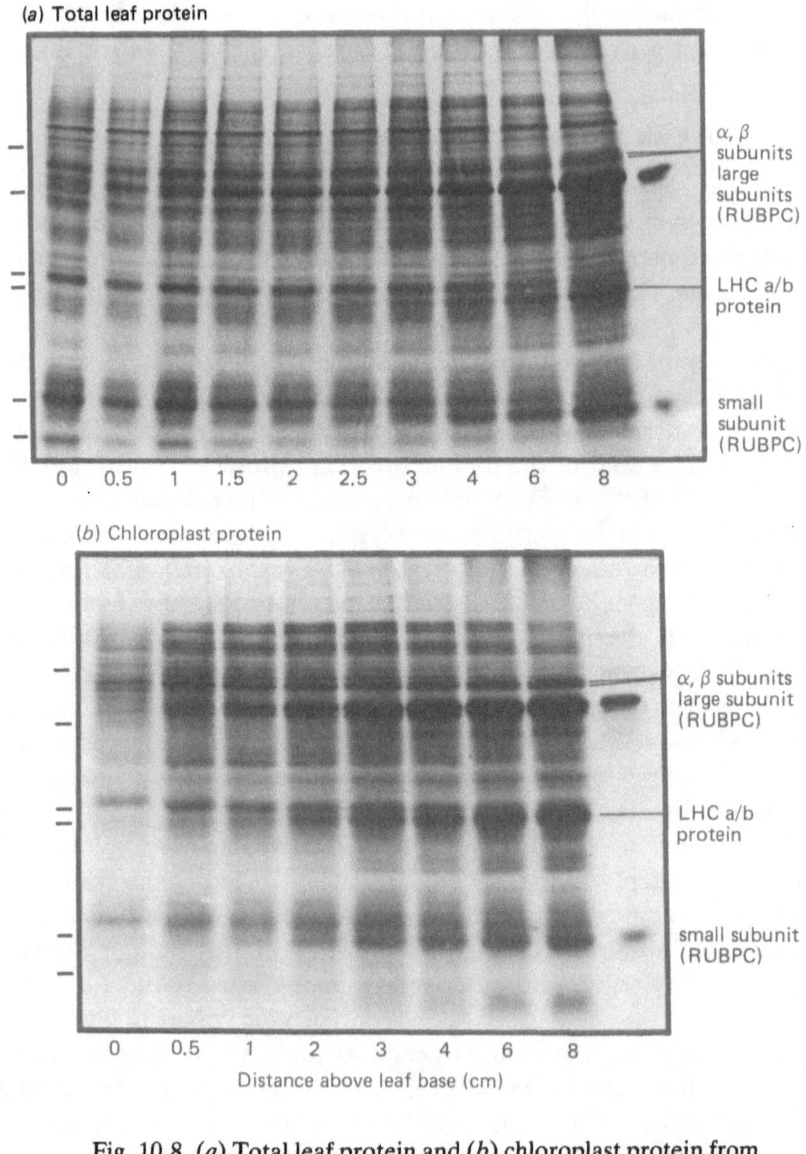


Fig. 10.8. (*a*) Total leaf protein and (*b*) chloroplast protein from different developmental stages of the first leaf of wheat (*Triticum aestivum* var. Maris Dove). Each track corresponds to a sample from a 5 mm leaf slice taken at the height above the leaf base indicated. Separation by SDS-polyacrylamide gel electrophoresis. Samples were loaded onto 15% slab gels so that the volumes in each track represented equal numbers of cells (*a*) or equal numbers of plastids (*b*). Molecular weight markers: bovine serum albumin (68 000), ovalbumin (43 000), chymotrypsin (25 000), trypsin (23 000),

more intriguingly, some decrease. The control of the synthesis of those proteins which are identifiable is amenable to analysis using the techniques of molecular genetics. Some of the remarkable conclusions of these studies will be reviewed in the following section in which the assembly of the photosynthetic apparatus is considered.

### *The formation of macromolecular assemblies*

As the molecular components of the photosynthetic system are synthesized they become incorporated into macromolecular assemblies which are the 'structural' components of the chloroplast. In Fig. 10.9 some of the components involved in the control of the development of the functional photosynthetic apparatus are represented diagrammatically. The assembly of the major lipid and protein components of the photosynthetic apparatus occurs in the chloroplast and when photosynthesis commences its products are used for synthesis of additional cellular and chloroplast components. The transcription and translation of the gene products which give rise to the polypeptides, pigments and lipids of the photosynthetic apparatus are processes which involve the collaborative biochemical function of several other cellular components in addition to the developing chloroplasts. Many aspects of the synthesis and assembly of the components of the chloroplast are controlled by light and by a complex series of interactions at the genetic level.

The leaf cell possesses three distinct genetic systems – the conventional 'nuclear genetic system' and two others, one found in the mitochondria and the other in the chloroplasts. Four interactive components in each genetic system collaborate in the expression of genetic information i.e. DNA, DNApolymerase, mRNA and a protein synthesizing system. The development of chloroplasts depends on a complex series of interactions between the nuclear genetic system and the plastid one. Some chloroplast polypeptides are synthesized inside the developing chloroplast; many others are made in the cytoplasm. Those chloroplast

Caption to Fig. 10.8 (*cont.*)
myoglobin (17 000) and haemoglobin (15 500). The large (49 000) and small (16 000) subunits of purified wheat RUBPC (Fraction I protein) were electrophoresed alongside the leaf and chloroplast proteins to identify the stained bands. The bands corresponding to the $\alpha$ and $\beta$ subunits (61 000, 59 000) and the light-harvesting a/b protein complex (LHCP 25 000) are also identified. (C. Dean and R. M. Leech, unpublished data.)

polypeptides synthesized within the organelle appear to be encoded in the chloroplast DNA while the chloroplast proteins synthesized outside the chloroplast are coded in the nuclear genome and synthesized on cytoplasmic ribosomes.

We have now some knowledge of the synthesis and assembly of a few of the quantitatively most significant chloroplast components, in particular Fraction I protein, the ATP synthase complex, chloroplast ribosomes and the light-harvesting chlorophyll a/b protein complex (Ellis, 1981). The mode of synthesis and assembly of the *ca* 200 additional chloroplast proteins is unknown.

The best-known example of the interaction of nuclear and chloroplast genomes is their role in the control of synthesis of Fraction I protein. This protein has a molecular weight of approximately $5.6 \times 10^5$ daltons and consists of eight large (MW $5.2–6 \times 10^4$ daltons) and eight small (MW $1.2–1.8 \times 10^4$ daltons) subunits. The small subunit is encoded in the nuclear DNA and its precursor, of molecular weight 20 000 daltons, is synthesized on free 80 S cytoplasmic ribosomes; the large subunit is coded for by the chloroplast genome and synthesized on 70 S ribosomes within the plastid. The DNA sequence of the large subunit in maize and the small subunit clone in pea have been established (McIntosh, Poulson & Bogorad, 1980, Bedbrook, Smith & Ellis, 1980). Recently

Fig. 10.9. Diagram showing some of the controls operating during the synthesis of the photosynthetic apparatus. For details see the text. $h\nu$ = photons.

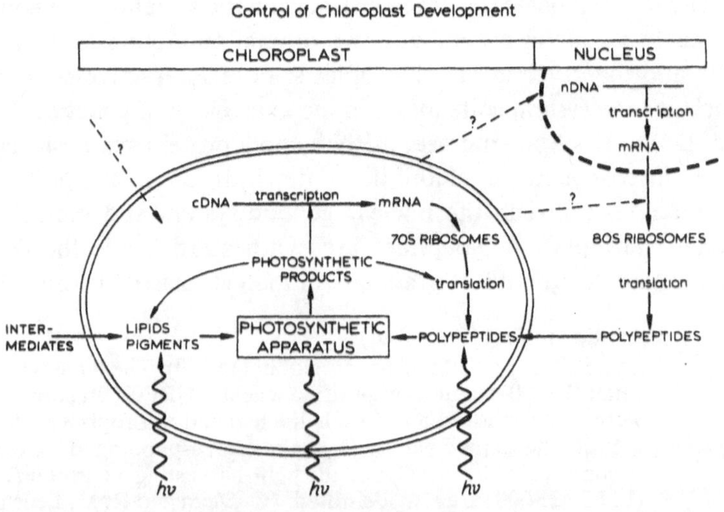

the expression of the maize and wheat chloroplast genes for the large subunit in *E. coli* has been reported (Gatenby, Castleton & Saul, 1980). The mRNA for the large subunit is non-polyadenylated and the gene for the large subunit is located as a single copy per chloroplast DNA circle. The precursor molecules of the small subunit move across the chloroplast envelope, a processing enzyme removes a small polypeptide of low molecular weight and the processed small subunit and the large subunit assemble inside the chloroplast.

The biogenesis of chloroplast ribosomes is another example of the co-operation between the nuclear and chloroplast genomes in development. The genes for the 5 S, 16 S and 23 S rRNA components of the chloroplast ribosome are all found in the chloroplast DNA and are transcribed as a single unit to give an RNA precursor of MW $2.7 \times 10^6$ daltons. At least some of the chloroplast tRNAs are also encoded in the chloroplast DNA. Chloroplast ribosomes contain about 90 distinct polypeptides and although a few of these are coded in the chloroplast genome the genes for the majority of the chloroplast ribosomal proteins are present in the nuclear genome and are inherited in a Mendelian fashion.

The mode of assembly of two thylakoid components, the ATP-synthase complex and the light-harvesting chlorophyll a/b protein complex has been the subject of extensive experimentation which has yielded interesting and surprising results. The ATP-synthase complex of spinach is made up of at least eight non-identical subunits. The extrinsic coupling factor ($CF_1$), which is visualized as small (*ca* 11 nm) particles on the surface of unstacked thylakoid membranes after freeze-fracture, is composed of five subunits of which at least two, the $\alpha$, $\beta$, also are synthesized within the chloroplast. The three subunit components of $CF_0$ are embedded in the membrane and two of the subunits are synthesized in the chloroplast but the third $CF_0$ subunit is synthesized on the cytoplasmic ribosomes. Assembly of the ATP-synthase complex occurs in isolated chloroplasts so a pool of cytoplasmically synthesized polypeptides appears to be contained within them.

The polypeptides of the light-harvesting chlorophyll a/b protein complex, like the small subunit of Fraction I protein, are synthesized as high molecular weight precursors in the cytoplasm and transported to the chloroplast where they are processed to smaller proteins (25 000 daltons MW) and assembled into the developing thylakoid membrane in association with chlorophylls a and b (see Fig. 10.11). The synthesis of one large molecular weight (29 500 daltons) precursor has been demonstrated in wheat germ extracts primed with cytoplasmic polyA-containing RNA (Apel, 1979) and this has been shown to enter isolated chloroplasts. The mRNA level for these polypeptides is controlled by light (Tobin, 1978; Apel, 1979). The transport of the polypeptide chain is post-translational after the chains have been released from the ribosomes and transport does not depend on concomitant protein synthesis. The extra sequence in the precursor is presumed to be involved in a specific interaction with the chloroplast envelope providing for its recognition, but it is not known whether the extra sequences are similar in dissimilar polypeptides. Continuous light is required to initiate translation, uptake, processing and membrane binding of the light-harvesting chlorophyll a/b protein complex. This is probably connected to the requirement for continuous light in chlorophyll synthesis. In the absence of chlorophyll synthesis LHCP turns over. Although the light-harvesting chlorophyll a/b protein precursor(s) synthesized in wheat germ *in vitro* translation systems has not been further fractionated, in the assembled thylakoid the complex consists of two light-harvesting chlorophyll a/b proteins which can be distinguished by their molecular weights of 25 000 and 26 000 daltons respectively. These are phosphoproteins and become heavily labelled when $^{32}P$-orthophosphate is fed to isolated chloroplasts. The phosphoprotein complex has recently been shown to perform a major regulatory role in the transfer of excitation energy within the thylakoid. In its phosphorylated state, the complex transfers excitation energy preferentially to PSI whereas in the dephosphorylated state the complex transfers excitation energy preferentially to PSII. Full photofunction of the light-harvesting and energy transduction phases of photosynthesis is dependent on the efficient operation of the phosphorylation and dephosphorylation system which itself is dependent on a phosphorylating enzyme, protein kinase, and a dephosphorylating phosphatase. The characteristics of the two enzyme systems are not yet known but the protein kinase is thylakoid-bound and activated by light-driven electron transport. Thus the synthesis and assembly and activation of the proteins of the LHCP complex play a key role in the attainment of photosynthetic function in the chloroplast and further knowledge of their characteristics is currently being actively pursued.

### *The assembly of macromolecular complexes into a photosynthetically functional chloroplast*

The regulation of macromolecular synthesis offers a number of potentially important control points in chloroplast development but assembly of the macromolecular complexes into a photosynthetically competent matrix is essential before photofunction can commence. The complexity of this task is exemplified here by considering aspects of the development of thylakoid photochemical functioning and photosynthetic enzyme systems.

### *Development of thylakoid photofunctions*

The photochemical reduction of $NADP^+$ and production of ATP by the thylakoid is dependent upon the absorption of photons by chlorophyll molecules and subsequent transfer of the trapped excitation energy to specific photochemically active chlorophyll molecules (P690 and P700), which then drive the transfer of electrons from water to $NADP^+$. All chlorophyll molecules in the mature light-harvesting apparatus are bound to membrane proteins, with three specific chlorophyll-protein complexes comprising the bulk of the apparatus; these are (i) a complex containing about 30% of the total chlorophyll and associated with PSI activity, chlorophyll-protein complex I (CPI), (ii) one containing up to 20% of the total chlorophyll and associated with PSII activity, and (iii) a light-harvesting chlorophyll a/b protein complex, often termed LHCP, which serves to transfer excitation energy to the complexes associated with PSI and PSII, and whose synthesis has already been considered above (Thornber, Markwell & Reinman, 1979). A simple diagrammatic representation of how these components interact in light harvesting is shown in Fig. 10.10. It is not possible here to provide extensive details of the complex mechanisms and numerous components involved in the photo-induced transfer of electrons from water to NADP (see Hinkle & McCarty, 1978 for details), but a simplified electron transport scheme depicting the major components and their spatial organization within the membrane is shown in Fig. 10.11. As water is photo-oxidized and electrons transferred from PSII to PSI through plastoquinone, (PQ), protons are pumped into the intrathylakoid space with the result that a proton electro-chemical gradient is established across the membrane. On chemiosmotic considerations, it is thought that this proton electrochemical gradient can be

Fig. 10.10. A model for the light-harvesting apparatus of the photosynthetic membrane showing the three major chlorophyll-protein complexes. $h\nu$ = photons.

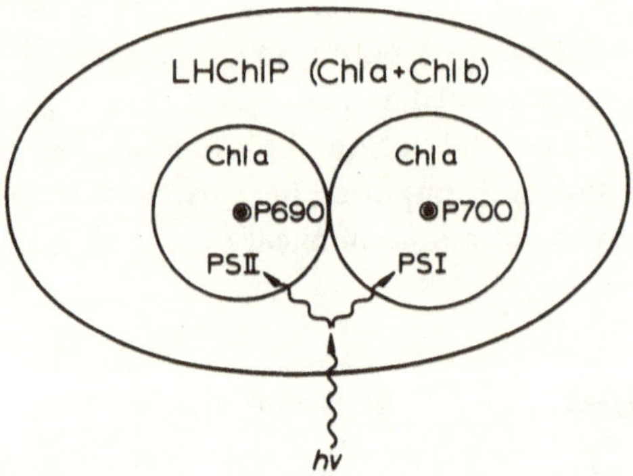


Fig. 10.11. Simplified scheme for photosynthetic electron transport within the thylakoid membrane and the chemiosmotic coupling to ATP synthesis. Key to symbols: ADP, adenosine diphosphate; ATP, adenosine triphosphate; $CF_0$, component of coupling factor responsible for directing protons across the membrane; $CF_1$, component of coupling factor responsible for ATP synthesis; cyt f, cytochrome f; Fd, ferredoxin; Mn, manganese; $NADP^+$, nicotinamide adenine dinucleotide phosphate; P690, reaction centre of photosystem II; P700, reaction centre of photosystem I; PC, plastocyanin; Pi, inorganic phosphate; PQ, plastoquinone; PSI, photosystem I; PSII, photosystem II; Q, primary electron acceptor of photosystem II; X, primary electron acceptor of photosystem I. For additional detail see Fig. 10.1.

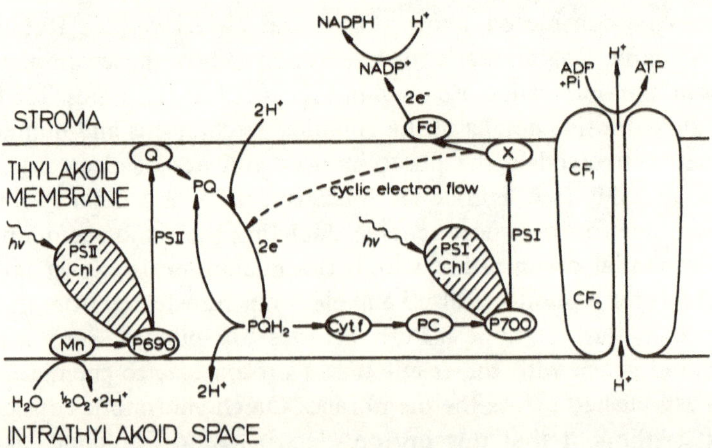

dissipated to produce ATP via the coupling factor, which is a proton translocating ATPase system consisting of two components, $CF_0$ and $CF_1$ (Jagendorf, 1977). $CF_1$ is an extrinsic membrane component containing the active site for ATP synthesis, whilst $CF_0$ is located within the membrane and is responsible for directing protons across the thylakoid (McCarty, 1979). Photophosphorylation can also occur as a result of proton pumping induced by a cyclic electron flow around PSI (Fig. 10.11).

In order to achieve competence in these photochemical processes, a very specific organization of membrane components is required. Much research has concentrated on the development of photofunction in greening etiolated systems and we will consider some important findings of these studies. However, it must be emphasized that although many similarities exist between thylakoid photofunctional development in both greening etiolated and 'naturally' grown leaves, for example in the development of reaction centre photochemistry, significant differences in the synchrony of thylakoid component synthesis and organization are found in plastids recovering from etiolation when compared with normal development. The sequence of development of photochemical function in greening etiolated leaves is given in Fig. 10.12.

Photochemically active reaction centres of PSI and PSII appear rapidly after the onset of chlorophyll synthesis (Baker & Butler, 1976;

Fig. 10.12. Proposed time sequence for the development of chloroplast structural and photochemical characteristics in greening etiolated leaves.

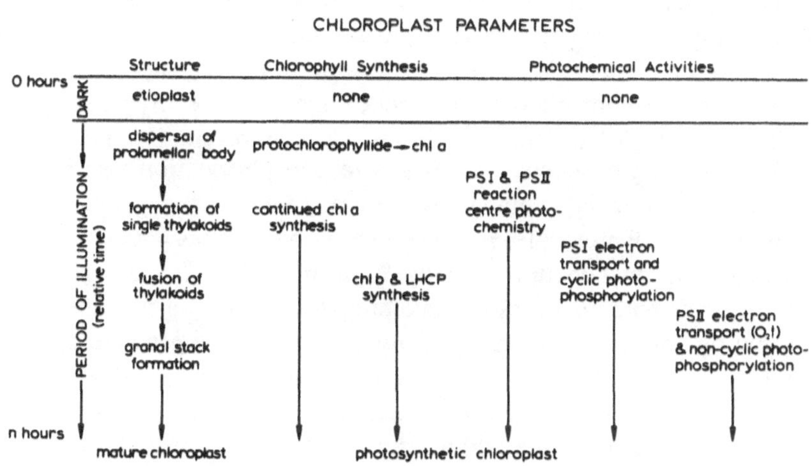

Egnéus,Selldén & Andersson, 1976; Baker & Miranda, 1982), thus implying their insertion into the membrane and activity at an early stage of development. However, for a period after the PSI and PSII reaction centres have become photochemically active the thylakoids cannot photo-oxidize water and thus cannot reduce $NADP^+$. It appears that the insertion of $Mn^{2+}$ into the water splitting enzyme complex is required to facilitate the splitting of water by PSII (Inoue, Kobayashi, Sakamoto & Shibata, 1975; Strasser, 1975; Strasser & Butler, 1976). The onset of water photo-oxidation coincides with the appearance of non-cyclic photophosphorylation in leaves, but it should be noted that PSI driven cyclic photophosphorylation appears before the non-cyclic process (Oelze-Karow & Butler, 1971) and generates a supply of ATP, but not NADPH, for the chloroplast. Presumably this ATP is utilized for the synthesis of chloroplast components since the chloroplast will also require a source of reducing power if it is to photo-reduce $CO_2$. Although all the photochemical activities of the thylakoid membrane become evident after water photo-oxidation appears, the efficiency of the photochemical processes continues to increase as the chloroplast develops. Accumulation of PSI, PSII and the light-harvesting chlorophyll-protein complexes increases the efficiency of photon capture by the membrane so that there is a decrease in the light energy required to saturate the photochemical reactions (Egnéus, Selldén & Andersson, 1976; Baker & Leech, 1977; Akoyunoglou, 1978). The appearance of the light-harvesting chlorophyll a/b protein complex confers an important control on the distribution of excitation energy to PSI and PSII reaction centres and thus on the balance between cyclic and non-cyclic electron transport. It will be remembered (see above) that this protein, in the unphosphorylated state, transfers energy preferentially to PSII and, when phosphorylated, to PSI; the phosphorylation of the protein appears to be controlled by the redox state of the thylakoid plastoquinone pool (Allen, Bennett, Steinbeck & Arntzen, 1981).

The coupling of electron transport to phosphorylation also becomes more efficient with greening, as demonstrated by the increase in the number of ATP molecules produced per two electrons passing along the electron transport chain (Weistrop & Stern, 1977). This may be a result of the thylakoid membrane becoming less leaky to protons and so becoming able to maintain a proton electrochemical gradient sufficient to drive ATP synthesis (Baker, 1982). Large changes in the polypeptide composition of the developing thylakoid membrane, especially the massive increase in the chlorophyll-protein content, will markedly modify membrane organization and so affect photochemical efficiency. Neither thylakoid nor chloroplast development are synchronous during leaf greening. Synthesis of new thylakoid membranes will occur throughout leaf development and give rise to changes in the concentration of photochemical centres per unit area of leaf. Both the concentration of active photochemical centres and the size of the light-harvesting apparatus associated with each centre, i.e. the average number of chlorophyll molecules per reaction centre, will determine the overall efficiency of light energy transduction of the leaf. As the number of reaction centres increases per unit area of leaf, the maximum rate of photosynthesis will increase. Under the majority of natural environmental conditions developing leaves are not exposed to saturating light levels, but an increase in the efficiency of light capture may be achieved by increasing the number of chlorophyll molecules associated with each reaction centre. This point is illustrated in Fig. 10.13 which gives the light saturation curves for four hypothetical situations found in developing leaves when variations occur in both the concentration of reaction centres and the size of their light-harvesting apparatus.

### *Development of carboxylating enzyme systems*

It has been suggested that carbon assimilation during leaf ontogeny may be regulated by the concentration and specific activity of the enzymes involved in the 'dark' reactions of photosynthesis since changes in the activity of enzyme systems per unit leaf area would result in changes in $r_x$ (see Fig. 10.3) of the leaf. Unfortunately, as yet, no comprehensive study has been made of all the C3 cycle enzymes during leaf ontogeny for a single species: studies have been restricted to a single enzyme or at best a few 'key' enzymes. Ribulose bisphosphate carboxylase is the most widely studied of the photosynthetic enzymes and has generally been found to show an increase in activity per unit leaf area during early phases of leaf development (e.g. Steer, 1971, 1973; Callow, 1974; Baker & Hardwick, 1973; O'Toole, Ludford & Ozbun, 1977), presumably as a result of increased synthesis of the enzyme, since no major change in the enzyme activity per unit of Fraction I protein has been observed (Callow, 1974; Steer, 1973). It is difficult to determine from *in vivo* measurements of $CO_2$ assimilation and *in vitro* enzyme assays whether an enzyme may limit the photosynthetic rate of a leaf because of the problems associated with differential efficiency of enzyme extraction and assay at different developmental stages. In cocoa, an increase in the *in vitro* ribulose bisphosphate carboxylase activity/leaf photosynthetic carbon assimilation rate ratio during development suggests that the enzyme was not a limiting factor in the development of photosynthetic capacity (Baker & Hardwick, 1973).

The situation in C4 plants, is more complex than for C3 species, because not only do the enzymes of the C3 cycle have to be considered, but also the carboxylating system of the mesophyll cells, the carbon translocation mechanisms between the mesophyll and bundle sheath cells and the decarboxylation system of the bundle sheath cells (see Fig. 10.14 for details). It is interesting that three key photosynthetic enzyme

Fig. 10.13. Models of photosynthetic unit assembly. Four hypothetical situations, (*a*) to (*d*), are considered in which the concentration of reaction centres per leaf area and the size of the light-harvesting apparatus per reaction centre are modified. Theoretical plots of the maximum photochemical activity per unit area, $V_{max}$, against light intensity are shown for each situation. Note that systems (*a*) and (*b*) have the same light saturation point but different $V_{max}$ values, as is the case for (*c*) and (*d*), whilst (*a*) and (*c*) have the same $V_{max}$ values, as is also the case with (*b*) and (*d*).

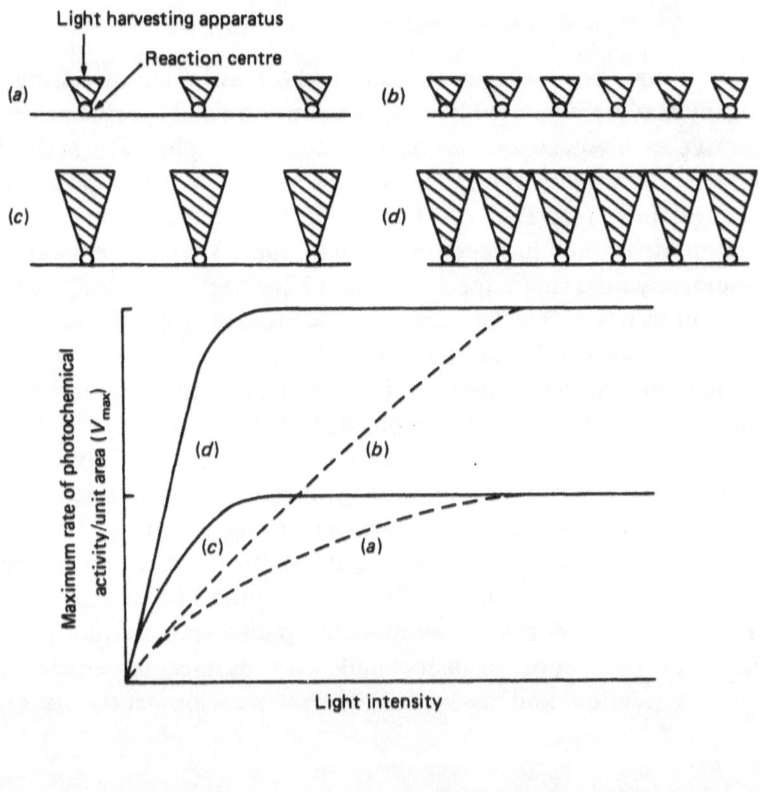

systems of the *Zea mays* leaf, i.e. PEP carboxylase, NADP-malic enzyme and the R5P system, show different patterns of development (Miranda, Baker & Long, 1981b). In another C4 plant, *Portulaca oleracea*, different photosynthetic carbon compounds are produced at different stages of ontogeny (Kennedy & Laetsch, 1973). Such changes may reflect a requirement by the leaf for different carbon skeletons during development, rather than simply changes in enzyme concentrations and activities, and introduce added complexity to the problem.

## Conclusions

The synthesis and assembly of the components of the photosynthetic apparatus in higher plants and the development of individual physiological functions have been discussed; some of the possible

Fig. 10.14. Pathway of photosynthetic C4 carbon assimilation in the mature *Zea mays* leaf. Key to symbols: PEP, phosphoenol pyruvate; OAA, oxaloacetate; PGA, 3-phosphoglycerate; (1), site of PEP carboxylase action; (2), site of NADP-malic enzyme action.

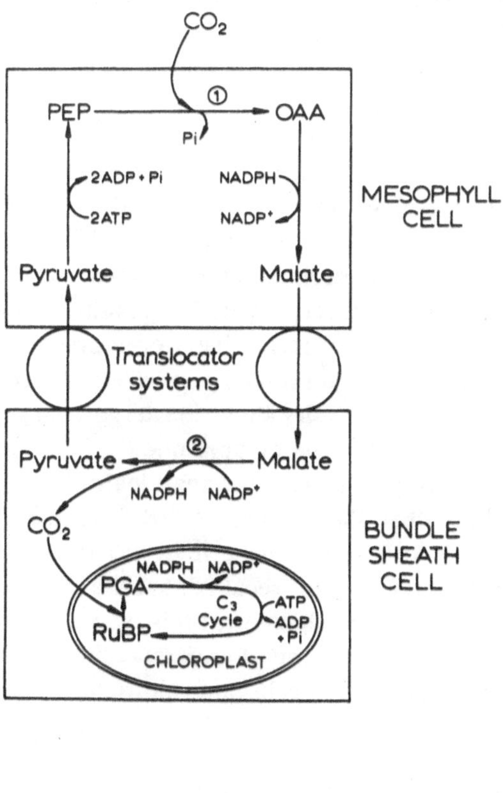

mechanisms which may operate to control discrete developmental processes have also been considered. Understanding of the development of photofunction has been greatly extended by studies of isolated systems such as enzymes, membranes and organelles and by the detailed analysis of individual activities such as energy transduction and oxygen evolution. But a comprehensive description of the development of photofunction in the whole leaf requires the consideration of another order of complexity, that is the heterogeneity in structure and development of the tissues of the leaf itself. Indeed the morphological and anatomical patterns of leaf development determined genetically and environmentally are reflected in the patterns of photosynthetic unit and plastid development. The distinctive patterns of cellular differentiation in developing leaves have been carefully described in several elegant studies (Chapter 6). Not all cells will become photosynthetic and it is clear that a leaf may have cells at different stages of development, some at the zenith of photosynthetic productivity while others are still maturing. Furthermore, chloroplasts in different cells are also at a variety of stages of differentiation from proplastids to chloroplasts, and the rates of membrane assembly, even within a single differentiating chloroplast, are also heterogeneous. The development of the photosynthetic apparatus of the whole leaf therefore reflects a myriad of stages of differentiation of its individual cells and its plastids which themselves are not necessarily developing at the same rates.

Although most investigators can only study one or two aspects of photofunction development at any one time, it is important not to lose sight of the fact that the leaf develops as an integrated whole and not as numerous discrete components. The careful choice of several parameters for the expression of experimental results may often reveal relationships which have not been suspected. The results shown in Fig. 10.15 illustrate well the additional facets of leaf development which are revealed when changes in ribosomal number *per cell* and *per plastid* are measured simultaneously. The graphs show the large changes in the numbers of chloroplast (70 S) and the cytoplasmic (80 S) ribosomes in cells of different ages within the same wheat leaf. At a stage in development (2 days old) when the number of 70 S ribosomes/plastid is declining, the number of 70 S ribosomes/cell is still increasing. This apparent paradox is resolved when it is recalled that the plastids are still replicating at this time. In relating changes of this type to the development of the whole leaf, a further degree of understanding can be added by examining the total populations of cells and chloroplasts in the whole leaf. Unfortunately there is no single internal standard to which all leaf developmental changes can be related since no single unchanging component has been identified. In a changing situation the more parameters which can be measured the greater will be the understanding of the integration of cellular and plastid processes, although some parameters may be more useful than others. The classical standards for measuring photosynthetic function, the chlorophyll content and unit leaf area, are clearly inadequate for developmental studies since the rate of chlorophyll synthesis is one of the most rapidly changing plastid

Fig. 10.15. Changes in ribosome number during cell and chloroplast development in leaf cells of wheat (*Triticum aestivum* var. Maris Dove). (C. Dean and Rachel M. Leech, unpublished data.)

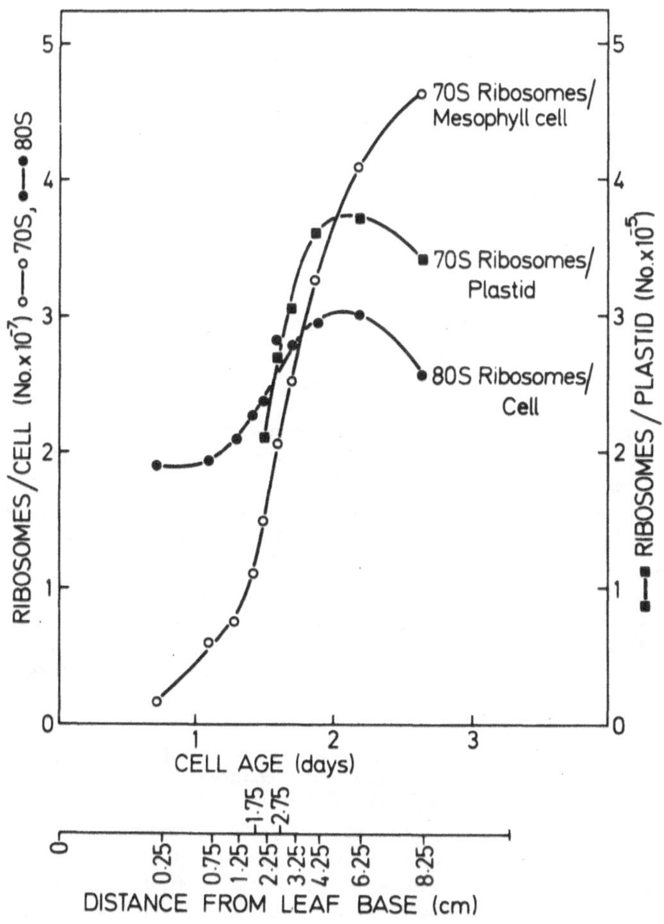

characteristics and measurements of cell and organelle volumes are more important than areas in considerations of cellular function.

What of the future? The need is to devise experimental procedures for the examination of the integration and control of the many facets of development and for the study of populations of organelles maturing at different rates within the same leaf. As more information becomes available we shall be able to move away from the present rather rigid concept of linear development (i.e. which process happens first, which second etc.) and consider the application of mathematical techniques of population analysis to aid understanding of photofunction development.

## References

Akoyunoglou, G. (1978). Growth of the PSII Unit as monitored by fluorescence measurements, the photo-induced absorbance change at 518 nm, and photochemical activity. In *Chloroplast Development*, ed. G. Akoyunoglou, pp. 355–66. Amsterdam: Elsevier.

Allen, J. F., Bennett, J., Steinbeck, K. E. & Arntzen, C. J. (1981). Chloroplast protein phosphorylation couples plastoquinone redox state to distribution of excitation energy between photosystems. *Nature*, **291**, 25–9.

Apel, K. (1979). Phytochrome-induced appearance of mRNA activity for the apoprotein of the light-harvesting chlorophyll a/b protein of barley *Hordeum vulgare*. *European Journal of Biochemistry*, **97**, 183–8.

Baker, N. R. (1983). Development of chloroplast photochemical function. In *Chloroplast Biogenesis*, ed. N. R. Baker & J. Barber. Amsterdam: Elsevier (in press).

Baker, N. R. & Butler, W. L. (1976). Development of the primary photochemical apparatus of photosynthesis during greening of etiolated bean leaves. *Plant Physiology*, **58**, 526–9.

Baker, N. R. & Hardwick, K. (1973). Biochemical and physiological aspects of leaf development in cocoa (*Theobroma cacao*). I. Development of chlorophyll and photosynthetic activity. *New Phytologist*, **72**, 1315–24.

Baker, N. R. & Leech, R. M. (1977). Development of photosystem I and photosystem II activities in leaves of light-grown maize (*Zea mays*). *Plant Physiology*, **60**, 640–4.

Baker, N. R. & Miranda, V. (1982). Development of primary photosynthetic processes in leaves grown under a diurnal light regime. In *Proceedings of 5th International Photosynthesis Congress*, ed. G. Akoyunoglou.

Barber, J. (1980). An explanation for the relationship between salt-induced thylakoid stacking and chlorophyll fluorescence changes associated with changes in spillover of energy from photosystem II to photosystem I. *Federation of European Biochemical Society Letters*, **118**, 1–10.

Bedbrook, J. R., Smith, S. M. & Ellis, R. J. (1980). Molecular cloning and sequencing of cDNA encoding the precursor to the small subunit of chloroplast ribulose 1,5-bisphosphate carboxylase. *Nature*, **287**, 692–7.

Boffey, S. A., Ellis, J. R., Selldén, G. & Leech, R. M. (1979). Chloroplast division and DNA synthesis in light-grown wheat leaves. *Plant Physiology*, **64**, 502–4.

Boffey, S. A., Selldén, G. & Leech, R. M. (1980). Influence of cell age on chlorophyll formation in light-grown and etiolated wheat seedlings. *Plant Physiology*, **65**, 680–4.

Brangeon, J. & Mustardy, L. (1979). The autogenetic assembly of intrachloroplastic lamellae viewed in 3-dimension. *Biologie Cellulaire*, **36**, 71–80.

Butterfass, T. (1979). Patterns of chloroplast reproduction: a developmental approach to protoplasmic plant anatomy. In *Cell Biology Monographs*, vol. 6, pp. 205. Vienna: Springer-Verlag.

Callow, J. A. (1974). Ribosomal RNA, fraction I protein synthesis and ribulose diphosphate carboxylase activity in developing and senescing leaves of cucumber. *New Phytologist*, **73**, 13–20.

Čatský, J., Tichá, I. & Solárová, J. (1976). Ontogenetic changes in internal limitations to bean leaf photosynthesis. I. Carbon dioxide exchange and conductances for carbon dioxide transfer. *Photosynthetica*, **10**, 394–402.

Edwards, G. E. & Walker, D. A. (1982). *C3, C4 – Some Aspects of Photosynthetic Carbon Assimilation*. Oxford University Press (in press).

Egnéus, H., Selldén, G. & Andersson, L. (1976). Appearance and development of P700 oxidation and photosystem I activity in etio-chloroplasts prepared from greening barley leaves. *Planta*, **133**, 47–52.

Ellis, R. J. (1979). The most abundant protein in the world. *Trends in Biochemical Sciences*, **4**, 241–4.

Ellis, R. J. (1981). Chloroplast proteins: synthesis, transport and assembly. *Annual Review of Plant Physiology*, **32**,111–37.

Gatenby, A. A., Castleton, J. A. & Saul, M. W. (1980). Expression in *E. coli* of maize and wheat chloroplast genes for large subunit of ribulose bisphosphate carboxylase. *Nature*, **291**, 117–21.

Griffiths, W. T. (1978). Light modulation of the activity of protochlorophyllide reductase. *Biochemical Journal*, **174**, 681–92.

Hiller, R. G. & Goodchild, D. J. (1981). Thylakoid membrane and pigment organisation. In *The Biochemistry of Plants. A Comprehensive Treatise*, vol. 8, *Photosynthesis*, ed. M. D. Hatch and N. K. Boardman, pp. 1–49. New York: Academic Press.

Hinkle, P. C. & McCarty, R. E. (1978). How Cells Make ATP. *Scientific American*, **238**, 104–23.

Inoue, Y., Kobayashi, Y., Sakamoto, E. & Shibata, K. (1975). Multiple flash activation of the water photolysis system in intermittently illuminated wheat leaves. *Plant and Cell Physiology*, **16**, 327–36.

Jagendorf, A. T. (1977). Photophosphorylation. In *Encyclopedia of Plant Physiology, New Series, Vol. 5: Photosynthesis I*, ed. A. Trebst & M. Avron, pp. 307–37. Berlin: Spinger-Verlag.

Kennedy, R. A. & Laetsch, W. M. (1973). Relationship between leaf development and primary photosynthetic products in the C4 plant *Portulaca oleracea* L. *Planta*, **115**, 113–24.

Kirk, J. T. O. & Tilney-Bassett, R. A. (1978). *The Plastids: Their Chemistry, Structure, Genetics and Inheritance*, second edition. Amsterdam: Elsevier/North-Holland.

Körner, C. H., Schael, J. A. & Bauer, H. (1979). Maximum leaf diffusive conductance in vascular plants. *Photosynthetica*, **13**, 45–82.

Leech, R. M. (1976). The replication of plastids in higher plants. In *Cell Division in Higher Plants*, ed. M. M. Yeoman, pp. 135–59. London: Academic Press.

Leech, R. M., Rumsby, M. G. & Thomson, W. W. (1972). Plastid differentiation, acyl lipid and fatty acid changes in developing green maize leaves. *Plant Physiology*, **52**, 240–5.

Ludlow, M. & Wilson, G. L. (1971). Photosynthesis of tropical pasture plants. III. Leaf age. *Australian Journal of Biological Sciences*, **24**, 1077–87.

Lurie, S. (1977a). Stomatal development in etiolated *Vicia faba*: relationship between structure and function. *Australian Journal of Plant Physiology*, **4**, 61–8.

Lurie, S. (1977b). Stomatal opening and photosynthesis in greening leaves of *Vicia faba*. *Australian Journal of Plant Physiology*, **4**, 69–74.

McCarty, R. E. (1979). Roles of a coupling factor for photophosphorylation in chloroplasts. *Annual Review of Plant Physiology*, **31**, 79–104.

McIntosh, L., Poulson, C. & Bogorad, L. (1980). Chloroplast gene sequence for the large subunit of ribulose bisphosphate carboxylase of maize. *Nature*, **288**, 556–90.

Mapleston, R. E. & Griffiths, W. T. (1980). Light modulation of the activity of protochlorophyllide reductase. *Biochemical Journal*, **189**, 125–33.

Miller, K. R. & Staehelin, L. A. (1976). Analysis of the thylakoid outer surface: coupling factor is limited to unstacked membrane regions. *Journal of Cell Biology*, **68**, 30–47.

Miranda, V., Baker, N. R. & Long, S. P. (1981a). Anatomical variation along the length of the *Zea mays* leaf in relation to photosynthesis. *New Phytologist*, **88**, 595–665.

Miranda, V., Baker, N. R. & Long, S. P. (1981b). Limitations of photosynthesis in different regions of the *Zea mays* leaf. *New Phytologist*, **89**, 165–78.

Murphy, D. J. & Leech, R. M. (1981). Photosynthesis of lipids from $CO_2$ in *Spinacia oleracea*. *Plant Physiology*, **68**, 762–5.

Mustardy, L. A. & Brangeon, J. (1978). 3-dimensional chloroplast infrastructure: Development aspects. In *Chloroplast Development*, ed. G. Akoyunoglou & J. H. Angyroudi-Akoyunoglou, pp. 489–94. Amsterdam: Elsevier/North-Holland.

Nobel, P. S. (1974). *Introduction to Biophysical Plant Physiology*. San Francisco: W. H. Freeman & Co.

Nobel, P. S., Zaragoza, L. J. & Smith, W. K. (1975). Relation between mesophyll surface area, photosynthetic rate and illumination level during development for leaves of *Plectranthus parviflorus* Henckel. *Plant Physiology*, **55**, 1067–70.

Oelze-Karow, H. & Butler, W. L. (1971). The development of photophosphorylation and photosynthesis in greening bean leaves. *Plant Physiology*, **48**, 621–5.

Ogawa, T. (1979). Stomatal responses to light and $CO_2$ in greening wheat leaves. *Plant and Cell Physiology*, **20**, 445–52.

Ohlrogge, J. B., Kuhn, D. N. & Stumpf, P. K. (1979). Subcellular localization of acyl carrier protein in leaf protoplasts of *Spinacia oleracea*. *Proceedings of the National Academy of Sciences, USA*, **76**, 1194–8.

O'Toole, J. C., Ludford, P. M. & Ozbun, J. L. (1977). Gas exchange and enzyme activity during leaf expansion in *Phaseolus vulgaris* L. *New Phytologist*, **78**, 565–71.

Paolillo, D. J. (1970). The 3-dimensional arrangement of intergranal lamellae in chloroplasts. *Journal of Cell Science*, **6**, 243–55.

Platt-Aloia, K. A. & Thomson, W. W. (1977). Chloroplast development in young sesame plants. *New Phytologist*, **78**, 599–605.

Possingham, J. V. (1980). Plastid replication and development in the life cycle of higher plants. *Annual Review of Plant Physiology*, **31**, 113–29.

Rawson, H. M. & Hackett, C. (1974). An exploration of the carbon economy of the tobacco plant. III. The gas exchange of leaves in relation to position on stem, ontogeny and nitrogen content. *Australian Journal of Plant Physiology*, **1**, 551–60.

Rawson, H. M. & Woodward, R. G. (1976). Photosynthesis and transpiration in dictoyledonous plants. I. Expanding leaves of tobacco and sunflower. *Australian Journal of Plant Physiology*, **3**, 247–56.

Reid, R. A. & Leech, R. M. (1980). *Structure and Function of Subcellular Organelles*. Glasgow: Blackie.

Robertson, D. & Laetsch, W. M. (1974). Structure and function of developing barley plastids. *Plant Physiology*, **54**, 148–60.

Roughan, P. G., Holland, R. & Slack, C. R. (1980). The role of chloroplasts and microsomal fractions in polar lipid synthesis from 1-$^{14}C$ acetate by cell free preparations from spinach (*Spinacia oleracea*) leaves. *Biochemical Journal*, **188**, 17–24.

Samsuddin, Z. & Impens, I. (1979). Relationship between leaf age and some carbon dioxide exchange characteristics of four *Hevea brasiliensis* Muell. Arg. clones. *Photosynthetica*, **13**, 208–10.

Selldén, G. & Selstam, E. (1976). Changes in chloroplast lipids during the development of photosynthetic activity in barley etiochloroplasts. *Physiologia Plantarum*, **37**, 35–41.

Šesták, Z., Čatský, J., Solárová, J., Strnadová, H. & Tichá, I. (1975). Carbon dioxide transfer and photochemical activities as factors of photosynthesis during ontogenesis of primary bean leaves. In *Genetic Aspects of Photosynthesis*, ed. Yu. S. Nasyrov & Z. Sesták, pp. 159–66. The Hague: W. Junk.

Staehelin, L. A., Armond, P. A. & Miller, K. R. (1977). Chloroplast membrane organisation at the supramolecular level and its functional implications. In *Chlorophyll Proteins, Reaction Centers and Photosynthetic Membranes*, ed. J. M. Olson & G. Hind, *Brookhaven Symposium in Biology*, *28*, pp. 278–315.

Staehelin, L. A. & Arntzen, C. J. (1979). Effects of ions and gravity forces on supra molecular organizations and excitation energy distribution in chloroplast membranes. In *Chlorophyll Organisation and Energy Transfer in Photosynthesis. CIBA Foundations Symposium*, *61*, pp. 147–69.

Steer, B. T. (1971). The dynamics of leaf growth and photosynthetic capacity in *Capsicum frutescens* L. *Annals of Botany*, **35**, 1003–15.

Steer, B. T. (1973). Control of ribulose – 1,5-diphosphate carboxylase during expansion of leaves of *Capsicum frutescens* L. *Annals of Botany*, **37**, 823–9.

Strasser, R. J. (1975). Studies on the oxygen evolving system in flashed leaves. In *Proceedings of 3rd International Congress on Photosynthesis*, ed. M. Avron, pp. 497–503. Elsevier: Amsterdam.

Strasser, R. J. & Butler, W. L. (1976). Correlation of absorbance changes and thylakoid fusion with the induction of oxygen evolution in bean leaves greened by brief flashes. *Plant Physiology*, **58**, 371–6.

Thomson, W. W. & Whatley, J. M. (1980). Development of nongreen plastids. *Annual Review of Plant Physiology*, **31**, 375–94.

Thornber, J. P. & Markwell, J. P. (1981). Photosynthetic pigment protein complexes in plant and bacterial membranes. *Trends in Biochemical Sciences*, **6**, 122–5.

Thornber, J. P., Markwell, J. P. & Reinman, S. (1979). Plant chlorophyll-protein complexes: recent advances. *Photochemistry and Photobiology*, **29**, 1205–16.

Tichá, I. & Čatský, J. (1977). Ontogenetic changes in the internal limitations to bean leaf photosynthesis. III. Leaf mesophyll structure and intercellular conductance for carbon dioxide transfer. *Photosynthetica*, **11**, 361–6.

Tobin, E. M. (1978). Light regulation of specific mRNA species in *Lemna gibba*. *Proceedings of the National Academy of Sciences, USA*, **75**, 4749–53.

Treffry, T. (1978). Biogenesis of the photochemical apparatus. *International Review of Cytology*, **52**, 159–96.

Weier, T. E. & Brown, D. L. (1970). Formation of the prolamellar body in 8-day dark-grown seedlings. *American Journal of Botany*, **57**, 267–75.

Weistrop, J. S. & Stern, A. I. (1977). Appearance of photochemical activity in isolated chloroplasts from far-red illuminated leaves of *Phaseolus vulgaris*. *Journal of Experimental Botany*, **28**, 354–65.

Wellburn, A. R., Hampp, R. & Wellburn, F. A. M. (1981). Interaction between plastids, mitochondria and cytoplasm during chloroplast development. In *Proceedings of 5th International Photosynthesis Congress*, ed. G. Akoyunoglou.

Whatley, J. M. (1977). Variations in the basic pathway of chloroplast development. *New Phytologist*, **78**, 407–20.

Woodward, R. G. & Rawson, H. M. (1976). Photosynthesis and transpiration in dicotyledonous plants. II. Expanding and senescing leaves of soybean. *Australian Journal of Plant Physiology*, **3**, 257–67.

# Summary and discussion

J. MOORBY

The general patterns of leaf growth have been known for some time. In the leaves of dicotyledonous plants there is generalized growth over the blade with basipetal differentiation whereas in many monocotyledonous leaves growth is from a basal meristem and the tissues are essentially mature when they emerge from the subtending sheath. This session was concerned with the details of these processes and how they related to and were affected by the external and internal environment. There was also some discussion of the development of the leaf into an organ capable of photosynthesis and the export of the assimilates so formed to other parts of the plant.

There was much discussion by Harte, Dengler and Poethig of the relative importance of cell division and expansion in the growth of leaves of several species. It became clear that it was of vital importance to make actual measurements of changes in cell numbers. Lyndon emphasized the heterogeneity of the growing dicotyledonous leaf and pointed out that even in leaf primordia some of the cells near the tip did not divide. Dale agreed and suggested that the situation was simpler in grass leaves where cell division was confined to the basal meristem. Kirby pointed out that in cereals leaves were initiated faster then they emerged, leading to an accumulation of primordia at the apex; i.e. the phyllochron was not identical to the plastochron. This made it impossible to use the concept of leaf plastochron index in cereals. Wardlaw discussed the effect of shading on leaf production and growth and pointed out that it could result in the compensatory growth of other leaves. Woolhouse added that there were also effects on stomata and the rate of water loss, and on phytochrome and growth substances; a point returned to in the discussion of the paper by Vince-Prue.

There was general agreement on the necessity for more information on changes in cell number as leaves differentiate, although Jarvis thought that it was necessary to develop a conceptual framework to accommodate the results obtained. Dale agreed with the proposal, but thought that there was still insufficient data, especially in the primordial growth phase, to permit the construction of a realistic model of the cellular basis of leaf development.

Much of the discussion of Terry's paper was concerned with the wide range of effects of environmental factors. The effect of light on cell wall extensibility was mentioned by Cleland but Terry maintained that, in terms of the total amount of growth made, the most important effect of light was its role in photosynthesis. Woolhouse questioned the generality of the constant allometric relationship between shoot and root weight found by Terry in sugar beet, saying that in grasses shoot growth was favoured at low light although the effect was less apparent in woodland species. Erickson pointed out that the effects of drought varied with leaf age, being less apparent in young leaves, and Starck added that water stress tended to have smaller effects on leaf expansion than on photosynthesis and translocation. (Many workers would not agree with this contention.) A more general point was made by Blacklow in querying the relevance to field conditions of experiments in controlled environments at 50 W $m^{-2}$ and Ritchie thought the temperature optima in leaf cell division and expansion would disappear at high irradiances. There was little discussion of the effects of mineral nutrients on leaf growth, this being surprising in view of their importance, particularly of nitrogen, in stimulating leaf growth in the field.

The discussion on the effects of growth substances was opened by Wightman, who amplified some of the results presented by Goodwin. This was followed by discussion on the effects of gibberellins on leaf shape. The effects of gibberellin on heteroblastic development was mentioned and Peacock pointed out that the response of grasses to gibberellin differed before and after flowering. Further discussion suggested that cell division was less affected than was cell expansion, but in dicotyledonous plants the heterogeneity of leaves both in tissue type and degree of differentiation complicated interpretation. Blacklow queried the link between temperature effects and growth substances. Goodwin replied that any stress tended to produce changes in the concentrations of growth substances and cited the relationship between water stress and the accumulation of abscisic acid. Wightman added that in winter wheat, exposure to low temperatures also caused an increase in the concentration of abscisic acid. Palmer took up a point made earlier by Wardlaw saying that leaves were part of a correlative system with other leaves and buds and that interference with one part of the system could lead to changes in the growth substances in, and growth of, other parts.

Jarvis challenged the view that the development of sun and shade leaves was influenced more by irradiance than by the ratio of red to far-red light. Vince-Prue replied that the work quoted by Jarvis was concerned with stem extension growth rather than with leaf growth. She agreed that there were effects of the red:far-red ratio, but thought that this was not as important in the development of sun and shade leaves as was irradiance. She added that it was possible to produce effects on, for example, specific leaf area and leaf area ratio by changing irradiance and not quality. Kemp asked about the effects of light on gibberellins and other growth substances. Vince-Prue replied that the situation was too complex to give a general answer. However, there were some specific situations where there were unambiguous responses – the effect of light on, for example, ethylene production in peas and on very fast changes in gibberellins and leaf unrolling in barley. King raised the question of the penetration of light into leaves, leaf sheaths and buds. Vince-Prue replied that more information was needed, but that the general effect was to lower the ratio of red:far-red light and that this could have major effects on the growth of primordia and young leaves. In addition, similar changes were found as light penetrated the leaf canopy and these might be important in the control of leaf senescence.

Silsbury raised the question of the most suitable lamp for use in controlled environment rooms and asked whether mercury halide lamps could be used alone. Warrington replied that the best light theoretically was one which produced the closest spectrum to natural sunlight, but this had to be balanced against installation costs, longevity, etc. Further, differences in plant growth in controlled environments could not always be attributed to light quality. He added that the phytotron at Palmerston North had successfully used mercury halide lamps for several years. He thought that some of the abnormal effects sometimes found with these lamps could be attributed to effects on leaf temperature which resulted from the lack of an efficient thermal barrier between the plants and lights.

Leech's paper described the development of the photosynthetic system in wheat leaves. Silk agreed with her that it was important to be sure whether the changes were expressed in terms of events in the same parts of a leaf at different times or events in different parts of the leaf. In discussion of the onset of photosynthesis Leech said that $O_2$ evolution could be detected in very low light in young tissues close to the base of the leaf. Terry asked about changes with leaf age in components of the electron transport system. Leech agreed that it was important to follow such changes but said she had concentrated on changes which were measured more readily. Lyndon queried the origin of the starch commonly found in young chloroplasts. Leech presumed that it originated from materials transported to the leaf early in development.

In a short presentation Ho described the movement of C, N, P and K into and out of a cucumber leaf over a period of 23 days and showed how the concentrations of C, N, P and K changed throughout the period. There was little specific discussion of the transport of materials to and from leaves although these processes were referred to in several papers. There is increasing understanding of the relationships between the transport processes and leaf growth, especially in the monocotyledons, but the heterogeneity of the leaves of dictoyledonous plants and the complexity of the vasculature makes the details less clear.

Maksymowych pointed out that a major gap in this section was the lack of any adequate treatment of petiole development. He described an extensive series of observations on the development of the petiole of *Xanthium*; these cannot be adequately summarized here and the interested reader is invited to contact Dr Maksymowych directly.

# PART III

## The mature leaf and its significance

The leaf on reaching maturity has developed its full capacity as a physiological organ. It is now a major member of the population of other leaves, ranging from primordial to senescent, borne on a population of stems of plants of like and unlike species. Here, we keep to our restricted theme of discussing the story of a leaf and its role in the functioning of the plant as a whole. It finds itself in an essentially hostile environment endeavouring to preserve sufficient water in its tissues to allow the other processes to proceed. Hence, the role of water is sufficiently removed functionally from that of other environmental components to deserve a separate chapter – it is a matter of maintenance rather than of simply influencing rates of reaction. Other important environmental factors are taken up in Chapter 12 and looked at from immediate, short-term (acclimation), and very long-term (adaptation) points of view. Emphasis here is on the two major processes – photosynthesis and respiration – and our understanding of these is brought together in the next chapter.

But the leaf is not an isolated organ and the whole plant depends very much on the transport of materials – between organelles within a cell, between cells and tissues, and between organs (from old to young developing leaves, to fruits and roots). This also requires a full treatment (Chapter 14), being perhaps no more important than when transferring fragments of previously functioning material from old to young leaves, leading to their senescence and eventual death. As pointed out in Chapter 15, much is known about the decline in leaf functioning but understanding is still meagre – underlying all of this story is the continuing recycling and use of many basic materials.

It is with the death of the leaf that our basic story ends but we have appended two further aspects. One of these discusses approaches and means of assembling the vast array of findings from reductionist experimentation, as also discussed in Chapter 4, in a way that can allow the whole to be understood and to direct our attention to those aspects which really matter. The other takes us into the real world and, in the restricted space we have allowed, looks at the growth and functioning of leaves in the field. This provides an appraisal of the importance of leaves in crop production but for further details and for extension of the system the reader is referred to texts of crop physiology.

# 11

## Water relations of the mature leaf

E. W. R. BARLOW

### Introduction

It is convenient to view the water relations of the mature leaf from an evolutionary standpoint of how it accumulates carbon without excessive loss of water. Many of the anatomical and morphological adaptations of the terrestrial plant leaf result from the inability of evolution to provide a perfect semi-permeable membrane that is permeable to carbon dioxide and not to water. However, in addition to the inevitable transpiration, water plays definite structural, solvent, transport and substrate roles. It is important that these roles of water are not ignored in a leaf experiencing a water deficit, because changes in leaf function may be controlled by alternations to one of these subsidiary roles.

The evolutionary pressures on terrestrial plants to reach a 'transpiration compromise' have resulted in the development of the leaf as a complex organ comprised of several distinctly different cell types. The arrangement and relative abundance of each of these tissue types must be kept in a clear perspective when the water relations of the leaf are considered. For example, the transverse section of a 'typical' C3 leaf shown in Fig. 11.1, illustrates how the mesophyll dominates the volume of the leaf, with the epidermis being quantitatively the least important and the vascular tissue being somewhat more so. While the structural and functional specialization of each of these cell types enable the leaf to respond to its environment in a quite sophisticated manner, they are a nightmare for experimentalists trying to unravel the mechanisms of leaf function.

The difficulty in adequately separating the various leaf tissues, physically or chemically, means that most analyses of leaf water relations and metabolites are bulk leaf values and reflect a volume average of all cells. In this chapter the discussion will be limited to mesophytic species, mostly crop plants, where any bulk leaf value usually represents the mesophyll.

### Water relations of mature cells

The state of water in plant tissues is most usefully described in terms of water potential or the chemical potential of water. By describing the state of water in this way, the direction and rate of water movement between different parts of the soil-plant-atmosphere continuum can be predicted (Slatyer & Taylor, 1960).

In thermodynamic terms, the water potential, $\psi$, of a plant cell is most commonly written as the sum of the pressure (turgor) potential, $P$, osmotic potential, $\pi$, and matric potential, $\tau$, components (Dainty, 1976).

$$\psi = P + \pi + \tau \qquad (11.1)$$

Units of pressure are now commonly used (MPa).

While the use of water potential as an expression of the energy state of water is thermodynamically correct and measurable, there are some

Fig. 11.1 Transverse section of a 'typical' C3 leaf illustrating the small proportion of total leaf volume that is occupied by the epidermis.

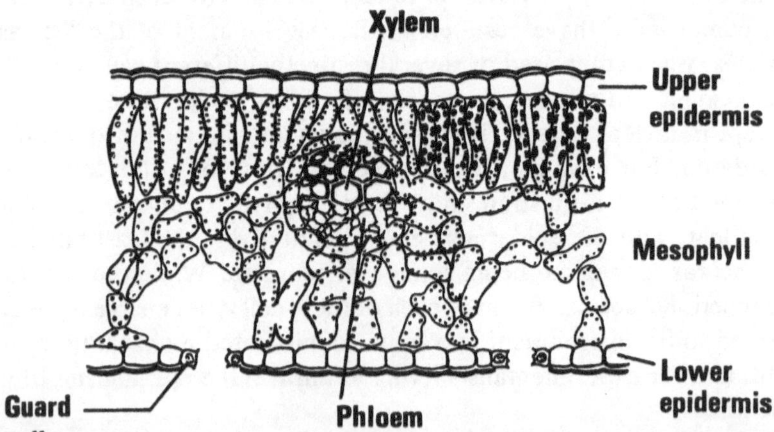

doubts regarding the splitting of water potential into the above components (Briggs, 1967; Weatherley, 1970; Passioura, 1980). The problem is that $\psi$ cannot be split into $P$, $\pi$ and $\tau$ components if there is any interaction between them. It is now clear that neither $P$ or $\pi$ is independent of $\tau$ and all $\tau$ effects are already implicitly accounted for in $P$ and $\pi$ (Passioura, 1980). Consequently, the water potential of a cell can be more conveniently described as

$$\psi = P + \pi \qquad (11.2)$$

Although equations (11.1) and (11.2) have been defined in terms of single cells, they hold for bulk leaf tissue if $\psi$, $P$ and $\pi$ are regarded as volume-averaged, or more precisely weight-averaged, values for all compartments of all cells in the tissue (Tyree & Jarvis, 1982). The inter-relationships between $\psi$, $\pi$ and $P$ and tissue hydration are best illustrated by the classical Höfler diagram (Fig. 11.2).

Fig. 11.2. A classical Höfler diagram illustrating the relationship of $P$, $\pi$, $\psi$, and relative water content (RWC) of wheat leaves at three development stages. The 75, 110 and 220 mm leaves contained 85, 75 and 25% immature elongating cells, respectively. (R. Vallance & E. W. R. Barlow, unpublished.)

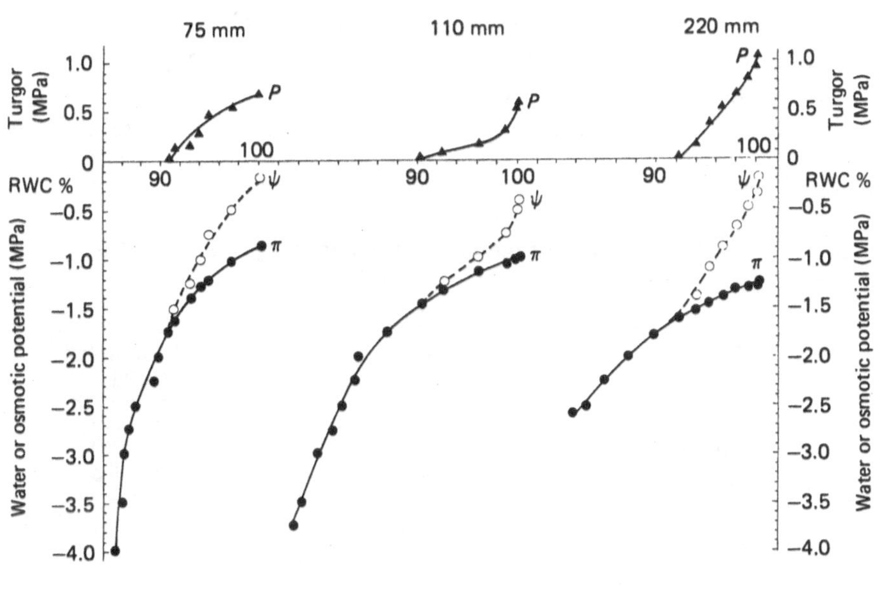

Both $P$ and $\pi$ are functions of cell volume ($v$) which in turn depends on the elasticity of the cell wall, such that

$$dP = \varepsilon \, dv/v \qquad (11.3)$$

and

$$d\pi = dv/v \qquad (11.4)$$

where $\varepsilon$ is the elastic modulus of the cell wall and has units of pressure. For small changes in volume it can be shown (Dainty, 1976) that

$$\frac{dv}{d\psi} = \frac{v}{(\varepsilon + \pi)} \qquad (11.5)$$

In tissues $\varepsilon$ can be defined as a bulk elastic modulus and is a weight-averaged modulus of all cells in the tissues. Cell volume is very difficult to measure and in tissues the weight of symplastic water, $W_s$, can be correctly used for cell volume. In practice, the relative water content (RWC) of the tissue can be used as a reasonable approximation of cell volume, because $W_c = W_a + W_s$ and the weight of apoplastic water, $W_a$, remains relatively constant (Wilson, Ludlow, Fisher & Schulze, 1980).

The weight-averaged nature of bulk leaf-water parameters, such as the elastic modulus, is well illustrated by the 5–10 fold variation in $\varepsilon$ between epidermis, subsidiary and mesophyll cells of *Tradescantia virginiana*, as measured with the pressure probe (Husken, Zimmermann & Schulze, 1979) (Table 11.1). A measurement of the bulk elastic modulus of this leaf would represent that of the mesophyll rather than that of the epidermal tissues.

The significance of the bulk elastic modulus in leaf water relations is

Table 11.1. *Water relation characteristics of* Tradescantia virginiana *cells.*

| Tissue | Average cell volume, $V$(nl) | Elastic modulus, $\varepsilon$ (MPa) |
|---|---|---|
| Epidermis | 0.17–0.53 | 4.0–10.0 |
| Subsidiary | 0.05–0.18 | 4.0– 8.0 |
| Mesophyll | 0.05–0.08 | 0.9– 1.4 |

From Husken, Zimmermann & Schulze (1979).

that it determines the capacitance of the leaf, or the decrease in $\psi$ for each increment of water lost from the tissue. This is illustrated in Fig. 11.2 where changes in the RWC with decreasing $\psi$ are compared for leaves of different bulk elastic moduli. As $\varepsilon$ is a cell wall characteristic it changes during the growth and maturity of the leaf. In this diagram the elastic modulus of wheat leaves, shown by the slope of the *P*/RWC curves, increased as the proportion of mature cells in the leaf increased. Although it is not known whether $\varepsilon$ changes during the senescence phase of the leaf, it does not appear to change in response to water stress (Wilson *et al.*, 1980). Consequently, $\varepsilon$ is probably a function of cell age, tissue type, species and growth environment. For further information on the determination of $\varepsilon$ and tabulated values of $\varepsilon$, the reader is referred to Zimmermann & Steudle (1978).

In the preceding discussion the static water relations of cells and tissues have been described in terms of leaf ontogeny and changes in water status. In the next section the factors and forces controlling leaf water status are considered under the heading of dynamic leaf water relations.

## Water relations of leaf tissue

The water potential of a leaf is very dependent on external factors, especially the supply of water from roots and the evaporative 'demand' of the leaf environment. Further, the small capacitance or water-holding capacity of most mesophytic leaves dictates that any small imbalance in water supply or demand will result in a rapid change in the $\psi$ of the leaf. For example, a rapidly transpiring wheat leaf transpires as much as 3% of its symplastic volume each minute and therefore a 50% decrease in supply can cause a decrease in $\psi$ of 1 MPa in 15 minutes.

The dynamic nature of leaf water relations results in considerable spatial and temporal variations in $\psi$ within a single herbaceous plant. There are perhaps even larger variations in $\psi$ within plant canopies. The spatial and temporal patterns of $\psi$ in the leaf are best understood by considering the mechanics of water flow into and out of the leaf.

### *Water flow to evaporative sites*

Under steady state conditions, the flux of water through each segment of the soil-plant-atmosphere continuum is identical and is governed by the decrease in $\psi$ across the segment and the resistance within the segment. Using the electrical analogy of van den Honert (1948), this relationship may be expressed in the general equation

$$E = (\psi_2 - \psi_1)/R \qquad (11.6)$$

where $E$ is the transpirational flux, $(\psi_2 - \psi_1)$ is the difference in water potential between the ends of the pathway and $R$ is the resistance of the pathway.

The liquid phase resistance of the plant, $R_{PL}$, can be conveniently divided into the flow resistances of the roots, $R_r$, stem, $R_s$, and leaves, $R_l$. For instance, under steady state conditions, then

$$R_l = (\psi_l - \psi_s)/E \qquad (11.7)$$

where $\psi_l$ and $\psi_s$ are the water potentials of the stem and leaf, respectively. $R_r$ is generally regarded as the major resistance to transpirational water movement in plants, whereas the resistance in the shoot, particularly in the stem, is frequently relatively low (Boyer, 1969, 1971). However, several recent reports suggest that the leaf resistance $R_l$ can be the largest potential resistance in some species (Jarvis, 1975) and may also vary with the transpirational flux (Boyer, 1974; Black, 1979).

The flux-dependent nature of the leaf resistance in sunflower is illustrated in Fig. 11.3, where a number of experiments are summarized; large changes in transpirational flux resulted in little or no measurable change in $\psi_l$. The location of the resistance in the leaf is unknown, although the major pathway of liquid water flow is probably through the cell walls to the sub-stomatal cavity after it has left vascular tissue (Sheriff & Meidner, 1974). The alternate protoplasmic pathway has a much higher resistance because of the low hydraulic conductivity of cell membranes. It has been suggested that at high transpirational flux the $R_l$ may increase due to the decrease in $\psi_l$ and subsequent emptying of some larger pore spaces in the cell wall. However, this effect is unlikely to be significant at leaf water potentials encountered by mesophytic species ($> -5$ MPa) because the mean diameter of these pore spaces ranges from 3 to 6 nm (Capita, Sabularse, Montezinos & Delmer, 1979) which corresponds to an air entry value below $-25$ MPa. Further, the $R_l$ tends to decrease at high transpirational fluxes.

A more important aspect of liquid water flow in the leaf is the site of its evaporation in the sub-stomatal cavity. The rate of evaporation per unit area of exposed wall is greater from guard cells and subsidiary cells than mesophyll cells because they are closest to the stomatal pore

(Tanton & Crowdy, 1972; Tyree & Yianoulis, 1980). In terms of supply, the hydraulic conductivity of the epidermis is sufficient to maintain a substantial proportion of evaporation within the leaf (Meidner, 1975). However, this does not necessarily imply that the total evaporative flux from the mesophyll is small, because the surface area of the mesophyll in contact with air spaces is of the order of 10 times the superficial area of the leaf and 100 times the surface area of the guard cells (Cowan, 1977).

The relative proportion of the total leaf transpirational flux that is being evaporated from the epidermis and mesophyll, respectively, is variable and dependent upon the degree of attachment of the mesophyll to the epidermis and the ambient environmental conditions. Temperature differences in the order of ±0.1 °C between the epidermis and mesophyll can result in $\psi$ differences of up to 0.6 MPa and therefore large changes in the relative rates of evaporation between the two

Fig. 11.3. Influence of transpiration rate on the leaf water potential of sunflower leaves in a number of studies where different environmental variables were used to regulate transpiration. (1) Light intensity. (2) Light intensity and windspeed. (3) (4) (5) Humidity. Note that in most studies $\psi_l$ is not proportional to transpiration rate indicating the flux dependence of leaf resistance. After Cowan (1977).

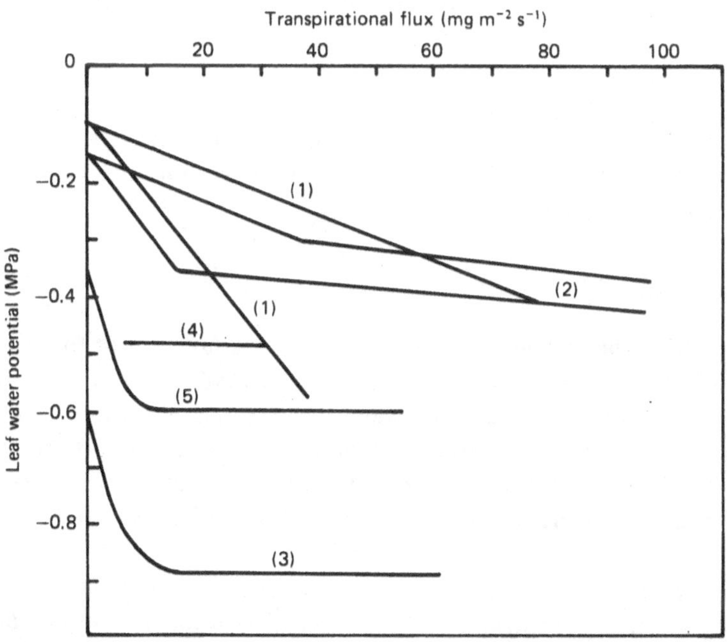

surfaces (Cowan, 1977). Such $\psi$ differences indicate the tenuous relationship between bulk leaf $\psi$, which is dominated by the mesophyll and the $\psi$ of the localized areas such as guard cells and subsidiary cells. If the environmental conditions necessary to initiate changes in the leaf transpiration flux, also cause a relative shift in evaporative sites within the leaf, the bulk leaf $\psi$ may be a very insensitive parameter to use in the calculation of $R_l$. Consequently, the decrease in the apparent liquid phase resistance at high transpirational flux may be more apparent than real. The significant reduction in $\psi$ at evaporation micro-sites in the epidermis also has implications for the reaction of stomata which control the vapour phase movement of water out of the leaf.

### *Vapour flow out of the leaf*

The flux density of water, $E$, lost from the leaf is controlled primarily by the resistance to vapour movement out of the leaf and the vapour pressure gradient between the leaf evaporative surfaces and the atmosphere; i.e.

$$E = (e_l - e_a)/R \qquad (11.8)$$

where $e_l$ and $e_a$ are the vapour pressure of air at the leaf evaporating surface and atmosphere, respectively, and $R$ is the total vapour phase resistance. The most important components of $R$ are the resistances due to stomata ($r_s$) and boundary layer outside the leaf ($r_a$), which are in series such that $R = r_a + r_s$.

A more rigorous analysis of all vapour phase resistances is to be found in Burrows & Milthorpe (1976). These resistances are more conveniently expressed as conductances such that

$$\frac{1}{g_l} = \frac{1}{g_a} + \frac{1}{g_s} \qquad (11.9)$$

The boundary layer conductance, $g_a$, which is a function of the wind speed and the geometry of the surface, is usually not under the short-term control of the leaf. However, turgor-mediated leaf movements such as leaf rolling, heliotropic and paraheliotropic movements (p. 330) which are initiated by the position of the sun and/or changes in leaf water status can change either leaf geometry or effective wind speed (Begg, 1980; Ehleringer & Forseth, 1980).

The stomatal conductance is the most powerful means the plant has of regulating water loss from the leaf. As such the stomatal conductance directly influences the leaf $\psi$ by determining 'leaf demand' for transpiration. However, the bulk leaf $\psi$ also influences stomatal conductance particularly as the leaf experiences water stress. These inter-relationships are illustrated by Fig. 11.4 (Cowan, 1977).

It is the response of the stomata, through the regulation of stomatal conductance, to environmental parameters such as visible radiation, ambient $CO_2$, vapour pressure deficit, and temperature that determine the environmental demand on the leaf to supply water for evaporation. The detailed responses of the stomata to these parameters are described in excellent reviews by Raschke (1979), Burrows & Milthorpe (1976) and Cowan (1977).

### *Dynamic leaf water relations*

#### *Leaf $\psi$ gradients*

The preceding discussion of water flow into and vapour movement out of the mature leaf indicates that $\psi$ is most unlikely to be equal in all parts of a mesophytic leaf. This is dictated by the leaf resistance to water movement and variations in the depth of the boundary layer along the leaf (see Monteith, 1973). In spite of these predictions, measurements of $\psi$ gradients along leaves have been limited to controlled environments for good methodological reasons. The pressure chamber is not suited to these measurements and *in situ* hygrometers used by Wiebe & Prosser (1977) in the study of $\psi$ gradients in corn leaves (Table 11.2) have not been used successfully in the field. Nevertheless, the $\psi$ gradients shown are likely to be more pronounced in the field, where radiation levels are higher and the curvature of individual leaves,

Fig. 11.4. Inter-relationships of leaf conductance to vapour transfer, $g_1$, transpiration rate, $E$, and the water potential of the leaf, $\psi_1$.

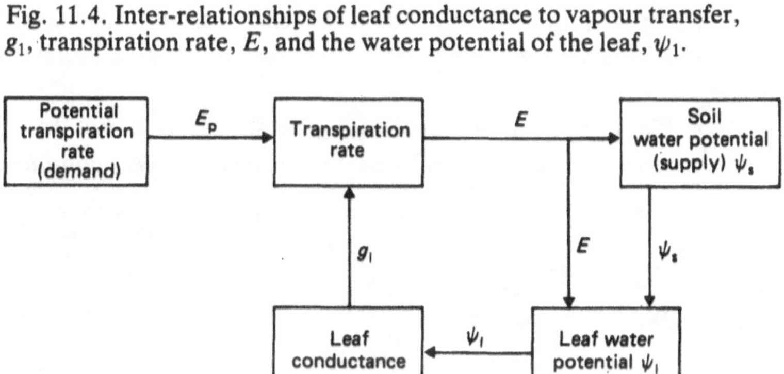

particularly monocots, results in differences in radiation interception along the leaf.

These $\psi$ gradients in individual leaves underline the heterogeneity of the leaf both longitudinally and radially in terms of the different tissues. There are few data available on the longitudinal variation in $\pi$ and $P$, although $P$ does seem to decrease towards the tip of wheat leaves (R. Vallance & E. W. R. Barlow, unpublished). The significance of these $\psi$ and $P$ gradients in leaf function awaits thorough evaluation.

### *Canopy $\psi$ gradients*

Complex profiles of $\psi$ exist within canopies because in addition to the variations in the length of the water transport pathways, environmental factors such as incident radiation, humidity, wind speed and temperature change rapidly within a canopy. In field crops and herbaceous communities where plants rarely reach a height of 3 m, the $\psi$ gradient needed to overcome gravity (0.01 MPa $m^{-1}$) does not make a significant contribution to the vertical $\psi$ gradients measured.

The vertical gradients of leaf $\psi$ in crop canopies are directly related to the transpirational flux through individual leaves and the resistances to liquid water flow encountered by water moving to the evaporation sites. In most situations the transpirational flux through a leaf is controlled principally by the stomatal conductance and the vapour pressure deficit between the leaf and its surrounding environment (equation 11.8). Stomata of non-senescent, well-watered leaves respond strongly to incident radiation. In dense crops with high leaf area index (LAI), transpiration from lower leaves is much reduced by the low radiation levels, lower wind speeds and high humidity conditions at the bottom of the canopy. These low transpiration rates result in high leaf $\psi$ and sharp $\psi$ gradients through the crop as illustrated by the tobacco crop with a LAI of 8.5 (Fig. 11.5). In contrast, the corn crop with a much lower LAI

Table 11.2. *Water potential gradient along a mature maize leaf held horizontally under controlled environment conditions.*

| | Position of leaf | | | |
|---|---|---|---|---|
| | base | midbase | midtip | tip |
| Water potential (MPa) | −0.31 | −0.43 | −0.55 | −0.64 |

From Wiebe & Prosser (1977).

allows radiation to penetrate deeper into the canopy and results in a very shallow vertical gradient. Although not reported in this study, it is likely that the vapour pressure gradients between leaf and environment are smaller at the bottom of the dense crop (Monteith, 1976).

Leaf $\psi$ profiles in crop canopies are also likely to be strongly influenced by liquid flow resistances due to the vascular anatomy of the particular species. Further, Turner & Begg (1973) have shown that considerable vertical $\pi$ gradients can occur in maize and tobacco. In general, $\pi$ is more negative in the younger upper leaves and tends to increase with leaf age as a particular leaf proceeds closer to senescence.

*Diurnal $\psi$ profiles*

Diurnal $\psi$ profiles in well-watered plants are a function of incident radiation acting through stomatal conductance. The diurnal $\psi$, $\pi$ and $g_l$ profiles shown in Fig. 11.6 illustrate the close relationship between incident radiation, leaf conductance, transpiration and the consequent inverse relationship with the leaf $\psi$. Comprehensive diurnal data on vertical $\psi$ gradients in crop canopies are not available, but the limited profiles published (Turner & Begg, 1973) suggest that gradients follow the pattern of radiation penetration into the canopy. The diurnal decrease in leaf $\psi$ at the top of a canopy, in response to evaporative demand, can result in low bulk leaf $P$ values being experienced in the middle of the day. However, these low values do not appear to influence stomatal function as indicated by leaf conductance. This is not always so in other crops in which leaf $\pi$ decreases in unison with leaf $\psi$ resulting in

Fig. 11.5. Vertical profile of (*a*) leaf $\psi$ (*b*) irradiance at midday in a maize (closed circles) and tobacco (open circles) crop. The leaf area index of the maize and tobacco crops was 3.2 and 8.5 respectively (redrawn from Turner & Begg, 1973).

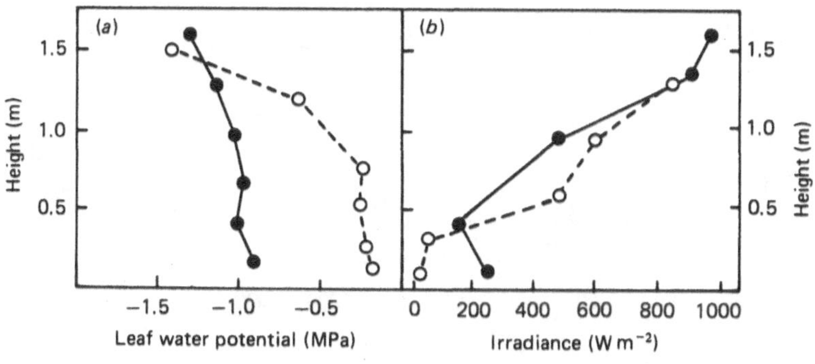

the maintenance of *P* throughout the day (Hsiao *et al.*, 1976). The maintenance of *P* by osmotic adjustment will be discussed in more detail below.

*Leaf ψ profiles during water deficits*

When the soil ψ decreases and water supply becomes limiting, leaf ψ tends to decrease in order to maintain the gradient in water potential between soil and the evaporating surface in the leaf. On a diurnal basis the daily minimum ψ becomes lower each day and takes longer to recover each night. This tends to skew the daily ψ profile to the afternoon (Fig. 11.7). Eventually, diurnal osmotic adjustments are not sufficient to maintain *P* and the plant may lose turgor for an appreciable portion of the day. The loss of bulk leaf turgor parallels adjustments to stomatal conductance resulting in, at first, decreases in leaf conductance in the middle of the day and, finally decreases for much of the day following a brief early morning opening. These midday adjustments to leaf conductance do not result in any increase in leaf ψ, but rather prevent further decreases by modulating water loss.

Fig. 11.6. Diurnal pattern of ψ and π in the upper leaves of a field grown sorghum crop in relation to solar radiation and leaf conductance (redrawn from Turner *et al.*, 1978).

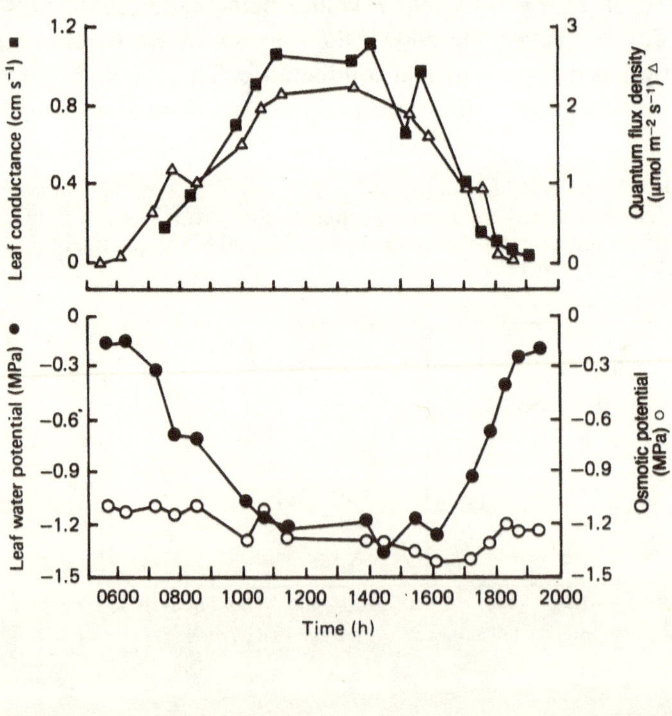

The vertical gradient of leaf $\psi$ becomes much smaller during periods of water stress (Turner, 1974a) due to the lowering of root $\psi$ and the stomata-induced lowering of water flow through the upper leaves. Consequently, water stress tends to result in the $\psi$ of all leaves becoming uniformly low.

*Osmotic adjustment*

With the development of a water deficit within the plant, and the resultant decrease in leaf water content, the bulk leaf turgor potential ($P$) may fall to very low levels or even zero (Fig. 11.2). This loss of turgor can initiate important changes in plant function such as cessation of leaf expansion (Hsiao, 1973) stomatal closure (Turner, 1974b) and an increase in the endogenous level of abscisic acid (Pierce & Raschke, 1980). The leaf water potential at which zero turgor is reached is determined by the osmotic potential ($\pi$) and the bulk modulus of elasticity of the tissues ($\varepsilon$). The osmotic potential of leaves may vary diurnally, weekly or seasonally, while the elasticity of mature cells is less variable (Tyree & Jarvis, 1982). In the leaves of some

Fig. 11.7. Diurnal changes in net photosynthesis ($F'$), soil water potential ($\psi_s$), leaf water potential ($\psi_l$) and leaf osmotic potential ($\pi$) of soybean at four different times during a long stress period. The hatched portion gives leaf turgor pressure (adapted from Turner & Burch, 1981).

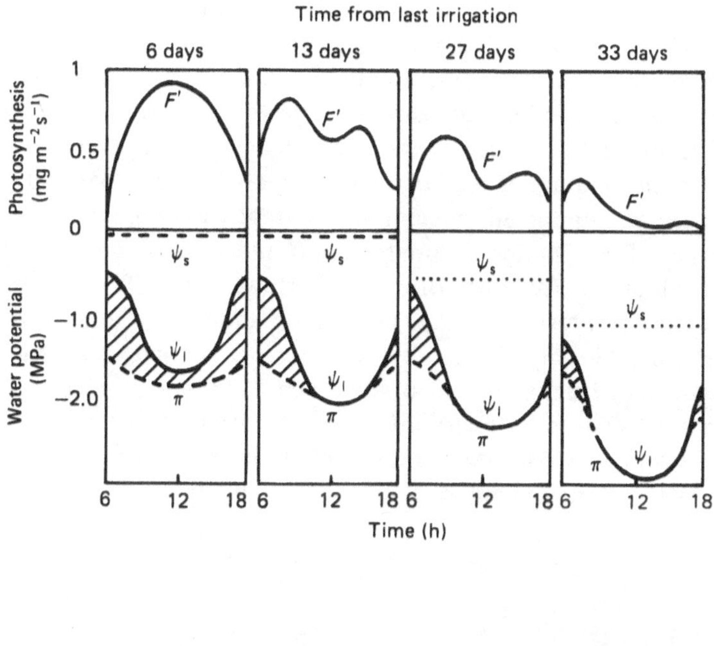

species, turgor is maintained at lower leaf $\psi$ by a lowering of the osmotic potential during periods of water stress. In higher plants this process is termed osmotic adjustment and involves a net increase in the number of solute molecules in each cell. It should be distinguished from the lowering of the osmotic potential by the passive concentration of existing solutes during tissue dehydration. The terms osmoregulation and turgor regulation are reserved for use in microorganisms and lower plants, particularly giant algae, where ion transport across the cell membrane has been shown to respond to changes in the internal osmotic and/or turgor potentials (Cram, 1976). The operation of turgor-dependent transport processes has not been demonstrated in terrestrial plants, although some experiments involving salinity and water stress suggest the presence of some turgor-regulating mechanism (Cram, 1976; Smith & Milburn, 1980). Further, the situation in the mature leaves of terrestrial plants is inherently more complex, because of the need for integrated reaction between all tissues and organs within the plant. For example, any turgor-regulating mechanism in the leaf involving ion transport would have to include long-distance transport to the leaf as well as short-distance transport across membranes, because the apoplastic pool of inorganic ions within the leaf is limited. Thus, whereas the giant algal cell may sequester inorganic ions from its immediate environment, the aerial leaf must rely on the plant roots to sequester ions from its environment and transport them to the leaf.

Osmotic adjustment frequently occurs in the mature leaves of terrestrial plants during periods of water stress and is usually greater when stress develops slowly (Turner & Kramer, 1980). These osmotic adjustments allow turgor to be maintained over a wider range of leaf water potentials. The extent of osmotic adjustment varies between species and cultivars within species and is usually within the −0.1 to −1 MPa range in the mature leaves of mesophytes. Within a particular plant the extent of osmotic adjustment varies greatly between different organs (Barlow, Lee, Munns & Smart, 1980) and organs at different stages of development (Munns, Brady & Barlow, 1979). The solutes that are usually the most important in osmotic adjustment are potassium, sometimes chloride, the soluble sugars (sucrose and to a lesser extent glucose and fructose), and organic acids (Munns *et al.*, 1979; Jones, Osmond & Turner, 1980; Ford & Wilson, 1981). It is likely that most of these solutes are located in the vacuole in the well vacuolated leaf cells. The soluble nitrogen compounds (amino acids, amides, proline, and glycine-betaine) are often quantitatively less important on a bulk tissue basis, but nevertheless equally important in osmotic adjustment because they are probably confined to the cytoplasm. At this stage it is not clear whether the accumulation of solutes during osmotic adjustment is an active adaptation to water stress or a passive process resulting from water-stress-induced changes in other plant functions such as expansive growth. However it is clear that osmotic adjustment does result in the maintenance of turgor at lower leaf potentials which can alter the $\psi$ threshold for processes such as stomatal closure (Ludlow, 1980).

## Influence of water deficits on leaf function

Much of the whole plant response to water deficits has been analysed in terms of the response of the mature leaf as it is the principal carbon assimilating organ. This section will deal exclusively with the response of the mature leaf to water deficits in terms of carbon assimilation and water loss. However, it should not be concluded from the attention accorded to these mature leaf responses to water deficits, that this response is the initial or the major plant response to water stress. It is likely that the whole plant response to water stress is highly integrated; moreover, our present knowledge of this response is limited to the physiological responses such as leaf growth, photosynthesis and transpiration that we can readily measure. However, until methods for monitoring responses, such as root growth, water relations of individual tissues such as the epidermis, and the flow of chemical messengers and regulators around the plant, are available, our view of the plant stress response will be, in a sense, dictated by what we are able to measure. The responses described in this section must be considered only part of the integrated plant response to water stress.

Leaf responses to water stress have been reviewed regularly over the last decade (Hsiao, 1973; Hsiao *et al.*, 1976; Boyer, 1976; Lange, Kappen & Schulze, 1976). More recently, much attention has been focused on crop adaptation to water stress and these reviews have necessarily considered the leaf response to water deficits (Boyer & McPherson, 1975; Begg & Turner, 1976; Mussell & Staples, 1979; Turner & Kramer, 1980; Teare & Peet, 1981). These reviews and treatises contain more comprehensive treatments than is possible here.

### *Leaf movement*

Although leaf movements were recognized for many centuries and were commented on by Darwin, their significance in terms of drought avoidance and adaptation is only beginning to be recognized (Begg, 1980; Ehleringer & Forseth, 1980). Leaf movements in response to water stress can be considered in two categories: active or paraheliotropic movements and wilting responses. Paraheliotropic movements are more complex because they involve active movements in response to water stress, whereas leaf wilting and rolling are both passive responses involving a loss of turgor from various leaf and/or petiole cells. All these leaf movements result in 'radiation shedding' or a reduction in the radiation intercepted by the leaf. Consequently, these movements result in both decreased carbon accumulation as well as water loss from the leaf.

#### *Active leaf movements in response to leaf water deficits*

A number of dicotyledonous plants exhibit diaheliotropism (solar tracking) or movement of leaf blades in such a way to remain perpendicular to direct solar radiation. These movements involve light-dependent volume and therefore turgor potential changes in the pulvinus at the base of each leaflet. These turgor changes appear to be mediated by phytochrome and involve the movement of potassium and possibly chloride ions across the membranes (Satter, 1979; Begg, 1980). These diaheliotropic movements which are exhibited by crop and pasture species such as Townsville stylo, sunflower, cotton and soybean and also many native herbs and shrubs of semi-arid and arid regions (Ehleringer & Forseth, 1980; Begg, 1980) result in an increased amount of solar radiation being intercepted by individual leaves (Fig. 11.8). In some solar-tracking species such as Townsville stylo, water stress results in a change from diaheliotropic to paraheliotropic leaf movements, where leaf orientation becomes parallel rather than perpendicular to direct solar radiation (Begg & Torssell, 1974). This results in large decreases in the daily radiation load and probably water loss from the leaf. Although it is assumed that the change from diaheliotropism to paraheliotropism is triggered by some threshold leaf $\psi$ or $P$ this has yet to be conclusively demonstrated. Consequently, the relationship between water stress induced stomatal regulation, leaf movements and the water use efficiency of the leaf awaits quantitative evaluation. In a diaheliotropic species, vertical wilting reduces light interception, and consequently heat load, by 50–60% (Rawson, 1979). However, as the stomata do not close completely in response to the decrease in bulk leaf *P*, the photosynthetic rate of the vertical wilted leaf was always about 50% of that of the horizontal leaf. The water use efficiency of a vertical wilted leaf is increased by a combination of leaf temperature changes and reductions in leaf conductances without corresponding changes in residual conductances (Rawson, 1979).

*Passive leaf movements in response to leaf water deficits*

The characteristic diurnal pattern of leaf $\psi$ during the development of a water deficit (see Fig. 11.7) illustrated a loss of bulk leaf *P* in the middle of the day, and this becomes progressively more protracted each day as the stress cycle continued. In many mesophytic species this loss of bulk leaf *P* results in changes in leaf position and/or morphology.

In many mesophytic dicotyledons the loss of bulk leaf *P* results in wilting which leads to a change in leaf angle from the horizontal to the vertical. In sunflower vertical wilting reduces the light interception and

Fig. 11.8. Interception of solar radiation between 400 and 700 nm by three leaf types during the course of a clear day (after Ehleringer & Forseth, 1980).

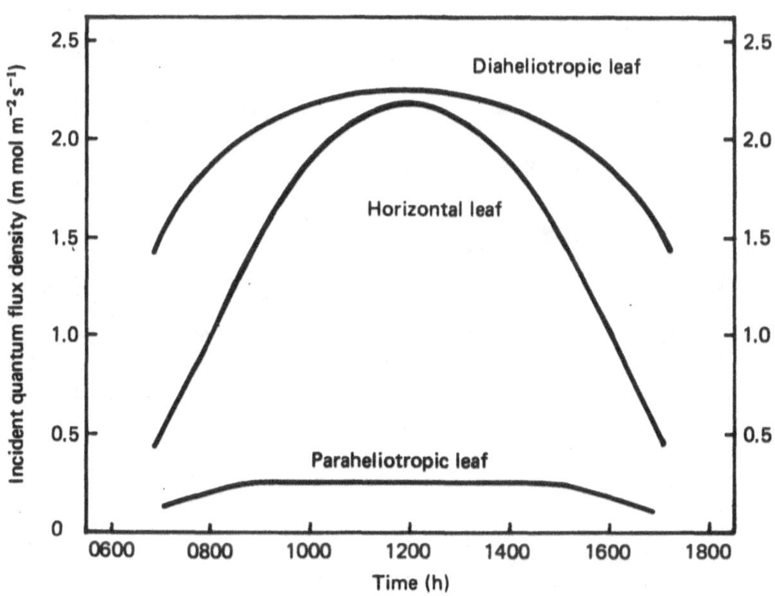

consequently heat load by as much as 50–60% (Rawson, 1979). However, the photosynthetic rate is reduced much less than heat load and transpiration demand and a large improvement in water use efficiency occurs (Fig. 11.9). The stomata did not close completely when the sunflower leaf wilted and photosynthesis remained about 50% of the non-stressed rate. Further, it is likely that wilting in the vertical position will change the boundary layer characteristics of the leaf. In this study the improvements to water use efficiency were aided by reductions in leaf conductance in the absence of appreciable changes in residual conductance.

Many monocotyledons, particularly grasses with fibrous leaves, exhibit a wilting response which results in leaf rolling or folding rather than vertical wilting. Leaf rolling or folding is common but by no means universal in mesophytic cultivated species such as rice (O'Toole & Moya, 1978), sorghum (Begg, 1980), temperate (Johns, 1978) and tropical grasses (Wilson *et al.*, 1980). Loss of turgor in enlarged epidermal cells with thin anticlinal walls, bulliform cells, and/or similarly

Fig. 11.9. Influence of vertical wilting on the photosynthetic rate (●), leaf conductance (○) and water use efficiency (WUE) (■) of a sunflower leaf (redrawn from Rawson, 1979). Leaves were alternately held horizontally and then allowed to assume a natural orientation.

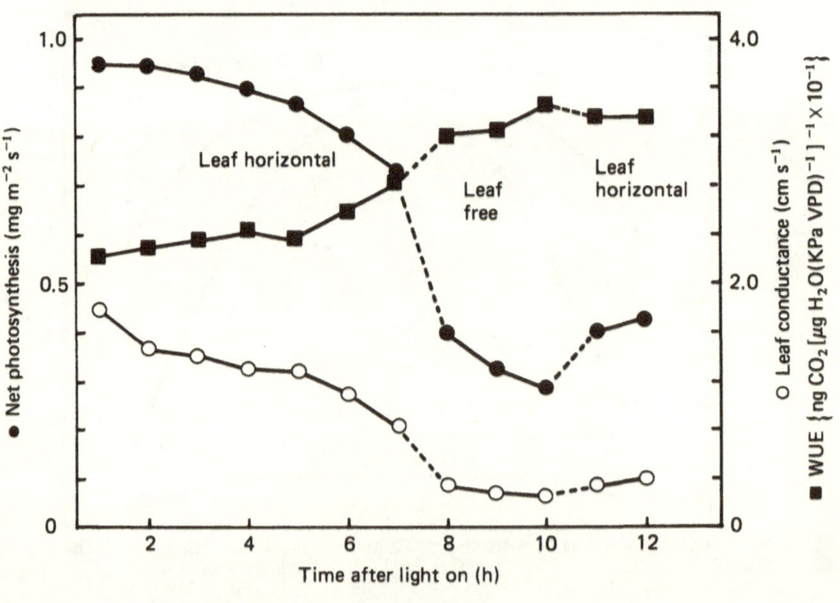

large mesophyll cells, hinge cells (Esau, 1977) is responsible for this response. The rolling is dependent on the leaf $\psi$ and in sorghum exhibits a threshold-type response similar to that of stomata (Turner, 1974b; Begg, 1980). Although no data are available, it is likely that leaf rolling correlates well with bulk leaf *P* because the large bulliform and hinge cells make up a significant proportion of the leaf volume (Esau, 1977). As with wilting in the vertical position, stomatal regulation is independent of leaf rolling, and indeed Begg (1980) has reported leaf rolling to precede stomatal closure. Leaf rolling results in a marked reduction in the effective leaf area and a more vertical leaf orientation, which can reduce transpiration by 50–70% and increase water use efficiency (Johns, 1978).

While these transient wilting responses are in many ways analogous to midday closure of the stomata, they have the added advantage of reducing the radiation load on each leaf. This may be an important mechanism in protecting the leaf's photosynthetic apparatus from photooxidation during water stress.

### *Photosynthesis and photorespiration*

No physiological response to water stress has received more attention than photosynthesis. In 1900 Pfeffer wrote that photosynthesis was reduced in wilted leaves and attributed this to stomatal closure. Since that time much effort has been expended in attempting to partition the photosynthetic response to reduced leaf $\psi$ into stomatal and non-stomatal factors. In this section the approach of Osmond, Winter & Powles (1980) will be followed in trying to link rather than separate stomatal and non-stomatal factors involved in the integrated leaf response to water stress.

The rate and severity of the water stress treatments to which plants are subjected can have a great effect on the photosynthetic rate. For example, when mesophytic plants are stressed rapidly the stomata close very effectively and the depression of photosynthesis can be almost totally ascribed to decreases in stomatal conductance (Osmond *et al.*, 1980). Similarly, when Barrs (1968) induced stomata to cycle and studied photosynthesis and transpiration through a number of dehydration–rehydration cycles, he found photosynthesis and transpiration to change in synchrony and concluded that stomatal movements completely accounted for all photosynthetic effects. Troughton & Slatyer (1969) investigated the importance of non-stomatal factors in water stress by forcing air through cotton leaves and found no effect on internal metabolism until the relative water content of the leaf decreased to 56%.

The net effect of any decrease in stomatal conductance, in the absence of non-stomatal changes during water stress, would be to lower the intercellular concentration of $CO_2$. A close approximation to intercellular $CO_2$ concentration, $C_i$, can be obtained from (Cowan & Farquhar, 1977)

$$C_i = C_a - \frac{F'}{g'_s}$$

relating ambient $CO_2$ concentration ($C_a$), assimilation ($F'$) and stomatal conductance ($g'_s$). It follows from this equation that $C_i$ remains constant only if the ratio of $F'/g'_s$ remains constant and the plot of $F'$ against $g'_s$ is linear. It can be seen from the experiments of Lawlor & Fock (1978) that $F'/g'_s$ is not constant following a rapid imposition of stress (Fig. 11.10*a*). However, when stress is applied slowly in a manner analogous to the development of water stress in the field, $F'$ tends to remain constant over a wide range of leaf $\psi$ in sorghum (Jones & Rawson, 1979) and soybean (Rawson, Turner & Begg, 1978). In this slow stress situation photosynthetic capacity and stomatal conductance decline in a coordinated way and $C_i$ tends to remain constant. These and other studies

Fig. 11.10. Relationship between $CO_2$ fixation and stomatal conductance for (*a*) *Zea mays* stressed at −0.1 to −1 MPa h$^{-1}$ (after Lawlor & Fock, 1978). (*b*) *Sorghum bicolor* plants stressed at −0.15 MPa d$^{-1}$ (Jones & Rawson, 1979). (*c*) *Glycine max* stressed in the field (Rawson *et al.*, 1978).

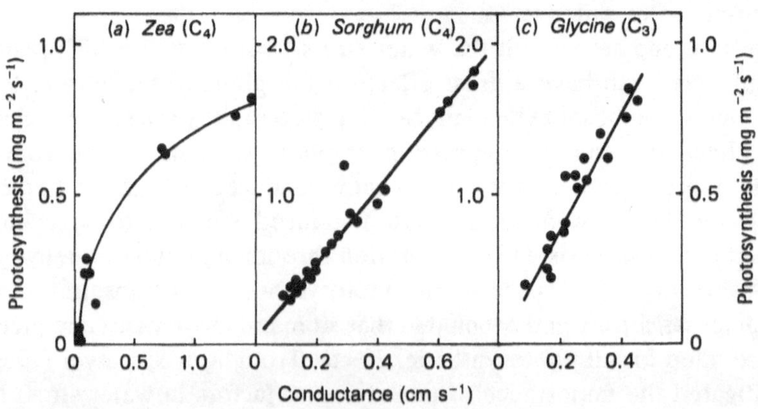

(Collatz *et al.*, 1976) when plants were stressed slowly emphasize the co-ordinated leaf response to water stress where concerted adjustments to stomatal conductance and the internal $CO_2$ fixation capacity are made.

The nature of the changes to the internal $CO_2$ fixation capacity during long-term stress and the manner in which internal $CO_2$ fixation capacity and stomatal conductance are co-ordinated are matters in which our interest is greater that our understanding at the present time. In the broad sense decreases in internal $CO_2$ fixation capacity could be due to photochemical and/or biochemical pathways within the leaf. There is evidence for both.

Boyer and co-workers have shown that chloroplasts isolated from water-stressed sunflower leaves have reduced photosystem II activity in proportion to the stress-induced decrease in leaf photosynthesis (Boyer & Bowen, 1970; Boyer, 1976). More recently, Powles & Osmond (1978) have demonstrated illumination of leaves at zero $CO_2$, which inhibits the operation of both the photorespiratory carbon oxidation (PCO) and photosynthetic carbon reduction (PCR) cycles, can lead to over-energization of the photochemical apparatus with resulting damage to photosystem II. This damage, which is largely reversible, is known as photo-inhibition. When similar plants are exposed to full sunlight at ambient $CO_2$ concentrations close to compensation (70 p.p.m.) no photo-inhibition occurs, illustrating a protective role for the PCO cycle in plants after the stomata have closed (Osmond *et al.*, 1980). Under these conditions Lawlor (1976) has demonstrated an increase in the pool size of some PCO cycle intermediates. The significance of these observations, in regard to the likelihood of photo-inhibition causing decreases in the $CO_2$-fixation capacity of water-stressed plants at ambient $CO_2$ levels, is not yet clear; however, several reported non-stomatal responses to water stress are remarkably similar to those observed following photo-inhibition and it may be that these responses have a common mechanistic basis and result from inhibitory processes (Osmond *et al.*, 1980).

Non-stomatal decreases in $CO_2$-fixation capacity may result from biochemical changes as well as photochemical damage. Water stress is known to decrease the activity of RuBP carboxylase (Jones, 1973; O'Toole, Crookston, Treharne & Ozbun, 1976) but it is not clear whether this is due to a change in the specific activity of the enzyme or the amount of the protein present. Brady, Scott & Munns (1974) have shown that a short water-stress period, or the supply of exogenous abscisic acid (ABA) results in a decline in the synthesis of RuBP carboxylase protein relative to the synthesis of other proteins. Although these observations are consistent with a decreased RuBPCase activity playing a role in the decrease in the internal $CO_2$ fixation capacity of the leaf during water stress, much more work is required to establish the quantitative relationship between RuBPCase and photosynthesis in both the short and long term.

The initiation and co-ordination of the leaf's response to decreasing $\psi$ deserves further comment in the light of the water relations of the stressed leaf. As argued above, the epidermal $\psi$ and particularly the guard cell $\psi$ may differ from the bulk leaf $\psi$ by as much as 0.8 MPa and the potential of the guard cells to make solute adjustments may result in large differences in $P$ as well. However, stomatal conductance is non-linearly related to bulk leaf $P$ rather than $\psi$ and usually exhibits a threshold response which may vary according to environmental conditions (Turner, 1974b; Jordan & Ritchie, 1971). It is possible that leaf function is co-ordinated by intercellular $CO_2$ in non-stress situations and a combination of intercellular $CO_2$ and ABA in stress situations (Raschke, 1975). Recently Pierce & Raschke (1980) have shown that ABA accumulation in a number of mesophytic species is initiated by a threshold turgor response in a manner analogous to stomatal closure. At this stage the ABA messenger represents an attractive hypothesis, which also explains the after-effect of stress on photosynthesis that is frequently observed (Kriedemann & Loveys, 1974; Boyer, 1976). However, in the short term, the midday closure phenomenon raises some questions regarding the ability of the guard cells to metabolize ABA in time for a late afternoon opening (Fig. 11.7).

### *Translocation and metabolism*

The influence of water stress on the translocation of photosynthate out of the mature leaf cannot be viewed in isolation because translocation can never be truly independent of photosynthate supply and particularly sink strength. It appears that the initial strength of the sink and its ability to continue accepting photosynthate during stress has a large influence on the translocation response to water stress. When the major sink is very sensitive to water stress, such as expansive growth of developing leaves (Hsiao, 1973), translocation can be more affected than photosynthesis (Wardlaw, 1969) and photosynthate accumulates in both the source and sink regions (Barlow & Boersma, 1976). In potatoes where tuber growth is about as sensitive to water stress as photosynthesis, the translocation of photosynthate out of water stressed leaves is proportional to the decline in photosynthesis (Munns & Pearson, 1974). Where the major sink is relatively unaffected by water stress, such as the cereal grain (Barlow, Lee, Munns & Smart, 1980), translocation is less inhibited than photosynthesis (Sung & Krieg, 1979) or even maintained (Wardlaw, 1967).

The above observations support the concept of sink control of translocation during water stress, with no direct effect of water stress on translocation. As phloem translocation is a turgor-dependent process, continued translocation during water stress will require some osmotic adjustment in the phloem pathway to maintain turgor. Smith & Milburn (1980) have shown that the rate of phloem loading in *Ricinus* actually increased during water stress when $\pi$ was decreasing and concluded that phloem loading is turgor regulated. This would be consistent with sink control of translocation during water stress.

As noted already water stress can lead to changes in the photosynthetic metabolism in the leaf when stomatal conductance is reduced and the proportion of carbon flowing through the PCO and PCR cycles respectively changes (Lawlor, 1979). In addition to decreasing the amount of carbon assimilated, these water-stress-induced changes in metabolism result in decreased flow of carbon in free hexoses and sucrose, in favour of glyceric acid, alanine, glycine and serine (Lawlor, 1976). On the other hand when vegetative plants, particularly grasses, are water stressed and translocation is reduced by decreased sink activity, soluble sugars, such as sucrose and to a lesser extent fructose and glucose, accumulate (Jones, Osmond & Turner, 1980; Ford & Wilson, 1981). At this stage in our knowledge, it would appear that water-stress-induced changes in the carbon metabolism of the mature leaf are largely indirect changes resulting from prior effects of water stress on the flow of $CO_2$ into, and/or photosynthate out of, the leaf.

Water stress also causes significant changes in the nitrogen metabolism of the leaf (Naylor, 1972). The most common observations are a decrease in protein synthetic capacity and the accumulation of the soluble nitrogen compounds, such as the imino acid proline, the amides asparagine and glutamine (Naylor, 1972) and the quaternary ammonium compound glycine betaine (Hanson & Nelsen, 1978). However, the inter-relationships between these changes and the mechanisms underlying them are not clear, due partly to the experimental problems of introducing a radioactively labelled substrate or intermediate without excising and/or rehydrating the water-stressed tissue. Reports of water stress-induced changes in nitrogen metabolism must also be evaluated carefully in terms of the age of the tissue and the likely influence of a prior cessation of growth in young tissues (see Chapter 7).

A spot estimate of the level of protein synthesis in water-stressed tissue can be made, without rehydrating the tissue, by comparing the percentage of polyribosomes in control and stressed tissue. Scott, Munns & Barlow (1979) found that fully expanded wheat leaves suffered no loss of polyribosomes, compared with controls, when water stressed to a level that caused large reductions in the polyribosomes of growing leaves (Fig. 11.11). However, during the 7-day stress period of this experiment, both control and stressed leaves lost RNA at the same rate, in accordance with the normal ageing of the wheat leaf (Brady & Scott, 1977). Therefore, it is probable that mild to moderate water stress does not seriously affect protein synthesis in mature leaf tissue. However, as mentioned above, there is good evidence that there may be changes in the types of proteins synthesized under stress conditions

Fig. 11.11. The proportion of polyribosomes, expressed as a percentage of the proportion in a well-watered control, in the apex (triangles), leaf 6 (squares) and the mature blade of leaf 5 (closed circles) in water-stressed wheat plants (after Scott, Munns & Barlow, 1979).

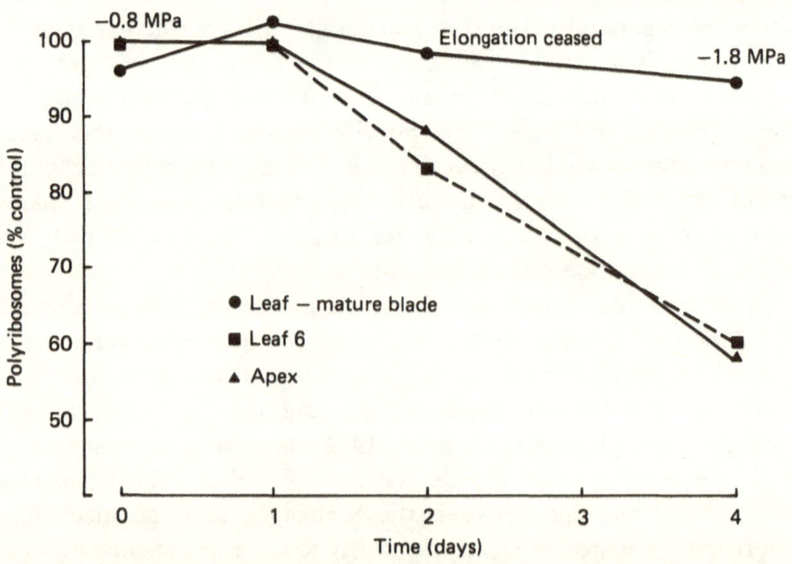

(Todd, 1972) and these changes may not be directly attributable to water stress. For example, Shaner & Boyer (1976) have been able to show that the decrease in the level of nitrate reductase during stress is due to the decreased flux of nitrate required to stimulate nitrate reductase synthesis rather than a direct effect on protein synthesis.

The accumulation of free proline in water-stressed leaves, first described by Kemble & MacPherson in 1954, is the best documented of all nitrogen metabolism responses (Stewart & Hanson, 1980). Proline accumulation results from increased proline synthesis and decreased proline oxidation in water-stressed leaves. In expanding tissues, where protein synthesis falls in response to decreases in growth, decreased incorporation of proline into protein will also contribute to its accumulation. However, as protein synthesis is not as severely affected in mature leaves, the accumulation of free proline is usually less in mature than expanding leaves (Munns, Brady & Barlow, 1979). Free proline does appear to be a very versatile nitrogen storage compound, in that it is neutral, very soluble (in comparison with some amides), easily translocated, readily metabolized after stress, and non-toxic to other cytoplasmic processes such as protein synthesis. Further, if located entirely in the cytoplasm, it can provide the solute molecules necessary to osmotically adjust the cytoplasm during stress (Ford & Wilson, 1981). However, the accumulation of pure proline should be regarded as a result of stress-induced perturbation of nitrogen metabolism rather than the cause of it.

More recently, another nitrogen-containing compound, glycine betaine, has been found to accumulate in leaves in response to water stress (Hanson & Nelsen, 1978; Wyn Jones & Storey, 1978). Glycine betaine levels in mature wheat leaves may equal those of proline (Pheloung & Barlow, unpublished), but unlike proline, glycine betaine is not readily metabolized after stress is released, and is therefore not a good nitrogen storage compound (Ladyman, Hitz & Hanson, 1980). In terms of solubility, translocation and cytoplasmic compatibility glycine betaine is very similar to proline.

In general, the water-stress-induced changes in the nitrogen metabolism of the mature leaf appear to be largely indirect. Changes in protein synthesis appear to depend on the level of growth inhibition caused by water stress, and the resulting accumulations of soluble nitrogen compounds initiate shifts in nitrogen metabolism which allow nitrogen to flow to compatible cytoplasmic storage compounds.

## Leaf adaptation to the environment

The scope of this chapter has been necessarily limited to mesophytic crop species and therefore the morphological adaptations of many non-cultivated plants to light and water availability have been neglected. Nevertheless, after dissecting the complex leaf functions into various categories it is instructive to attempt to formulate an overall strategy for leaf function. Cowan & Farquhar (1977) have recently put forward a hypothesis which attempts to link water loss and carbon gain into a strategy of leaf response to the environment. They propose that the stomatal mechanism not only acts as a device which prevents desiccation of plant tissue, but has, in principle, the capability of varying the conductances to diffusion so as to maximize the ratio of the mean rate of carbon assimilation to the mean rate of evaporation in a fluctuating environment. Implicit in this hypothesis is the assumption that stomatal aperture responds to changes in the state of water in the leaf directly or indirectly. This hypothesis is consistent with the daily patterns of photosynthesis during the development of water deficits shown in Fig. 11.7 and indeed the vertical wilting of sunflower (Fig. 11.9). While the hypothesis awaits further verification, it must be appreciated that it provides a strategy, but not a mechanism of how the stomata are controlled by the bulk of the leaf, the mesophyll. The recent data of Pierce & Raschke (1980), demonstrating a rapid ABA response to decreases in bulk leaf turgor, may provide a mechanism for communication between the mesophyll and epidermis of the mature leaf. Such a mechanism also provides a vital link between the biophysical and biochemical functions of the leaf and is consistent with the absence of direct short-term effects of water deficits on leaf metabolism. Consequently, the elucidation of the kinetics of growth substance (particularly ABA) synthesis, metabolism and translocation within the leaf may be a fertile area of future research into leaf function. The emphasis, over the last 15 years, on the biophysical aspects of leaf water relations and function, has failed to provide a biophysical mechanism of co-ordinated leaf function.

## References

Barlow, E. W. R. & Boersma, L. (1976). Interaction between leaf elongation, photosynthesis and carbohydrate levels of water-stressed corn seedlings. *Agronomy Journal*, **68**, 923–6.

Barlow, E. W. R., Lee, J. W., Munns, R. & Smart, M. G. (1980). Water relations of the developing wheat grain. *Australian Journal of Plant Physiology*, **7**, 519–25.
Barrs, H. D. (1968). Effect of cyclic variations on gas exchange under constant environmental conditions on the ratio of transpiration to net photosynthesis. *Physiologia Plantarum*, **21**, 918–29.
Barrs, H. D. (1973). Controlled environment studies of the effects of variable atmospheric water stress on photosynthesis, transpiration and water status of *Zea mays* L. and other species. In *Plant Response to Climatic Factors*. Proceedings Uppsala Symposium, ed. R. O. Slatyer, pp. 249–58. Paris: UNESCO.
Begg, J. E. (1980). Morphological adaptations of leaves to water stress. In *Adaptation of Plants to Water and High Temperature Stress*, ed. N. C. Turner & P. J. Kramer, pp. 33–43. New York: Wiley Interscience.
Begg, J. E. & Torssell, B. W. R. (1974). Diaphotonastic and paraphotonastic leaf movement in *Stylosanthes humilis H. B. K.* (Townsville stylo). In *Mechanisms of Regulation of Plant Growth*, ed. R. L. Bieleski, A. R. Ferguson & M. M. Cresswell, pp. 277–83. Wellington: Royal Society of New Zealand.
Begg, J. E. & Turner, N. C. (1976). Crop water deficits. *Advances in Agronomy*, **28**, 161–217.
Black, C. R. (1979). The relative magnitude of the partial resistances to transpirational water movement in sunflower (*Helianthus annuus* L.). *Journal of Experimental Botany*, **30**, 245–53.
Boyer, J. S. (1969). Free energy transfer in plants. *Science*, **163**, 1219–20.
Boyer, J. S. (1971). Resistances to water transport in soybean, bean and sunflower. *Crop Science*, **11**, 403–7.
Boyer, J. S. (1974). Water transport in plants: Mechanism of apparent changes in resistance during absorption. *Planta*, **117**, 187–207.
Boyer, J. S. (1976). Water deficits and photosynthesis. In *Water Deficits and Plant Growth*, vol. IV, ed. T. T. Kozlowski, pp. 154–91. New York: Academic Press.
Boyer, J. S. & Bowen, B. L. (1970). Inhibition of oxygen evolution in chloroplasts isolated from leaves with low water potentials. *Plant Physiology*, **45**, 612–15.
Boyer, J. S. & McPherson, H. G. (1975). Physiology of water deficits in cereal crops. *Advances in Agronomy*, **27**, 1–23.
Brady, C. J. & Scott, N. S. (1977). Chloroplast polyribosomes and synthesis of Fraction 1 Protein in the developing wheat leaf. *Australian Journal of Plant Physiology*, **4**, 327–35.
Brady, C. J., Scott, N. S. & Munns, R. (1974). The interaction of water stress with the senescence pattern of leaves. In *Mechanisms of Regulation of Plant Growth*, ed. R. L. Bieleski, A. R. Ferguson & M. M. Cresswell. Wellington: The Royal Society of New Zealand.
Briggs, G. E. (1967). *Movement of Water in Plants*. Oxford: Blackwell.
Burrows, F. J. & Milthorpe, F. L. (1976). Stomatal conductance in the control of gas exchange. In *Water Deficits and Plant Growth*, vol. IV, ed. T. T. Kozlowski, pp. 103–54. New York: Academic Press.
Capita, N., Sabularse, D., Montezinos, D. & Delmer, D. P. (1979). Determination of the pore size of cell walls of living plant cells. *Science*, **205**, 1144–7.
Collatz, J., Ferrar, P. J. & Slatyer, R. O. (1976). Effects of water stress and differential hardening treatments on photosynthetic characteristics of xeromorphic shrub *Eucalyptus socialis*, F. Muell. *Oecologia*, Berlin, **23**, 95–105.
Cowan, I. R. (1977). Stomatal behaviour and environment. *Advances in Botanical Research*, **4**, 117–228.
Cowan, I. R. & Farquhar, G. D. (1977). Stomatal function in relation to leaf metabolism and environment. In *Integration of Activity in the Higher Plant. Society for Experimental Biology Symposium 31*, ed. D. H. Jennings. Cambridge University Press.

Cram, W. J. (1976). Negative feedback regulation of transport in cells. The maintenance of turgor, volume and nutrient supply. In *Encyclopedia of Plant Physiology* (new series), vol. IIA, *Transport in Plants*, ed. U. Lüttge & M. G. Pitman, pp. 284–316. Berlin: Springer-Verlag.

Dainty, J. (1976). Water relations of plant cells. In *Encyclopedia of Plant Physiology* (new series), vol. IIA, *Transport in Plants*, ed. U. Lüttge & M. G. Pitman, pp. 12–35. Berlin: Springer-Verlag.

Ehleringer, J. & Forseth, I. (1980). Solar tracking by plants. *Science*, **210**, 1094–8.

Esau, K. (1977). *Anatomy of Seed Plants*. New York: John Wiley.

Ford, C. W. & Wilson, J. R. (1981). Changes in levels of solutes during osmotic adjustment to water stress in leaves of four tropical pasture species. *Australian Journal of Plant Physiology*, **8**, 77–91.

Hanson, A. D. & Nelsen, C. E. (1978). Betaine accumulation and $^{14}C$ formate metabolism in water-stressed barley leaves. *Plant Physiology*, **62**, 305–12.

Honert, T. H. van den (1948). Water transport in plants as a catenary process. *Discuss. Faraday Society*, **3**, 146–53.

Hsiao, T. C. (1973). Plant response to water stress. *Annual Review of Plant Physiology*, **24**, 519–70.

Hsiao, T. C., Acevedo, E., Ferres, E. & Henderson, D. W. (1976). Water stress, growth and osmotic adjustment. *Philosophical Transactions of the Royal Society of London*, **B273**, 479–500.

Husken, D., Zimmermann, U. & Schulze, E.-D. (1979). Water relations of leaves of *Tradescantia virginiana*: direct turgor pressure measurement. In *Plant Membrane Transport: Current Conceptual Issues*, ed. R. M. Spanswick, W. J. Lucas & J. Dainty, pp. 469–70. Oxford: Pergamon Press.

Jarvis, P. G. (1975). Water transfer in plants. In *Heat and Mass Transfer in the Biosphere. I. Transfer Processes in the Plant Environment*, ed. D. A. de Vries & N. H. Afgan, pp. 369–94. Washington: Halsted Press.

Johns, G. G. (1978). Transpirational leaf area, stomatal and photosynthetic responses to gradually induced water stress in four herbage species. *Australian Journal of Plant Physiology*, **5**, 113–25.

Jones, H. G. (1973). Moderate-term water stresses and associated changes in some photosynthetic parameters in cotton. *New Phytologist*, **72**, 1095–105.

Jones, M. M., Osmond, C. B. & Turner, N. C. (1980). Accumulation of solutes in leaves of sorghum and sunflower in response to water deficits. *Australian Journal of Plant Physiology*, **7**, 193–205.

Jones, M. M. & Rawson, H. M. (1979). Influence of rate of development of leaf water deficits upon photosynthesis, leaf conductance, water use efficiency, and osmotic potential in sorghum. *Physiologia Plantarum*, **45**, 103–11.

Jordan, W. R. & Ritchie, J. T. (1971). Influence of soil water stress on evaporation, root absorption and internal water status of cotton. *Plant Physiology*, **48**, 783–8.

Kemble, A. R. & MacPherson, H. T. (1954). Nitrogen metabolism of wilting ryegrass. *Biochemistry Journal*, **58**, 46–9.

Kriedemann, P. E. & Loveys, B. R. (1974). Hormonal mediation of plant responses to environmental stress. In *Mechanisms of Regulation of Plant Growth*, ed. R. L. Bieleski, A. R. Ferguson & M. M. Cresswell, pp. 461–5. Wellington: Royal Society of New Zealand

Ladyman, J. A. R., Hitz, W. O. & Hanson, A. D. (1980). Translocation and metabolism of glycine betaine by barley plants in relation to water stress. *Planta*, **150**, 191–6.

Lange, O. L., Kappen, L. & Schulze, E.-D. (eds.) (1976). *Water and Plant Life: Problems and Modern Approaches*. Berlin: Springer-Verlag.

Lawlor, D. W. (1976). Assimilation of carbon into photosynthetic intermediates of water-stressed wheat. *Photosynthetica*, **10**, 431–9.

Lawlor, D. W. (1979). Effects of water and water stress on carbon metabolism of plants with C3 and C4 photosynthesis. In *Stress Physiology in Crop Plants*, ed. H. Mussell & R. C. Staples, pp. 303–26. New York: Wiley Interscience.

Lawlor, D. W. & Fock, H. (1978). Photosynthesis, respiration, and carbon assimilation in water-stressed maize at two oxygen concentrations. *Journal of Experimental Botany*, **29**, 579–93.

Ludlow, M. M. (1980). Adaptive significance of stomatal responses to water stress. In *Adaptation of Plants to Water and High Temperature Stress*, ed. N. C. Turner & P. J. Kramer, pp. 123–39. New York: Wiley Interscience.

Meidner, H. (1975). Water supply, evaporation and vapour diffusion in leaves. *Journal of Experimental Botany*, **26**, 666–73.

Monteith, J. L. (1973). *Principles of Environmental Physics*. London: Edward Arnold.

Monteith, J. L. (1976). *Vegetation and the Atmosphere*, vol. 2, *Case Studies*. London: Academic Press.

Munns, R., Brady, C. J. & Barlow, E. W. R. (1979). Solute accumulation in the apex and leaves of wheat during water stress. *Australian Journal of Plant Physiology*, **6**, 379–89.

Munns, R. & Pearson, C. J. (1974). Effect of water stress on translocation of carbohydrate in *Solanum tuberosum*. *Australian Journal of Plant Physiology*, **1**, 529–37.

Mussell, H. & Staples, R. C. (eds.) (1979). *Stress Physiology of Crop Plants*. New York: Wiley Interscience.

Naylor, A. W. (1972). Water deficits and nitrogen metabolism. In *Water Deficits and Plant Growth*, vol. III, ed. T. T. Kozlowski, pp. 241–55. New York: Academic Press.

Neumann, H. H., Thurtell, G. W. & Stevenson, K. R. (1974). *In situ* measurements of leaf water potential and resistance to water flow. *Canadian Journal of Plant Science*, **54**, 175–84.

Osmond, C. B., Winter, K. & Powles, S. B. (1980). Adaptive significance of carbon dioxide cycling during photosynthesis in water-stressed plants. In *Adaptation of Plants to Water and High Temperature Stress*, ed. N. C. Turner & P. J. Kramer, pp. 139–55. New York: Wiley Interscience.

O'Toole, J. C., Crookston, R. K., Treharne, J. & Ozbun, J. L. (1976). Mesophyll resistance and carboxylase activity – a comparison under water stress conditions. *Plant Physiology*, **57**, 465–8.

O'Toole, J. C. & Moya, T. B. (1978). Genotypic variation in the maintenance of leaf water potential in rice. *Crop Science*, **18**, 873–6.

Passioura, J. B. (1980). The meaning of matric potential. *Journal of Experimental Botany*, **31**, 1161–9.

Pierce, M. & Raschke, K. (1980). Correlation between loss of turgor and accumulation of ABA in detached leaves. *Planta*, **148**, 174–82.

Powles, S. B. & Osmond, C. B. (1978). Inhibition of the capacity and efficiency of photosynthesis in bean leaflets illuminated in a $CO_2$-free atmosphere at low oxygen: a possible role for photorespiration. *Australian Journal of Plant Physiology*, **5**, 619–29.

Raschke, K. (1975). Stomatal action. *Annual Review of Plant Physiology*, **26**, 309–40.

Raschke, K. (1979). 4. Movements using turgor mechanisms. 4.1. Movements of stomata. In *Encylopedia of Plant Physiology* (new series), vol. 7, *Physiology of Movements*, ed. W. Haupt & M. E. Fenleib, pp. 383–441. Berlin: Springer-Verlag.

Rawson, H. M. (1979). Vertical wilting and photosynthesis, transpiration and water use efficiency of sunflower leaves. *Australian Journal of Plant Physiology*, **6**, 109–20.

Rawson, H. M., Turner, N. C. & Begg, J. E. (1978). Agronomic and physiological responses of soybean and sorghum crops to water deficits. IV. Photosynthesis, transpiration and water use efficiency of leaves. *Australian Journal of Plant Physiology*, **5**, 195–209.

Satter, R. L. (1979). Leaf movements and tendril curling. In *Encyclopedia of Plant*

*Physiology* (new series), vol. VII, *Physiology of Movements*, ed. W. Haupt & M. E. Feinleib, pp. 442–85. Berlin: Springer-Verlag.

Scott, N. S., Munns, R. & Barlow, E. W. R. (1979). Polyribosome content in young and aged wheat leaves subjected to drought. *Journal of Experimental Botany*, **30**, 905–11.

Shaner, D. L. & Boyer, J. S. (1976). Nitrate reductase activity in maize (*Zea mays* L.) leaves. II. Regulation by nitrate flux at low leaf water potential. *Plant Physiology*, **58**, 505–9.

Sheriff, D. W. & Meidner, H. (1974). Water pathways in leaves of *Hedera helix* L. and *Tradescantia virginiana* L. *Journal of Experimental Botany*, **26**, 1147–56.

Slatyer, R. O. & Taylor, S. A. (1960), Terminology in plant-soil-water relations. *Nature, London*, **187**, 922–4.

Smith, J. A. C. & Milburn, J. A. (1980). Water stress and phloem loading. *Berichte der deutschen botanischen Gesellschaft*, **93**, 269–80.

Stewart, C. R. & Hanson, A. D. (1980). Proline accumulation as a metabolic response to water stress. In *Adaptation of Plants to Water and High Temperature Stress*, ed. N. C. Turner & P. J. Kramer, pp. 173–91. New York: Wiley Interscience.

Stoker, R. & Weatherley, P. E. (1971). The influence of the root system on the relationship between the rate of transpiration and depression of leaf water potential. *New Phytologist*, **70**, 547–54.

Sung, F. J. M. & Krieg, D. R. (1979). Relative sensitivity of photosynthetic assimilation and translocation of $^{14}$carbon to water stress. *Plant Physiology*, **64**, 852–6.

Tanton, T. W. & Crowdy, S. H. (1972). Water pathways in higher plants. *Journal of Experimental Botany*, **23**, 600–18.

Teare, I. D. & Peet, M. M. (eds.) (1981). *Crop-Water Relations.* New York: Wiley Interscience.

Todd, G. W. (1972). Water deficits and enzyme activity. In *Water Deficits and Plant Growth*, vol. III, ed. T. T. Kozlowski, pp. 177–217. New York: Academic Press.

Troughton, J. H. & Slatyer, R. O. (1969). Plant water status, leaf temperature and the calculated mesophyll resistance to carbon dioxide of cotton leaves. *Australian Journal of Biological Sciences*, **22**, 815–27.

Turner, N. C. (1974a). Stomatal behaviour and water status of maize, sorghum and tobacco under field conditions. II. At low soil water potential. *Plant Physiology*, **53**, 360–5.

Turner, N. C. (1974b). Stomatal response to light and water under field conditions. In *Mechanisms of Regulation of Plant Growth*, ed. R. L. Bieleski, A. R. Ferguson & M. M. Cresswell, pp. 423–32. Wellington: Royal Society of New Zealand.

Turner, N. C. & Begg, J. E. (1973). Stomatal behaviour and water status of maize, sorghum and tobacco under field conditions. I. At high soil water potential. *Plant Physiology*, **51**, 31–6.

Turner, N. C., Begg, J. E., Rawson, H. M., English, S. D. & Hearn, A. B. (1978). Agronomic and physiological responses of soybean and sorghum crops to water deficits. III. Components of leaf water potential, leaf conductance, $^{14}CO_2$ photosynthesis, and adaptation to water deficits. *Australian Journal of Plant Physiology*, **5**, 179–94.

Turner, N. C. & Burch, G. J. (1981). The role of water in plants. In *Crop-Water Relations*, ed. I. D. Teare & M. M. Peet. New York: Wiley Interscience. (In press.)

Turner, N. C. & Kramer, P. J. (eds.) (1980). *Adaptation of Plants to Water and High Temperature Stress.* New York: Wiley Interscience.

Tyree, M. T. & Jarvis, P. G. (1982). Water in tissues and cells. In *Encyclopedia of Plant Physiology* (new series), vol. B, *Physiological Plant Ecology*, ed. O. Lange & B. Osmond. Berlin: Springer-Verlag. (In press.)

Tyree, M. T. & Yianoulis, P. (1980). The site of water evaporation from sub-stomatal cavities, liquid path resistances and hydroactive stomatal closure. *Annals of Botany*, **46**, 175–93.

Wardlaw, I. F. (1967). The effect of water stress on translocation in relation to photosynthesis and growth. I. Effect during grain development in wheat. *Australian Journal of Biological Sciences*, **20**, 25–39.

Wardlaw, I. F. (1969). The effect of water stress on translocation in relation to photosynthesis and growth. II. Effect during leaf development in *Lolium temulentum*. *Australian Journal of Biological Sciences*, **22**, 1–16.

Weatherley, P. E. (1970). Some aspects of water relations. *Advances in Botanical Research*, **3**, 171–206.

Wiebe, H. H. & Prosser, R. J. (1977). Influence of temperature gradients on leaf water potential. *Plant Physiology*, **59**, 256–8.

Wilson, J. R., Ludlow, M. M., Fisher, M. J. & Schulze, E.-D. (1980). Adaptation to water stress of the leaf water relations of four tropical forage species. *Australian Journal of Plant Physiology*, **7**, 207–20.

Wyn Jones, R. G. & Storey, R. (1978). Salt stress and comparative physiology in the Gramineae. II. Glycinebetaine and proline accumulation in two salt and water-stressed barley cultivars. *Australian Journal of Plant Physiology*, **5**, 801–16.

Zimmermann, U. & Steudle, E. (1978). Physical aspects of water relations of plant cells. *Advances in Botanical Research*, **5**, 45–117.

# 12

# External factors influencing photosynthesis and respiration

MERVYN M. LUDLOW

## Introduction

Rates of photosynthesis and respiration of leaves result from an interplay of internal (Chapter 13) and external factors. Although internal factors determine the ultimate capacity of these processes, external factors influence the extent to which this capacity is achieved. The external factors discussed here (light, temperature, humidity, wind and atmospheric gases) are the main components of the aerial environment of leaves.

Environmental conditions under which leaves develop and to which they have been previously exposed indirectly influence current performance because of their effects on both structure and function. Prevailing environmental conditions affect function and, hence, directly influence current performance. Both the direct and indirect effects are modified by adaptation of the leaf and its capacity to acclimate to changing conditions. Adaptation is *heritable* modification to structure or function that increases the probability of an organism surviving and reproducing in a particular environment. On the other hand, acclimation is *non-heritable* modification to characteristics or processes caused by exposure of organisms to new environmental conditions such as warmer, cooler or drier weather (Kramer, 1980).

External factors affect leaf function by influencing the environment at the reaction sites within the leaf. For example, dark respiration rate is determined by the temperature of the mitochondria rather than the temperature of the air surrounding the leaf. Organelle temperature may be similar to leaf temperature but the latter is determined not only by prevailing environmental conditions but also by characteristics of the leaf itself (Grace, 1977).

**Nature of photosynthesis and respiration**

Photosynthesis involves transfer of atmospheric $CO_2$ to the chloroplast, photochemical processes driven by radiant energy which yield protons, and biochemical processes which use the protons to reduce $CO_2$ to carbohydrate in the photosynthetic carbon reduction (PCR) cycle. As $CO_2$ molecules diffuse along a concentration gradient from the atmosphere to the chloroplast (Fig. 12.1), they encounter resistances associated with the boundary layer ($r_a'$), stomata and intercellular spaces ($r_s'$) and the photochemical and biochemical processes (intracellular resistance, $r_i'$). The $r_a'$ results from the still air near

Fig. 12.1. Schematic diagram of the resistance to carbon dioxide movement into and within leaves. $C_a$, $C_i$, $C_c$ and $C_r$ are the carbon dioxide concentrations (or partial pressures) in the ambient air, intercellular spaces, site of carboxylation in the chloroplast and at respiratory sites within the cell, respectively; $r'_a$, $r'_s$ and $r'_i$ are resistances to carbon dioxide transfer associated with the boundary layer of the leaf, stomata and intracellular processes of photosynthesis, respectively; P and R represent sites of photosynthesis and respiration.

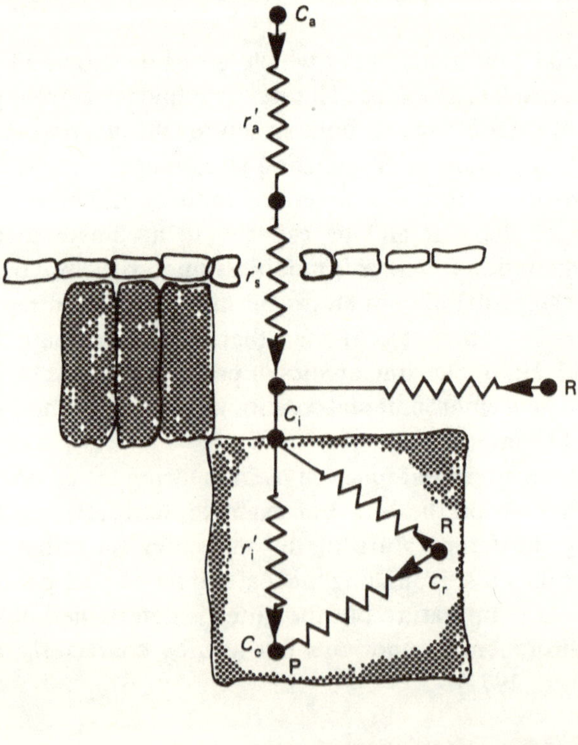

the leaf surface and its magnitude decreases with narrower leaves and with higher wind speed (Cowan & Milthorpe, 1968). Because $r_s'$ is calculated from the equivalent resistance for water vapour flux out of the leaf ($r_s$) and the ratio of the diffusion coefficients of $CO_2$ and $H_2O$ ($r_s' = 1.65r_s$), stomatal resistance is the sum of the resistances of the stomatal pore and of the intercellular spaces up to the mesophyll cell wall. The $r_i'$ is a complex of resistances for $CO_2$ transfer from the mesophyll cell wall to the site of fixation in the chloroplast, and the biochemical and photochemical processes at the site of fixation. The transfer part of $r_i'$ is mainly physical but it may have a biochemical component, especially if carbonic anhydrase or any other active transport process is involved.

Applying Fick's law of diffusion, the relationship between these resistances and leaf net photosynthetic rate ($F'$) through the whole or part of the pathway is (Fig. 12.1):

$$F' = \frac{C_a - C_c}{r_a' + r_s' + r_i'} = \frac{C_a - C_i}{r_a' + r_s'} = \frac{C_i - C_c}{r_i'} \quad (12.1)$$

where $C_a$, $C_i$ and $C_c$ are the carbon dioxide partial pressures (Pa) in ambient air, in intercellular spaces and at photosynthetic sites respectively. If, as commonly assumed, $C_c = 0$ the limitation to net $CO_2$ uptake by respiration in the light is included in $r_i'$. The influence of this respiratory flux can be eliminated if $C_c = \Gamma$, the carbon dioxide compensation concentration (Lake, 1967).

When the relative limitations to photosynthesis by the boundary layer, stomata, or intracellular processes are being compared in series they should be expressed as resistances. However, when studying the relationship between any one of these limitations and $F'$, or between them and an environmental factor such as temperature, it is less confusing and more convenient to use conductances ($g_a'$, $g_s'$ and $g_i'$; where $g$ = 1/resistance) because fluxes are proportional to conductances but inversely proportional to resistances (Burrows & Milthorpe, 1976).

Details of the photochemical and biochemical processes of photosynthesis and respiration are discussed in Chapters 10 and 13.

Two types of respiration are found in leaves; dark respiration and photorespiration. Release of $CO_2$ in the dark can arise from both the Krebs cycle and the pentose phosphate pathway. The latter operates only in the dark but there is evidence, though not unequivocal, that the Krebs cycle continues in the light (Graham & Chapman, 1979). In this chapter I assume that dark respiration continues in the light at the same rate as in the dark.

Photorespiration is the light-dependent process of oxygen uptake and $CO_2$ release which results from competition between atmospheric $CO_2$ and $O_2$ on the active sites of RuBP carboxylase/oxygenase (Chollet & Ogren, 1975; Andrews & Lorimer, 1978). Carbon dioxide is released when phosphoglycolate is metabolized by the photorespiratory carbon oxidation (PCO) cycle. $C_3$ plants exhibit photorespiration, but it is apparently absent in $C_4$ plants under normal atmospheric conditions because the increased $CO_2$ concentration in the bundle sheath increases the rate of carboxylation of RuBP, suppressing oxygenation activity (Hatch & Osmond, 1976).

No current method gives precise estimates of the rate of photorespiration ($R_L$; Ludlow & Jarvis, 1971a) because photosynthesis, photorespiration, and possibly also the Krebs cycle, operate simultaneously in the light. Moreover, the amount of $CO_2$ refixed cannot be calculated because we are uncertain about the concentration gradients and the resistances to $CO_2$ transfer within both the cell and leaf (but see Chapter 13).

## Response, acclimation and adaptation to external factors

### *Light*

#### *Nature of light*

Photosynthetically active radiation (PAR) is the radiant flux density in the 400–700 nm waveband of the electromagnetic spectrum. Normally flux density of PAR is 50% of the short-wave radiation of sunlight, although the proportion can vary between 40 and 60% depending upon location, time of day and atmospheric conditions. However, since photosynthesis depends upon quanta rather than energy, light for photosynthesis is best defined as quantum flux density ($\mu$mole of quanta $m^{-2} s^{-1}$). Sunlight contains about 4.6$\mu$mole $J^{-1}$ PAR. Photochemical processes in leaves are influenced by *absorbed* rather than *incident* light. The amount of light lost by reflection (about 10%) and transmission (about 10%) varies with leaf anatomy and morphology and environmental conditions.

#### *Response to light*

Rates of net photosynthesis ($F'$), dark respiration ($R_D$) and photorespiration ($R_L$) respond to the amount, direction, frequency and quality of light received. The relationship between $F'$ and $Q_v$, often called the light response curve, is fundamental to understanding the photosynthetic performance of leaves and of canopies (Fig. 12.2). When $Q_v = 0$, the flux of $CO_2$ from the leaf is $R_D$. In the initial linear phase at low $Q_v$ the supply of electrons limits photosynthesis, while in the subsequent curvilinear phase $F'$ is limited by both the electron transport and the carboxylation reactions. Beyond the point of light saturation $F'$ is limited by the capacity for carboxylation and is independent of $Q_v$. The slope of the linear phase is the maximum light utilization efficiency (based on incident $Q_v$) or the quantum yield (based on absorbed $Q_v$). Quantum yield for most $C_3$ and $C_4$ plants at 20–30 °C is 0.06–0.08 mole $CO_2 \cdot$ mole quanta$^{-1}$ which is equivalent to a quantum requirement of 12–16 quanta per mol of fixed $CO_2$. Because of the constancy of this value among higher plants, the light compensation point is determined primarily by $R_D$, but it varies with environmental conditions, environmental history and biological factors such as leaf age.

Fig. 12.2. Relationship between net photosynthetic rate ($F'$) and quantum flux density ($Q_v$) of leaves of two $C_4$ species (broken line; ●, *Pennisteum purpureum*; ◆, *Panicum maximum*) and two $C_3$ species (solid line; ■, *Camissonia claviformis*; ×, *Calopogonium mucunoides*). The leaf of *P. maximum* was suffering water stress (leaf water potential = −0.85 MPa).

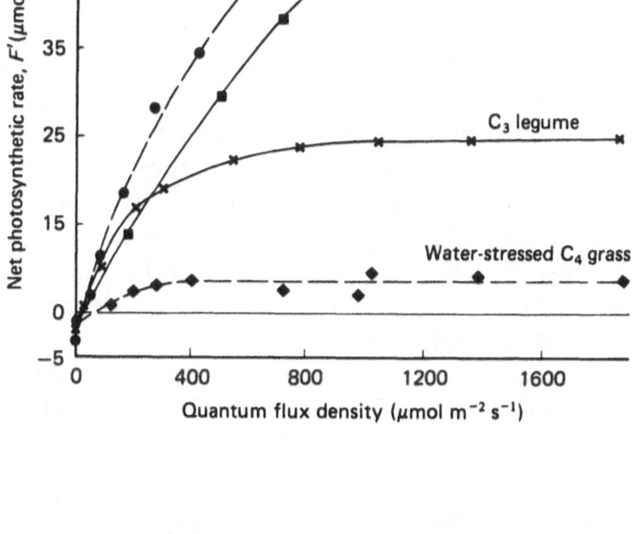

Light saturation of photosynthesis is said to be a characteristic of $C_3$ plants and its absence a characteristic of $C_4$ plants (Fig. 12.2). While this is usually true under optimal conditions of growth and measurement, light saturation is absent in many $C_3$ species with high photosynthetic capacities (e.g. sunflower, *Camissonia*) and commonly absent in leaves from plants grown under field conditions (Mooney, Ehleringer & Berry, 1976; Ehleringer & Björkman, 1978; Fig. 12.2). In leaves of $C_4$ plants, light saturation can occur when conditions are sub-optimal (e.g. severe water deficits; Fig. 12.2) or when internal factors such as leaf age reduce photosynthetic capacity. Absence of light saturation in both $C_3$ and $C_4$ plants is associated with high values of $g_s'$ and $g_i'$.

The shape of the light response curve of $C_4$ plants and of $C_3$ plants for $Q_v$ values below the light saturation point can be described by a non-rectangular hyperbola:

$$m(F')^2 - \left(\alpha Q_v + \frac{C_c}{r_x}\right)(F')^2 + \alpha Q_v \frac{C_c}{r_x} = 0$$

where $\alpha$ is the light utilization efficiency, $Q_v$ is the absorbed energy, $C_c$ is the $CO_2$ concentration at the carboxylation site and $r_x$ is the carboxylation resistance (Prioul & Chartier, 1979). This equation describes a range of curves between a rectangular hyperbola ($m = 0$) and a hyperbola reduced to its two asymptotes ($m = 1$). Most other mechanistic or empirical relationships which have been used to describe light response curves (Tenhunen, Weber, Yocum & Gates, 1979; Charles-Edwards, 1981; Farquhar & von Caemmerer, 1981) reduce to this relationship.

Dorsiventral leaves, such as sunflower (*Helianthus annuus*), tobacco (*Nicotiana tabacum*) and some pasture legumes, have a lower rate of net photosynthesis when illuminated on the abaxial surface than on the adaxial surface with the same $Q_v$. The lower rate is associated mainly with lower $g_i'$ which results from a higher reflectivity to incident light and less efficient light harvesting by the underlying spongy mesophyll cells compared with the adaxial surface. Normal plant structure and also diaheliotropic leaf movements in many well-watered dicotyledons (Begg & Torssell, 1974; Ehleringer & Forseth, 1980) ensure that the adaxial surface of dorsiventral leaves receives high $Q_v$, thus maximizing net photosynthesis of individual plants under the prevailing environmental conditions. Isobilateral leaves of grasses and desert holly (*Perezia nana*) are often illuminated on both surfaces in nature. They have the same rate of net photosynthesis when illuminated with the same $Q_v$ on either leaf surface (Moss, 1964, 1965; Ludlow & Wilson, 1971a; Syvertsen & Cunningham, 1979). However, if both surfaces are illuminated simultaneously, the rate is greater than when either surface is illuminated separately with double the $Q_v$ (Moss, 1964, 1965).

It is sometimes claimed that light received as intermittent flashes is used more efficiently in photosynthesis than continuous light, because the limitation of the dark reactions is reduced (e.g. Kriedemann, Torokfalvy & Smart, 1973). However, when results are expressed on a common energy or photon flux basis, this claim is not supported by the data available (Sagar & Giger, 1980). Similarly, there seems little evidence of daylength (photoperiod) *per se* having a *direct* effect on $F'$ or $R$, although inductive daylengths alter $g_s$ in the short-day plant *Xanthium pennsylvanicum* compared with plants in non-inductive long days (Krizek & Milthorpe, 1966; Meidner, 1970). However crassulacean acid metabolism (CAM) is induced in *Kalanchoe blossfeldiana* by short days (Osmond, 1978).

In general, stomata of $C_3$ and $C_4$ plants open in the light and close in the dark, whereas those of water-stressed CAM plants do the opposite. Stomatal conductance of leaves of $C_4$ plants increases continuously with quantum flux, in the same manner as net photosynthesis, up to values equivalent to full sunlight (Gifford, 1971; Ludlow & Wilson, 1971a). On the other hand, stomatal conductance of leaves of most $C_3$ plants, like net photosynthesis, becomes light saturated at quantum fluxes between one-third and one-half of full sunlight. There are basically two mechanisms which cause stomata to open with increasing $Q_v$ (Sheriff, 1979). One is sensitive only, or predominantly, to blue light, and does not appear to be closely linked with photosynthesis. The second operates via the decrease in $C_i$ (probably the intracellular guard cell carbon dioxide concentration) caused by photosynthesis as $Q_v$ increases. Indeed, there is evidence that stomatal conductance can be determined by the rate of photosynthesis because there is a linear relationship between $F'$ and $g_s'$ and because $C_i$ is maintained approximately constant at about 22 Pa in $C_3$ species, and 10 Pa in $C_4$ plants with quantum fluxes greater than about 200 $\mu$mol m$^{-2}$ s$^{-1}$ (Wong, Cowan & Farquhar, 1979; Warritt, Landsberg & Thorpe, 1980). This proposal has recently been challenged by Jarvis & Morison (1981) who quote many instances when there is little or no relationship between $F'$, $g_s'$ and $C_i$.

By definition, photorespiration occurs only in the light. Notwithstanding the difficulty of measuring $R_L$, it appears that under

$CO_2$-limiting conditions $R_l$ increases with $Q_v$ in much the same way as $F'$ (Jackson & Volk, 1970; Ludlow & Wilson, 1971a).

Blue (425–450 nm) and red (575–675 nm) are more effective in photosynthesis than either green light (*ca* 675 nm) or the longer (>675 nm) and shorter (<425 nm) wavelenths (McCree, 1971; Fig. 12.3). There are only minor variations in this response amongst herbaceous higher plants and these arise from different carotenoid: chlorophyll ratios which give differences in quantum yield in the blue region (Björkman, 1973). Blue light is even less effective in conifers because of the greater reflection by their waxy blue-green needles and because of the higher proportions of carotenoid which compete with chlorophylls for unreflected blue light. Within the 400–700 nm waveband the enhancement of net photosynthesis by light of different quality is so small that the effects of different wavelengths can be treated as independent and additive (McCree, 1972). On the other hand, it appears that blue light is usually more effective in stomatal opening than is red light (Sheriff, 1979).

Because the PCR and PCO cycles are intimately linked via RuBP carboxylase/oxygenase, the action spectrum of photorespiration is similar to that of photosynthesis (Bulley, Nelson & Tregunna, 1969). However there are some reports of photorespiration being stimulated by blue compared with red or white light (Jackson & Volk, 1970).

Fig. 12.3. Mean quantum yield of net photosynthesis as a function of wavelength of absorbed light for leaves of 22 higher plant species. (Redrawn from McCree, 1971.)

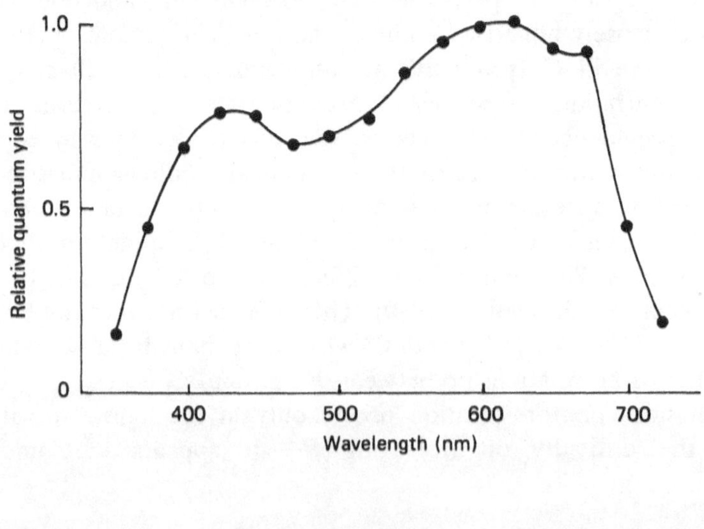

*Acclimation and adaptation to different light environments*

The light environment under which leaves develop has a profound effect on their structure and function (Björkman, 1973, 1981; Boardman, 1977; Patterson, 1980). Moreover, some changes to structure and many changes in function occur even in mature leaves when their light environment is altered. These effects vary among species and ecotypes of the same species. Some species appear to be adapted to grow only at high quantum fluxes (obligate 'sun' plants), others only at low quantum fluxes (obligate 'shade' plants), whereas some species have the capacity to grow in both environments (facultative 'shade' plants). For brevity, I shall use the terms 'sun' and 'shade' plants to represent the extreme types of behaviour. The characteristics of these

Fig. 12.4. Relationship between quantum flux density and net photosynthetic rate of leaves of *Atriplex patula* which developed under high, intermediate and low light levels, and of *Alocasia macrorhiza* growing on the floor of a rainforest in south-east Queensland. The inset, which is an expanded version of that part of the figure near the origin, shows that the quantum yield was unaffected by light environment under which the leaves developed and the low dark respiration rate of *Alocasia* (arrow). (Adapted from Osmond *et al.*, 1980).

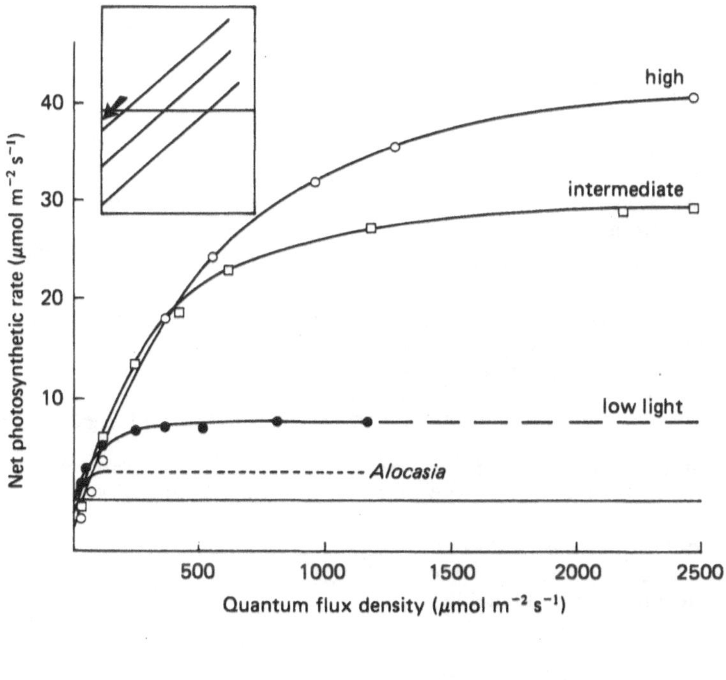

plants and their response to changes in light environment are summarized in Table 12.1 and Fig. 12.4. Characteristics of sun plants grown in low light are similar, though not as extreme, as those of shade plants shown in the table.

Morphological, anatomical and physiological characteristics of leaves from shade plants (called shade leaves) maximize light-harvesting capacity and minimize the synthetic and maintenance costs of soluble proteins and components of the electron transport chain which would be superfluous at low light (Fig. 12.4); for example, the ratio of chlorophyll

Table 12.1. *Differences in morphological and physiological characteristics of leaves from 'sun' and 'shade' plants, and changes in these characteristics when 'sun' and 'shade' plants are grown under or transferred to, respectively, low and high quantum fluxes ($Q_v$)*

| Characteristic | Sun versus shade | Sun plant – low $Q_v$ | Shade plant – high $Q_v$ |
|---|---|---|---|
| *Morphological* | | | |
| leaf mesophyll thickness | $\gg$ or $<$[a] | –[c] | + |
| cell surface area : leaf area | $>$ | – | + |
| specific leaf weight ($g\,m^{-2}$) | $\gg$ or $<$ | – | + |
| leaf area ratio ($m^2\,g^{-1}$) | $<$ | + | – or 0 |
| *Photosynthetic and respiratory* | | | |
| maximum net phot. rate[b] | $>$ | – | – or 0 |
| light saturation point | $>$ | – | + or 0 |
| light compensation point | $>$ | – | + |
| light utilization efficiency | 0 | 0 | 0 |
| stomatal frequency[b] | 0 | – | + |
| stomatal conductance | $>$ | – or 0 | – |
| intracellular conductance | $>$ | – | – |
| dark respiration rate[b] | $>$ | – | + |
| photorespiration rate[b] | NA[d] | 0 or + | 0 |
| *Biochemical and photochemical* | | | |
| carboxylase activity[b] | $>$ | – | 0 |
| chlorophyll content | $\gg$ or $<$ | – | – |
| electron transport capacity | $<$ | – | + |
| PSU size (chlorophyll/P700) | 0 or $<$ | – or + | NA |
| PSU density[b] | NA | – | + |

Adapted from Björkman (1973), Boardman (1977) and Patterson (1980).

[a] $>$ sun greater than shade, $\gg$ sun much greater than shade, $<$ sun less than shade.

[b] rate or content per unit leaf area.

[c] + = increase, – = decrease, 0 = no change.

[d] NA = information not available.

to soluble protein is seven times higher in shade than in sun leaves (Björkman, Troughton & Nobs, 1974). Shade leaves are thinner and of lower specific leaf weight than sun leaves, and shade plants invest a larger proportion of their dry weight in leaves.

The reduced mesophyll thickness in thin shade leaves results in a smaller cell volume and surface area per unit leaf area. The lower light saturation point and light-saturated rate of photosynthesis arise mostly from lower mesophyll conductances, although stomatal density and stomatal conductance are usually also lower in shade leaves. Quantum yield is usually not affected by light environment and consequently the lower light compensation point of shade leaves results from the greatly reduced dark respiration rate (Ludlow & Wilson, 1971a; Björkman, 1973; Patterson, 1980; Fig. 12.4). Electron transport capacity and carboxylase activity are reduced, but the size and density of photosynthetic units are normally similar to those of sun leaves, although in a recent study the mean size of photosynthetic unit in a range of shade plants was twice that of a range of sun species (Malkin & Fork, 1981). Whereas the chlorophyll concentration per unit leaf area in shade leaves may be higher, lower, or similar to sun leaves, the distribution of chlorophyll is quite different. Shade leaves have larger chloroplasts with larger grana and less stroma lamellae, and more chlorophyll per chloroplast volume and per cell (Boardman, 1977). This highly efficient light-harvesting arrangement is necessary for survival of shade species such as those that occur on the floor of rainforests where the daily $Q_v$ is less than 0.5% of that at the top of the canopy (Björkman & Ludlow, 1972).

The capacity for acclimation to low quantum fluxes varies greatly between obligate sun species, such as most crop and pasture plants, and facultative shade plants such as *Alocasia macrorhiza* and *Cordyline rubra* which occur both in full sunlight and the darkest places on the rainforest floor. Nevertheless, the qualitative changes in anatomy and physiology which occur are basically similar in sun and shade plants and they may have a common basis. The lower rates of leaf net photosynthesis (on an area basis) of sun plants which have developed in low light (Fig. 12.4) can, in many instances, be explained completely by the lower cell volume and cell surface area per unit leaf area of the thinner leaves compared with sun leaves (Charles-Edwards & Ludwig, 1975; Nobel, 1980; Patterson, 1980). Consequently there is a higher resistance to carbon dioxide uptake by, and transport within, mesophyll tissues and less photochemical and biochemical machinery per unit leaf area.

However, in other instances, differences in anatomy assessed from cell dimensions (Nobel, 1977, 1980) or eliminated by use of cell suspensions (Harvey, 1980), could not account for differences in net photosynthesis between sun and shade leaves. In these examples reduced rates of electron transport and carboxylation were probably responsible for the lowered rate of net photosynthesis of shade leaves.

Because of the large numbers of processes influenced by the light environment under which leaves develop, it is unlikely that there will be one factor responsible for the reduced photosynthetic capacity in all situations. In practice, there seems to be an integrated adjustment of the various processes involved in light harvesting and carbon dioxide fixation to suit the available $Q_v$.

Similar types of change occur in fully expanded leaves when the light environment changes although the extent and rate of change may differ (Patterson, 1980; Table 12.1). For example, leaves of facultative shade plants which developed under low light become thicker when transferred to high light, and specific leaf weight and cell surface area to leaf area increase. In addition, there are increases in $F'$ and its component processes. Increases in enzyme activities, PSU size and $F'$ are evident within one day of transfer; other properties, such as specific leaf weight, chlorophyll content per unit area and the density of photosynthetic units, alter more slowly. However, in mature leaves of shade plants and in leaves of some sun plants that have developed in low light, even though many of these structural and metabolic adjustments occur, $F'$ and associated processes often decrease rather than increase in high light. The reduced capacities for electron transport and carbon dioxide fixation cannot utilize the higher $Q_v$, and photoinhibition results. This effect can be so severe in some obligate shade plants that they die in high-light regimes (Björkman, 1975).

The scope for structural change in mature leaves is, obviously, much more limited than alterations to metabolism, although sometimes most of the changes in $F'$ of fully expanded leaves can be explained by changes in leaf anatomy and its associated characteristics. However, as with the difference between sun and shade leaves, it appears that there are reversible integrated adjustments to both structure and function, and no one single controlling factor can be identified for all situations.

Most of the information on acclimation to different light environments comes from comparisons of characteristics of leaves from natural environments with different peak $Q_v$, or changes in characteristics when the $Q_v$ has been altered in both natural and artificial environments.

However, it appears that anatomical and photosynthetic characteristics of leaves are determined by the daily integral of $Q_v$, rather than $Q_v$ itself under constant conditions or peak $Q_v$ when there is diurnal variation, even though all three measures are inter-related (Ludlow & Ng, 1976; Chabot, Jurik & Chabot, 1979; Nobel, 1980). Even this finding may be due to correlated factors because recently Nobel & Hartsock (1981) found that leaf thickness, and hence subsequent photosynthetic peformance, is more dependent upon the amount of photosynthesis than upon the daily $Q_v$ integral during leaf formation.

### *Ultra-violet radiation*

Photosynthesis is progressively inhibited as the wavelength decreases below 400 nm. However, because of absorption by stratospheric ozone, there is essentially no solar radiation below 295 nm received at the Earth's surface (Robberecht & Caldwell, 1978). Therefore, interest has been on the effect of UV-B radiation (280–315 nm) on plants and the consequences of increasing UV-B if the stratospheric ozone is depleted (Caldwell, 1971, 1977).

Inhibition of $F'$ by UV-B is cumulative and proportional to the total amount received during leaf ontogeny and $R_D$ is enhanced by exposure to UV-B. In contrast, leaf growth is only influenced during the first two or three days of expansion (Caldwell, 1977). The degree of inhibition of $F'$ varies with species (Caldwell, 1971, 1977; Van, Garrard & West, 1976; Sisson & Caldwell, 1976; Robberecht & Caldwell, 1978) and results from damage to the photosynthetic apparatus rather than from lowered $g_s'$. The thylakoid membranes of the chloroplasts dilate after only 15 minutes exposure and electron transport capacity (PS-II activity) is reduced (Brandle, Campbell, Sisson & Caldwell, 1977).

The reasons for the variation in adaptation among species and their ability to acclimate to changes in UV-B are unknown (Caldwell, 1977). Differences or changes in leaf reflectivity, absorption by epidermal tissues and photoreactivity, and ability to repair damaged structures or functions may be involved.

### *Temperature*

#### *Nature of leaf temperature*

Most leaves are sufficiently thin that the temperature gradient between the ad- and abaxial surfaces is small compared with spatial and temporal variations. Consequently, the temperature of cell organelles is probably similar to leaf temperature. Leaf temperature depends primarily on radiant energy absorbed by the leaf, the convective heat transfer between the leaf and the surrounding air, the loss of latent heat from the leaf by transpiration, and re-radiation of heat from the leaf surfaces. These processes are influenced by leaf morphological factors such as thickness, size, reflectivity, angle and orientation, and by environmental factors such as air temperature, incident radiant flux density, humidity and windspeed.

Leaves are exposed to temperatures from −40 to 60 °C. Depending upon the species, and the preceding and current conditions, water in leaves changes from the liquid phase to the solid phase at some temperature below 0 °C. This change of state has a profound effect on survival and on both current and future leaf activity. Important, though less dramatic, effects result when chilling-sensitive plants from tropical and sub-tropical areas are exposed to non-freezing temperatures below 15 °C.

*Response to temperature*

The primary photochemical reactions of photosynthesis are generally considered independent of temperature as long as the functional integrity of the chloroplast membrane is maintained. In contrast, like most metabolic processes, the other component reactions of photosynthesis and the reactions of both dark and photorespiration are strongly dependent on temperature. For temperatures up to the optimum, the $Q_{10}$ of these metabolic processes is about 2, but it can vary between 1.5 and 2.5 (Fig. 12.5). Respiration in the dark seems to have a larger $Q_{10}$ than respiration in the light (Hofstra & Hesketh, 1969); for example, the $Q_{10}$ was 1.8 for $R_L$ and 2.3 for $R_D$ in *Rumex acetosa* (Holmgren & Jarvis, 1967). The responsiveness of $F'$ to temperature increases with $Q_v$ (Ludlow & Wilson, 1971a) and with $CO_2$ concentration (Osmond, Björkman & Anderson, 1980).

Rates of photosynthesis and respiration begin to decline at an ever-increasing rate above their optimum temperature which is lower for $F'$ than for $R_D$ and $R_L$. Moreover, the dark respiration processes are much more resistant to high temperature denaturation than those of photosynthesis (Berry & Björkman, 1980). At very high temperatures the rates of all photosynthetic and respiratory processes suffer a time-dependent and largely irreversible decline as vital components of the metabolic processes are inactivated and the functional integrity of cellular organelles such as mitochondria is disrupted. Sensitivity to high temperature varies with species (adaptation) and growth temperature (acclimation).

The lower $F'$ at temperatures below the optimum is primarily due to the limitation of the biochemical processes because stomatal conductance does not decrease until photosynthesis is severely inhibited (Ludlow & Jarvis, 1971b; Williams, 1974; Neilson & Jarvis, 1975). As temperature falls, the activity of rate-limiting enzymes declines in both $C_3$ and $C_4$ plants (Berry & Björkman, 1980). At even lower temperatures, conformational changes result in denaturation of proteins, and phase changes in lipids disrupt membranes and membrane-associated processes, especially in chilling-sensitive plants (Osmond *et al.*, 1980).

Where $g_s'$ does respond to temperature, it usually increases up to the optimum temperature for $F'$ (Pearcy, 1977; Sheriff, 1979; Berry & Björkman, 1980). While the opening may be a response to decreasing $C_i$, there does not seem to be as close a relation between $g_s'$, $C_i$ and $F'$ as there is with $Q_v$. It seems likely that opening is due to a direct effect of increasing temperature on guard cell metabolism because it also occurs in the dark (Sheriff, 1979; Muchow, Ludlow, Fisher & Myers, 1980a). The response of stomata to temperature is very complex and poorly understood; for example $g_s'$ of kenaf (*Hibiscus cannabinus*) decreases as

Fig. 12.5. Relationship between leaf temperature and net photosynthetic rate of a $C_4$ grass (*Cenchrus ciliaris*) and a $C_3$ legume (*Calopogonium muconoides*) at high quantum flux densities and ambient carbon dioxide and oxygen concentrations.

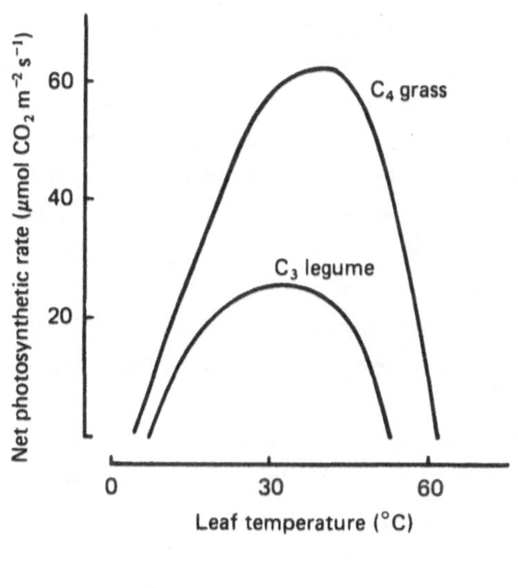

temperature falls from 30 °C, reaches a minimum between 10 and 15 °C and *increases* again as temperature falls to 5 °C (Muchow *et al.*, 1980a).

Stomata close at supra-optimal temperatures (e.g. van den Driessche, Connor & Tunstall, 1971; Williams & Kemp, 1976), because of increasing $C_i$ due to enhanced respiration or because of increasing leaf-air vapour pressure difference and water deficits (Sheriff, 1979). If leaf-air vapour pressure difference is kept low, $g_s'$ continues to increase up to temperatures which cause damage to the photosynthetic apparatus (Berry & Björkman, 1980). But the main causes of the decline of $F'$ as temperatures increase above the optimum are firstly that respiration increases at a greater rate than photosynthesis (Larcher, 1980) and then secondly that $g_i'$ decreases as irreversible heat inactivation disrupts the functional integrity of the photosynthetic apparatus in the chloroplast. The stimulation of $R_L$ arises because the oxygenase activity of RuBP carboxylase/oxygenase increases more with temperature than does the carboxylation activity (Farquhar, von Caemmerer & Berry, 1980). For example, the lower optimum and maximum temperatures of $C_3$ plants compared with $C_4$ plants (Fig. 12.5) can be eliminated if photorespiration is suppressed by low $O_2$ or high $CO_2$ concentrations (Osmond *et al.*, 1980). Loss of functional integrity arises from inactivation of both thylakoid membrane reactions and enzymes of photosynthetic carbon metabolism. Photosystem II is particularly sensitive to heat but the mechanism is not understood. Inactivation of PS-II-driven electron transport results in inhibition of non-cyclic photophosphorylation and uncoupling of phosphorylation from electron transport.

Certain soluble enzymes located outside the thylakoid membrane in either the stroma region of the chloroplast or in the cytoplasm are also inactivated by temperatures that cause irreversible inhibition of photosynthesis. Three enzymes (NADP glyceraldehyde-3P dehydrogenase, ribulose-5P kinase and NADP malate dehydrogenase) are more heat sensitive than other photosynthetic enzymes including RuBP carboxylase, and show changes similar to those found in $F'$. The heat inactivation of these 'light-activated' enzymes may be linked with inhibition of PS-II activity. Heat treatment also causes denaturation and precipitation of other soluble proteins, but the nature of the proteins involved and the consequences of this loss for net photosynthesis have not been determined.

Short-term exposure of chilling-sensitive plants to temperatures below 10 °C can depress subsequent $F'$ (Crookston *et al.*, 1974). It is uncertain, however, how important this is under field conditions when plants usually have time to acclimate to declining seasonal temperatures. A reduction of canopy photosynthesis could not be detected following exposure of *Setaria anceps* and *Macroptilium atropurpureum* to 11 °C for the whole night during the summer when mean daily maximum and minimum temperatures were 29 and 18 °C respectively (M. M. Ludlow & R. Davis, unpublished). The after-effects of more extreme temperatures are marked. Inhibition of $F'$ and the rate of its recovery depends upon the maximum or minimum temperature and the time of exposure (Bauer, Larcher & Walker, 1975). Exposure of chilling-sensitive plants to sub-zero but non-freezing temperatures (Ludlow & Taylor, 1974) and of frost-tolerant plants to freezing temperatures (Larcher, 1980) depresses $F'$ severely during the following day and progressively less so in subsequent days. Similarly, recovery from short-term exposure to high temperatures, which damage thylakoid membranes and reduce non-cyclic photophosphorylation and PS-II activity, may take from days to several weeks.

*Adaptation and acclimation to temperature*

This topic has been reviewed extensively (Berry & Björkman, 1980; Osmond *et al.*, 1980; Patterson, 1980; Turner & Kramer, 1980). Plants occupying thermally contrasting habitats generally exhibit photosynthetic and respiratory characteristics that reflect adaptation to the temperature regimes of their respective habitats (Berry & Björkman, 1980; Larcher, 1980). Cardinal temperatures for $F'$ from a range of plants are given in Table 12.2. In general, species or ecotypes from (or grown at) low temperatures have higher $F'$ measured at low temperatures and lower optimum temperature than those from warm areas or regimes. The converse applies when cool- and heat-adapted species are grown and measured at high temperatures. Similar differences in cardinal temperatures for $F'$ result if the same species or ecotype is grown at a range of temperatures under controlled conditions or is measured at a number of occasions during the normal seasonal change of temperature in the field. The temperature response curve shifts in the direction of the growth temperature. By Kramer's (1980) definition this is acclimation. However, Berry & Björkman (1980) believe that the $F'$ at the new temperature must increase to confer an ecological advantage. For example, the optimum temperature for $F'$ of *Atriplex sabulosa*, a native of cool coastal habitats of California, increases with growth temperature but values of $F'$ measured at all temperatures are lower than those from the preferred cool temperature regime. Thus this plant is at a photosynthetic disadvantage when grown at high temperatures even though the optimum temperature has increased.

The extent of acclimation varies with species or ecotype and the size of the temperature change. Some of the largest seasonal shifts of optimum temperature for $F'$ between winter and summer are: 10 and 11 °C for two altitudinal ecotypes of *Eucalyptus pauciflora* in south-eastern Australia; 12 °C for *Hammada scorparia* in the Negev Desert, Israel (Fig. 12.6); and 14 °C for the shrub *Larrea divaricata* in Death Valley, California (Berry & Björkman, 1980). Plants from habitats with large seasonal changes of temperature tend to possess a greater potential for acclimation than do plants from habitats with relatively stable temperatures. Moreover, plants restricted in distribution to cool environments tend to have a relatively low preferred temperature and a limited potential for acclimation to high temperatures, whereas the converse is usually true for plants from hot environments, especially $C_4$ species. When a species grows in a wide range of thermal environments it may be either composed of a number of genetically different ecotypes each with a limited potential for acclimation or of genetically identical plants with a high potential for photosynthetic acclimation to a wide range of temperatures (Berry & Björkman, 1980).

Acclimation to changes in temperature regime can arise from changes in structure or function of new leaves which develop under the new temperature regime. However, it also occurs in mature leaves with

Table 12.2. *Cardinal (minimum, optimum and maximum) temperatures for leaf net photosynthetic rate measured at ambient carbon dioxide and oxygen levels and high quantum fluxes*

| | Minimum (°C) | Optimum (°C) | Maximum (°C) |
|---|---|---|---|
| Winter annuals, spring-flowering and alpine plants | −7 to −2 | 10 to 20 | 30 to 40 |
| Evergreen conifers | −5 to −3 | 10 to 25 | 35 to 42 |
| Dwarf shrubs or heath and tundra; winter deciduous trees of temperate zones | −3 to −1 | 15 to 25 | 40 to 45 |
| Herbaceous $C_3$ plants | −2 to 0 | 15 to 30 | 35 to 45 |
| Desert plants and shrubs; sclerophyllous trees | −5 to 5 | 15 to 35 | 42 to 55 |
| Evergreen trees in tropics and sub-tropics | 0 to 5 | 25 to 30 | 45 to 50 |
| Tropical $C_4$ grasses | 5 to 10 | 35 to 45 | 45 to 60 |

Adapted from Bauer *et al.* (1975), Ludlow (1976) and Larcher (1980).

alterations to function and, to a lesser extent, to structure. The time for acclimation in mature leaves varies from days to weeks. Acclimation of tropical grass and tropical legume leaves which developed at 20 °C to a new temperature of 30 °C was complete after 15 hours overnight (Ludlow & Wilson, 1971b), and the thermal stability of *Nerium oleander* leaves which developed at 20 °C increased to that of leaves grown at 45 °C within one or two days (Osmond *et al.*, 1980). With other species changes in $F'$ and $g_s'$ are evident within a few days, but in some circumstances it can take up to two weeks until all the photochemical, biochemical and structural alterations cease. For example, in *Nerium oleander* complete acclimation took 12 days after a temperature change from 45 to 20 °C and vice versa (Björkman, Badger & Armond, 1980). Even this rate would be sufficient in most situations to accommodate the relatively slow seasonal march of temperature.

The mechanisms of adaptation to particular thermal regimes and mechanisms of acclimation to altered temperature are basically similar. Consequently, they will be discussed collectively. Growth temperature affects innumerable structural and functional characteristics of leaves (see Patterson, 1980 for a comprehensive list). However, I am going to restrict my discussion to structural features such as leaf thickness,

Fig. 12.6. Acclimation of the temperate response of leaf net photosynthetic rate in *Hammada scorparia* during seasonal changes in temperature in the Negev Desert, Israel. (Adapted from Lange *et al.*, 1974.)

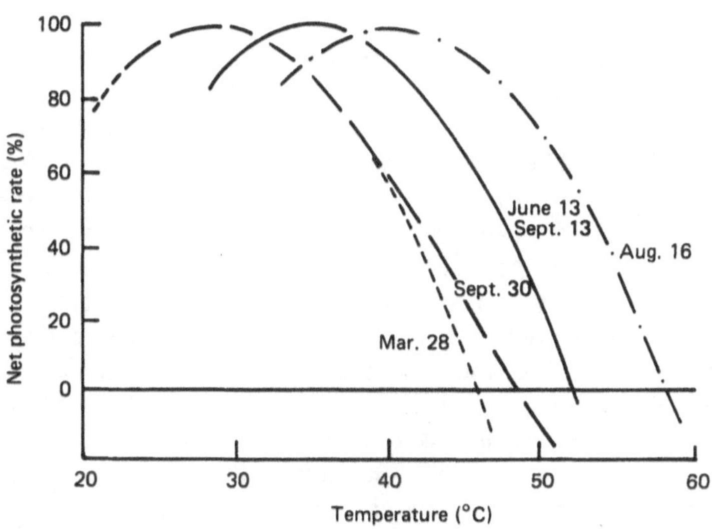

specific leaf weight and ratio of cell surface area to projected leaf area ($A^{mes}/A$), and the functional features $R_D$, $R_L$, $g_s'$ and the intrinsic biochemical properties of photosynthesis.

Both $R_D$ and $R_L$, and activities of respiratory enzymes per unit leaf area, usually increase as growth temperature decreases (Patterson, 1980), presumably because of the increase in leaf thickness and specific leaf weight. The few studies made on the role of respiration in photosynthetic acclimation to temperature suggest that it is not significant (Berry & Björkman, 1980; Osmond *et al.*, 1980). Obviously the absence of photorespiration in $C_4$ plants aids adaptation and acclimation to high temperatures, whereas photorespiration is a burden which $C_3$ plants must bear at all temperatures and which is accentuated at high temperatures.

Differences in $g_s'$ account for some of the adaptation of photosynthesis to various temperature regimes and changes in $g_s'$ contribute to changes in $F'$ during acclimation to temperature. However, its importance varies from moderate in tropical $C_4$ pasture grasses and legumes (Ludlow & Wilson, 1971b) to minor in species native to various regions of California (Berry & Björkman, 1980; Osmond *et al.*, 1980). Overall, while stomata could be responsible for the more rapid changes in $F'$ when growth temperature is altered, they appear to exercise only a minor role in the adaptation and acclimation responses to temperature. Therefore, we must look at structural features which modify the size of the photosynthetic machinery per unit leaf area and at the intrinsic photosynthetic characteristics of mesophyll cells to find explanations.

Leaf thickness and specific leaf weight decrease as growth temperatures increase in many species, but they may both increase again at supra-optimal temperatures (Chabot & Chabot, 1977; Nobel, 1977, 1980; Patterson, 1980). The decrease in leaf thickness does not lead to lower $A^{mes}/A$ as would be expected, because of the compensatory effect of decreasing cell size (Wilson & Ludlow, 1968; Nobel, 1980). Moreover, specific leaf weight may not be a good measure of the amounts of biochemical components per unit leaf area because starch accumulation increases at lower temperatures (Chabot & Chabot, 1977). There is not always a good correlation between $F'$ and leaf structure in leaves formed at different temperatures, although there is a good association between leaf thickness, specific leaf weight and increasing $R_D$ as growth temperature decreases (Chabot & Chabot, 1977; Nobel, 1977; Björkman *et al.*, 1980; Patterson, 1980). Structural differences thus appear relatively unimportant for temperature acclimation and are not major determinants of differences in adaptation to different temperature regimes. It is unlikely that photosynthetic acclimation to temperature in mature leaves can result from changes in structure because the scope for change is limited and because of the rapidity of change (Ludlow & Wilson, 1971b; Berry & Björkman, 1980; Patterson, 1980).

Consequently, $g_i'$ must be the major factor determining the photosynthetic adaptation to low and high temperatures and in temperature acclimation, because stomata only play a minor role (Ludlow & Wilson, 1971b; Williams, 1974). The rapidity (overnight) of the increase in $g_i'$ when the temperature was increased from 20 to 30 °C in tropical grasses and legumes suggests changes in biochemical processes are responsible. Acclimation to low temperature involves increase in the amount of photosynthetic enzymes, such as RuBP carboxylase/oxygenase and fructose-1,6-bisphosphate phosphatase, which increase photosynthetic *capacity*. In addition, changes to the physical properties and composition of membrane lipids contribute to the low temperature or chilling *tolerance* (Berry & Björkman, 1980). It has been suggested, though not proved beyond doubt, that chilling injury results from a phase separation of membrane lipids resulting in disruption of membrane-associated reactions such as PS-II and photophosphorylation. On the other hand, changes in the physical properties of chloroplast membrane lipids leading to increased heat stability of these membranes are a key factor in acclimation to high temperature, together with increased heat stability of soluble enzymes. It has been suggested that heat stability arises from a decrease in fluidity of the polar lipids of the membranes due in part to a decrease in unsaturation of their component fatty acids. These changes can occur within one or two days which is more than adequate to accommodate the seasonal increase in temperature.

## *Humidity*

### *Nature of humidity*

Humidity is a loose term used to describe the moisture content of air. It can be expressed more precisely as relative humidity – the ratio of the actual vapour pressure of water to the saturated vapour pressure at the particular dry bulb temperature. However, relative humidity is of limited value for describing the moisture content of air as it influences plant processes such as transpiration and leaf characteristics such as stomatal conductance. Water vapour moves from the intercellular spaces of leaves to the ambient air along gradients of vapour pressure (or partial pressures). This air–leaf vapour pressure difference ($\Delta e$) equals the saturated vapour pressure at leaf temperature, less the vapour pressure of the air. If the air and leaf are at the same temperature $\Delta e$ is identical to saturation deficit of the air ($\delta e$), i.e. the difference between the actual and saturated vapour pressure at the particular dry bulb temperature.

Whereas $\Delta e$ defines the gradient for transpiration and is the most appropriate measure to which $g_s$ should be related, we are often forced to use $\delta e$ especially in field studies. $\Delta e$ and $\delta e$ will usually be similar except when leaves are significantly cooler than air during radiation frosts, or in well-watered crops growing in dry environments, or significantly warmer because stomata are closed, or the leaf is subjected to high radiation loads such as those found close to the soil surface in hot, dry and sunny environments.

*Response to humidity*

Respiration and $g_i'$ do not appear to respond to $\Delta e$ *or* $\delta e$. On the other hand, $\Delta e$ influences $F'$ by affecting $g'_s$ (Ludlow & Wilson,

Fig. 12.7. Relationship between leaf–air vapour pressure difference, and net photosynthetic rate ($F'$ closed symbols) and stomatal conductance for carbon dioxide ($g_s'$, open symbols) of leaves of a $C_4$ grass (***Pennisetum purpureum***, circles) and a $C_3$ legume (***Vigna luteola***, squares).

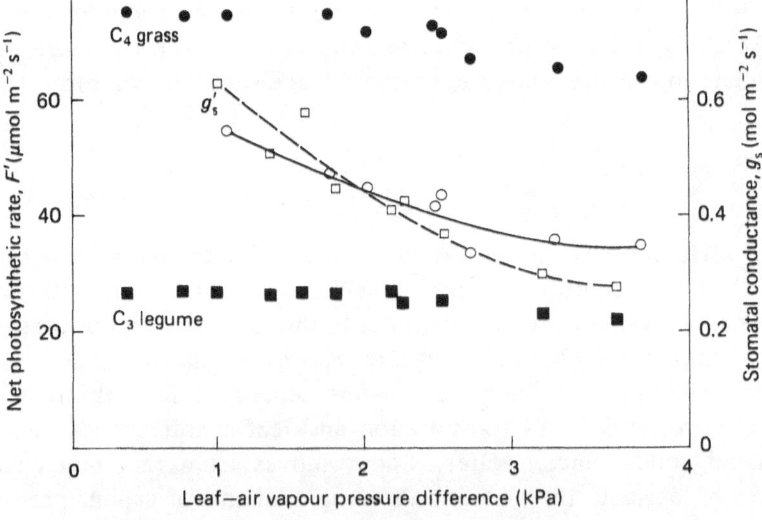

1971a; Hall, Schulze & Lange, 1976; Fig. 12.7). Increasing $\Delta e$ can *indirectly* cause stomatal closure by lowering bulk leaf water potential. However, stomata of a wide range of species from different habitats appear to respond *directly* to humidity and $g'_s$ decreases as $\Delta e$ increases, independent of bulk leaf water potential (Hall *et al.*, 1976; Sheriff, 1979; Jarvis, 1980; Ludlow, 1980). The direct response is most evident at high leaf water potentials (Ludlow, 1980; M. M. Ludlow, M. J. Fisher & R. Davis, unpublished data; cf. Schulze & Küppers, 1979), low quantum fluxes (Hall *et al.*, 1976), and either at low (Hall *et al.*, 1976; Jarvis, 1980) or between low and optimum temperatures (Muchow, Fisher, Ludlow & Myers, 1980b).

Whereas stomata respond to bulk leaf water potential in a classical feed-back manner, they appear to respond to $\Delta e$ in a feed-forward manner; local water deficits develop in the epidermis which induce stomatal closure before bulk leaf water potential falls appreciably (Ludlow, 1980). This prevents harmful deficits from developing in the bulk of the leaf and reduces the possibility of permanent injury. The mechanism of the response to $\Delta e$ is not understood. However, it has been suggested that deficits in the epidermis are the result of increased evaporation from guard cells or adjacent hypodermal cells *within* the leaf (the so-called peristomatal transpiration) in some species, and from guard cells and subsidiary cells, *external* to the stomatal pore in others (Sheriff, 1979; Jarvis, 1980).

The influence of increased $\Delta e$ on $F'$ is smaller than on $g_s'$; for example increasing $\Delta e$ from 1.6 to 3.3 kPa only depressed $F'$ by 20% in $C_4$ pasture grasses and $C_3$ legumes (Ludlow & Wilson, 1971a; Fig. 12.7). The influence of $\Delta e$ on $F'$ will depend upon the responsiveness of $g_s'$ to $\Delta e$ and relative sizes of $g_s'$ and $g_i'$ such as in $C_4$ ($g_i' > g_s'$) and $C_3$ ($g_i' < g_s'$) species.

### *Adaptation and acclimation to humidity*

Stomata of many species from contrasting habitats respond to $\Delta e$ but with differing degrees of sensitivity. Many species native to arid areas and a number of agricultural crops grown in semiarid regions demonstrate the response, but there are many exceptions (Hall *et al.*, 1976; Rawson, Begg & Woodward, 1977; Jarvis, 1980; Ludlow, 1980). Response to humidity is not associated with ecological distribution (Sheriff, 1977) or phylogeny. Therefore the ecological significance for survival, and the agronomic significance for crop production, of stomatal responsiveness to humidity are unclear. Moreover, even though species with large sub-stomatal cavities seem more responsive (Sheriff, 1977), the anatomical and physiological characteristics associated with, and necessary for, the stomatal response to humidity remain to be described.

Stomata of some species can acclimate to $\Delta e$ in that $g'_s$ of leaves exposed to large $\Delta e$ become more sensitive (Maier-Maercker, 1979). Moreover, the response of $g'_s$ to $\Delta e$ changes in some species with the degree of water stress they have experienced, resulting from both soil and atmospheric deficits. Other species show no adjustment (Ludlow, 1980). Of the responsive species, some become more sensitive and others become less sensitive to $\Delta e$ (Ludlow, 1980; M. M. Ludlow, M. J. Fisher & R. Davis, unpublished). Whereas reduced sensitivity may reflect a passive response to water stress, and not acclimation, the increased sensitivity to $\Delta e$ of stomata of plants previously exposed to water stress (Osonubi & Davies, 1980) suggests acclimation. For example, Schulze *et al.* (1972) and Schulze *et al.* (1975) found with apricot in the Negev Desert of Israel that the slope of the $g_o'/\Delta e$ relationship was steeper in unirrigated than in irrigated plants, and that the relationship became steeper in unirrigated plants as soil water declined and as leaf water potential decreased from spring to summer. This behaviour has clear adaptive advantages in reducing water loss, maintaining plant water status and water use efficiency, and assisting survival in an arid environment. However, it is not an obligatory characteristic because other plants that do not exhibit this response grow and survive there. In less arid areas increased stomatal sensitivity to $\Delta e$ would assist survival of species unable to withstand severe leaf water deficits, although it may have no particular advantage and may be a disadvantage for growth of species able to tolerate such deficits and with access to adequate soil water (Ludlow, 1980).

### *Atmospheric gases*

#### *Response to oxygen and carbon dioxide concentrations*

Dark respiration is not influenced by oxygen concentration in the ambient atmosphere ($O_a$) above 2% (Hofstra & Hesketh, 1969; Canvin, 1979), nor by $CO_2$ concentration or partial pressure ($C_a$). Similarly, at normal $O_2$ levels, photorespiration is either independent of $C_a$ or it may decline slightly at high $C_a$ (Canvin, 1979; Farquhar *et al.*, 1980). However, as discussed later, photorespiration is strongly influenced by $O_a$.

Stomata open as $C_a$ falls and close as $C_a$ increases, but only in the presence of abscisic acid (ABA) in some species (Raschke, 1975; Burrows & Milthorpe, 1976; Sheriff, 1979; Fig. 12.8). Stomata appear to respond to the $CO_2$ concentration in guard cells which is apparently related to $C_i$, but the mechanism of the response is unknown.

At normal $CO_2$ levels $F'$ of $C_4$ species is unaffected when $O_a$ is reduced from 21 to <1% (Canvin, 1979), because the $CO_2$ concentrating mechanism of the $C_4$ pathway in the bundle sheath suppresses the oxygenase activity of RuBPc. However, $F'$ of $C_4$ plants is depressed progressively as $O_a$ increases above ambient levels and $O_2$ molecules effectively compete with $CO_2$ molecules for active sites on the enzyme. Stomatal conductance falls at $O_a$ >40% but this is probably a result rather than a cause of the depressed photosynthesis.

On the other hand, $F'$ of $C_3$ plants is stimulated by up to 50% when $O_a$ is lowered from 21 to 1–2% (Hesketh, 1967; Ludlow, 1970; Canvin, 1979). The increase in $F'$ is due to suppression of the oxygenase and stimulation of the carboxylase activity of RuBP, and to inhibition of the PCO cycle. As with $C_4$ plants, $g_s{}'$ is unaffected by $O_a$ over this range (Ludlow, 1970; Burrows & Milthorpe, 1976; Akita & Tanaka, 1979).

Fig. 12.8. Relationships between ambient (open symbols) and intracellular (closed symbols) carbon dioxide partial pressure, and net photosynthetic rate ($F'$) and stomatal conductance for carbon dioxide ($g_s{}'$) of (*a*) a $C_4$ grass (*Sorghum almum*) and (*b*) a $C_3$ legume (*Vigna luteola*).

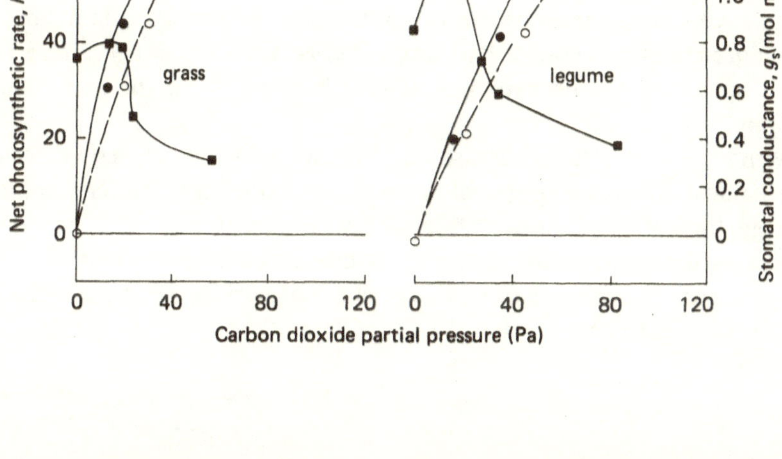

The close association between $g_s'$ and $C_i$ (Wong *et al.*, 1979) does not apply to the influence of oxygen concentration because $g_s'$ of $C_3$ legumes was unaltered when $C_i$ fell 2 Pa as $O_a$ was reduced from 21 to 0.2% (Ludlow, 1970). $F'$ is progressively reduced at normal $CO_2$ concentrations as $O_a$ rises above 21% but the reduction is less at higher $CO_2$ concentrations. For example, in rice, $F'$ at 80% $O_2$ was only 16% of that at 21% $O_2$ when $C_a$ was 30 Pa, whereas it was 75% at 220 Pa (Akita & Tanaka, 1979).

At normal $O_2$ concentrations, $F'$ of both $C_3$ and $C_4$ plants increases with $C_a$ (Ludlow & Wilson, 1971a; Osmond *et al.*, 1980), $C_4$ species being more responsive up to concentrations of 30 Pa. However, $F'$ of $C_4$ plants saturates at lower $CO_2$ concentrations than $C_3$ plants because of the lowered responsiveness of RuBP carboxylase in the bundle sheath where the $CO_2$ concentration is 2–5 times that of ambient (Hatch & Osmond, 1976). There are no differences in $F'$ between $C_3$ and $C_4$ plants at saturating $CO_2$ concentrations. A model of $C_3$ photosynthesis (Farquhar *et al.*, 1980) predicted that the slope of the $F'/C_i$ relationship would change as the limitation by carboxylation efficiency at low $CO_2$ concentration gave way to limitation by electron transport capacity. Subsequent careful experimentation has verified this prediction (von Caemmerer & Farquhar, 1981).

## *Wind*

### *Nature of wind*

Air is never still in nature, it being usually quite windy and turbulent in plant communities and close to leaf surfaces. Wind speeds of over 30 $m\,s^{-1}$ are recorded in violent storms and cyclones (Grace, 1977), and values of 15 $m\,s^{-1}$ have been measured at alpine timberlines (Caldwell, 1970a). However, at the top of agricultural crops 1–2 m high, wind speed rarely exceeds 3 $m\,s^{-1}$, and values between 0.5 and 1 $m\,s^{-1}$ are more common (Day & Parkinson, 1979). Within plant communities wind speed decreases exponentially with depth from the top of the canopy.

Wind influences the structure and function of leaves *directly* through mechanical effects and physical damage, and *indirectly* by affecting their energy budget. The direct effects of wind result from the frictional forces as the momentum of the air is dissipated over the leaf surface. Macro- and microscopic abrasion of the leaf surface occurs as well as mechanical buffeting (Grace, 1975, 1977).

The energy balance of leaves is influenced by effects on the convective and radiative heat fluxes between leaves and the surrounding atmosphere (Grace, 1977). The effects on these fluxes depend upon leaf characteristics such as size, shape, and surface properties, energy incident upon the leaf, and environmental conditions such as the temperature and humidity of the air. As wind speed increases the size of the boundary layer at the leaf surface decreases, the conductances for heat, water vapour and $CO_2$ increase, and leaf temperature approaches air temperature; hence the temperature of leaves under high radiation loads or with closed stomata will fall, whereas it will rise in leaves exposed to clear skies at night. Wind has the largest effect on $g_a$ in broad leaves but the angle of the leaf and whether it flutters or not seem of little importance (Grace, 1975, 1977). Increased wind velocity also reduces the humidity close to the leaf surface because water vapour is moved away from the leaf more rapidly into the surrounding air. Depending upon whether leaf temperature rises or falls, this will usually influence $\Delta e$ and hence water loss from the leaf.

*Response to wind*

Dark respiration rate increases with wind velocity, presumably as a result of mechanical damage to leaves (Grace, 1977; Russell & Grace, 1978b). Both dark respiration and photorespiration are sensitive to changes in leaf temperature, usually associated with variation in wind speed.

The increase in the boundary layer conductance with increasing wind velocity will usually have a greater effect on water loss than on $CO_2$ uptake, because of the additional limitations on the $CO_2$ pathway (Fig. 12.1). However, the influence of the enhanced boundary layer conductance will depend upon the sizes of the other conductances in both pathways. For example, it will have the greatest effect on $H_2O$ when $g_s$ is large and humidity is high enough to prevent stomatal closure (Yabuki & Miyagawa, 1970), and the least effect on $CO_2$ uptake of $C_3$ species where $g_i'$ is low or of both $C_3$ and $C_4$ species when $(g_s' + g_i')$ is small.

It is often difficult to determine separately water loss through the cuticle and loss through stomatal pores. However, it seems that the increase in leaf conductance ($g_1$ = stomatal plus cuticular loss) recorded at high wind speeds in some species (Russell & Grace, 1978b) is due both to increase in $g_s$ and enhanced cuticular loss associated with abrasion to the leaf surface (Grace, 1977). On the other hand, increasing wind speed usually decreases $g_s$ indirectly by lowering the humidity at the leaf surface, causing water deficits in the epidermis and the bulk of the leaf, and by lowering leaf temperature (Burrows & Milthorpe, 1976). Therefore despite the possible increase in cuticular loss from abraded leaf surfaces in some species, transpiration rate usually decreases as wind speed increases because stomata close (Grace, 1975, 1977).

Although wind speed influences transpiration more than photosynthesis, the changes are in the same direction. If wind only affects $g_a'$, a small increase in $F'$ with increasing wind velocity is expected from theory (Grace, 1977); for example, $F'$ of *Cucumis* increased with wind speeds up to $2\,m\,s^{-1}$ when humidity was high enough to prevent stomatal closure, but it reached a maximum at $0.5\,m\,s^{-1}$ at a lower humidity because stomata closed progressively (Yabuki & Miyagawa, 1970). However, the stomatal closure associated with the indirect effects of wind normally result in depressed $F'$ as wind velocity increases. There is sometimes no depression of $F'$ by wind (Caldwell, 1970a; Russell & Grace, 1978b).

Exposure to wind has an after-effect on subsequent gas exchange. Because wax regenerates slowly in mature leaves any abrasion of the epidermis and cuticle will persist (Grace, 1977). Exposure of *Festuca arundinacea* to high winds for several days depressed $F'$ by 25%, due to decreased $g_i'$ despite an increase in $g_l'$. The reduced $g_i'$ was attributed to leaf water deficits (Grace & Thompson, 1973) and later to a direct effect of mechanical shaking (Grace, 1977).

### *Adaptation and acclimation to wind*

Species differ markedly in the extent to which increased wind speed depresses transpiration and photosynthesis (Grace, 1975, 1977); for example, *Pinus cembra*, a species from exposed sites, was unaffected by wind speeds of $15\,m\,s^{-1}$, whereas *Rhododendron ferrugineum*, a species from protected sites, was very sensitive (Caldwell, 1970a). Differences in the response of $F'$ to wind are partly due to differences in $g_s'$, in mechanical properties of leaves, and in unknown factors. The ability of leaves of some species to oscillate in the wind and cause air to move through them by mass flow does not, as previously thought, assist gas exchange of most agricultural species, but it could be important in tall trees in situations where wind speeds are high (Day & Parkinson, 1979).

Plants seem to acclimate to wind by altering both the structure and function of their leaves. Leaves of some species developed in strong winds have xeromorphic characteristics such as thicker, stronger leaves with thick cuticles and a larger number of small stomata (Grace, 1975; Grace & Russell, 1977; cf. Russell & Grace, 1978a). Partial closure of stomata on the current year's flush of needles but not of those of older needles of *Pinus cembra* when exposed to high wind speeds (Caldwell, 1970b) indicates acclimation of stomatal behaviour to wind.

## Conclusion

Although much was learned prior to 1970 on the effects of external factors on photosynthesis and respiration, I believe there was substantial progress in the 1970s. Towards the end of that decade modelling of leaf photosynthesis and respiration emerged as potentially a most fruitful new approach for research (see Chapter 16). Models such as those by Berry & Farquhar (1977) for some aspects of $C_4$ photosynthesis and by Farquhar *et al.* (1980) and Farquhar & von Caemmerer (1981) for $C_3$ photosynthesis based on simplified biochemical schemes of the processes are able to simulate most of the observed responses to external factors. In some cases they predicted behaviour prior to its discovery. More recently, some models (e.g. Tenhunen & Westrin, 1979) have included acclimation to light and temperature but much more needs to be done before adaptation and acclimation behaviour can be effectively described and predicted.

## References


Akita, S. & Tanaka, I. (1979). Studies on the mechanism of differences in photosynthesis among species. V. Stomatal response in high oxygen concentrations and its effect on the rate of apparent photosynthesis. *Japanese Journal of Crop Science*, **48**, 470–4.

Andrews, T. J. & Lorimer, G. H. (1978). Photorespiration – still unavoidable? *FEBS Letters*, **90**, 1–9.

Bauer, H., Larcher, W. & Walker, R. B. (1975). Influence of temperature stress on $CO_2$-gas exchange. In *Photosynthesis and Productivity in Different Environments*, ed. J. P. Cooper, pp. 557–86. Cambridge University Press.

Begg, J. E. & Torssell, B. W. R. (1974). Diaphotonastic and parahelionastic leaf movements in *Stylosanthes humilis* H.B.K. (Townsville stylo). In *Mechanisms of Regulation of Plant Growth*, ed. R. L. Bieleski, A. R. Ferguson & M. M. Cresswell, pp. 277–83. Wellington: Royal Society of New Zealand.

Berry, J. A. & Björkman, O. (1980). Photosynthetic response and adaptation to temperature in higher plants. *Annual Review of Plant Physiology*, **31**, 491–543.

Berry, J. A. & Farquhar, G. D. (1977). The $CO_2$ concentrating function of $C_4$ photosynthesis, a biochemical model. In *Proceedings of the Fourth International Congress on Photosynthesis*, ed. D. O. Hall, J. Coombs & T. W. Goodwin, pp. 119–31. London: Biochemical Society.

Björkman, O. (1973). Comparative studies on photosynthesis in higher plants. In *Current Topics in Photobiology, Photochemistry and Photophysiology*, vol. 8, ed. A. C. Giese, pp. 1–63. New York: Academic Press.
Björkman, O. (1975). Environmental and biological control of photosynthesis: inaugural address. In *Environmental and Biological Control of Photosynthesis*, ed. R. Marcelle, pp. 1–16. The Hague: Dr W. Junk.
Björkman, O. (1981). Ecological adaptation of the photosynthetic apparatus. In *Proceedings of the Fifth International Congress on Photosynthesis, Photosynthesis VI, Photosynthesis and Productivity, Photosynthesis and Environment*, ed. G. Akoyunoglou, pp. 191–202. Philadelphia: Balaban Int. Sci. Services.
Björkman, O. & Ludlow, M. M. (1972). Characterisation of the light climate on the floor of a Queensland rainforest. *Yearbook of the Carnegie Institution of Washington*, **71**, 85–94.
Björkman, O., Troughton, J. H. & Nobs, M. (1974). Photosynthesis in relation to leaf structure. In *Basic Mechanisms in Plant Morphogenesis, Brookhaven Symposia in Biology*, No. 25, pp. 206–26. New York: Brookhaven National Laboratory.
Boardman, N. K. (1977). Comparative photosynthesis of sun and shade plants. *Annual Review of Plant Physiology*, **28**, 355–77.
Brandle, J. R., Campbell, W. F., Sisson, W. B. & Caldwell, M. M. (1977). Net photosynthesis, electron transport capacity and ultrastructure of *Pisum sativa* L. exposed to ultraviolet-B radiation. *Plant Physiology*, **60**, 165–9.
Bulley, N. R., Nelson, C. D. & Tregunna, E. B. (1969). Photosynthesis: action spectrum for leaves in normal and low oxygen. *Plant Physiology*, **44**, 678–84.
Burrows, F. J. & Milthorpe, F. L. (1976). Stomatal conductance in the control of gas exchange. In *Water Deficits in Plant Growth*, vol. 4, ed. T. T. Kozlowski, pp. 103–52. New York: Academic Press.
von Caemmerer, S. & Farquhar, G. D. (1981). Some relationships between the biochemistry of photosynthesis and the gas exchange of leaves. *Planta (Berlin)*, **153**, 376–87.
Caldwell, M. M. (1970a). Plant gas exchange at high wind speeds. *Plant Physiology*, **46**, 535–7.
Caldwell, M. M. (1970b). The effect of wind on stomatal aperture, photosynthesis, and transpiration of *Rhododendron ferrugineum* L. and *Pinus cembra* L. *Centralblatt für das Gasamte Forstwesen*, **87**, 193–201.
Caldwell, M. M. (1971). Solar UV irradiation and the growth and development of higher plants. In *Photophysiology*, vol. 6, ed. A. C. Giese, pp. 131–77. New York: Academic Press.
Caldwell, M. M. (1977). The effects of solar UV-B radiation (280–315 nm) on higher plants: implications of stratospheric ozone reduction. In *Research in Photobiology*, ed. A. Castellani, pp. 597–607. New York: Plenum.
Canvin, D. T. (1979). Photorespiration: comparison between $C_3$ and $C_4$ plants. In *Encyclopedia of Plant Physiology* (new series), vol. 6, *Photosynthesis II, Photosynthetic Carbon Metabolism and Related Processes*, ed. M. Gibbs & E. Latzko, pp. 368–96. Berlin: Springer-Verlag.
Chabot, B. F. & Chabot, J. F. (1977). Effects of light and temperature on leaf anatomy and photosynthesis in *Fragaria vesca*. *Oecologia (Berlin)*, **26**, 363–77.
Chabot, B. F., Jurik, T. W. & Chabot, J. F. (1979). Influence of instantaneous and integrated light-flux density on leaf anatomy and photosynthesis. *American Journal of Botany*, **66**, 940–5.
Charles-Edwards, D. A. (1981). *The Mathematics of Photosynthesis and Productivity*. London: Academic Press.
Charles-Edwards, D. A. & Ludwig, J. L. (1975). The basis of expression of leaf photosynthetic activities. In *Environmental and Biological Control of Photosynthesis*,

ed. R. Marcelle, pp. 37–44. The Hague: Dr W. Junk.
Chollet, R. & Ogren, W. L. (1975). Regulation of photorespiration in $C_3$ and $C_4$ species. *Botanical Review*, **41**, 137–79.
Cowan, I. R. & Milthorpe, F. L. (1968). Plant factors influencing the water status of plant tissues. In *Water Deficits and Plant Growth*, vol. 1, ed. T. T. Kozlowski, pp. 137–93. New York: Academic Press.
Crookston, R. K., O'Toole, J., Lee, R., Ozbun, J. L. & Wallace, D. H. (1974). Photosynthetic depression in beans after exposure to cold for one night. *Crop Science*, **14**, 457–64.
Day, W. & Parkinson, K. J. (1979). Importance to gas exchange of mass flow of air through leaves. *Plant Physiology*, **64**, 345–6.
van den Driessche, R., Connor, D. J. & Tunstall, B. R. (1971). Photosynthetic response of brigalow to irradiance, temperature and water potential. *Photosynthetica*, **5**, 210–17.
Ehleringer, J. R. & Björkman, O. (1978). A comparison of photosynthetic characteristics of *Encelia* species possessing glabrous and pubescent leaves. *Plant Physiology*, **62**, 185–90.
Ehleringer, J. R. & Forseth, I. (1980). Solar tracking by plants. *Science*, **210**, 1094–8.
Farquhar, G. D. & von Caemmerer, S. (1981). Modelling of photosynthetic response to environmental conditions. In *Encyclopedia of Plant Physiology* (new series), vol. 12B, *Physiological Plant Ecology: Water Relations and Photosynthetic Productivity*, ed. O. L. Lange, P. S. Nobel, C. B. Osmond & H. Zeigler. Berlin: Springer-Verlag. (In press.)
Farquhar, G. D., von Caemmerer, S. & Berry, J. A. (1980). A biochemical model of photosynthetic $CO_2$ assimilation in leaves of $C_3$ species. *Planta (Berlin)*, **149**, 78–90.
Gifford, R. M. (1971). The light response of $CO_2$ exchange: on the source of differences between $C_3$ and $C_4$ species. In *Photosynthesis and Photorespiration*, ed. M. D. Hatch, C. B. Osmond & R. O. Slatyer, pp. 51–6. New York: Wiley.
Grace, J. (1975). Wind damage to vegetation. *Current Advances in Plant Science*, **6**, 883–94.
Grace, J. (1977). *Plant Response to Wind*. London: Academic Press.
Grace, J. & Russell, G. (1977). The effect of wind on grasses. III. Influence of continuous drought or wind on anatomy and water relations in *Festuca arundinacea* Schreib. *Journal of Experimental Botany*, **28**, 268–78.
Grace, J. & Thompson, J. R. (1973). The after-effect of wind on the photosynthesis and transpiration of *Festuca arundinacea*. *Physiologia Plantarum*, **28**, 541–7.
Graham, D. & Chapman, E. A. (1979). Interactions between photosynthesis and respiration in higher plants. In *Encyclopedia of Plant Physiology* (new series), vol. 6, *Photosynthesis II, Photosynthetic Carbon Metabolism and Related Processes*, ed. M. Gibbs & E. Latzko, pp. 150–62. Berlin: Springer-Verlag.
Hall, A. E., Schulze, E.-D. & Lange, O. L. (1976). Current perspectives of steady-state stomatal responses to environment. In *Water and Plant Life, Problems and Modern Approaches*, ed. O. L. Lange, L. Kappen & E.-D. Schulze, pp. 169–88. Berlin: Springer-Verlag.
Harvey, G. W. (1980). Photosynthetic performance of isolated leaf cells from sun and shade plants. *Yearbook of the Carnegie Institution of Washington*, **79**, 160–4.
Hatch, M. D. & Osmond, C. B. (1976). Compartmentation and transport in $C_4$ photosynthesis. In *Encyclopedia of Plant Physiology* (new series), vol. 3, ed. C. R. Stocking & U. Heber, pp. 143–84. Berlin: Springer-Verlag.
Hesketh, J. D. (1967). Enhancement of photosynthetic $CO_2$ assimilation in the absence of oxygen, as dependent upon species and temperature. *Planta (Berlin)*, **76**, 371–4.
Hofstra, G. & Hesketh, J. D. (1969). Effects of temperature on the gas exchange of leaves in the light and dark. *Planta (Berlin)*, **85**, 228–37.
Holmgren, P. & Jarvis, P. G. (1967). Carbon dioxide efflux from leaves in light and darkness. *Physiologia Plantarum*, **20**, 1045–51.

Jackson, W. A. & Volk, R. J. (1970). Photorespiration. *Annual Review of Plant Physiology*, **21**, 385–432.
Jarvis, P. G. (1980). Stomatal response to water stress in conifers. In *Adaptation of Plants to Water and High Temperature Stress*, ed. N. C. Turner & P. J. Kramer, pp. 105–22. New York: Wiley.
Jarvis, P. G. & Morison, J. I. L. (1981). The control of transpiration and photosynthesis by the stomata. In *Stomatal Physiology*, ed. P. G. Jarvis & T. A. Mansfield, pp. 247–79. Cambridge University Press.
Kramer, P. J. (1980). Drought, stress and the origin of adaptations. In *Adaptation of Plants to Water and High Temperature Stress*, ed. N. C. Turner & P. J. Kramer, pp. 7–20. New York: Wiley.
Kriedemann, P. E., Torokfalvy, E. & Smart, R. E. (1973). Natural occurrence and photosynthetic utilisation of sunflecks by grapevine leaves. *Photosynthetica*, **7**, 18–27.
Krizek, D. T. & Milthorpe, F. L. (1966). Effect of photoperiodic induction on the transpiration rate and stomatal behaviour of debudded *Xanthium pennsylvanicum* plants. *Journal of Experimental Botany*, **24**, 76–86.
Lake, J. V. (1967). Respiration of leaves during photosynthesis. II. Effects on the estimation of mesophyll resistance. *Australian Journal of Biological Sciences*, **20**, 495–9.
Lange, O. L., Schulze, E.-D., Evenari, M., Kappen, L. & Buschbom, U. (1974). The temperature-related photosynthetic capacity of plants under desert conditions. I. Seasonal changes of the photosynthetic response to temperature. *Oecologia (Berlin)*, **17**, 97–110.
Larcher, W. (1980). *Physiological Plant Ecology*, 2nd edition. Berlin: Springer-Verlag.
Ludlow, M. M. (1970). Effect of oxygen concentration on leaf photosynthesis and resistances to carbon dioxide diffusion. *Planta (Berlin)*, **91**, 285–90.
Ludlow, M. M. (1976). Ecophysiology of $C_4$ grasses. In *Water and Plant Life, Problems and Modern Approaches*, ed. O. L. Lange, L. Kappen & E.-D. Schulze, pp. 364–86. Berlin:Springer-Verlag.
Ludlow, M. M. (1980). Adaptive significance of stomatal responses to water stress. In *Adaptation of Plants to Water and High Temperature Stress*, ed. N. C. Turner & P. J. Kramer, pp. 123–38. New York: Wiley.
Ludlow, M. M. & Jarvis, P. G. (1971a). Methods for measuring photorespiration on leaves. In *Plant Photosynthetic Production: Manual of Methods*, ed. Z. Sestak, J. Catsky & P. G. Jarvis, pp. 294–315. The Hague: Dr W. Junk.
Ludlow, M. M. & Jarvis, P. G. (1971b). Photosynthesis in sitka spruce (*Picea sitchensis* (Bong.) Carr.). 1. General characteristics. *Journal of Applied Ecology*, **8**, 925–53.
Ludlow, M. M. & Ng, T. T. (1976). Photosynthetic light response curves of leaves from controlled environment facilities, glasshouses or outdoors. *Photosynthetica*, **10**, 457–62.
Ludlow, M. M. & Taylor, A. O. (1974). Effect of sub-zero temperatures on the gas exchange of buffel grass. In *Mechanisms of Regulation of Plant Growth*, ed. R. L. Bieleski, A. R. Ferguson & M. M. Cresswell, pp. 513–18. Wellington: Royal Society of New Zealand.
Ludlow, M. M. & Wilson, G. L. (1971a). Photosynthesis of tropical pasture plants. 1. Illuminance, carbon dioxide concentration, leaf temperature and leaf-air vapour pressure difference. *Australian Journal of Biological Sciences*, **24**, 449–70.
Ludlow, M. M. & Wilson, G. L. (1971b). Photosynthesis of tropical pasture plants. 2. Temperature and illuminance history. *Australian Journal of Biological Sciences*, **24**, 1065–75.
McCree, K. J. (1971). The action spectrum, absorptance and quantum yield of photosynthesis in crop plants. *Agricultural Meteorology*, **9**, 191–216.
McCree, K. J. (1972). Significance of enhancement for calculations based on the action spectrum for photosynthesis. *Plant Physiology*, **49**, 704–6.
Maier-Maercker, U. (1979). Peristomatal transpiration and stomatal movement: a

controversial view. II. Observation of stomatal movements under different conditions of water supply and demand. *Zeitschrift für Pflanzenphysiologie*, **91**, 157–72.

Malkin, S. & Fork, D. C. (1981). Photosynthetic units of sun and shade plants. *Plant Physiology*, **67**, 580–3.

Meidner, H. (1970). Effects of photoperiodic induction and debudding in *Xanthium pennsylvanicum* and of partial defoliation in *Phaseolus vulgaris* on rates of net photosynthesis and stomatal conductances. *Journal of Experimental Botany*, **21**, 164–9.

Mooney, H. A., Ehleringer, J. & Berry, J. A. (1976). High photosynthetic capacity of a winter annual in Death Valley. *Science*, **194**, 322–4.

Moss, D. N. (1964). Optimum lighting of leaves. *Crop Science*, **4**, 131–6.

Moss, D. N. (1965). Capture of radiant energy by plants. *Meteorological Monographs*, **6**, 90–108.

Muchow, R. C., Fisher, M. J., Ludlow, M. M. & Myers, R. J. K. (1980b). Stomatal behaviour of kenaf and sorghum in a semiarid tropical environment. II. During the day. *Australian Journal of Plant Physiology*, **7**, 621–8.

Muchow, R. C., Ludlow, M. M., Fisher, M. J. & Myers, R. J. K. (1980a). Stomatal behaviour of kenaf and sorghum in a semiarid tropical environment. I. During the night. *Australian Journal of Plant Physiology*, **7**, 609–19.

Neilson, R. E. & Jarvis, P. G. (1975). Photosynthesis in sitka spruce (*Picea sitchensis* (Bong.) Carr.). VI. Response of stomata to temperature. *Journal of Applied Ecology*, **12**, 879–81.

Nobel, P. S. (1977). Internal leaf area and cellular $CO_2$ resistance: photosynthetic implications of variations with growth conditions and plant species. *Physiologia Plantarum*, **40**, 137–44.

Nobel, P. S. (1980). Leaf anatomy and water use efficiency. In *Adaptation of Plants to Water and High Temperature Stress*, ed. N. C. Turner & P. J. Kramer, pp. 43–55. New York: Wiley.

Nobel, P. S. & Hartsock, T. L. (1981). Development of leaf thickness for *Plectranthus parviflorus* – influence of photosynthetically active radiation. *Physiologia Plantarum*, **51**, 163–6.

Osmond, C. B. (1978). Crassulacean acid metabolism: a curiosity in context. *Annual Review of Plant Physiology*, **29**, 379–414.

Osmond, C. B., Björkman, O. & Anderson, D. J. (1980). *Physiological Processes in Plant Ecology*. Ecological Studies 36. Berlin: Springer-Verlag.

Osonubi, O. & Davies, W. J. (1980). The influence of plant water stress on stomatal control of gas exchange at different levels of atmospheric humidity. *Oecologia (Berlin)*, **46**, 1–6.

Patterson, D. T. (1980). Light and temperature adaptation. In *Predicting Photosynthesis for Ecosystem Models*, vol. 2, ed. J. D. Hesketh & J. W. Jones, pp. 205–35. Boca Raton: CRC Press.

Pearcy, R. W. (1977). Acclimation of photosynthetic and respiratory carbon dioxide exchange to growth temperature in *Atriplex lentiformis* (Torr.) Wats. *Plant Physiology*, **59**, 795–9.

Prioul, J. L. & Chartier, P. (1979). Partitioning of transfer and carboxylation components of intracellular resistance to photosynthetic $CO_2$ fixation: a critical analysis of the methods used. *Annals of Botany*, **41**, 789–800.

Raschke, K. (1975). Stomatal action. *Annual Review of Plant Physiology*, **26**, 309–40.

Rawson, H. M., Begg, J. E. & Woodward, R. G. (1977). The effect of atmospheric humidity on photosynthesis, transpiration and water use efficiency of leaves of several plant species. *Planta (Berlin)*, **134**, 5–10.

Robberecht, R. & Caldwell, M. M. (1978). Leaf epidermal transmittance of ultraviolet radiation and its implications for plant sensitivity to ultraviolet-radiation induced injury. *Oecologia (Berlin)*, **32**, 277–87.

Russell, G. & Grace, J. (1978a). The effect of wind on grasses. IV. Some influences of drought or wind on *Lolium perenne*. *Journal of Experimental Botany*, **29**, 245–55.

Russell, G. & Grace, J. (1978b). The effect of wind on grasses. V. Leaf extension, diffusive conductance, and photosynthesis in the wind tunnel. *Journal of Experimental Botany*, **29**, 1249–58.

Sagar, J. C. & Giger, W. (1980). Re-evaluation of published data on the relative photosynthetic efficiency of intermittent and continuous light. *Agricultural Meteorology*, **22**, 289–302.

Schulze, E.-D. & Küppers, M. (1979). Short-term and long-term effects of plant water deficits on stomatal responses to humidity in *Corylus avellana* L. *Planta (Berlin)*, **146**, 319–26.

Schulze, E.-D., Lange, O., Buschbom, U., Kappen, L. & Evenari, M. (1972). Stomatal responses to changes in humidity in plants growing in the desert. *Planta (Berlin)*, **108**, 259–70.

Schulze, E.-D., Lange, O. L., Kappen, L., Evenari, M. & Buschbom, U. (1975). The role of air humidity and leaf temperature in controlling stomatal resistance of *Prunus armeniaca* L. under desert conditions. II. The significance of leaf water status and internal carbon dioxide concentration. *Oecologia (Berlin)*, **18**, 219–33.

Sheriff, D. W. (1977). The effect of humidity on water uptake by, and viscous flow resistance of, excised leaves of a number of species: physiological and anatomical observations. *Journal of Experimental Botany*, **28**, 1399–1407.

Sheriff, D. W. (1979). Stomatal aperture and the sensing of the environment by guard cells. *Plant, Cell and Environment*, **2**, 15–22.

Sisson, W. B. & Caldwell, M. M. (1976). Photosynthesis, dark respiration, and growth of *Rumex patienta* L. exposed to ultraviolet irradiance (288–315 nanometers) simulating a reduced atmospheric ozone column. *Plant Physiology*, **58**, 563–8.

Syvertsen, J. P. & Cunningham, G. L. (1979). The effects of irradiating adaxial and abaxial leaf surface on the rate of net photosynthesis of *Perezia nana* and *Helianthus annuus*. *Photosynthetica*, **13**, 287–93.

Tenhunen, J. D., Weber, J. A., Yocum, C. S. & Gates, D. M. (1979). Solubility of gases and the temperature dependency of whole leaf affinities for carbon dioxide and oxygen, an alternative perspective. *Plant Physiology*, **63**, 916–23.

Tenhunen, J. D. & Westrin, S. S. (1979). Development of a photosynthesis model with an emphasis on ecological applications. IV. Wholephot – whole leaf photosynthesis in response to four independent variables. *Oecologia (Berlin)*, **41**, 145–62.

Turner, N. C. & Kramer, P. J. (Eds) (1980). *Adaptation of Plants to Water and High Temperature Stress*. New York: Wiley.

Van, T. K., Garrard, L. A. & West, S. H. (1976). Effects of UV-B radiation on net photosynthesis of some crop plants. *Crop Science*, **16**, 715–18.

Warritt, B., Landsberg, J. J. & Thorpe, M. R. (1980). Responses of apple leaf stomata and environmental factors. *Plant, Cell and Environment*, **3**, 13–22.

Williams, G. J. (1974). Photosynthetic adaptation to temperature in $C_3$ and $C_4$ grasses, a possible role in the shortgrass prairie. *Plant Physiology*, **54**, 709–11.

Williams, G. J. & Kemp, P. R. (1976). Temperature relations of photosynthetic response in populations of *Verbascum thapsus* L. *Oecologia (Berlin)*, **25**, 47–54.

Wilson, G. L. & Ludlow, M. M. (1968). Bean leaf expansion in relation to temperature. *Journal of Experimental Botany*, **19**, 309–21.

Wong, S. C., Cowan, I. R. & Farquhar, G. D. (1979). Stomatal conductance correlates with photosynthetic capacity. *Nature*, **282**, 424–6.

Yabuki, K. & Miyagawa, H. (1970). Studies on the effect of wind speed upon photosynthesis. 2. The relation between wind and photosynthesis. *Journal of Agricultural Meterology*, **26**, 21–6.

# 13

## Internal factors influencing photosynthesis and respiration

J. D. HESKETH, E. M. LARSON, A. J. GORDON AND D. B. PETERS

General issues of photosynthesis and respiration discussed in Chapters 10 and 12 serve as background to this chapter. Here we will be principally concerned with attempting to describe these processes at the cell-chloroplast-molecular level, to draw together current concepts in a simplified, possibly controversial, quantitative conceptual model, along the lines given in the lucid summary by Nobel (1974), and then to identify possible potential limiting factors, such as might occur in natural ecosystems.

### The capture and conversion of light energy

#### *The photosynthetic unit*

In a typical leaf, light is captured and converted to excited electrons by some 400 'antenna' chlorophyll molecules of a photosynthetic unit. The electromagnetic energy of these excited electrons is transferred among antenna chlorophylls to two reaction-centre 'trap' chlorophylls: $P_{700}$ (for photosystem I) and $P_{680}$ (for photosystem II). ($P_{680}$ is designated sometimes as $P_{690}$). The reaction-centre molecules, with excited electrons, participate in a series of oxidation-reduction reactions, which result in the flow of electrons through numerous intermediates of the photosystems. Such photosystem intermediates in series are collectively called the Z-scheme (Govindjee & Govindjee, 1975; Golbeck, Lien & San Pietro, 1977). The energy of an excited Z-scheme electron is used to reduce $NADP^+$ to NADPH in the stroma and to transport protons from the stroma across the thylakoid membrane into the thylakoid space. The various Z-scheme intermediates are embedded in the thylakoid membrane, but some are quite mobile within the membrane. The proton gradient across the thylakoid membrane drives another set of reactions through a complex of enzymes embedded and attached to the membrane, resulting in the production of ATP from ADP and inorganic P in the stroma. The proton gradient involves an electrical charge and a chemical activity component (or pH gradient), both of which drive the ADP/ATP reactions through a chemi-osmotic mechanism, first proposed by Mitchell (cf. Trebst & Avron, 1977 and chapters therein; Hinkle & McCarty, 1978; Mitchell, 1979).

Forty-five to 70% of the insoluble leaf protein contains thylakoid membranes complexed to chlorophylls, as well as components of nuclei, mitochondria and endoplasmic reticulum (Wildman & Kwanyuen, 1978; Morita, 1980). Subjecting extracts by anionic detergents of the insoluble-protein fraction to polyacrylamide gel electrophoresis resulted in two chlorophyll-protein groups as two separate bands on a gel column (Thornber, 1975; Thornber & Alberte, 1977). A third, 'free pigment', band in the gel column was postulated to have been complexed originally to two other proteins.

Measurements of chlorophyll a and b, as well as of $P_{700}$ found in polyacrylamide gel bands, from many plant species (as well as chlorophyll-deficient mutants) grown under different conditions, have led to an oversimplified conceptual model of the chlorophyll composition of a photosynthetic unit (Fig. 13.1). In deriving and using the model, it has been assumed that the ratio of $P_{700}$ to photosynthetic units is 1 : 1. One band in the gel typically contained one $P_{700}$ chlorophyll molecule to 40 chlorophyll a molecules, and was designated the '*$P_{700}$-chlorophyll a-protein*' complex (Thornber, 1975; Thornber *et al.*, 1977; Thornber & Alberte, 1977). This band contained about 10% of the leaf chlorophyll (40/400 chlorophyll molecules, with 400 molecules in a photosynthetic unit of the hypothetical 'typical' leaf). A second band, containing about 55% (220/400 molecules) of the leaf chlorophyll, was composed of equal quantities of chlorophyll a and b, and was named the *'light-harvesting chlorophyll a/b protein'* complex (Thornber & Highkin, 1974). A third free pigment band on the column, composed of 35% (140/400 molecules) of the leaf chlorophyll, contained only chlorophyll a. 10% of the total leaf chlorophyll (40/400 molecules), which ended up in the free pigment band, was postulated to be part of a *'reaction-centre II chlorophyll-protein'* complex, with one $P_{680}$ to 40 molecules like the $P_{700}$-chlorophyll a-protein complex. The remaining leaf chlorophyll found in the free pigment band (100/400 molecules) was thought to be associated with a '*photosystem I light-harvesting chlorophyll-protein*' complex (Thornber *et al.*, 1977).

A chlorophyll b-deficient pea mutant, with a leaf chlorophyll a/b ratio of 10–18 and containing one-third of the amount of chlorophyll of a normal leaf, played a role in the development of the model. This mutant was capable of 73% of the $CO_2$ exchange rate of a normal pea leaf at about 20 W $m^{-2}$ and 93% of that of a normal leaf at about 1000 W $m^{-2}$ (Highkin, Boardman & Goodchild, 1969). Such mutant/wildtype comparisons illustrate the role of the a/b accessory chlorophylls in collecting light energy under weak and intense light levels.

Analyses of chlorophyll-protein complexes and $P_{700}$ have contributed significantly to ecological studies of light-temperature acclimation and adaptation, the greening of etiolated leaves, water stress, mineral nutrition, the $C_4:C_3$ phenomenon, chlorophyll-deficient mutants, and various aspects of genotype variation (cf. Chapters 10 and 12). Diverse results from many such studies can easily be interpreted using the photosynthetic unit model described, proving at least its utility. However, one recent exception (Malkin, Armond, Mooney & Fork, 1981) challenges the 1:1 ratio for photosystem I and II reaction centres in some species.

Fig. 13.1. Model of the chlorophyll-protein composition of a photosynthetic unit. Approximate relative proportions are indicated in brackets but these vary with species and growing conditions. (Adapted from Thornber *et al.*, 1977.)

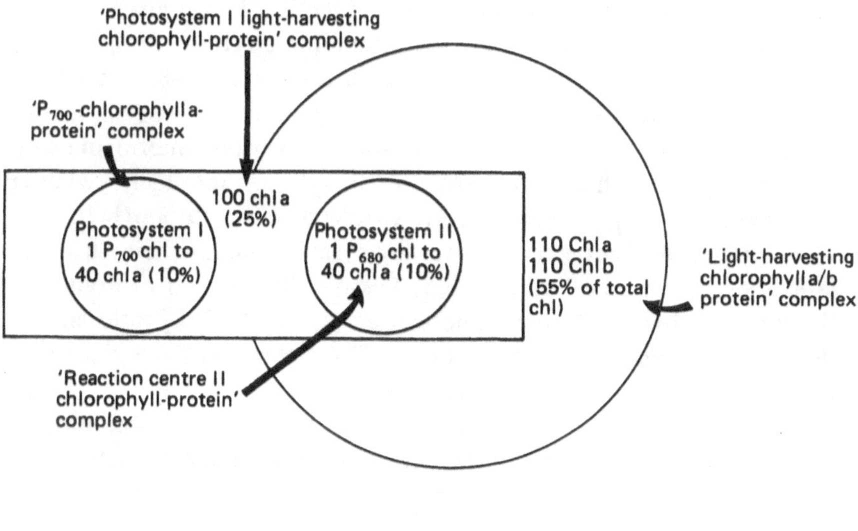

### *Membrane translocators*

The trans-membrane fluxes of inorganic phosphate, triosephosphates, dicarboxylates, and ATP/ADP influence photosynthetic rates. We list in Table 13.1 some of the translocators ('carriers') proposed for the light and dark reactions. With the exception of the Z-scheme which is a highly complex system, it is assumed that the flux of each metabolite follows Michaelis-Menten kinetics, with the restriction that something usually exchanges for something else across the membrane. Differential pH levels set up by the various proton fluxes can affect the $CO_2:HCO_3^-$ ratio and, secondarily, the $CO_2$ diffusion rates. Values of $V_{max}$ and $K_m$ for the various translocators are being currently estimated (Heber & Heldt, 1981). Their major importance in our context is in the control of translocation of fixed carbon out of the cell compared with its storage as starch in the chloroplast, in $C_4$-acid transport in $C_4$ plants, and in the transport of metabolites between chloroplasts, peroxisomes and mitochondria in photorespiring cells.

### *ATP and NADPH production rates*

The rate at which a photosynthetic unit can process electrons to produce ATP and NADPH has been measured from $0_2$ evolution rates of broken chloroplasts and the number of $P_{700}$ chlorophyll molecules (or photosynthetic units) in the reaction mixture. Four electrons are assumed to pass through the photosynthetic unit for every $0_2$ evolved and such a system must be free of respiratory $0_2$ consumption. Half-times for dark reduction of components of the Z-scheme, after exposure to saturating light, are also used for estimating electron transport rates by a complex kinetic analysis (Whitmarsh & Cramer, 1980). Z-scheme components prior to the rate-limiting step, which seems to involve electron transfer from plastoquinone through various intermediates to $P_{700}$, are in a reduced (electron-rich) state in the light, whereas components after plastoquinone are oxidized (electron-poor). In other words, in strong light electrons flowing through the Z-scheme accumulate prior to the rate-limiting step. Passive diffusion of protons back through the thylakoid membrane will affect the $ATP:0_2$ ratio and the quantum efficiency. Moreover, the ATP/ADP ratio affects dark reaction kinetics of ATP-mediated reactions.

Estimates of the rates of production and consumption may be attempted. A soybean leaf in the field has 0.6 g chlorophyll per $m^2$ leaf

Table 13.1. *Translocators in photosynthetic cell membranes*

| Membrane | Translocator and metabolic behaviour | Reviewer |
|---|---|---|
| Thylakoid | *Z-scheme*: for 8 photons and 4 electrons processed and each $O_2$ evolved, 6 protons are taken up outside the thylakoid membrane and 8 appear inside. 2 NADP are reduced to NADPH, yielding $2H^+$. The 8 protons produced can yield 2.67 ATP | Hinkle & McCarty (1978) |
| | *ADP/ATP proton*: 3 protons move back into the stroma for every ATP produced from ADP in the stroma | |
| Inner chloroplastic | *Phosphate*: 1:1 bidirection exchange of divalent $P_i$, 3-phosphoglyceric acid (3-PGA) and dihydroxyacetone phosphate (DHAP). The ratio of divalent to trivalent 3-PGA is pH dependent at pH 7 to 8 | Walker (1976), Heber & Heldt (1981) |
| | *Dicarboxylate*: involving L-malate, oxaloacetate, α-ketoglutarate, L-aspartate and L-glutamate | Heber & Heldt (1981) |
| | *ATP/ADP*: slow | Heber & Heldt (1981) |
| Inner mitochondrial | *Respiratory chain*: for 2 NADH oxidized and 1 $O_2$ evolved, 12 protons are transferred from inside to outside the membrane | Hinkle & McCarty (1978) |
| | *ATP/ADP proton*: 1 ATP from ADP inside the membrane for 2 protons transferred to the inner side of the membrane | |
| | *ATP/ADP*, $P_i/OH^-$, *Succinate*/$P_i$, $Na^+/H^+$, $Ca^{2+}$, *NADPH/NADP proton*: 1 NADPH for 2 protons transferred | Tolbert (1979) |
| Peroxisomes | *Dicarboxylate*: see above | Huber & Moreland (1981) |
| Plasmalemma | *ADP/ATP*, $K^+$, *sucrose*: also in phloem parenchyma plasmalemma | Heber & Heldt (1981) |

$CO_2$, $NH_3$, serine, glycine, glycolate, glyoxylate, and some amino acids are believed to move by diffusion.

area and approximately 400 chlorophyll molecules per Z-scheme or photosynthetic unit (Bunce, Patterson, Peet & Alberte, 1977; Hesketh, Ogren, Hageman & Peters, 1981) or $10^{18}$ photosynthetic units $m^{-2}$ [(0.6 g $\times$ 6.02 $\times$ $10^{23}$ molecules $mol^{-1}$)/ (893.5 g $mol^{-1}$ $\times$ 400 molecules per photosynthetic unit)]. Given 10 ms to process an electron through a Z-scheme of a photosynthetic unit (Nobel, 1974) and one ATP and one NADPH generated per two electrons processed, we get 5 $\times$ $10^{21}$ molecules each of ATP and NADPH $m^{-2}$ $s^{-1}$ formed in saturating light. A typical photorespiring leaf requires 4 ATP and 2.9 NADPH for every $CO_2$ molecule fixed (Ogren, 1977). For a very high net flux density, $F'$ = 50 $\mu$mole $CO_2$ $m^{-2}$ $s^{-1}$, 3 $\times$ $10^{19}$ molecules $CO_2$ $m^{-2}$ $s^{-1}$ (5 $\times$ $10^{-5}$ mol $m^{-2}$ $s^{-1}$ $\times$ 6.02 $\times$ $10^{23}$ molecules $mol^{-1}$) are fixed, requiring 1.2 $\times$ $10^{20}$ ATP and 9 $\times$ $10^{19}$ NADPH molecules $m^{-2}$ $s^{-1}$. That is, less than 2% of the ATP and NADPH generated is used in net $CO_2$ fixation.

### *Quantum yield and efficiency*

Radmer & Kok (1977) reviewed the basic principles associated with measuring light energy which can be used for photosynthesis and the Warburg-Emerson controversy; current evidence supports Emerson's measurements of 8–10 photons per $O_2$ molecule evolved.

The number of protons generated inside the thylakoid membrane per quantum absorbed or electron transferred through the Z-scheme and the ratio of protons/ATP produced (Table 13.1), the ATP-NADPH requirement for $CO_2$ fixation, the amount of photo- and dark-respiration and the associated ATP-NADPH cost, $NO_2$ reduction, and other energy-requiring steps also influence the observed quantum yield. Light quality and the photosynthetically usable energy per quantum, together with the numbers, sizes, and distribution of photosynthetic units and efficiency of capture of photons by 'antennae' chlorophyll molecules and the transmission of electrons to the reaction centres within the photosynthetic unit also affect quantum efficiency in a leaf. Leaf and chloroplast morphology control numbers and distribution of photosynthetic units.

### *The reduction of carbon dioxide*

Gaastra (1959) presented a major advance in understanding and quantifying photosynthetic $CO_2$-flux in and out of a leaf. $CO_2$-flux theory based upon an 'electrical resistance analogue' had existed for some time (Rabinowitch, 1951), but Gaastra developed experimental methods for estimating some of the major resistances involved. Since 1959, considerable experimental evidence has accumulated about such resistances (Šesták, Čatský & Jarvis, 1971; Burrows & Milthorpe, 1976; Korner, Scheel & Bauer, 1979; Pospíšilova & Solárová, 1980). Nobel (1974) gives a lucid discussion and additional information can be found in Rabinowitch (1951).

In Tables 13.2 to 13.5 we use this approach (cf. Figs. 10.3 and 12.1) to attempt to estimate $CO_2$ concentrations inside a leaf exposed to full sunlight under optimal conditions, particularly in the stroma near the fixation sites; this analysis should then apply to maximum rates of photosynthesis. Conductances were estimated from the pathlength, $\Delta x_i$, increased by a tortuosity factor, the actual cross-sectional area being traversed, $S$, and the diffusivity of $CO_2$, $D_i$, in the medium (i) involved. The ratio of cell wall area exposed to air to the projected leaf area is about 30:1 (Longstreth & Nobel, 1980). About 50% of the volume of the cell wall is void, with an estimated tortuosity factor of 1.3; hence, $K$, the fraction of the wall through which $CO_2$ can diffuse, equals 0.35. Other values can be found in Nobel (1974), with the exception of those for membranes, which we took from Guthnecht, Bisson & Tosteson (1977). To estimate the reduction in $CO_2$ concentration along a pathway, the $CO_2$ flux is divided by the conductance involved (Table 13.2). We recommend the reader make similar calculations, after challenging all our assumptions. Once respiration is introduced into the model, the situation becomes more complicated (see Nobel, 1974, for possible solutions to such a problem). However, the crudeness of the model is such that we can ignore such complications in the simple exercise undertaken here.

The $CO_2$ concentration, $C_{CO_2}$, and light intensity, $Q$, at the reaction site, and the $K_m$ (stroma $C_{CO_2}$ or $Q$ at $F' = F'_{max}/2$, $F'_{max} = F'$ as stroma $C_{CO_2}$ or $Q \rightarrow \infty$) of the dark and light reactions, also control the $CO_2$ flux rate, as discussed below. As the $CO_2$ concentrations estimated in Tables 13.2 and 13.3 approach zero, the assumed flux becomes impossible for the flux conductances considered.

$CO_2$ may diffuse in the liquid phase as $CO_2$, $HCO_3^-$ or $H_2CO_3$, each of which diffuse independently of each other along their respective gradients, to complicate the resistance analogue. Carbonic anhydrase enhances the rate of the $CO_2$:$HCO_3^-$ interchange. The air:water partition coefficient is controlled by temperature. The conductance of membranes for $CO_2$ flux is much greater than that for $HCO_3^-$. pH

Table 13.2. *Assumptions and equations used in estimating values shown in Tables 13.3–13.5*

| Assumptions | Symbol | Value-units[a] |
|---|---|---|
| Diffusion surface to leaf area ratio: (cell wall, plasmalemma, cytoplasm, chloroplast) | $S$ | 20, 10–40 |
| Void volume: | $vv_i$ | |
| cell wall (0.5 void × 1.3 tortuosity) | | 0.66 of total |
| cytoplasm, chloroplast, stromal | | 0.33 of total |
| Pathlengths for $CO_2$ flux: | $\Delta x_i$ | |
| boundary layer | $\Delta x_a$ | $1 \times 10^{-3}$ m |
| intercellular air space | $\Delta x_c$ | $5 \times 10^{-5}$ m |
| cell wall | $\Delta x_{cw}$ | $5, 10-15 \times 10^{-7}$ m |
| cytoplasmic | $\Delta x_n$ | $2 \times 10^{-7}$ m |
| inner membrane space of chloroplast | $\Delta x_\mu$ | $1 \times 10^{-7}$ m |
| stromal | $\Delta x_t$ | $3 \times 10^{-6}$ m |
| $CO_2$ diffusivity: | $D_i$ | |
| in air | $D_{air}$ | $1.6 \times 10^{-5}$ m$^2$s$^{-1}$ |
| in water at 10 °C | $D_{water}$ | $1.17 \times 10^{-9}$ m$^2$s$^{-1}$ |
| 15 °C | | $1.37 \times 10^{-9}$ m$^2$s$^{-1}$ |
| 20 °C | | $1.70 \times 10^{-9}$ m$^2$s$^{-1}$ |
| 25 °C | | $1.82 \times 10^{-9}$ m$^2$s$^{-1}$ |
| 30 °C | | $2.06 \times 10^{-9}$ m$^2$s$^{-1}$ |
| 35 °C | | $2.26 \times 10^{-9}$ m$^2$s$^{-1}$ |
| 40 °C | | $2.75 \times 10^{-9}$ m$^2$s$^{-1}$ |
| in membranes | $D_{mem}$ | $1.15 \times 10^{-11}$ m$^2$s$^{-1}$ |
| Water/air $CO_2$ partition coefficient: (Bunsen coefficient for solubility of $CO_2$ at a specified temperature) | $\alpha_i$ | |
| 10 °C | $\alpha_{10}$ | 1.19 |
| 15 °C | $\alpha_{15}$ | 1.02 |
| 20 °C | $\alpha_{20}$ | 0.88 |
| 25 °C | $\alpha_{25}$ | 0.76 |
| 30 °C | $\alpha_{30}$ | 0.67 |
| 35 °C | $\alpha_{35}$ | 0.59 |
| 40 °C | $\alpha_{40}$ | 0.53 |
| Conductances of $CO_2$ through: | | |
| the membrane | $g'_{mem}$ | [a]$3, 1-2 \times 10^{-3}$ m$^{-2}$s$^{-1}$ |
| air (0.64 $g'_l$) | $g'_l$ | [a]$1.03, 0.33-0.5 \times 10^{-3}$ m$^{-2}$s$^{-1}$ |
| Photosynthetic $CO_2$ exchange rate: | $F'$ | [a]30, 10–50 $\mu$mol m$^{-2}$s$^{-1}$ |
| Equations: | $K_i^{CO_2} = 1 - vv_i$ | |
| | $g'_i = (SK_i^{CO_2}D_i)/\Delta x_i$ | |
| | $(\Delta CO_2)_i = F'/g'_i$ | |

[a] Value before comma used in Table 13.3, values after comma are manipulations used in Tables 13.4 and 13.5.

controls the ratio of $HCO_3^-$ to $CO_2$, which can be considerable (at pH = 8.2, $HCO_3^-/CO_2 = 76$). RuBP carboxylase reacts with $CO_2$; PEP carboxylase seems to react with $HCO_3^-$.

$CO_2$ flux ceases in $C_4$ plants once $CO_2$ is fixed through PEP carboxylase into a $C_4$-acid in the cytoplasm of the mesophyll cells. Because of the 'concentrating' effect of the $C_4$ transport system, $CO_2$ may even diffuse outwards from the bundle sheath cells to the mesophyll cells (Hatch & Osmond, 1976). The rates of $C_4$-flux and the kinetics of $CO_2$ fixation and release then become important in limiting the leaf $CO_2$-exchange rate (see below: the $C_4$:$C_3$ Phenomenon, Photorespiration, and Associated Limiting Factors). Transport of $CO_2$ across membranes is critical; an active transport mechanism would greatly increase the conductance but involves the expenditure of metabolic energy.

### *The carboxylating enzyme*

Aspects of ribulose bisphosphate carboxylase/oxygenase (RuBP carboxylase/oxygenase) behaviour were reviewed in many chapters in Siegelman & Hind (1978); that by Lane and Miziorko on the enzyme chemistry is particularly relevant. Because of its oxygenase behaviour and its effect on photosynthetic rates, this enzyme is being studied in great detail. Two groups (Bedbrook, Smith & Ellis, 1980; McIntosh, Poulson & Bogorad, 1980) have sequenced the DNA for the large (chloroplast DNA) and small (nuclear DNA) subunits of the enzyme (see also Chapter 10). Zima & Šesták (1979) have surveyed the literature concerned with activities of these and other enzymes during leaf ontogeny.

In Table 13.3, estimates of $CO_2$ concentrations at the photosynthetic site or at some point 'upstream' in the diffusion network are presented; these indicate that with an ambient concentration of 330 p.p.m., the concentration at the fixation site would be 99 p.p.m. (4 $\mu$M). Reaction kinetics, accounting for the competition between $O_2$ and $CO_2$ for the fixation site on the carboxylation enzyme, give the following (Laing, Ogren & Hageman, 1974):

$$F' = F'_{max}\, C_{CO_2}\, [K_m + C_{CO_2}\, (1 + C_{O_2}/K_o)]^{-1}$$

where $F'_{max} = F'$ as $C_{CO_2} \rightarrow \infty$, $C_{O_2}$ = the oxygen concentration at the reaction site and $K_m$ and $K_o$ are the Michaelis-Menten constants for carboxylase and oxygenase activity, respectively. We may assume that

Table 13.3. *Estimate of $CO_2$ concentrations for maximum photosynthesis rates inside a hypothetical leaf; various assumptions about the 'electrical conductance analog' for $CO_2$ flux from bulk air to the stroma are made. Air phase conductances were estimated for one surface of the leaf. Leaf conductances are given on a projected leaf area basis*

| Leaf component traversed | $k_i^{CO_2}$ | $D_i$ ($m^2\ s^{-1}$) | $\Delta x_i$ (m) | $g'_i$[(a)] ($cm\ s^{-1}$) | $S$[(b)] | $Sg'_i$[(c)] ($cm\ s^{-1}$) | $F'/S$[(d)] ($\mu mol\ m^{-2}\ s^{-1}$) | $C^{CO_2}$[(e)] ($mmol\ m^{-3}$) | $C^{CO_2}$[(e)] (p.p.m.) | Symbol for conductance |
|---|---|---|---|---|---|---|---|---|---|---|
| Bulk air | | | | | | | | 13.3 | 330 | |
| Boundary layer | 1.0 | $1.6\times10^{-5}$ | $1\times10^{-3}$ | 1.6 | 2.0 | 3.2 | 15 | 12.3 | 305 | $g'_a$ |
| Stomatal | | | | 1.0 | 2.0 | 2.0 | 15 | 10.8 | 268 | $g'_s$ |
| Interair space | 1.0 | $1.6\times10^{-5}$ | $5\times10^{-4}$ | 3.2 | 2.0 | 6.4 | 15 | 10.4 | 258 | $g'_c$ |
| Interface partition coefficient ($\alpha_{30} = 0.67$)[(f)] | | | | 0.044 | 20.0 | 0.88 | 1.5 | 6.9 | 171 | $g'_\alpha$ |
| Wall water [(g)] | 0.33 | $2.06\times10^{-9}$ | $5.0\times10^{-7}$ | 0.136 | 20.0 | 2.72 | 1.5 | 5.8 | 144 | $g'_{cw}$ |
| Plasmalemma[(h)] | 1.3 | $1.15\times10^{-11}$ | $5.0\times10^{-9}$ | 0.30 | 20.0 | 6.00 | 1.5 | 5.4 | 134 | $g'_m$ |
| Cytoplasm | 0.67 | $2.06\times10^{-9}$ | $2.0\times10^{-7}$ | 0.690 | 20.0 | 13.8 | 1.5 | 5.1 | 126 | $g'_n$ |
| Outer chloroplast membrane | 1.3 | $1.15\times10^{-11}$ | $5.0\times10^{-9}$ | 0.30 | 20.0 | 6.00 | 1.5 | 4.6 | 144 | $g'_o$ |
| Inner membrane space | 0.67 | $2.06\times10^{-9}$ | $1.0\times10^{-7}$ | 1.38 | 20.0 | 27.6 | 1.5 | 4.5 | 112 | $g'_u$ |
| Inner chloroplast membrane | 1.3 | $1.15\times10^{-11}$ | $5.0\times10^{-9}$ | 0.30 | 20.0 | 6.00 | 1.5 | 4.0 | 100 | $g'_q$ |
| Stroma[(i)] | 50.7 | $2.06\times10^{-9}$ | $3.0\times10^{-6}$ | 3.48 | 20.0 | 69.6 | 1.5 | 4.0 | 99 | $g'_t$ |

Notes to Table 13.3 (*contd.*)

(a) Conductance $g'_i = (D_i K_i^{CO_2})/\Delta x_i$.
(b) Surface involved in $CO_2$ flux divided by the projected leaf area (the area of one side of a leaf).
(c) $(1/g'_l) = (1/g'_a + 1/g'_s + 1/g'_c)$.
(d) $F'$ always equalled 30 $\mu$mol $CO_2$ $m^{-2}$ $s^{-1}$ on a projected leaf area basis. The values given were divided by the $S$ value.
(e) $CO_2$ concentration just beyond the leaf component traversed.
(f) Not a true conductance. The $g'_\alpha$ value calculated is dependent on $g'_l$, leaf temperature and $F'$, and is valid for these conditions only.
(g) $K_i^{CO_2}$ includes the void volume adjustment of 0.5 (50% wall material) and a tortuosity factor of 1.3 (0.5/1.3 = 0.33).
(h) See Guthnecht *et al.* (1977) for discussion and references.
(i) Assuming carbonic anhydrase facilitated diffusion with 76 mol $HCO_3^-$ to 1 mol $CO_2$, pH = 8.2. $K_i^{CO_2} = (1 - 0.33) \times (76) = 50.7$.

$C_{CO_2}$ $(1 + C_{O_2}/K_o) = 1.66$, but there is some uncertainty about values of $K_m$ (Tenhunnen, Hesketh & Gates, 1980). Jones & Slatyer (1972), Ku & Edwards (1977a, b) and Tenhunnen *et al.* (1979) proposed values ranging from 0.1 to 2 $\mu$M. Mooney, Björkman & Collatz (1977) and Malkin *et al.* (1981) reported values of 5 $\mu$M in the wall water. Biochemists usually assume values of 10–20 $\mu$M for fixation by chloroplasts or cells suspended in liquid media but these differ appreciably from the intact system. Simple substitution in the Michaelis-Menten relation shows that, with $K_m = 1$, $F'/F'_{max}$ varies from 0.45 to 0.57 and, with $K_m = 10$, from 0.15 to 0.38, as $C_{CO_2}$ in the stroma varies from 2 to 10 $\mu$M. With $K_m = 5$ and $C_{CO_2} = 4$ (Table 13.3), $F'/F'_{max}$ would be 0.34. Estimates of $F'_{max}$ may be made as follows: a soybean leaf with 60 g dry weight $m^{-2}$ and 4% N has approximately 7 g RuBP carboxylase/oxygenase $m^{-2}$. Taking $5.6 \times 10^5$ as the molecular weight and 8 reaction sites per molecule we find $6 \times 10^{19}$ sites $m^{-2}$ or $1.25 \times 10^{-5}$ mol $m^{-2}$. With a specific activity of 16.7 mol $CO_2$ per mol enzyme per second, $F'_{max} = 210$ $\mu$mol $CO_2$ fixed $m^{-2}$ $s^{-1}$. Taking $F'/F'_{max}$ as 0.34, we would expect a maximum net $CO_2$ exchange, $F'$, of about 70 $\mu$mol $m^{-2}$ $s^{-1}$. Recorded values from $C_3$ plants range up to about 60 (Ludlow & Wilson, 1971; Mooney, Ehleringer & Berry, 1976; Longstreth & Nobel 1980).

Effects of variation in the size of various components in the pathway on the estimated stromal concentration of $CO_2$ are shown in Tables 13.4 and 13.5.

### *Light and dark respiration*

3 ATP and 2 NADPH molecules are required in the Calvin-Benson cycle for each molecule of $CO_2$ fixed. This assumes no photorespiration. For each photorespiratory event resulting in the uptake of 1 $O_2$ molecule and the production of a $C_3$ and a $C_2$ acid, 2 ATP and 2.5 NADPH are required (G. H. Lorimer, T. J. Andrews, unpublished) and 0.5 'fixed' carbon molecules are converted to $CO_2$. $C_4$ plants require 5 ATP and 2 NADPH per carbon fixed. From the proportion of $CO_2$ and $O_2$ fixed, one can estimate the ATP-NADPH requirement for a photorespiring leaf.

It is generally considered that dark respiration proceeds in the light at about the same rate as in the dark (Graham & Chapman, 1979) but better evidence is needed for conclusive proof. The cost of loading sucrose into the phloem for long-distance transport (Chapter 14) is small (about 1 ATP per ATP equivalent loaded).

**What factors inside leaves control photosynthesis?**

In the above section we attempted to quantify the many steps in the photosynthetic processes occurring in a leaf, with the aim of delineating their relative importance. Every assumption and point made, using that crude model, is subject to critical discussion and further research, especially the estimation of parameter values over a much wider range of conditions and leaf ontogeny. Here we will attempt

Table 13.4. *Analysis of the effects of variable wall area ratios and membrane conductances on the stroma* $CO_2$ *concentration in strong light. The air phase* ($g'_l$) *and pseudo conductance due to the partitioning of* $CO_2$ *into water from air* ($g'_\alpha$) *are maintained constant. The parameters are those described in Tables 13.2 and 13.3 except for the changes shown here. Three probable membrane conductances have been matched with four wall ratios. The* $g'_{mem}$ *is assumed to apply to the three cell membranes, while the area ratio (S) applies to all liquid phase components*

| Conductance or component | Membrane conductances for $CO_2$ $g'_{mem}$ (m s$^{-1}$) | | |
|---|---|---|---|
| | $3\times10^{-3}$ | $2\times10^{-3}$ | $1\times10^{-3}$ |
| $g'_l$, air phase (cm s$^{-1}$) | 1.03 | 1.03 | 1.03 |
| $g'_\alpha$, air/water interface (cm s$^{-1}$)[(a)] | 0.88 | 0.88 | 0.88 |
| $g'_{liq}$, liquid phase (cm s$^{-1}$)[(a)] | 0.05 | 0.04 | 0.025 |
| due to water | 0.101 | 0.101 | 0.101 |
| due to membranes | 0.100 | 0.067 | 0.033 |
| $g'$, total leaf (cm s$^{-1}$)[(b)] | | | |
| $S$, area ratio = 10 | 0.24 | 0.22 | 0.16 |
| 20 | 0.32 | 0.30 | 0.24 |
| 30 | 0.30 | 0.34 | 0.29 |
| 40 | 0.38 | 0.37 | 0.32 |
| $C_{CO_2}$ in stroma ($\mu$M)[(c)] | | | |
| $S$, area ratio = 10 | 0.8 | −0.3 | −5.4 |
| 20 | 3.9 | 3.3 | 0.8 |
| 30 | 5.0 | 4.5 | 2.9 |
| 40 | 5.4 | 5.2 | 3.9 |

(a) A pseudo conductance that is applicable only for the $g'_l$, $F'$ (30 $\mu$mol m$^{-2}$ s$^{-1}$) and leaf temperature (30 °C) used.
(b) $(1/g') = 1/g'_l + 1/g'_\alpha + 1/g'_{liq}$.
(c) $13.3 - 3/g' = C_{CO_2}$ in the stroma.

Table 13.5. *Effect of changes in various parameters of the* $CO_2$ *conductance network on stroma* $CO_2$ *concentrations. The conditions are the same as in Table 13.3, with stroma* $C_{CO_2} = 4.0\,\mu M$

| Parameter changed | Old value | New value | Units | Stroma $C_{CO_2}$ (mmol m$^{-3}$) ($\mu$M l$^{-1}$) | Comments |
|---|---|---|---|---|---|
| $g'_1$ | 1.03 | 0.5 | cm s$^{-1}$ | 1.9 | Air phase conductance cut in half |
| | 1.03 | 0.33 | | −0.2 | cut in third |
| Leaf $T$, $g'_\alpha$[(a)] | 30, 0.88 | 15, | °C | 7.3 | Eliminates $g'_\alpha$ |
| | 30, 0.88 | 35, 0.72 | cm s$^{-1}$ | 3.1 | Increase $g'_\alpha$ |
| | 30, 0.88 | 25, 1.18 | | 4.9 | Slight decrease $g'_\alpha$ |
| | 30, 0.88 | 20, 2.31 | | 6.2 | Moderate decrease $g'_\alpha$ |
| Wall thickness | 0.5 | 1.0 | $\mu$m | 2.9 | Doubles $\Delta x_{cw}$ |
| | 0.5 | 1.5 | $\mu$m | 1.8 | Triples $\Delta x_{cw}$ |
| Wall area ratio | 20 | 10 | | 0.8 | One-half $S$ |
| | 20 | 40 | | 5.4 | Double $S$ |
| $g'_{mem}$ | 0.30 | 0.20 | cm s$^{-1}$ | 3.3 | See Table 13.4 |
| | 0.30 | 0.10 | | 0.8 | |
| $F'$ | 30 | 50 | $\mu$mol m$^{-2}$ s$^{-1}$ | 0.7 | 1.7 × $F'$ |
| | 30 | 10 | | 7.3 | 0.33 × $F'$ |
| $g'_t$ | 69.6 | 1.37 | cm s$^{-1}$ | 1.8 | Eliminates carbonic anhydrase facilitated diffusion |

(a) $g'\alpha$ is a pseudo conductance applicable only for a $F'$ value of 30 $\mu$mol m$^{-2}$ s$^{-1}$. Note that temperature changes modulate the bulk air $CO_2$ level and the water phase $CO_2$ diffusion coefficient.

to review the more direct experimental evidence relating to the assumptions made and relevant to the role of the many internal factors in controlling photosynthesis under natural conditions. In this exercise, much care must be exercised in interpreting simple observations and correlations made on a very complex system without much understanding of how that system works.

### *Implications of the leaf area index*

Leaves situated deep in a canopy supporting a leaf area : ground area ratio of 3 to 8 or more are limited in their ability to capture and convert light into reductant. Such leaves have large photosynthetic units and fewer of them per unit area (e.g. Patterson, 1980). A large photosynthetic unit has more antennae chlorophyll molecules for capture of incoming photons. An improvement in quantum efficiency of the light reactions in weak light would enhance the photosynthetic rate in a large fraction of leaves in the natural ecosystem. Research on the behaviour of light, especially the capture and funnelling of electrons to the Z-scheme reaction sites and the efficiency of the chemi-osmotic system in producing ATP-reactions in weak light, therefore is vital for improving primary productivity. In strong light, the limiting factor seems to be the rate of processing electrons through the Z-scheme. It is important to note that the apparent excess of leaf area serves as storage for reduced nitrogen, prior to the appearance of large N-demands during reproductive growth.

### $CO_2$-*enrichment*

Leaves in light-rich environments and at a high temperature (frequently 30 to 40 °C) can exhibit a doubling or tripling in photosynthetic $CO_2$ exchange rates when $CO_2$ in the air is increased. Here, there appears to be excess reductant at normal levels of $CO_2$, but the suppression of photorespiration and its energy requirement as $CO_2$ is increased contributes to the responses (Radmer & Kok, 1977). The activity of the carboxylase may also be increased by higher $CO_2$ concentrations. The photosynthetic response to $CO_2$ interacts considerably with other factors such as temperature and light and, in a canopy, the fraction of leaves in a light-rich environment becomes important.

Plants exposed to optimal conditions for both growth and photosynthesis during the reproductive stage will respond to $CO_2$ enrichment with an increase in dry matter production in the reproductive parts. Such plants must be capable of establishing 'sinks' for usage of the enhanced supply of photosynthate; for example, the plant's nitrogen status, as influenced by supply from the roots or from nodules, may restrict the responses.

### *Behaviour of chlorophyll b deficient mutants*

Mutants deficient in chlorophyll b have been used for some time as genetic markers by plant breeders. With these the initial slope of the light response curve is depressed, but photosynthetic rates in intense light are similar to those of normal leaves. Growth rates and yields are also similar to those of the normal plant type. Robert Keck (personal communication) has suggested that b-deficient leaves, which transmit more light than normal leaves (cf. Gabrielsen, 1948) should perform well in a canopy consisting of a large leaf area packed into a short depth. Better transmission of light to lower leaves in such a canopy should result in better overall canopy performance. Unfortunately, Keck's hypothesis, as well as the precise comparative yielding ability of b-deficient crops, have yet to be tested experimentally.

The b-deficient mutants tend to have as many photosynthetic units per unit leaf area (Alberte *et al.*, 1974), but the antennae chlorophyll a/b shell (Fig. 13.1) is missing. Comparison would allow the importance of this shell for generating reductant to be examined. Other pigments are involved in photon capture, and the size of the a/b shell in these mutants depends upon the environment – particularly the irradiance level. More study is required before one can generalize about the role of light reactions in controlling primary productivity in nature.

### *The correlation between* $CO_2$ *exchange and leaf nitrogen*

Studies listed in Table 13.6, and others cited by Yoshida & Coronel (1976), show strong correlations between photosynthetic $CO_2$ exchange and leaf nitrogen status. These correlations are convincing evidence that enzyme levels limit the dark reaction rates, although a correlation between dark reaction enzyme level and light reaction protein level has not been ruled out, or even tested (but see Friedrich & Huffaker, 1980). It is likely that in many of these studies an increase in nitrogen was also associated with an increase in chlorophyll and the number of photosynthetic units – possibly with more chloroplasts and thicker leaves.

There have been suggestions that the specific activity of the RuBP carboxylase enzyme, or its $K_m$ value for $CO_2$, varies among genotypes or with stage of growth. McNeil *et al.* (1981) concluded that difficulties in assaying for carboxylase activity make such comparisons problematical; enzymes of the genotypes of *Lolium perenne* they compared had similar activity.

Studies have shown an effect of leaf nitrogen status on stomatal conductance, in addition to an expected effect on the residual conductance (Table 13.7). Much has been made of the relative invariance of the photosynthesis : transpiration ratio in such studies, but there are exceptions in the studies listed.

Oritani *et al.* (1979) and Hesketh *et al.* (1981) reported an inverse correlation between photosynthesis rate and area per leaf, where the latter was varied by selecting late-maturing types of genotypes known to have large areas per leaf. Evans & Dunstone (1970) and Khan & Tsunoda (1970a, b) reported similar correlations among the wild and cultivated wheat species. Photosynthetic area per plant is an important determinant of plant total photosynthesis; plants with large leaves but with low percentage N and low photosynthetic rates may nevertheless show larger total plant photosynthesis.

Energy is required for reducing nitrogen whether applied as fertilizer or fixed in nodules. Energy (and dollar) costs of applied N are now of concern. These factors, together with effects on protein yields, influence the maintenance of high N-levels in leaves. Improvements in enzyme activities would benefit N-fixing plants and low protein crops.

### *The correlation between $CO_2$ exchange and leaf chlorophyll*

Gabrielsen (1948), in comparing some 17 species, showed that the quantum efficiency, as determined in weak light, was about 90% the maximum value at chlorophyll concentrations of 300–400 mg $m^{-2}$ with chlorophyll levels ranging up to 900 mg $m^{-2}$. Theoretically, the amount of chlorophyll should not be limiting in natural conditions. Plants with higher photosynthetic rates than those he studied are now known to exist.

Photorespiration has an apparent effect on quantum yields at low light intensities. Temperature (see Table 13.2) affects resistances to $CO_2$ flux (Tenhunnen *et al.*, 1979), $CO_2$ and $O_2$ concentrations and their competition for the reaction sites (Ku & Edwards, 1978), and enzyme activities (Ogren, 1977). The $CO_2$ concentration external or internal to

Table 13.6. *Literature survey of studies showing correlations between photosynthetic* $CO_2$ *exchange (F′) and leaf nitrogen status*

| Authors | Species | Experimental factors varied | Kind of N measured | Notes |
|---|---|---|---|---|
| Murata (1961) | *Oryza sativa* L. | plant age | % N | Correlated also with $P_2O_5$, $K_2O$, chlorophyll, SLW, and respiration rate |
| Björkman (1968) | *Solidago virgaurea* | sun and shade ecotypes | RuBPcase | |
| Nevins & Loomis (1970) | *Beta vulgaris* L. | withheld N | % N | Resistance analysis |
| Khan & Tsunoda (1970a,b) | Wheat species and cultivars | genotype | % N | Inverse correlation with area per leaf |
| Treharne (1972) | *Lolium* species | genotype | RuBPcase | |
| Sato & Tsuno (1975) | *Oryza sativa* L.<br>*Zea mays* L.<br>*Sorghum bicolor* L. | leaf parts | % N | |
| Wilson (1975) | *Panicum maximum*<br>*Lolium perenne* | N-fertility | % N | |

| | | | | |
|---|---|---|---|---|
| Yoshida & Coronel (1976) | *Oryza sativa* L. | | | Lists 7 additional references showing correlations; also has a resistance analysis |
| Ludlow & Ng (1976) | *Panicum maximum* | N-fertility | % N | |
| Boote *et al*. (1978) | *Glycine max* (Merr.) L. | senescing leaves | % N | |
| Oritani *et al*. (1979) | *Oryza sativa* L. | maturity grouping | soluble N<br>total N | Inverse correlation with area per leaf |
| Ishihara *et al*. (1979) | *Oryza sativa* L. | N-fertility<br>leaf position | % N | Resistance analysis |
| Wittenbach (1979) | *Triticum aestivum* L. | senescing leaves | RuBPcase<br>soluble protein | |
| Bolton & Brown (1980) | *Festuca arundinacea*<br>*Panicum maximum* | N-fertility | organic N | Resistance analysis |
| Uchida *et al*. (1980) | *Oryza sativa* L. | leaf development | RuBPcase<br>soluble N<br>total N | Resistance analysis |
| Friedrich & Huffaker (1980) | *Hordeum vulgare* L. | senescing leaves | RuBPcase<br>total protein<br>non-RuBPcase | |
| Hesketh *et al*. (1981) | *Glycine max* L. | genotype<br>maturity group | RuBPcase<br>soluble N | Inverse correlation with area per leaf |

Table 13.7. *Literature survey of conductance analyses of photosynthesis: leaf N status experimental results*

| Author | Species | Relative response to treatment (high N value/low N value) $CO_2$ exchange ($F'$) ratio | Conductance ratios $g_l'$ | $g'_{residual}$ | $F'/g'_l$ ratio |
|---|---|---|---|---|---|
| Ryle & Hesketh (1969) | *Zea mays* L.[a] | 4.0 | 2.5 | 8.0 | 1.6 |
| | [b] | 2.4 | 2.0 | 3.6 | 1.2 |
| Nevins & Loomis (1970) | *Beta vulgaris* L. | 1.6 | 1.4 | 2.0 | 1.1 |
| Medina (1971) | *Atriplex hastata* | 2.2 | 0.9 | 1.8 | 2.5 |
| Yoshida & Coronel (1976) | *Oryza sativa* L. | 4.4 | 7.0 | | 1.0 |
| Adams *et al.* (1977) | *Simmondsia chinensis* | 1.9 | 1.0 | 2.3 | 1.9 |
| Wong *et al.* (1979) | *Zea mays* L. | | | | 1.0 |
| Goudriann & van Keulen (1979) | *Zea mays* L. | 2.0 | 2.0 | | 1.0 |
| Ishihara *et al.* (1979) | *Oryza sativa* L. | 2.9 | 3.7 | 3.0 | 0.8 |
| Bolton & Brown (1980) | *Panicum maximum* | 5.0 | 7.0 | 8.0 | 1.4 |
| | *Festuca arundinacea* | 3.0 | 3.5 | 5.0 | 1.1 |
| Longstreth & Nobel (1980) | *Gossypium hirsutum* L. | 2.3 | 1.0 | 2.9 | 2.3 |
| Radin & Ackerson (1981) | *Gossypium hirsutum* L. | 1.8 | 1.0 | 1.8 | 1.8 |

[a] Lower leaves were removed to create a 'sink' effect.
[b] Control.

Table 13.8. *A survey of literature indicating correlations between photosynthetic $CO_2$ exchange and chlorophyll*

| Authors | Species | Chlorophyll (mg $m^{-2}$) at % maximum $F'$ 50 | 90 | Maximum $F'$ ($\mu$mol $m^{-2}$ $s^{-1}$) | Saturated chlorophyll level (mg $m^{-2}$) |
|---|---|---|---|---|---|
| Patterson *et al.* (1977) | *Gossypium hirsutum* L. | 280 | 500 | 28 | 560 |
| Teeri *et al.* (1977) | *Zea mays* L. | 260 | 450 | 40 | 500 |
| Terry (1980) | *Beta vulgaris* L. | 300 | 550 | 69 | 600 |
| Alberte *et al.* (1976) | Barley mutants (1000 p.p.m. $CO_2$) | 170 | 380 | 19 | 430 |

the leaf affects the quantum yield in weak light and more data about possible mechanisms are needed before one can conclude exactly how quantum yields, temperature and $CO_2$ interact. $C_4$ and $C_3$ plants seem to have the same quantum efficiency in 21% $O_2$; $C_3$ leaves often have a higher quantum efficiency in low oxygen (Ku & Edwards, 1978).

Four studies (Table 13.8) show a good correlation between leaf photosynthetic rate and chlorophyll. When one includes the chlorophyll b deficient mutants in such an analysis, the correlation is even better. Alberte *et al.* (1976) have shown a strong correlation between $CO_2$ exchange and photosynthetic unit density among species, including $C_3$ and $C_4$ species.

### *The* $C_4$:$C_3$ *phenomenon,* $C_4$:$C_3$ *intermediates,* CAM-*aquatic plant variations, and associated limiting factors*

Hatch & Osmond (1976), Scharrenberger & Fock (1976), Ray & Black (1979), and many others have reviewed this subject. The $C_4$:$C_3$ story is a complex one; both the history of the discoveries and the complexities of the biochemistry, cells and organelles involved make fascinating reading. Plants exhibiting intermediate behaviour, as encountered naturally (including CAM plants and submersed aquatics such as 'sea grasses'), or bred from $C_4 \times C_3$ crosses, offer great potential for studying this phenomenon.

Somerville & Ogren (1981) have developed methods for isolating mutants of *Arabidopsis* with defective enzymes at various steps of the photorespiratory carbon pathway. The many mutants isolated to date do not survive in normal air where pathway intermediates pile up to a poisonous level. However, considerable information is accumulating about the characteristics of the pathway and hope of finding a mutant with an altered carboxylase that will only accept $CO_2$ onto its active site remains. A second approach is to alter an enzyme such as RuBP carboxylase by altering its DNA code (cf. Bedbrook *et al.*, 1980; McIntosh *et al.*, 1980). But for this to be successful, more needs to be known about the chemistry of the $CO_2$ fixation site.

There has been a massive search for photorespiration inhibitors (Zelitch, 1979; Servaites & Ogren, 1977). Here, the identification and incorporation of defective enzymes may be unsuccessful as, when the pathway is blocked, pathway metabolites accumulate to inhibitory levels for photosynthesis.

Hatch & Osmond (1976) presented an equation and associated

assumptions for quantifying fluxes of $C_4$ acids between mesophyll and bundle sheath cells. Their analysis is similar in nature to that given in Tables 13.2–13.4, including considerations of transport (plasmodesmata) surfaces, diffusivities, pathlengths, enzyme $K_m$ values, $CO_2$ concentrations, and cell anatomy. Berry & Farquhar (1978) presented a more sophisticated treatment of the same problem, with computer solutions to the equations involved. This system, however, needs more research, particularly on $CO_2$ concentrations in the bundle sheath cells.

Raghavendra & Das (1978) have updated the list of known $C_4$ plants and Teeri & Stowe (1976), Doliner & Jolliffe (1979), and Tieszen, Nenyimba, Imbamba & Troughton (1979) have analysed the geographic-altitudinal distribution of $C_4$ species and its ecological significance. The occurrence of $C_4$ species is favoured under conditions of high temperatures and intermittent water stress.

### *The $K_m$ for $CO_2$ and $CO_2$-concentration mechanisms*

Walker (1976) reviewed the history of attempts to equate dark reaction kinetics ($K_m$ and $V_{max}$ for carboxylation) of $CO_2$ fixation to leaf photosynthetic $CO_2$ fluxes. $CO_2$-concentrating mechanisms were frequently introduced to explain discrepancies between the two. As indicated in the first section of this chapter, there still is a major discrepancy between $CO_2$ concentrations in the stroma, $K_m$ values as reported by biochemists for $CO_2$ fixation, and known rates of photosynthesis.

Among the evidence for $CO_2$-concentrating mechanisms is the estimate by Hatch & Osmond (1976) of 13–15 $\mu$M $CO_2$ at pH 8 in the bundle sheath cells of $C_4$ plants. Kaplan, Badger & Berry (1980) and Badger, Kaplan & Berry (1980) reported that internal $CO_2$ concentrations (ignoring $HCO_3^-$, $H_2CO_3$) were some 40-fold of those in surrounding media in *Chlamydomonas* and *Anabaena* cells. They suggested an energy-dependent transport system linked to ATP, moving $HCO_3^-$. If carbonic anhydrase were involved and an active influx of $HCO_3^-$ existed in such cells, the carboxylation enzyme could be exposed to elevated $CO_2$ concentrations. The $K_m$ for $CO_2$ with cells grown in ambient $CO_2$ was less than 4 $\mu$M, yet the $K_m$ for the isolated enzyme was 30–40 $\mu$M. These results stress the need to know more about how $CO_2$ is fixed on the carboxylase, the role of $HCO_3^-$ near such a site, and more about extracting and measuring activities of the enzyme; certainly, the historical problem of equating the kinetics of the carboxylating enzyme and known leaf photosynthetic rates in nature is still far from resolved.

### $CO_2$ *conductances*

The importance of the various conductances for diffusion of $CO_2$ on photosynthesis is now widely recognized. Raschke (1979) has reviewed evidence for interactions between $CO_2$ fixation, the intercellular $CO_2$ concentration and guard cell behaviour. Cowan & Farquhar (1977) have postulated that stomata function so that the total loss of water is a minimum for the total amount of carbon taken up (see also Goudriaan & van Keulen, 1979). They, and Wong *et al.* (1979), show that the internal $CO_2$ concentration at the cell walls of a photosynthesizing leaf tends to remain constant, the stomatal conductance changing to achieve this constancy. This suggests that stomata are influenced by the supply of some substance from the mesophyll. Korner, Scheel & Bauer (1979) have surveyed the literature for measurements of leaf diffusive conductances and found high correlations between photosynthetic rates and the leaf and residual conductances.

Charles-Edwards & Ludwig (1975) and Nobel and coworkers (e.g. Longstreth & Nobel, 1980) have emphasized the importance of leaf thickness and of the ratio of the cell wall area in the leaf to the projected leaf area. Both factors are related to numbers of cells, chloroplasts, photosynthetic units and carboxylating enzyme molecules per unit leaf area. We have illustrated above (Table 13.3) the roles these surfaces play in determining internal flux conductances; the conductance, in effect, is multiplied by the ratio of the internal surface to the projected leaf surface. Methods have been proposed for estimating liquid $CO_2$-flux conductances (Ludlow & Wilson, 1972; Longstreth & Nobel, 1980; Tenhunnen *et al.*, 1979) but although these have yielded interesting estimates, all are conceptually questionable. The approach of Tenhunnen *et al.* (1979), involving computer estimation of parameters difficult to measure or estimate, is the most sound conceptually, but even then numerous 'guesses' must be made, as was done in the analysis of the $CO_2$-diffusion network earlier.

### *Plant growth regulators*

The role of plant growth regulators in influencing rates of photosynthesis is still most uncertain but readers may wish to refer to Chapter 8 and to Wareing (1979) and Guinn & Mauney (1980) for reviews of current concepts.

*Source: sink effects and feed-back control*

In a recent review of the relevant literature relating to possible mechanisms for regulation of photosynthesis by assimilate demand, the concluding sentence (translated) reads 'the mechanisms by which the flux and accumulation of assimilates are able to regulate photosynthesis, still remains confused' (PintoC, 1980). Certainly, the results of attempts to affect photosynthesis by increasing leaf carbohydrate levels have been conflicting. In some experiments, treatments designed to increase starch levels appear to have reduced photosynthetic rate while in others there has been no clear effect. Many of these treatments have been somewhat drastic including, for example, extended light periods, removal of fruits and girdling of petioles, thereby resulting in long periods during which high concentrations of starch were stored. Attempts to increase photosynthetic rates have involved partial defoliation, shading of all leaves but one, and grafting on additional fruits. A danger with these experiments is that the alterations in the plant's environment or the modification of the plant itself may indirectly affect photosynthesis so that the observed responses may have little causal relation with carbohydrate levels (Potter & Breen, 1980). However, there does seem to be potential for control of photosynthesis within the cell at the chloroplast/cytoplasm interface (Herold, 1981). Experiments using isolated intact chloroplasts, despite the difficulties in relating the results to the microenvironment of the normal leaf cell, are essential if we are ever to understand the relationships between assimilate supply and demand.

Only a limited number of metabolites are able to cross the chloroplast membrane and all of these are phosphorylated compounds. A strict counter exchange system operates mediated by the phosphate translocator (Heldt, 1976), such that as triose phosphate (the principal compound concerned) is exported from the chloroplast, inorganic phosphate (Pi) is imported. Triose phosphate is utilized in the synthesis of hexose phosphates and ultimately of sucrose which is then exported from the cell, and Pi is released and can again exchange for an additional molecule of triose phosphate.

The formation of starch within the chloroplast has been manipulated by alterations in concentrations of external Pi (Heldt *et al.*, 1977). The enzymes responsible for starch synthesis and degradation are controlled by Pi and PGA (Preiss & Kosuge, 1970). Photosynthesis by isolated chloroplasts may be reduced when the external concentration of Pi is low and the internal concentration of sugar phosphates is high (A. R.

Portis, personal communication). Photosynthesis is also inhibited at high Pi levels (Bamberger, Ehrlich & Gibbs, 1974) but this may be due to forced export of triose phosphate resulting in insufficient substrate for carboxylase. Isolated chloroplasts are very responsive to external concentrations of Pi and triose phosphates; hence, fluctuations in these compounds brought about by sink demands may provide one control of photosynthesis.

There is also a need for improved understanding concerning the compartmentation of various metabolites. Information is becoming available about the internal chloroplast concentrations of many key metabolites (Stitt, Wirtz & Heldt, 1980) but knowledge of the concentrations of metabolites in the cytoplasm is also required. A further need concerns the role of the vacuole where most of the very large pool of Pi appears to be located. If there are substantial barriers separated between the vacuole and the cytoplasm we must learn how the exchange is controlled.

## Conclusions

We have presented the ingredients, involving mathematical and experimental methods, for the development of a conceptual model describing how photosynthetic and other processes interact inside a leaf. Our understanding of these systems is imperfect and awaits further exploitation of available methodology. We must also ensure that some of the experimental material is representative of what occurs naturally under a range of environmental conditions. Short- and long-term effects, such as occur during a day or a growing season, of environment on plant behaviour need to be accounted for. Evolutionary effects should be included. Because of the present state of understanding, we can only raise questions as to mechanisms responsible for responses, such as those associated with sun and shade, cold, heat or water stress, $CO_2$ enrichment, the growth regulator-source-sink complex, and genotypes. Values for $CO_2$ conductances and dark reaction kinetics need to be established that will explain photosynthetic rates in nature.

## References

Adams, J. A., Johnson, H. B., Bingham, F. T. & Yermanos, D. M. (1977). Gaseous exchange of *Simmondsia chinensis* (Jojoba) measured with a double isotope porometer and related to water stress and nitrogen deficiency. *Crop Science*, **17**, 11–15.

Alberte, R. S., Hesketh, J. D., Hofstra, G., Thornber, J. P., Naylor, A. W., Bernard, R. L., Brim, C., Endrizzi, J. & Kohel, R. J. (1974). Composition and activity of the photosynthetic apparatus in temperature-sensitive mutants of higher plants. *Proceedings of the National Academy of Sciences*, USA, **71**, 2414–18.

Alberte, R. S., Hesketh, J. D., Thornber, J. P. & Kleinkops, A. (1976). Comparisons of photosynthetic activity and chloroplast lamellar characteristics of chlorophyll mutants of barley. *Plant Physiology*, *S*-72.

Badger, M. R., Kaplan, A. & Berry, J. A. (1980). Internal inorganic carbon pool of *Chlamydomonas reinhardtii*, evidence for a carbon dioxide-concentrating mechanism. *Plant Physiology*, **66**, 407–13.

Bamberger, E. S., Ehrlich, B. A. & Gibbs, M. (1974). The glyceraldehyde 3-phosphate and glycerate 3-phosphate shuttle and carbon dioxide assimilation in intact spinach chloroplasts. *Plant Physiology*, **55**, 1023–30.

Bedbrook, J. R., Smith, S. M. & Ellis, R. J. (1980). Molecular cloning and sequencing of cDNA encoding the precursor to the small subunit of chlorophyll ribulose 1,5-bisphosphate carboxylase. *Nature*, **287**, 692–7.

Berry, J. & Farquhar, G. (1978). The $CO_2$ concentrating function of $C_4$ photosynthesis, a biochemical model. In *Proceedings of the 4th International Congress on Photosynthesis*, ed. D. D. Hall, J. Combs & T. W. Goodwin, pp. 119–31. London: The Biochemical Society.

Björkman, O. (1968). Further studies on differentiation of photosynthetic properties in sun and shade ecotypes of *Solidago virgaurea. Physiologia Plantarum*, **21**, 84–99.

Bolton, J. K. & Brown, R. H. (1980). Photosynthesis of grass species differing in carbon dioxide fixation pathways. V. Responses of *Panicum millioides*, and tall fescue (*Festuca arundinacea*) to nitrogen nutrition. *Plant Physiology*, **66**, 97–100.

Boote, K. J., Gallaher, R. N., Robertson, W. K., Hinson, K. & Hammond, L. C. (1978). Effect of foliar fertilization on photosynthesis, leaf nutrition, and yield of soybeans. *Agronomy Journal*, **70**, 787–91.

Bunce, J. A., Patterson, D. T., Peet, M. M. & Alberte, R. S. (1977). Light acclimation during and after leaf expansion in soybean. *Plant Physiology*, **60**, 255–8.

Burrows, F. J. & Milthorpe, F. L. (1976). Stomatal conductance in the control of gas exchange. In *Water Deficits and Plant Growth*, ed. T. T. Kozlowski, vol. 4, pp. 103–52. New York: Academic Press.

Charles-Edwards, D. A. & Ludwig, L. J. (1975). The basis of expression of leaf photosynthetic activities. In *Environmental and Biological Control of Photosynthesis*, ed. R. Marcelle, pp. 37–44. The Hague: Junk Publishers.

Cowan, I. R. & Farquhar, G. D. (1977). Stomatal function in relation to leaf metabolism and environment. In *Integration of Activity in the Higher Plant*, ed. D. H. Jennings, *Symposium of the Society for Experimental Biology No. 31*, pp. 471–505. Cambridge University Press.

Doliner, L. H. & Jolliffe, P. A. (1979). Ecological evidence concerning the adaptive significance of the $C_4$ dicarboxylic acid pathway of photosynthesis. *Oecologia*, **38**, 23–34.

Evans, L. T. & Dunstone, R. L. (1970). Some physiological aspects of evolution in wheat. *Australian Journal of Biological Science*, **23**, 725–41.

Friedrich, J. W. & Huffaker, R. C. (1980). Photosynthesis, leaf resistances, and ribulose-1,5-bisphosphate carboxylase degradation in senescing barley leaves. *Plant Physiology*, **65**, 1103–7.

Gaastra, P. (1959). Photosynthesis of crop plants as influenced by light, carbon dioxide, temperature, and stomatal diffusion. *Mededelingen van de Landbouwhoogeschool te Wageningen*, **59**, 1–68.

Gabrielsen, E. K. (1948). Effects of different chlorophyll concentrations on photosynthesis in foliage leaves. *Physiologia Plantarum*, **1**, 5–37.

Golbeck, J. H., Lien, S. & San Pietro, A. (1977). Electron transport in chloroplasts. In *Encyclopedia of Plant Physiology*, (new series) vol. 5, *Photosynthesis, I. Photosynthetic Electron Transport and Photophosphorylation*, ed. A. Trebst & M. Avron. Berlin: Springer-Verlag.

Goudriaan, J. & van Keulen, H. (1979). The direct and indirect effects of nitrogen shortage on photosynthesis and transpiration in maize and sunflower. *Netherlands Journal of Agricultural Science*, **27**, 227–34.

Govindjee & Govindjee, R. (1975). Introduction to photosynthesis. In *Bioenergetics of Photosynthesis*, ed. Govindjee, pp. 1–50. New York: Academic Press.

Graham, D. & Chapman, E. A. (1979). Interactions between photosynthesis and respiration in higher plants. In *Encyclopedia of Plant Physiology*, (new series) vol. 6, *Photosynthesis II: Photosynthetic Carbon Metabolism and Related Processes*, ed. M. Gibbs & E. Latzko, pp. 150–62. Berlin: Springer-Verlag.

Guinn, G. & Mauney, J. R. (1980). Analysis of $CO_2$ exchange assumptions: feedback control. In *Predicting Photosynthesis for Ecosystem Models*, vol. II, ed. J. D. Hesketh & J. W. Jones, pp. 1–16. Boca Raton: CRC Press.

Guthnecht, J., Bisson, M. A. & Tosteson, F. D. (1977). Diffusion of carbon dioxide through bilayer membranes, effects of carbonic anhydrase, bicarbonate, and unstirred layers. *Journal of General Physiology*, **69**, 779–94.

Hatch, M. D. & Osmond, C. B. (1976). Compartmentation and transport in $C_4$ photosynthesis. In *Encyclopedia of Plant Physiology*, (new series) vol. 3, *Transport in Plants. III. Intracellular Interactions and Transport Processes*, ed. C. R. Stocking & U. Heber, pp. 144–84. Berlin: Springer-Verlag.

Heber, U. & Heldt, H. W. (1981). The chloroplast envelope: Structure, function and role in leaf metabolism. *Annual Review of Plant Physiology*, **32**, 139–68.

Heldt, H. W. (1976). Metabolite carriers of chloroplasts. In *Encyclopedia of Plant Physiology*, (new series) vol. 3, *Transport in Plants. III. Intracellular Interactions and Transport Processes*, ed. C. R. Stocking & U. Heber, pp. 137–43. Berlin: Springer-Verlag.

Heldt, H. W., Chon, C. J., Maronde, D., Herold, A., Stankovic, Z. S., Walker, D. A., Kraminer, A., Kirk, M. R. & Heber, U. (1977). Role of orthophosphate and other factors in the regulation of starch formation in leaves and isolated chloroplasts. *Plant Physiology*, **59**, 1146–55.

Herold, A. (1981). Regulation of photosynthesis by sink activity – the missing link. *New Phytologist*, **86**, 131–44.

Hesketh, J. D., Ogren, W. L., Hageman, M. E. & Peters, D. B. (1981). Correlations among leaf $CO_2$-exchange rates, areas and enzyme activities among soybean cultivars. *Photosynthesis Research* (in press).

Highkin, H. R., Boardman, N. K. & Goodchild, D. J. (1969). Photosynthetic studies on a pea-mutant deficient in chlorophyll. *Plant Physiology*, **44**, 1310–20.

Hinkle, P. C. & McCarty, R. E. (1978). How cells make ATP. *Scientific American*, **238**, 104–23.

Huber, S. C. & Moreland, D. E. (1981). Co-transport of potassium and sugars across the plasmalemma of mesophyll protoplasts. *Plant Physiology*, **67**, 163–9.

Ishihara, K., Iida, O., Hirasawa, T. & Ogura, T. (1979). Relationship between nitrogen content in leaf blades and photosynthetic rate of rice plants with reference to stomatal aperture and conductance. *Japanese Journal of Crop Science*, **48**, 543–50.

Jones, H. & Slatyer, R. (1972). Estimation of the transport and carboxylation components of the intracellular limitation to leaf photosynthesis. *Plant Physiology*, **50**, 283–8.

Kaplan, A., Badger, M. R. & Berry, J. A. (1980). Photosynthesis and the intracellular inorganic carbon pool in the blue green alga *Anabaena variabilis*: Response to external $CO_2$ concentration. *Planta*, **149**, 219–26.

Khan, M. A. & Tsunoda, S. (1970a). Evolutionary trends in leaf photosynthesis and related leaf characters among cultivated wheat species and its wild relatives. *Japan Journal of Breeding*, **20**, 133–40.

Khan, M. A. & Tsunoda, S. (1970b). Leaf photosynthesis and transpiration under different levels of air flow rate and light intensity in cultivated wheat species and its wild relatives. *Japan Journal of Breeding*, **20**, 305–13.

Korner, C., Scheel, J. A. & Bauer, H. (1979). Maximum leaf diffusive conductance in vascular plants. *Photosynthetica*, **13**, 45–82.

Ku, S. & Edwards, G. (1977a). Oxygen inhibition of photosynthesis. I. Temperature dependence and relation to $O_2/CO_2$ solubility ratio. *Plant Physiology*, **59**, 986–90.

Ku, S. & Edwards, G. (1977b). Oxygen inhibition of photosynthesis. II. Kinetic characteristics as affected by temperature. *Plant Physiology*, **59**, 991–9.

Ku, S. & Edwards, G. (1978). Oxygen inhibition of photosynthesis. III. Temperature dependence of quantum yield and its relation to $O_2/CO_2$ solubility ratio. *Planta*, **140**, 1–6.

Laing, W. A., Ogren, W. L. & Hageman, R. H. (1974). Regulation of soybean net photosynthetic fixation by the interaction of $CO_2$, $O_2$, ribulose-1,5-diphosphate carboxylase. *Plant Physiology*, **54**, 678–85.

Longstreth, D. J. & Nobel, P. S. (1980). Nutrient influences on leaf photosynthesis, effects of nitrogen, phosphorus, and potassium for *Gossypium hirsutum* L. *Plant Physiology*, **65**, 541–3.

Ludlow, M. M. & Ng, T. T. (1976). Effect of water deficit on carbon dioxide exchange and leaf elongation rate of *Panicum maximum* var. *Trichoglume*. *Australian Journal of Plant Physiology*, **3**, 401–13.

Ludlow, M. M. & Wilson, G. L. (1971). Photosynthesis of tropical pasture plants. I. Illuminance, carbon dioxide concentration, leaf temperature, and leaf-air vapour pressure difference. *Australian Journal of Biological Sciences*, **24**, 449–70.

Ludlow, M. M. & Wilson, G. L. (1972). Photosynthesis of tropical pasture plants. IV. Basis and consequences of differences between grasses and legumes. *Australian Journal of Biological Sciences*, **25**, 1133–45.

McIntosh, L., Poulson, C. & Bogorad, L. (1980). Chloroplast gene sequence for the large subunit of ribulose bisphosphate carboxylase of maize. *Nature*, **288**, 556–60.

McNeil, P. H., Foyer, C. H., Walker, D. A., Bird, I. F., Cornelium, M. J. & Keyes, A. J. (1981). Similarity of ribulose-1,5-bisphosphate carboxylases of isogenic diploid and tetraploid ryegrass (*Lolium perenne* L.) cultivars. *Plant Physiology*, **67**, 530–4.

Malkin, S., Armond, P. A., Mooney, H. A. & Fork, D. C. (1981). Photosystem II photosynthetic unit sizes from fluorescence induction in leaves. *Plant Physiology*, **67**, 570–9.

Mauney, J. R., Guinn, G., Fry, K. E. & Hesketh, J. D. (1979). Correlation of photosynthetic carbon dioxide uptake and carbohydrate accumulation in cotton, soybean, sunflower and sorghum. *Photosynthetica*, **13**, 260–6.

Medina, E. (1971). Effect of nitrogen supply and light intensity during growth on the photosynthetic capacity and carboxydismutase activity of leaves of *Atriplex patula* ssp. *hastata*. *Carnegie Institute of Washington Yearbook*, **70**, 551–9.

Mitchell, P. (1979). Keilin's respiratory chain concept and its chemiosmotic consequences. *Science*, **206**, 1148–59.

Mooney, H. A., Björkman, O. & Collatz, G. J. (1977). Photosynthetic acclimation to temperature and water stress in the desert shrub *Larrea divaricata*. *Carnegie Institution of Washington Yearbook*, **76**, 328–35.

Mooney, H. A., Ehleringer, J. & Berry, J. A. (1976). High photosynthetic capacity of a winter annual in Death Valley. *Science*, **194**, 322–3.

Morita, K. (1980). Release of nitrogen from chloroplasts during leaf senescence in rice (*Oryza sativa* L.). *Annals of Botany*, **46**, 297–302.

Murata, Y. (1961). Studies on the photosynthesis of rice plants and its culture significance. *Bulletin of the National Institute of Agricultural Sciences, Series D*, (*Nishigahara, Tokyo*), **9**, 1–170.

Nevins, D. J. & Loomis, R. S. (1970). Nitrogen nutrition and photosynthesis in sugar beet (*Beta vulgaris* L.). *Crop Science*, **10**, 21–5.

Nobel, P. S. (1974). *Introduction to Biophysical Plant Physiology*. San Francisco: W. H. Freeman.

Ogren, W. L. (1977). Increasing carbon fixation by crop plants. In *Proceedings of the 4th International Congress on Photosynthesis*, ed. D. D. Hall, J. Combs & T. W. Goodwin, pp. 721–33. London: The Biochemical Society.

Oritani, T., Enbutsu, T. & Hoshida, R. (1979). Studies on nitrogen metabolism in crop plants. VI. Changes in photosynthesis and nitrogen metabolism in relation to leaf area growth of several rice varieties. *Japanese Journal of Crop Science*, **48**, 10–6.

Patterson, D. T. (1980). Light and temperature adaptation. In *Predicting Photosynthesis for Ecosystem Models*, vol. 2, ed. J. D. Hesketh & J. W. Jones, pp. 205–35. Boca Raton: CRC Press.

Patterson, D. T., Bunce, J. A., Alberte, R. S. & Van Volkenburgh, E. (1977). Photosynthesis in relation to leaf characteristics of cotton from controlled and field environments. *Plant Physiology*, **59**, 384–7.

PintoC, M. (1980). Régulation de la photosynthèse par la demande d'assimilats: mécanismes possibles. *Photosynthetica*, **14**, 611–37.

Pospíšilova, J. & Solárová, J. (1980). Environmental and biological control of diffusive conductances of adaxial and abaxial leaf epidermes. *Photosynthetica*, **14**, 90–127.

Potter, J. R. & Breen, P. J. (1980). Maintenance of high photosynthetic rates during the accumulation of high leaf starch levels in sunflower and soybean. *Plant Physiology*, **66**, 528–31.

Preiss, J. & Kosuge, T. (1970). Regulation of enzyme activity in photosynthetic systems. *Annual Review of Plant Physiology*, **21**, 433–66.

Rabinowitch, E. I. (1951). *Photosynthesis*, vol. II, Part I. New York: Interscience Publishers.

Radin, J. W. & Ackerson, R. C. (1981). Water relations of cotton plants under nitrogen deficiency. III. Stomatal conductance, photosynthesis, and abscisic acid accumulation. *Plant Physiology*, **67**, 115–19.

Radmer, R. J. & Kok, B. (1977). Light conversion efficiency in photosynthesis. In *Encyclopedia of Plant Physiology*, (new series) vol. 5, *Photosynthesis. I. Photosynthetic Electron Transport and Photophosphorylation*, ed. A. Trebst & M. Avron, pp. 125–35. Berlin: Springer-Verlag.

Raghavendra, A. S. & Das, V. S. R. (1978). The occurrence of $C_4$-photosynthesis: A supplementary list of $C_4$ plants reported during late 1974–mid 1977. *Photosynthetica*, **12**, 200–8.

Raschke, K. (1979). 4. Movements using turgor mechanisms. 4.1 Movements of stomata. In *Encyclopedia of Plant Physiology*, (new series) vol. 7, *Physiology of Movements*, ed. W. Haupt & M. E. Fenleib, pp. 383–441. Berlin: Springer-Verlag.

Ray, T. B. & Black, C. C. (1979). The $C_4$ and crassulacean acid pathways. 6. The $C_4$ pathway and its regulation. In *Encyclopedia of Plant Physiology*, (new series) vol. 6, *Photosynthesis. II. Photosynthetic Carbon Metabolism and Related Processes*, ed. M. Gibbs & E. Latzko, pp. 77–101. Berlin: Springer-Verlag.

Ryle, G. J. A. & Hesketh, J. D. (1969). Carbon dioxide uptake in nitrogen-deficient plants. *Crop Science*, **9**, 451–4.

Sato, T. & Tsuno, Y. (1975). Studies on $CO_2$ uptake and $CO_2$ evolution in each part of crop plants. III. Variation of photosynthetic rate in different parts of leaves of rice, corn and sorghum plants. *Proceedings of the Crop Science Society of Japan*, **44,** 389–96.

Scharrenberger, C. & Fock, H. (1976). Interactions among organelles involved in photorespiration. In *Encyclopedia of Plant Physiology*, (new series) vol. 3, *Transport in Plants. III. Intracellular Interactions and Transport Processes*, ed. C. R. Stocking & U. Heber, pp. 185–234. Berlin: Springer-Verlag.

Servaites, J. C. & Ogren, W. L. (1977). Chemical inhibition of the glycolate pathway in soybean leaf cells. *Plant Physiology*, **60,** 461–6.

Šesták, Z., Čarský, J. & Jarvis, P. G. (eds.) (1971). *Plant Photosynthetic Production. Manual of Methods*. The Hague: W. Junk.

Siegelman, H. W. & Hind, G. (eds.) (1978). *Photosynthetic Carbon Assimilation*. New York: Plenum Press.

Somerville, C. R. & Ogren, W. L. (1981). Inhibition of photosynthesis in *Arabidopsis* mutants lacking leaf glutamate synthase activity. *Nature*, **286,** 257–9.

Stitt, M., Wirtz, W. & Heldt, H. W. (1980). Metabolite levels during induction in the chloroplast and extrachloroplast compartments of spinach protoplasts. *Biochimica Biophysica Acta*, **593,** 85–102.

Teeri, J. A., Patterson, D. T., Alberte, R. S. & Castleberry, R. M. (1977). Changes in the photosynthetic apparatus of maize in response to simulated natural temperature fluctuations. *Plant Physiology*, **60,** 370–3.

Teeri, J. A. & Stowe, L. G. (1976). Climatic patterns and the distribution of $C_4$ grasses in North America. *Oecologia*, **23,** 1–12.

Tenhunnen, J. D., Hesketh, J. D. & Gates, D. M. (1980). Leaf photosynthesis models. In *Predicting Photosynthesis for Ecosystem Models*, vol. I, ed. J. D. Hesketh & J. W. Jones, pp. 123–81. Boca Raton: CRC Press.

Tenhunnen, J. D., Weber, J. A., Yocum, C. S. & Gates, D. M. (1979). Solubility of gases and the temperature dependency of whole leaf affinities for carbon dioxide and oxygen, an alternative perspective. *Plant Physiology*, **63,** 916–23.

Terry, N. (1980). Limiting factors in photosynthesis. I. Use of iron stress to control photochemical capacity *in vivo*. *Plant Physiology*, **65,** 114–20.

Thornber, J. P. (1975). Chloroplast-proteins: light-harvesting and reaction centre components of plants. *Annual Review of Plant Physiology*, **26,** 127–58.

Thornber, J. P. & Alberte, R. S. (1977). The organization of chlorophyll *in vivo*. In *Encyclopedia of Plant Physiology*, (new series) vol. 5, *Photosynthesis I. Photosynthetic Electron Transport and Photophosphorylation*, ed. A. Trebst & M. Avron, pp. 574–82. Berlin: Springer-Verlag.

Thornber, J. P., Alberte, R. S., Hunter, F. A., Sciozawa, J. A. & Kan, K.-S. (1977). The organization of chlorophyll in the plant photosynthetic unit. *Brookhaven Symposium of Biology*, **28,** 132–58.

Thornber, J. P. & Highkin, H. R. (1974). Comparison of the photosynthetic apparatus of normal barley leaves and a mutant lacking chlorophyll b. *European Journal of Biochemistry*, **41,** 109–16.

Tieszen, L. L., Nenyimba, M. M., Imbamba, S. K. & Troughton, J. H. (1979). The distribution of $C_3$ and $C_4$ gasses and carbon isotope discrimination along an altitudinal and moisture gradient in Kenya. *Oecologia*, **38,** 337–50.

Tolbert, N. E. (1979). Glycolate metabolism by higher plants and algae. In *Encyclopedia of Plant Physiology*, (new series) vol. 6, *Photosynthesis II. Photosynthetic Carbon Metabolism and Related Processes*, ed. M. Gibbs & E. Latzko, pp. 338–52. Berlin: Springer-Verlag.

Trebst, A. & Avron, M. (1977). Introduction. In *Encyclopedia of Plant Physiology*, (new series) vol. 5, *Photosynthesis I. Photosynthetic Electron Transport and Photophosphorylation*, ed. A. Trebst & M. Avron, pp. 1–4. Berlin: Springer-Verlag.

Treharne, K. J. (1972). Biochemical limitations to photosynthetic rates. In *Crop Processes in Controlled Environments*, ed. A. R. Rees, K. E. Cockshull, D. W. Hand & R. G. Hurd, pp. 285–303. New York: Academic Press.

Uchida, N., Itoh, R. & Murata, Y. (1980). Studies on the changes in the photosynthetic activity of a crop leaf during its development and senescence. I. Changes in the developmental stage of a rice leaf. *Japanese Journal of Crop Science*, **49**, 127–34.

Walker, D. A. (1976). Plastids and intracellular transport. In *Encyclopedia of Plant Physiology*, (new series) vol. 3, *Transport in Plants. III. Intracellular Interactions and Transport Processes*, ed. C. R. Stocking & U. Heber, pp. 85–136, 144–84. Berlin: Springer-Verlag.

Wareing, P. F. (1979). Inaugural address: Plant development and crop yield. In *Photosynthesis and Plant Development*, ed. R. Marcelle, H. Clysters & M. Van Poucke, pp. 1–17. The Hague: Dr W. Junk.

Whitmarsh, J. & Cramer, W. A. (1980). Theoretical time dependence of oxidation-reduction reactions in photosynthetic electron transport: Reduction of a linear chain by the plastoquinone pool. *Methods in Enzymology*, **69**, 202–23.

Wildman, S. G. & Kwanyuen, P. (1978). Fraction I protein and other products from tobacco for food. In *Photosynthetic Carbon Assimilation*, ed. H. W. Siegelman & G. Hind, pp. 1–18. New York: Plenum Press.

Wilson, J. R. (1975) Comparative response to nitrogen deficiency of a tropical and temperate grass in the interrelation between photosynthesis, growth and the accumulation of non-structural carbohydrate. *Netherlands Journal of Agricultural Science*, **23**, 104–12.

Wittenbach, V. A. (1979). Ribulose bisphosphate carboxylase and proteolytic activity in wheat leaves from anthesis through senescence. *Plant Physiology*, **64**, 884–7.

Wong, S. C., Cowan, I. R. & Farquhar, G. D. (1979). Stomatal conductance correlates with photosynthesis capacity. *Nature*, **282**, 424–6.

Yoshida, S. & Coronel, U. (1976). Nitrogen nutrition, leaf resistance and leaf photosynthetic rate of the rice plant. *Soil Science and Plant Nutrition*, **22**, 207–11.

Zelitch, I. (1979). Photorespiration: Studies with whole tissue. In *Encyclopedia of Plant Physiology*, (new series) vol. 6, *Photosynthesis. II. Photosynthetic Carbon Metabolism and Related Processes*, ed. M. Gibbs & E. Latzko, pp. 353–67. Berlin: Springer-Verlag.

Zima, J. & Šesták, Z. (1979). Photosynthetic characteristics during ontogenesis of leaves. 4. Carbon fixation pathways, their enzymes and products. *Photosynthetica*, **13**, 83–106.

# 14

## Transport functions of leaves

U. LÜTTGE

### Introduction

*Survey of transport functions of leaves*

Transport plays an important role in the way in which organisms reflect and respond to their environment. But it is also essential within organisms, where division of labour has led to specialization, i.e. of organelles within cells, cells within tissues, tissues in organs and organs of whole plants.

All of these aspects of transport apply to leaves, which are dependent on import of material, such as inorganic ions and water, via the xylem from the roots, and in turn export materials, notably photosynthetic products, via the phloem. Aerial leaves communicate with the external environment mainly by gas exchange ($O_2$, $CO_2$, $H_2O$) via the stomata but also by other mechanisms – for instance, through the absorption of water from rain or dew, the leaching of solutes in the rain, or excretion or secretion of water and solutes by passive and active hydathodes, salt glands and nectaries. Submerged leaves of aquatic higher plants may take up ions over their entire surface. As in any other type of eukaryotic cell there is a vast number of transport processes across membrane boundaries of organelles within leaf cells. Most conspicuous, however, in green leaf cells are the sets of transport mechanisms at the envelopes of mitochondria and chloroplasts which allow control of dissimilatory and assimilatory reaction sequences and regulation of respiratory and photosynthetic energy metabolism on the transition from dark to light (see reviews in Stocking & Heber, 1976). In normal bifacial leaves there is compartmentation and transport between palisade parenchyma and spongy parenchyma (Outlaw & Fisher, 1975; Outlaw, Fisher & Christy,

1975), epidermal and vascular tissues. Even more specific relationships occur in the leaves of plants having Kranz anatomy, where mesophyll and bundle-sheath cells contain separated sets of enzyme machineries cooperating in $C_4$-photosynthesis (Hatch & Osmond, 1976).

Thus leaves have a more-or-less complex anatomy with division of labour between various cell types. In vascular plants the most uniform leaves are found among species which are secondarily adapted to aquatic life. Submerged leaves do not need stomata; instead they often have a large internal gas phase or aërenchyma made up of intracellular spaces serving for gas storage ($O_2$, $CO_2$). Vascular tissues, especially the xylem, are usually much reduced, there being no particular requirements for long-distance transport of water and inorganic ions. The most simple leaves of higher plants are those of *Elodea*, with only two cell layers, an upper one with larger cells and a lower one with smaller cells (Fig. 14.1*a*). But the cells of the lower layer have a complex structure. The outer cell wall forms protuberances towards the cell interior with the result that the plasmalemma area is increased substantially (Fig. 14.1*b*). Cells with such structures, the so-called transfer cells (Gunning, 1977), are usually found in tissues and organs of plants at locations with a particular requirement for solute transport across cell boundaries. In other words, the whole lower surface of *Elodea* leaves is a layer of transfer cells. Another example of the occurrence of transfer cells is in the hydropote glands of the lower surface of floating water lily leaves, which function in ion uptake (Fig. 14.1*c*).

Influx and efflux across the plasmalemma of cells and intracellular transport across membranes of subcellular compartments may be considered as 'short-distance' transport, intercellular transport in tissues and organs as 'medium-distance' and inter-organ transport in whole plants as 'long-distance' transport (Lüttge & Higinbotham, 1979). In the study of ionic relations of leaves it has proved useful to consider general structural models of the interactions between the various transport processes.

Fig. 14.1. Transfer cells of leaves of aquatic higher plants. (*a*) Cross section of a leaf of *Elodea callitrichoides*. (*b*) Electronmicrograph of the inner surface of the external wall of a cell from the lower layer of *E. callitrichoides* with a labyrinth formed by the cell wall protuberances. (*c*) Hydropote of the lower surface of a water lily leaf (*Nymphaea*), with a degenerating outer cell (O) and two gland cells (G), the outer one having an extremely complex cell wall labyrinth, the inner one possessing smaller wall protuberances. W, cell wall. (*a*) and (*b*) from Falk & Sitte (1963).

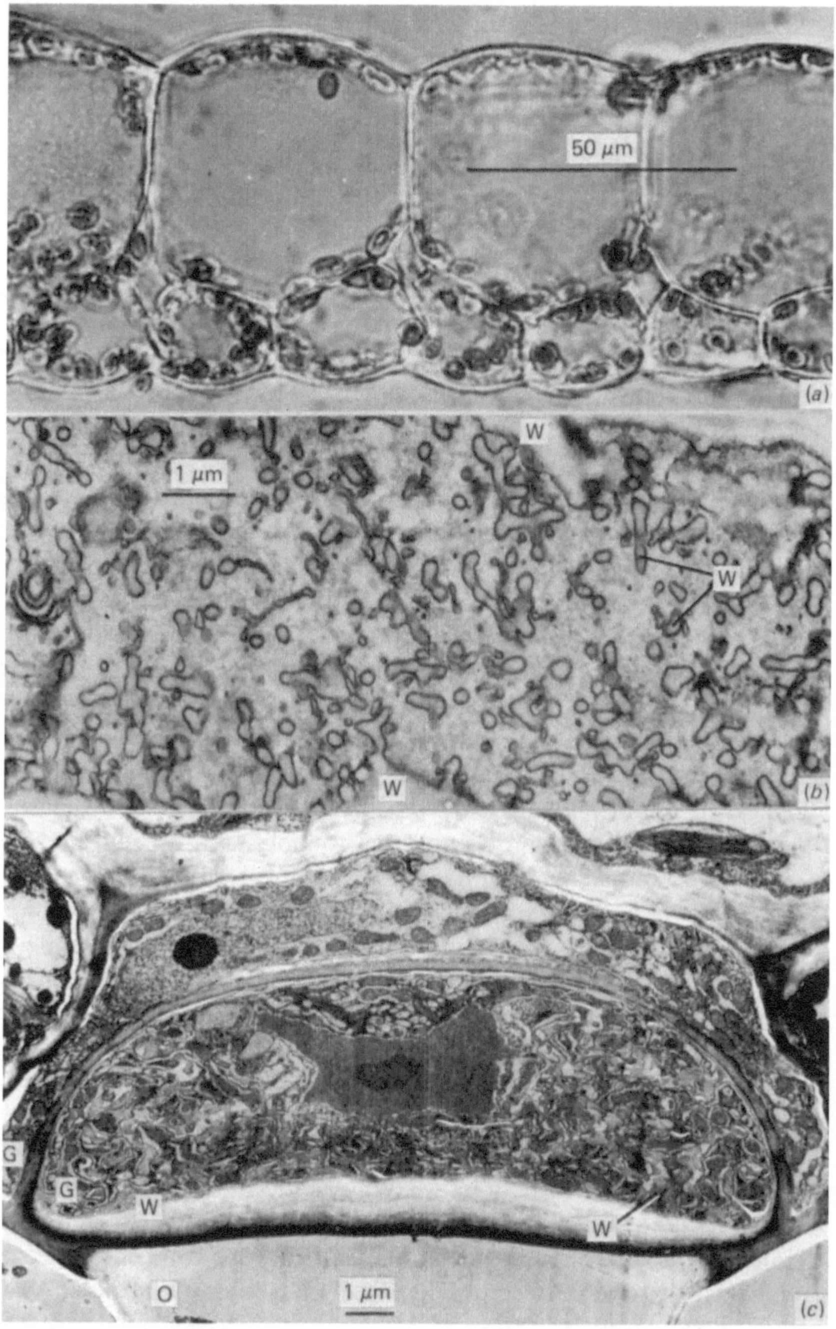
50 μm
(a)
1 μm
W
W
W
(b)
G
G
W
W
O
1 μm
(c)

In the simplest systems, as for example, in lengthy leaf strips of the aquatic macrophyte *Vallisneria* (Arisz, 1960), there is uptake and accumulation of ions across plasmalemma and tonoplast by all cells and there are apoplastic and symplastic pathways for intercellular transport. The situation is not too different in aerial leaves, if we envisage that water and solutes supplied via the xylem are distributed in the leaf apoplast via the tracheidal endings of leaf veins.

The systems become more complex when considering input of ions to the leaves, export from salt glands, salt bladders and hydathodes, or reabsorption from the xylem and retranslocation in the phloem. Salt-excreting systems are depicted in the left part of Fig. 14.2 and compared with the ion transport mechanisms operating across roots. The latter have been investigated very extensively (see pp. 323–36 of Lüttge & Higinbotham, 1979). One of the outstanding features in the root is the obstruction of the apoplastic pathway at the endodermal cell layer by the Casparian strip, an impregnation of the free spaces of the cell wall with cutin- and lignin-like polymers. Thus mechanisms of ion uptake and release at the plasmalemma of root hair, epidermis and cortex cells peripheral to the endodermis and of ion release or reabsorption across the plasmalemma of xylem parenchyma cells in the central cylinder are only coupled via the symplastic pathway. This principle is repeated in leaves with glands and salt bladders. There are apoplast barriers in most glands (Hill & Hill, 1976; Lüttge & Schnepf, 1976). Hence, there is apoplastic transport only up to a barrier which is bypassed by symplastic transport. There is transport across the tonoplasts of all cells and the vacuoles are used as compartments for accumulation and storage; and there are the two important transport mechanisms at the plasmalemma on both sides of the apoplast barrier. With a wealth of experiments Pitman and his colleagues have characterized the different properties of these 'pumps' in roots, using inhibitors of metabolism, phytohormones such as cytokinins and abscisic acid, inhibitors of protein synthesis, and amino acid analogues application of which leads to the synthesis of non-functional proteins (Pitman, 1977; Lüttge & Higinbotham, 1979). Guttation by gland-less passive hydathodes of leaves – for example, at the margins of many plants like lady's mantle, *Alchemilla*, or strawberry, *Fragaria*, or at the tips of young grass leaves – shows similar responses to such agents as does the export of ions from the xylem of excised whole roots. As shown in Fig. 14.2 no additional membrane system is traversed by transport in these hydathodes, unlike in salt glands (Dieffenbach, Lüttge & Pitman, 1980).

Fig. 14.2. Simplified structural scheme summarizing transport functions. Left part: ion transport with (from top to bottom) passive hydathode, bladder cell, salt gland and root. Right part: sugar transport with phloem loading (top) and nectary (bottom). The schemes show (i) *short-distance transport at membranes*, i.e. firstly, uptake into the symplast from the ambient apoplast; secondly, release into bladder vacuoles, to the leaf surface or into the xylem; and thirdly, fluxes into and out of the vacuoles; (ii) *symplastic transport* between cooperating cells of leaf and root tissues, respectively; (iii) *long-distance transport* and recycling in the xylem and phloem. Apoplastic transport can occur in the cell walls but not in regions of Casparian-strip like structures indicated by heavy black. The cytoplasm is dotted in gland-like cells and companion cells more heavily than in normal parenchyma cells; v, vacuoles; chl, chloroplasts.

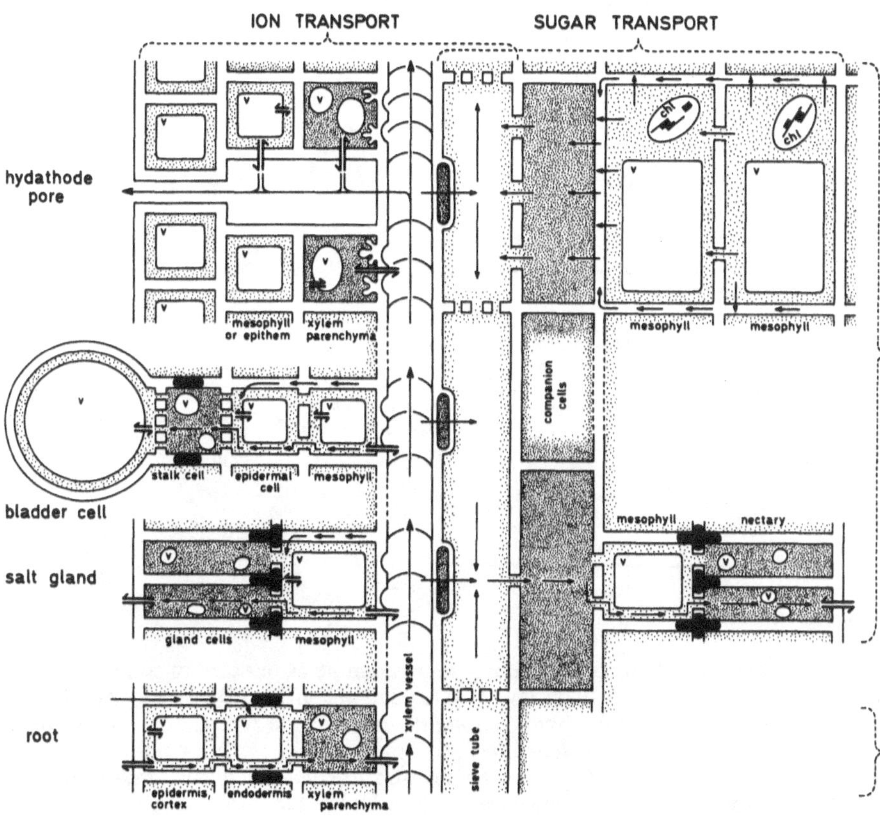

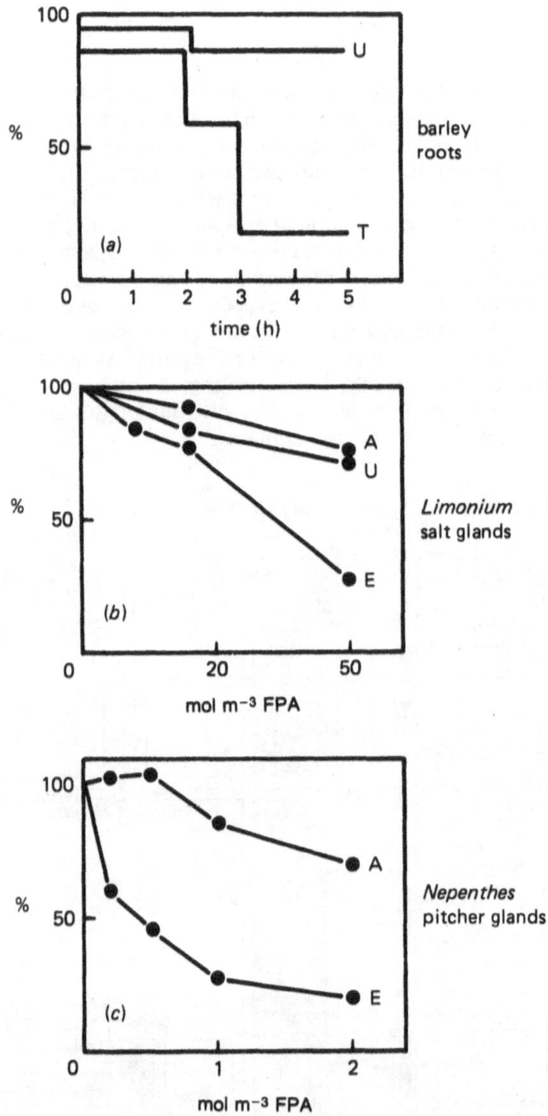


Fig. 14.3. Effects of *p*-fluoro-phenylalanine as per cent of controls on $Cl^-$ movements in barley roots (*a*) and in leaf-gland systems, (*b*) salt glands of the halophyte *Limonium vulgare* and (*c*) pitcher glands of the carnivorous plant *Nepenthes*. (*a*) Time course with 2mol $m^{-3}$ FPA added at time 0; (*b*) and (*c*) concentration dependence in long-term experiments extending over 15–20 h. U, uptake into tissue; T, transport out of the xylem from the cut end of excised roots; E, excretion from glands; A, accumulation in gland-bearing leaf tissue. In *Limonium* whole leaves were fed with FPA via the petioles (large

That leaf-gland systems, compared here structurally with roots, also function in a similar way is shown from experiments using the amino acid analogue *p*-fluoro-phenylalanine (FPA) (Fig. 14.3). Export of chloride into the xylem vessels of barley roots and excretion by the salt glands of *Limonium* or by the pitcher glands of *Nepenthes* is much more sensitive to FPA than uptake by and accumulation in the root tissue, the leaf tissue of *Limonium*, and the pitcher wall tissue of *Nepenthes* which morphologically is homologous to a leaf. Thus it appears that, in all of these cases, the proteins involved in the release of the ion from the symplast are more readily disabled by the incorporation of the amino acid analogue than those operating in uptake into the symplast.

In Fig. 14.2 one can also see how root-shoot interactions may operate in ionic relations of whole plants. The solution transported in the xylem to the shoots may be modified along the route. This is largely due to the activity of xylem parenchyma cells in the upper root zones, the hypocotyl, the stems and the leaves, and in their ultrastructure these cells are transfer cells (Läuchli, 1979). Such cells are also important for uptake of nitrogenous solutes from the xylem in leaves and distribution within the leaves (Pate, 1980). Reabsorption of salt from the xylem can also be mediated by transfer cells in the phloem (E. Winter-Sluiter, personal communication) and subsequently retranslocation of the ions to the roots occurs in the phloem.

To complete this survey similar schemes are considered, in the right side of Fig. 14.2, for transport of organic solutes, particularly of sugar. These schemes comprise loading and unloading of sieve tubes by companion cells and secretion of sugar by nectaries. Sucrose can move in the symplast and apoplast from the photosynthesizing mesophyll cells to the companion cells of sieve tubes. There is now much evidence that it must enter the apoplast prior to loading of the sieve tubes by active transport from the apoplast into the symplast of the sieve element-companion cell complex (Geiger, 1975, Giaquinta, 1980). In most cases of nectar secretion apoplast barriers are involved as in salt excretion. Several alternative pathways of transport can be discussed (see below), but after symplastic movement of sugar from the sieve tubes to the gland cells, an active transport out of the symplast is probably the most important step in secretion.

Caption to Fig. 14.3 (*cont.*)
concentrations needed), in *Nepenthes* pitcher wall discs were used. (*a*) after Fig. 4 of Schaefer, Wildes & Pitman (1975), (*b*) and (*c*) after Figs. 2 and 3 from Jung & Lüttge (1980).

## Interrelations in the whole plant

### *Source-sink concepts and signalling systems*

Source-sink concepts have long been used to explain sugar translocation in the phloem by a mass flow mechanism (Münch, 1930). With respect to photosynthetic products, during their ontogeny leaves change from net-consuming, importing sinks to net-producing, exporting sources. During this transition correlated physiological and structural changes occur (see Chapters 6–10). The source-sink concept has proved extraordinarily useful in considering interrelationships in the plant as a whole, including the distribution of inorganic nutrients. The rate of ion absorption by leaf cells is related to cell division and extension and thus is highly dependent on leaf age (Fig. 14.4). Source-sink relations in whole plants have been investigated using systems which can be easily handled, such as young barley seedlings (see below), or germinating seeds where the seminal roots and the young shoots compete for reserves of organic and inorganic nutrients from the seed (Sutcliffe, 1976).

In tissues and cells the salt status has considerable feedback effects on solute uptake. This can be very unspecific as in root parenchyma cells where high KCl loading inhibits the uptake of various kinds of other solutes (Cram, 1980). But it is often highly specific as shown in studies on N- and P-nutrition in many organisms, including yeast and algae. An example related to leaves is shown in Fig. 14.5. Phosphate content and growth of fronds of *Lemna paucicostata* are greatly reduced in a phosphate-free medium, but phosphate uptake by such fronds is considerably increased; in other words, the rate of phosphate uptake is inversely related to the phosphate content of the tissue. The fronds of *Lemna* (duckweeds) are morphogenetically homologous to leaves with a highly reduced shoot axis.

Clarkson, Sanderson & Scattergood (1978) showed that in barley plants most of the additional phosphate absorbed in response to phosphate deficiency is translocated to the shoot, and they suggested that phosphate transport in roots is under the control of phosphate levels in leaves. Over such large inter-organ distances information about the loading state of tissues can be conveyed by the rates of transport of materials themselves. Assimilate export from the shoots appears to be mainly regulated by the assimilation rate and not by the requirement of the roots (Lambers, 1979). More subtle signals must also be involved which may be electrical or hormonal.

By injecting a small electrical current into *Elodea* leaves, through glass capillaries inserted into cells and serving as micro-salt bridges, it was shown that there is intercellular electrical coupling via the symplast (Spanswick, 1972). Transient oscillations of membrane potential triggered by switching photosynthesis on or off through dark-light-dark transitions can also be used to demonstrate this. These signals (see for example Figure 14.8*c*) are only produced by photosynthetically active green cells, but not by pale cells of variegated leaves or by the

Fig. 14.4. Uptake of $^{22}Na^+$, $^{42}K^+$ and $^{86}Rb^+$ from 0.1 mol $m^{-3}$ and 5 mol $m^{-3}$ solutions of their chlorides by slices of primary bean leaves as a function of leaf age. (After Jacoby, Abas & Steinitz, 1973.)

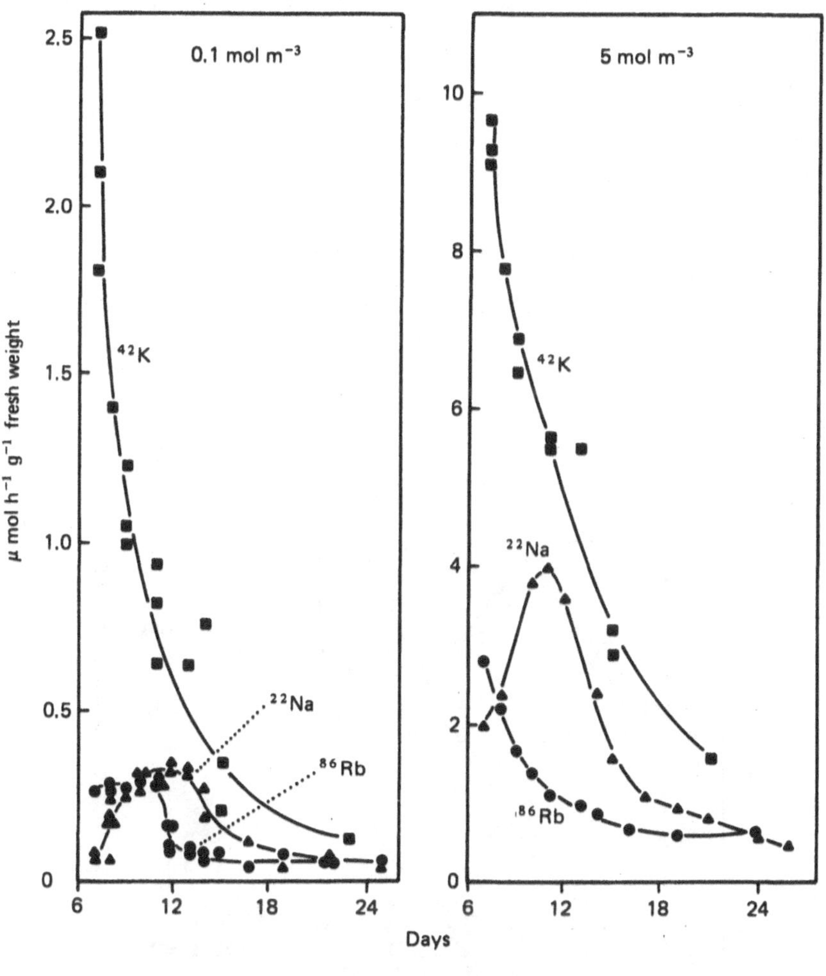

photosynthetically inactive bladder cells of *Atriplex* and *Chenopodium* leaves. However, when leaf tissue comprising both types of cells is subjected to light-dark transitions such signals are picked up with an electrode in photosynthetically inactive cell (see Lüttge & Higinbotham, 1979, pp. 296–301). Using a variegated leaf of *Oenothera*, the largest distance tested between the electrode in a pale cell and the closest green cells was 1 mm. There was no discernible lag phase before arrival of the signal at the electrode. The contact between the distant cells in the leaf could not have been due to symplastic transport of charged particles and electrotonic coupling involving membranes must be assumed (Brinckmann & Lüttge, 1974). Both the plasmalemma and the membranes of the endoplasmic reticulum extend, through the plasmodesmata, from cell to cell to form a continuous membrane system (Gunning & Robards, 1976).

Phytohormones have manifold effects on transport (van Steveninck, 1976). Synthesis of particular phytohormones may be restricted to specific locations in the plants, e.g. in roots or leaves, but all have been shown to be mobile in both directions between shoots and roots. The

Fig. 14.5. Growth, phosphate uptake and total phosphate content of fronds of *Lemna paucicostata* grown without (−P) and with 1.47 mol $m^{-3}$ phosphate (+P) in the medium. (After Ullrich-Eberius, Novacky, Fischer & Lüttge, 1981.)

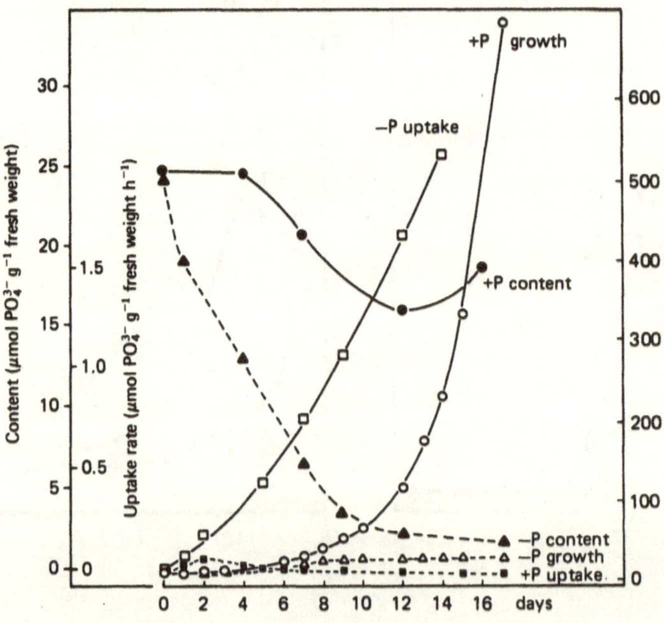

pathways of long-distance transport in xylem and phloem can serve this distribution of hormones (King, 1976). In this way phytohormones can play a role as signalling systems relating transport processes at the inter-organ level.

An arbitrary scheme illustrates the elements of the control that leaves can exert on their ionic relations (Fig. 14.6) (Pitman, 1972; Lüttge & Higinbotham, 1979: p. 368).

Fig. 14.6. Elements of a signalling system by which leaves control their ionic relations. P, phloem transport; X, xylem transport.

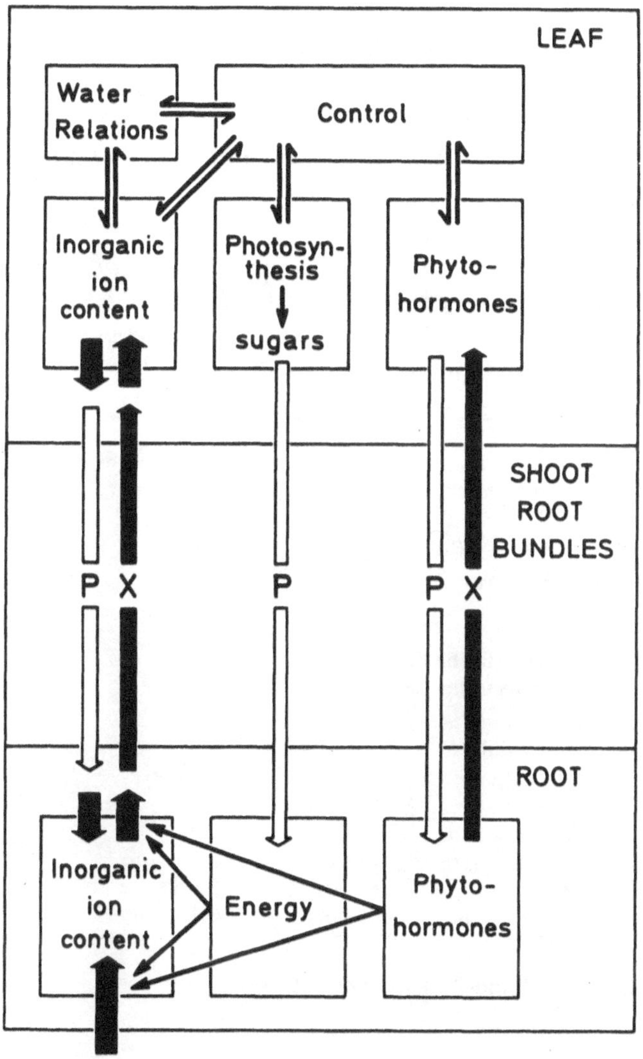

Experiments with week-old barley seedlings provide examples for the possible operation of the elements of the scheme of Fig. 14.6.

The role of photosynthetic substrate production and supply to the roots for regulation of ionic relations of leaves is shown in Table 14.1. In plants grown under a greatly reduced daily photoperiod (2 h L : 22 h D) relative growth rate was markedly decreased and physiological properties of leaves related to photosynthesis were highly affected. However, the ion content and transport processes of leaf cells were barely changed. The control of leaf ionic relations was accomplished by the roots which responded to the reduced demand from the leaves. In sunflower, it was shown by Graham & Bowling (1977) that the supply of metabolic substrates from shoot to root was also essential for the maintenance of electrical membrane potentials of root cells.

In other experiments roots of intact barley seedlings were sealed in a chamber with a water-saturated atmosphere so that there was no water deficit between the root tissue and the environment, but no liquid water

Table 14.1. *Ionic relations and physiological activities of leaves and roots of barley plants when daily photoperiod is reduced to 2 h, compared with those of plants under 16 h photoperiod; values are presented as ratios in short (2 h) to long (16 h) photoperiods.*

| | Photoperiod ratio: $\frac{2\,\text{h L} : 22\,\text{h D}}{16\,\text{h L} : 8\,\text{h D}}$ |
|---|---|
| *Relative growth rate:* | 0.25 |
| *Chlorophyll content of the leaves:* | 0.17 |
| *Ionic relations of the leaves:* | |
| $K^+$ content | 0.88 |
| $K^+$ uptake by leaf cells | 1.00 |
| $Cl^-$ uptake by leaf cells in the light | 1.22 |
| $Cl^-$ uptake by leaf cells in the dark | 0.94 |
| *Physiological activities of the leaves:* | |
| photosynthetic $O_2$ evolution | 0.67 |
| respiratory $O_2$ uptake | 0.99 |
| *Ionic relations of the roots:* | |
| active ion uptake by the root | 0.38 |
| active ion accumulation in the vacuoles | 0.16 |
| passive ion efflux | 0.91 |
| transport in the xylem towards the leaves | 0.13 |

After Pitman (1972) and Pitman *et al.* (1974b).

was available for transport to the shoot. In this way leaves were mildly wilted. Subsequently roots were submerged again in nutrient solution. Water content of leaves and physiological activities like photosynthesis, respiration, protein synthesis, and ion uptake by leaf cells recovered very rapidly. However, in roots cut from these plants an appreciable time after complete recovery, the ion export from the xylem was still inhibited, while ion accumulation by the roots was enhanced (Fig. 14.7).

Fig. 14.7. Water content (*triangles*) of barley leaves during mild wilting and during recovery; transpiration and ion accumulation of plants during recovery; export of ions from roots cut from prewilted plants after various times of recovery (i.e. at times indicated by *small arrows* on the abscissa: 15 min, ····; 45 min, –·–·–·–; 195 min, –––). Water content in % = 100 × (water content of leaves : water content of fully turgid leaves). Transpiration, ion accumulation and export are given as % of nonwilted controls. (From Lüttge, 1974.)

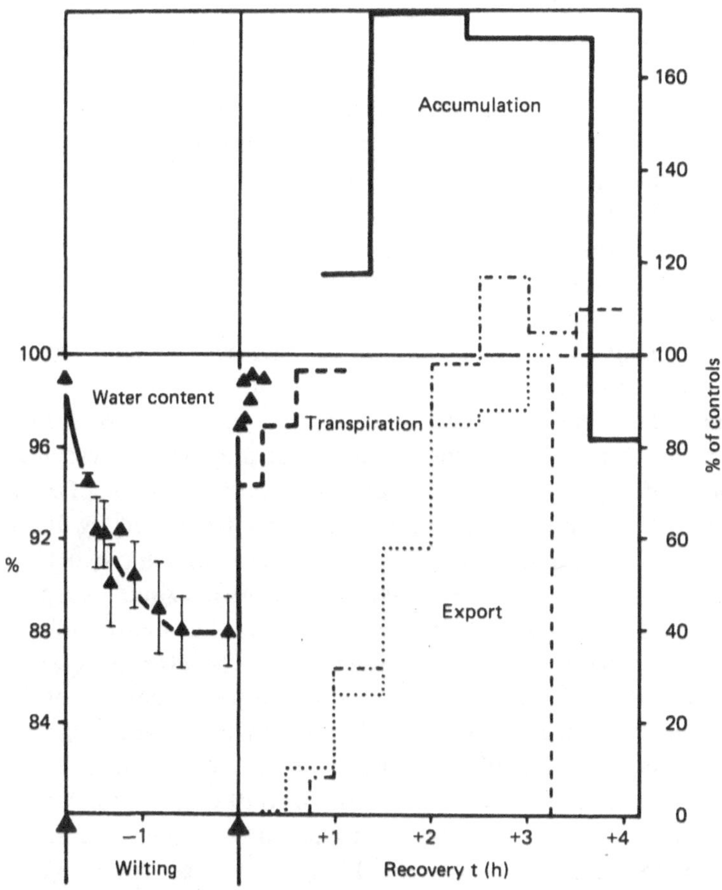

These long after-effects of wilting must be due to transport of some factor, presumably a hormone, from the leaves to the roots. The behaviour of the roots of the previously wilted plants is very similar to that of roots treated with abscisic acid (ABA) where, under certain conditions, xylem transport, but not accumulation, is inhibited (Cram & Pitman, 1972; Pitman, Lüttge, Läuchli & Ball, 1974a; Pitman & Wellfare, 1978; Dieffenbach *et al.*, 1980). It is well known that leaves respond to water stress by increasing ABA levels in their cells considerably (see Chapter 11) and that ABA can be transported to the roots in the phloem. This hormone could well be the factor involved in regulating the response of roots to wilting of leaves.

### Sources of metabolic energy available for ion uptake

It was known in the early 1930s that light can stimulate ion uptake by green cells. In 1965 MacRobbie suggested that in the Characean alga, *Nitella translucens*, $K^+$ uptake is driven by ATP and light-dependent $Cl^-$ uptake by a different mechanism directly consuming redox energy. This initiated a vast exploration of the possible sources of metabolic energy available for ion uptake by green leaves, including aquatic and aerial leaves. The latter had to be cut in thin slices to minimize limitations by diffusion when the tissue was submerged in experimental solutions (Smith & Epstein, 1964; Osmond, 1968). Problems remained, however, including the phenomena of adaptive ageing due to wounding after cutting (Pitman, Lüttge, Läuchli & Ball, 1974b), and predominant gas exchange via the stomata even in submerged leaf slices (Ullrich-Eberius, Lüttge & Neher, 1976).

Ion fluxes have been studied under very many different conditions: in the light and in the dark, using inhibitors of metabolism, varying the wavelength of irradiated light, manipulating the composition of external gas phases, and using photosynthetic mutants (reviewed by Jeschke, 1976; Lüttge & Higinbotham, 1979; pp. 216–34). In some experiments quite specific responses were observed, but two examples only are discussed here. *First*, experiments with some $C_4$ plants gave indications of the operation of mechanisms as suggested by MacRobbie (1965) for *Nitella*. The light-dependent fraction of $Cl^-$ uptake needed non-cyclic photosynthetic electron flow and appeared to be independent of, or even enhanced by, uncouplers of phosphorylation, so that the involvement of redox energy could be implied. In contrast, a cation transport mechanism in $C_4$ plants, and also $Cl^-$ uptake in $C_3$ plants, were inhibited by uncouplers, thus ATP was the likely source of energy (Lüttge,

Pallaghy & Osmond, 1970; Lüttge, Ball & von Willert, 1971). *Second*, and in contrast, in leaves of the water plant *Vallisneria* it seemed that only $Cl^-$ uptake could be powered by ATP originating from cyclic photophosphorylation. For cation uptake (measured with $^{86}Rb^+$ as a tracer) non-cyclic electron flow was needed; its action spectrum showed the characteristic 'red drop' similar to that of photosynthetic $O_2$ evolution, indicating the involvement of photosystem II in addition to photosystem I. The action spectrum of $Cl^-$ uptake declined more gradually in the far-red region suggesting that photosystem II, which is not excited above 705 nm, was not required for energy supply (Prins, 1973).

Dark-grown week-old barley ($C_3$) and maize ($C_4$) seedlings were used to relate ion uptake ($^{36}Cl^-$, $^{86}Rb^+$) to the development of photosynthetic capacity during greening, when the etiolated plants were illuminated. Leaf slices were obtained from etiolated and greening plants, and $O_2$ exchange and rates of ion uptake were measured in the light and in the dark, and in air and an atmosphere of nitrogen. Under the conditions applied, photosynthetic $O_2$ evolution reached a steady rate within 4 h after the commencement of illuminating the etiolated seedlings (Lüttge, 1973). Rates of ion uptake by leaves of greening plants in light + $N_2$ increased within 4 h with similar kinetics to photosynthetic $O_2$ evolution. In air, similar rates of ion uptake were obtained independently of illumination and irrespective of whether the tissue was etiolated or greened. Only in the dark + $N_2$ and with etiolated tissue also in the light + $N_2$, when glycolysis was the only energy source available, were rates of ion uptake much reduced (Lüttge & Ball, 1973). Table 14.2 gives data for etiolated and greened barley leaves, including levels of ATP, under various conditions. As long as respiratory or photosynthetic energy was available ATP content and rates of ion uptake were high, 40 nmol $g^{-1}$ and 4–6 $\mu$mol $g^{-1}$ $h^{-1}$, respectively, irrespective of the actual energy turnover, which ranged from 10 to 55 $\mu$mol ~P $g^{-1}$ $h^{-1}$. Only under conditions where cells were constrained to glycolysis were ATP levels and ion uptake very low.

In general, the type of work described in this section shows that the response of ion uptake by leaf cells to light depends very much on the nutritional and physiological state of the tissue and the nature of the leaves. Transport processes are not obligatorily linked to particular energy-providing systems, but can utilize energy produced by any of the various pathways of cellular energy metabolism. Transport stops only when all energy sources are impeded.

Table 14.2 *Energy relations and ion uptake in cells of green and etiolated barley leaves in light (L) and darkness (D) and in aerobic (air) and anaerobic ($N_2$) conditions. Phosphorylation rates were obtained from respiratory $O_2$ uptake (assuming a ~P/O ratio of 3) and photosynthetic $O_2$ evolution (assuming a ~P/O ratio of 1).*

| | L, air | D, air | L, $N_2$ | D, $N_2$ |
|---|---|---|---|---|
| *Green leaves:* | | | | |
| process of phosphorylation | non-cyclic photo-phosphorylation | respiration | cyclic photophosphorylation, limited non-cyclic photophosphorylation; limited respiration ($O_2$ production in light) | glycolysis |
| phosphorylation ($\mu$mol $g^{-1}$ $h^{-1}$)[a] | 55 | 50 | 10 | n.d. |
| ATP-level (nmol $g^{-1}$)[a] | 40 | 35 | 40 | 7 |
| ion uptake ($\mu$mol $g^{-1}$ $h^{-1}$)[a] | 6 | 6 | 6 | 1.5 |
| *Etiolated leaves:* | | | | |
| process of phosphorylation | respiration | respiration | glycolysis | glycolysis |
| phosphorylation ($\mu$mol $g^{-1}$ $h^{-1}$)[a] | 50 | 50 | n.d. | n.d. |
| ATP-level (nmol $g^{-1}$)[a] | n.d. | n.d. | 10 | 10 |
| ion uptake ($\mu$mol $g^{-1}$ $h^{-1}$)[a] | 4 | 4 | 1.5 | 1.2 |

After Lüttge & Ball (1976).
[a] Note that in this table and throughout this chapter weights refer to fresh weights.

In the second half of the 1970s the search for specific energy-providing reactions driving specific transport processes was replaced by the tendency to relate all transport processes to an electro-chemical gradient of protons ($\Delta\bar{\mu}_{H^+}$) at the plasmalemma. This unifying concept has developed from Mitchell's chemiosmotic theory of energy coupling. According to this theory a reversible ATPase can either use the $\Delta\bar{\mu}_{H^+}$ to synthesize ATP or establish a $\Delta\bar{\mu}_{H^+}$ by consuming ATP (Mitchell, 1967). The stimulus for investigations with green plant leaves again came from work on algae. Komor & Tanner (1980) investigated hexose uptake by an inducible transport process in the alga *Chlorella vulgaris* and demonstrated that sugar is taken up together with protons ($H^+$-sugar cotransport or symport) utilizing the energy of $\Delta\bar{\mu}_{H^+}$ established by an active $H^+$-extrusion pump in the plasmalemma.

## Electrical membrane potentials of leaf cells

### *Membrane potential in light and darkness*

When micro-salt bridges to measure intracellular electrical potentials are inserted into higher plant cells it is difficult to place the capillary tip in the cytoplasm because normally it slips immediately into the vacuole. Thus electrical potential is measured across two membranes, tonoplast and plasmalemma, in series. The problem can be overcome by gentle centrifugation of the material prior to the measurements, so that the capillary tip can be more easily inserted into the cytoplasm (Zurzycki, 1968; Lüttge & Zirke, 1974) or by slowly moving the capillary through the cell, which gives an 'electrical profile' with successive measurements in cell wall, cytoplasm, vacuole, cytoplasm and cell wall (Rona, Pitman, Lüttge & Ball, 1980). Where separate measurements of plasmalemma and tonoplast potentials have been possible, the potential at the tonoplast was found to be zero or slightly positive inside the vacuole. Such measurements have been made on oat coleoptiles (Etherton, 1970), thalli of the hornwort *Phaeoceros laevis* (Davis, 1974), leaves of the mosses *Funaria hygrometrica* (Zurzycki, 1968) and *Mnium cuspidatum* (Lüttge & Zirke, 1974), the aquatic plant *Potamogeton schweinfurthii* (Denny & Weeks, 1968) and the crassulacean acid metabolism plant *Kalanchoë daigremontiana* (Rona *et al.*, 1980). Thus, the highly negative potentials obtained with tips of electrodes in the vacuole are largely plasmalemma potentials. Inhibitor studies show that these potentials are composed of two components, a contribution due to passive diffusion of ions, the diffusion potential, and a contribution due to an active process consuming metabolic energy, the electrogenic pump.

There are conflicting observations on the effect of light on the membrane potential of green leaf cells. Transient oscillations of the membrane potential are observed in all green cells when there are transitions from darkness to light or vice versa (e.g. see Fig. 14.8*c*) They are due to intracellular regulation processes occurring after non-cyclic photosynthetic electron flow is switched on or off (Bentrup, 1974; Brinckmann & Lüttge, 1974). But the resting potential usually is similar in continuous light and darkness although there are reports of less negative, i.e. smaller, potential differences in the dark than in the light (Bentrup, 1974). Since the transient oscillations triggered by changes in illumination often take about one hour or even longer to settle, care must be taken that new steady levels are attained. But some of the reported light–dark differences of membrane potential are definitely genuine.

What are the reasons for these differences? It has been shown that in *Vallisneria* leaves the low potential observed in the dark is largely, or entirely, a diffusion potential (Bentrup, Gratz & Unbehauen, 1973). Prins, Harper & Higinbotham (1980) suggest that the diffusion potential in *Vallisneria spiralis* is about −110 mV; and since there is a small electrogenic component of −13 mV in the dark and a large electrogenic component of −80 mV in the light, potentials are about −125 mV and −190 mV in the dark and light, respectively. Conversely, however, M. Rohling and U. Lüttge (unpublished) found no difference due to illumination with resting potentials of $-189 \pm 9$ (SD, $n = 8$) and $-207 \pm 20$ (SD, $n = 10$) in the dark and light respectively. The diffusion potential in the presence of inhibitors was −90 to −95 mV, i.e. similar to that observed by Prins *et al.* (1980).

Differences in the light response of the resting potential of a given species may be due to variation in the ionic compositions of experimental solutions. Phytochrome effects could also be involved in control of electrical membrane potentials of green stems and leaves but, although significant, they are only of small magnitude (Löppert, Kronberger & Kandeler, 1978; Racusen & Galston, 1980). It is much more likely, however, that the large differences in resting potential obtained under certain conditions in light and darkness are due to nutritional and energy status. This is shown by detailed analyses of the membrane potential of duckweed cells (*Lemna gibba*) which can be readily grown in axenic cultures. Best growth is obtained under mixotrophic conditions in short days (8 h L : 16 h D) with 29 mol m$^{-3}$ sucrose added to the inorganic nutrient solution. These plants always have, in dark *and* in light, a highly negative membrane potential (around −200 mV) with a large metabolism-dependent component of −100 to −120 mV and a diffusion potential of −90 to −100 mV. When plants are kept for 8 days under long day conditions (16 h L : 8 h D) without sucrose in the medium, they grow very fast but presumably deplete their energy reserves because if this is followed by 1–2 days of treatment in inorganic medium in complete darkness, the energy-dependent component of the membrane potential is partially or totally abolished so that it is close to diffusion potential of −90 to −100 mV. In the dark addition of 50 mol m$^{-3}$ glucose to the medium, which can be taken up by diffusion and serve as a respiratory substrate, leads rapidly to a hyperpolarization of the cells as does illumination (Fig. 14.8*a*, *b*). ATP levels of the 'low-potential plants' are only 20–50 nmol g$^{-1}$ as compared to 80–140 nmol g$^{-1}$ in the 'high-potential plants' (Novacky, Ullrich-Eberius & Lüttge, 1978; Böcher, Fischer, Ullrich-Eberius & Novacky, 1980). It is evident that the membrane potential in a given organism may have different stable states depending on nutritional conditions and that the effect of light differs greatly depending on the physiological state of the cells. Transitions from one stable state to another may not only be triggered by provision of energy but also by other effectors like hormones (see Fig. 14.8*f*).

### *Electrogenic proton pumps in the plasmalemma*

Theoretically any exchange of two cations or two anions or any co-transport or symport of a cation with an anion across a membrane can cause electrogenicity if the coupling is not exactly 1:1 so that there is imbalance of charge. Since there are usually many indirect couplings of fluxes across a membrane, it is often very difficult to determine the transport of which ionic species is the primary active step leading to charge separation. Pursuing the Mitchell-type mechanism of a reversible membrane ATPase, which can utilize $\Delta\tilde{\mu}_{H^+}$ at a membrane to synthezise ATP, or consume ATP to establish $\Delta\tilde{\mu}_{H^+}$, has indicated that plant cells have plasmalemma ATPases extruding protons from the cells so that $H^+$-pumping is the primary active transport. Potassium is exchanged for $H^+$; but $\Delta\tilde{\mu}_{H^+}$ at the plasmalemma can also be utilized as an energy source for other processes at the membrane.

In leaf cells it is extraordinarily difficult to measure $\Delta\bar{\mu}_{H^+}$ at the plasmalemma and the effects of metabolism on it. $\Delta\bar{\mu}_{H^+}$ comprises an electrical and a chemical component:

$$\Delta\bar{\mu}_{H^+} = RT\ln\frac{c_i H^+}{c_o H^+} + F\Delta E = RT \times 2.3 \times \Delta pH + F\Delta E \qquad (14.1)$$

Fig. 14.8. Membrane potential differences (PD) of cells of *Lemna gibba* under various conditions. (*a*) Hyperpolarization of a low-PD cell in the dark after addition of 50 mol m$^{-3}$ glucose to the medium for about 13 min. 1X = purely inorganic medium. (*b*) Hyperpolarization of a low-PD cell by light(L). (*c*) Transient changes of PD of a high-PD cell after dark–light–dark transitions. (*d*) Hyperpolarization of a low-PD cell by 0.015 mol m$^{-3}$ fusicoccin (FC) in the dark. (*e*) Hyperpolarization of a high-PD cell by 0.03 mol m$^{-3}$ fusicoccin (FC) in the dark; EtOH = ethanol control (solvent for FC). (*f*) Hyperpolarization of a low-PD cell by 10$^{-3}$ mol m$^{-3}$ abscisic acid (ABA) in the dark. (*g*) Transient depolarization of a high-PD cell by 5 mol m$^{-3}$ tributylbenzylammonium (TBBA$^+$) in the dark. (*h*) Transient PD changes of a high-PD cell upon addition and removal of 1 mol m$^{-3}$ L-alanine (L-Ala) in the dark.

(*a, b*) from Novacky *et al.* (1978a); (*c, g*) M. Rohling unpublished; (*d, e, h*) from Lüttge *et al.* (1981); (*f*) from Hartung, Ullrich-Eberius, Lüttge, Böcher & Novacky (1980).

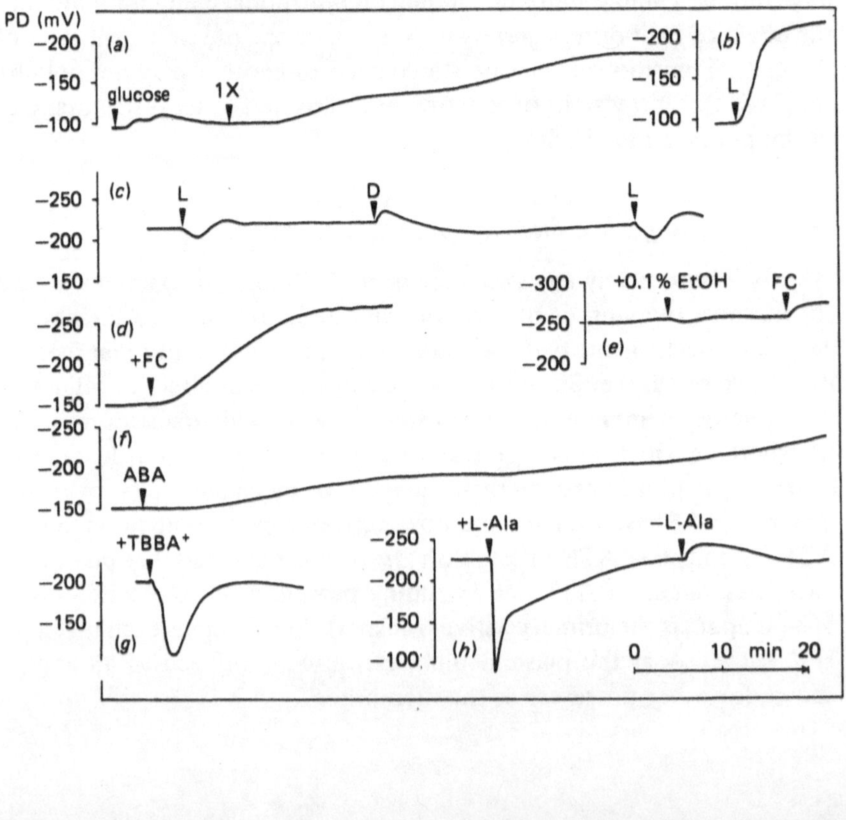

where *R*, *T* and *F* have the usual meanings, $c^{H^+}$ is proton concentration inside (i) and outside (o), respectively. While the electrical component ($\Delta E$) is readily measured (see above), the chemical component ($\Delta$pH) is difficult to determine. Measurements of cytoplasmic pH-values for green cells are only available so far from studies with characean algae, where a value close to 7.5 seems to be most likely (Spanswick & Miller, 1977; Smith & Raven, 1979; Keifer, 1980). It seems reasonable to assume a similar value in leaf cells of higher plants. The pH of the medium is often taken to be equivalent to the external pH but due to diffusion in the apoplast and buffering capacity of fixed ionic groups in the cell wall, the external pH in the vicinity of the plasmalemma is never known exactly (Cleland, 1976; Novacky, Ullrich-Eberius & Lüttge, 1980).

The use of more-or-less specific membrane effectors has been essential in the demonstration of $H^+$ co-transport. Of particular value have been carbonyl cyanide *m*-chlorophenyl-hydrazone (CCCP), which increases passive permeability of membranes for protons, and the fungal toxin fusicoccin (FC), which stimulates proton-extruding ATPases.

Fusicoccin has proved to be an extraordinarily useful tool since it stimulates very specifically the proton-extruding ATPase in the plasmalemma of higher plant cells (Marrè, 1979). This ATPase consists of two proteins, one of which has the ATPase activity and the other carries the FC binding site (Tognoli, Beffagna, Pesci & Marrè, 1979; Stout & Cleland, 1980). In cells of leaves of the CAM plant *Kalanchoe daigremontianum*, FC hyperpolarized the plasmalemma but had no effect on the tonoplast potential, either because it did not penetrate into the cells, or because the tonoplast lacks an FC-sensitive $H^+$-pumping ATPase (Rona *et al.*, 1980). In *Lemna gibba* the relative effects of FC depend on the state of the cells. The hyperpolarization of low potential cells by FC is considerable, that of the high potential cells smaller, but eventually in both types of cells similar highly negative potentials are obtained with FC (Fig. 14.8*d*, *e*). Fusicoccin does not increase the low ATP levels of the low potential plants (Böcher *et al.*, 1980) but it increases respiration (Lüttge, Jung & Ullrich-Eberius, 1981). Therefore it can be assumed that FC stimulates the membrane ATPase to an extent that it becomes a strong sink for ATP within the cell and thus can establish a high potential difference in spite of the low ATP levels. Hyperpolarization of the plasmalemma by FC can be taken as proof of operation of an $H^+$ extruding ATPase.

Abscisic acid, ABA, which has been inferred to interact with $H^+$–$K^+$

exchange in stomatal guard cells (Hsiao, 1976), also hyperpolarizes the low potential *Lemna* cells (Fig. 14.8*f*). Another interesting agent is the lipophilic cation tributylbenzylammonium ($TBBA^+$). When added at 5 mol m$^{-3}$ to the medium of leaf cells, it often leads to marked transient or persistent depolarizations (Lüttge & Ball, 1980). This may be due to ready uptake of positive charge by passive diffusion of $TBBA^+$, or to dissociation equilibria of the lipophilic cations (weak-base effects) altering cytoplasmic pH and hence affecting $\Delta\bar{\mu}_{H^+}$ at the plasmalemma (see Walker & Smith, 1975; Keifer & Spanswick, 1979). The repolarization of the membrane potential in the presence of $TBBA^+$ could be brought about by increased pumping of $H^+$ out of the cells thus removing positive charge (Fig. 14.8*g*).

## Proton-electrochemical gradients driving the transport of other solutes across the plasmalemma of leaf cells

It has been mentioned above that, in *Chlorella*, an electrochemical proton gradient powers sugar uptake via a proton-sugar co-transport or symport mechanism. There is now substantial evidence that a similar mechanism drives uptake of sugars and amino acids in green cells of the aquatic liverwort *Riccia fluitans* (Felle, Lühring & Bentrup, 1979; Felle & Bentrup, 1980), *Lemna* (Novacky *et al.*, 1978a, b, 1980; Lüttge *et al.*, 1981), coleoptiles (Etherton & Rubinstein, 1978) and mesophyll cells of aerial leaves of higher plants (Delrot, Despeghel & Bonnemain, 1980; Despeghel, Delrot, Bonnemain & Besson, 1980; Guy, Reinhold & Rahat, 1980; see also (*a*) in Fig. 14.9.)

Solute $H^+$ co-transport requires two entities in the membrane; a carrier, C, which binds $H^+$ and the solute transported across the membrane, and the proton extrusion pump, P, which establishes an $H^+$ electrochemical gradient at the membrane. The interplay between C and P results in changes of the $H^+$ electrochemical gradient ($\Delta\bar{\mu}_{H^+}$, equation 14.1) when solute transport commences or stops. The transient nature of these changes is due to a small delay in the response of P, pumping protons at increased and decreased rates respectively after co-transport begins or stops. Possibly these responses of P are regulated by cytoplasmic pH which begins to decrease when $H^+$ co-transport sets in and to rise when at increased pumping rates $H^+$ co-transport stops. Examples are shown in Fig. 14.8*h* of transients in the electrical component ($\Delta E$) of $\Delta\bar{\mu}_{H^+}$, a depolarization after commencement and a hyperpolarization after the cessation of solute uptake. Transients in the chemical component ($\Delta$pH) of $\Delta\tilde{\mu}_{H^+}$, an alkalinization of the medium after the addition of transportable sugar, have been demonstrated with cotyledons of *Ricinus* (Hutchings, 1978a; Komor *et al.*, 1980), pulvini of *Samanea* (Racusen & Galston, 1977) and fronds of *Lemna* (Novacky *et al.*, 1980). Stimulation of solute uptake by fusicoccin is also evidence for the existence of $H^+$ co-transport mechanisms.

Solutes taken up by *Lemna* cells by such mechanisms also include

Fig. 14.9. Secondary active transport processes powered via flux coupling by primary active proton efflux; and possible primary active $Cl^-$ excretion; (*a*) organic solute- and (*b*) inorganic anion-proton co-transport, (*c*) $K^+/H^+$ exchange, (*d*) putative sugar secretion with $H^+$ extruded, (*e*) $Cl^-$ pump, (*f*) malate synthesis.
PEP, phosphoenolpyruvate; PEP-C, phosphoenolpyruvate-carboxylase; OAA, oxaloacetate; MDH, malate-dehydrogenase.

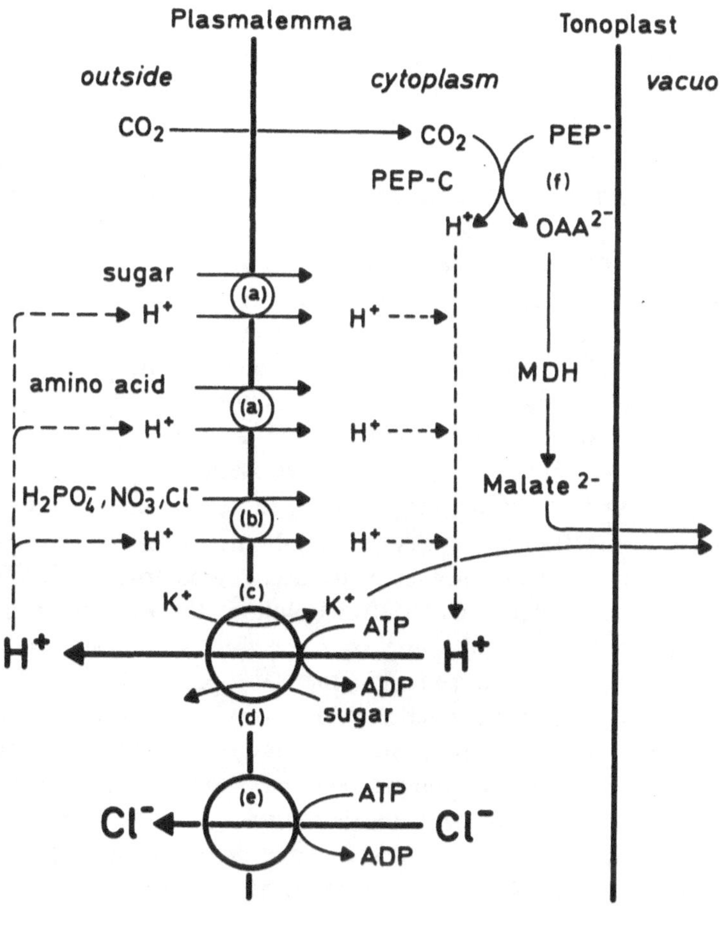

$NO_3^-$ (Novacky *et al.*, 1978a) and $H_2PO_4^-$ (Ullrich-Eberius, Novacky, Fischer & Lüttge, 1981). Since in these cases transient membrane-potential depolarizations are also observed after commencement of uptake, more than one proton must be transported with each of these inorganic anions.

## Functions of leaves utilizing the proton extruding ATPase of the plasmalemma

### *Phloem loading*

In aerial leaves of higher plants transport processes linked to the proton pump at the plasmalemma of leaf cells serve a variety of purposes. One of the most important is phloem loading. As already pointed out (see Fig. 14.2) sucrose is extruded from the symplast of the supplying mesophyll cells into the apoplast before it re-enters the symplast of the exporting complex of sieve tubes and companion cells (Geiger, 1975; Giaquinta, 1980). The latter transport process is specific for sucrose and must be considered to be the actual sugar-concentrating loading mechanism. Among the evidence that loading occurs at an apoplast/symplast interface is its specific inhibition by the sulfhydryl reagent *p*-chloromercuribenzene-sulfonic acid (PCMBS), which only penetrates the apoplast but is not taken up into the cells and has no effects on photosynthesis and respiration (Giaquinta, 1976).

There is much evidence that loading operates by proton-sugar co-transport (Hutchings, 1978a, b; Cho & Komor, 1980; Delrot *et al.*, 1980; Giaquinta, 1980; Heyser, 1980; Martin & Komor, 1980). PCMBS appears to act both on the proton-extrusion pump, P, and the co-transport carrier, C (Giaquinta, 1980). In cells of soybean cotyledons $2\,mol\,m^{-3}$ PCMBS depolarizes the membrane to the diffusion potential and inhibits the sucrose-induced transient depolarizations. When the tissue is washed with PCMBS-free solution, the membrane potential recovers to the control value, but the size of the sucrose-induced transients remains reduced (Lichtner & Spanswick, 1981). Similar results were found for the action $0.01\,mol\,m^{-3}$ $HgCl_2$ on glucose uptake by *Lemna gibba* (B. Golle and U. Lüttge, unpublished). Thus the effects of SH-reagents on the proton pump and the solute uptake carrier are different, and only with the latter does firm binding seem to be involved. Using another SH-reagent, *N*-ethylmaleimide, Komor, Weber & Tanner (1978) showed that the hexose carrier of *Chlorella* also needs an SH group for undisturbed function. In studies on phloem loading in *Vicia faba* leaves the use of PCMBS and *N*-ethylmaleimide led to the conclusion that the carrier but not the proton pump possesses functional SH-groups on the external side of the membrane (Delrot *et al.*, 1980). Application of two-substrate carrier kinetics describing the binding of $H^+$ and sucrose suggests that the sucrose-loaded carrier is half-protonated at pH 9.0, a value which is not too different from the pK of 10 characteristic of thiol groups (Delrot & Bonnemain, 1981).

### *Various special functions*

Other functions of leaves utilizing the proton-extruding ATPase of the plasmalemma are summarized in Fig. 14.9.

*Extension growth, volume increase of stomatal guard cells and motor cells of pulvini.*

According to the acid growth hypothesis, cell wall extensibility is brought about by acidification of the walls due to active $H^+$ extrusion; $K^+$ is being exchanged for $H^+$ (Fig. 14.9*c*). Since $H^+$ does not return into the cells as in $H^+$ co-transport (Fig. 14.9*a*, *b*) protons have to be provided for maintenance of cytoplasmic pH. This is accomplished by formation of new carboxyl groups through dark fixation of $CO_2$ (Fig. 14.9*f*). The malate synthesized is accumulated together with the $K^+$ in the vacuole, where it serves as osmotically active material for the establishment of turgor, which drives cell extension. This mechanism has been extensively investigated with cereal coleoptiles (Haschke & Lüttge, 1977), but it also operates in leaves (Volkenburgh & Cleland, 1980). Similarly, $K^+$ and malate accumulation in the vacuoles of guard cells driven by proton extrusion lead to the turgor increases required for stomatal opening (Hsiao, 1976; Raschke, 1979). In starch-free leaves of *Allium cepa* this mechanism does not work, however, because phosphoenolpyruvate (PEP), the precursor for malate by $CO_2$ dark fixation in stomatal regulation, must be derived from glucans. *Allium* guard cells instead need $Cl^-$ obligatorily to balance the charge of $K^+$ taken up (Schnabl & Ziegler, 1977; Schnabl & Raschke, 1980). In cells of other species uptake of $Cl^-$, depending on its availability, is an alternative to malate synthesis (Raschke & Schnabl, 1978). Chloride uptake is not necessarily a separate active mechanism; it could be due to a $H^+/Cl^-$ symport and thus linked to the proton extrusion pump without requiring malic acid synthesis for cytoplasmic pH regulation (Fig. 14.9*c* and *b*).

Similarly, the motor cells of pulvini involved in nyctinastic movements use KCl as osmoticum (Satter, 1979; Lüttge & Higinbotham 1979, pp. 241–5).

*Salt glands*

The leaves of some halophytes have salt glands excreting large amounts of NaCl thereby reducing the salt load of the mesophyll (Hill & Hill, 1976). The comparison of two mangrove species is informative in this respect. *Rhizophora mucronata* partially excludes salt from the transpiration stream at the root level. The $Cl^-$ concentration of the xylem sap is 17 mol $m^{-3}$. In contrast, the xylem sap of *Aegialitis annulata* contains 85–122 mol $Cl^-$ $m^{-3}$. As a consequence, a leaf of *R. mucronata* receives 17 $\mu$mol $Cl^-$ $d^{-1}$ and a leaf of *A. annulata* 100 $\mu$mol $d^{-1}$. Nevertheless, the amounts of $Cl^-$ and $Na^+$ found in the leaves are quite similar in both species, because *A. annulata*, but not *R. mucronata*, has salt glands excreting NaCl (Atkinson *et al.*, 1967).

Hill & Hill (1976) have investigated the salt glands of *Limonium vulgare* and found an inducible $Cl^-$-ATPase responsible for excretion (Fig. 14.9*e*) and Jung & Lüttge (1980) observed that fusicoccin inhibited net $Cl^-$ export by *Limonium* glands. In view of the very specific stimulating effect of FC on the proton-extruding ATPase (see above) an inhibition of the independent $Cl^-$-ATPase proposed by Hill & Hill is unlikely. However, the proton pump could drive reabsorption of $Cl^-$ by $H^+/Cl^-$ symport, this would be stimulated by FC and hence net $Cl^-$ excretion inhibited (combination of (*e*) and (*b*) in Fig. 14.9).

*Hydropotes and ion transport across water plant leaves*

Salt-absorbing hydropote glands of water plants with enormous transfer-cell-like wall labyrinths (Lüttge & Krapf, 1969; Lüttge, Pallaghy & von Willert, 1971) have already been mentioned (Fig. 14.1*c*). Perhaps the transfer-cell layers at the abaxial surfaces of *Elodea* and *Potamogeton* leaves (Fig. 14.1*b*) may be considered as equivalent to large hydropotes. These leaves in the light show a directed transport of $K^+$ from their abaxial to their adaxial surface and develop an $H^+$-electrochemical gradient across themselves, with the adaxial surface becoming alkaline and negative, and the abaxial surface becoming acidic and positive (Jeschke, 1976). It appears probable that proton-extrusion pumps lead to acidification in the apoplastic spaces of the cell wall protuberances at the abaxial surface. This would lead to formation of $CO_2$ from $HCO_3^-$ to be used in photosynthesis and neutralization of the remaining $OH^-$; $H^+$ would also be exchanged for $K^+$ (Fig. 14.9*c*). $K^+$ would move down its electrochemical gradient across the leaf to be released at the adaxial surface with perhaps passive exchange for $H^+$ (Prins, Snel, Helder & Zanstra, 1979).

*Glands of carnivorous plants*

Some of the transport functions of the glands on leaves of carnivorous plants, or on trapping organs morphogenetically derived from leaves, can also be explained by mechanisms shown in Fig. 14.9; these include $Cl^-$ excretion and acidification of the excreted sap in *Nepenthes* (Lüttge, 1981) and *Dionaea* (P. Rea, unpublished), and absorption of amino acids from the digested prey (P. Rea, unpublished). Others are due to vesicle-mediated secretion such as the extrusion of the content of Golgi vesicles during the secretion of the polysaccharide trapping mucilage (Lüttge & Schnepf, 1976). Cisternae of the endoplasmic reticulum in contact with the plasmalemma may serve as the secretion pathway of enzyme proteins in *Dionaea* (Robins & Juniper, 1980).

*Nectary glands*

Nectaries are usually located on leaves or parts of flowers such as sepals, petals, stamens and receptacles. Sugar transport in such systems can be visualized using the model of Fig. 14.2. A sugar-concentrating active transport process must be involved in the overall process (Lüttge & Schnepf, 1976). An intriguing question is whether or not $H^+$-sugar co-transport, which is so widely distributed in higher plants, could also be involved in nectar secretion. It appears unlikely that active phloem unloading uses this mechanism since the opposite process of phloem loading utilizes $H^+$-sugar co-transport; and moreover since the phloem sap is usually somewhat alkaline, the $H^+$ pump will work in the direction from sieve tube and companion cell cytoplasm to the apoplast. Presumably symplastic transport prevails from the sugar-providing phloem to the gland cells, and there may not be a pump concentrating sugars in the gland cells. Thus, it remains important to consider active secretion of sugars out of the gland cytoplasm. There are some indications that the nectar becomes acidic during secretion (Lüttge, 1977). Although more data are needed, this could indicate an involvement of $H^+$ transport in nectar secretion. It must be noted, however, that this would differ from the $H^+$ co-transport mechanisms discussed above. Sugar would be transported in the

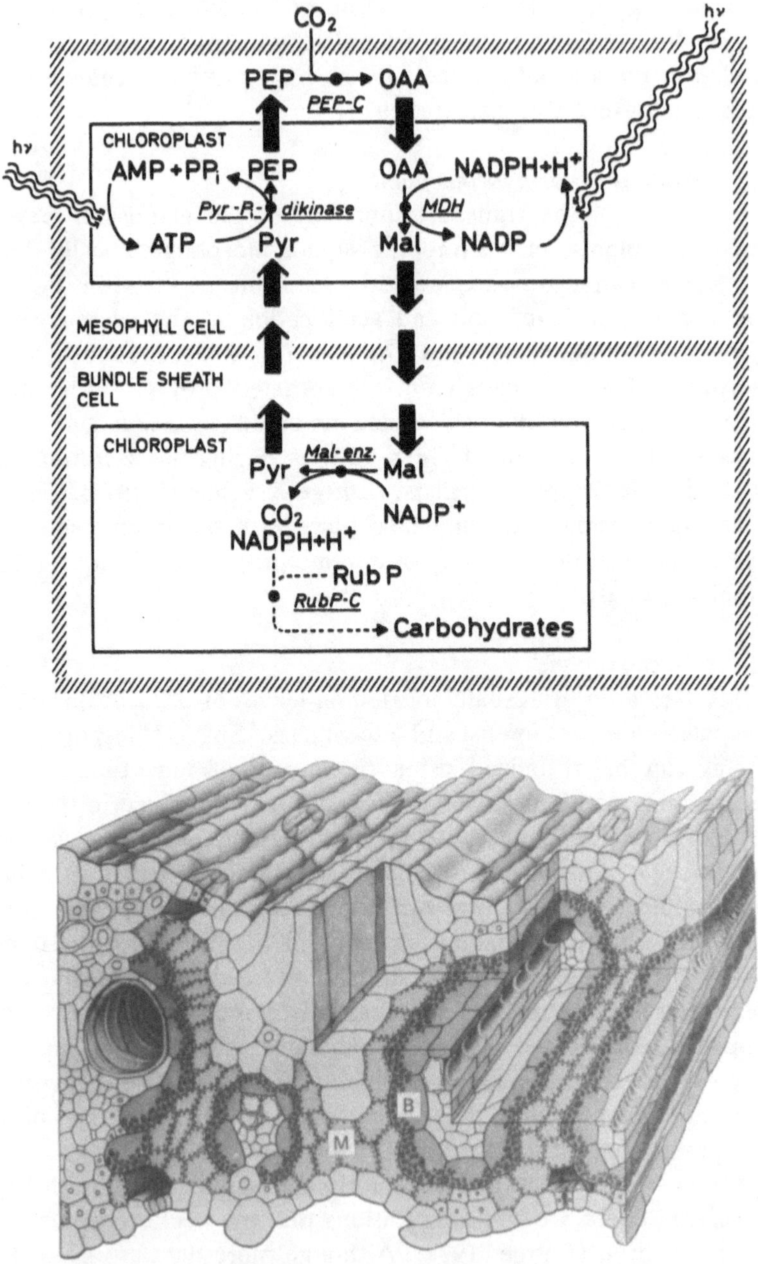


Fig. 14.10. Strongly simplified scheme of the $C_4$ pathway of photosynthesis with malate and pyruvate as transport metabolites. Thick black arrows: chloroplast membrane transport plus symplastic

direction of the proton pump (Fig. 14.9*d*) and not opposite to it with protons taken up back into the cells (Fig. 14.9*a*).

## The role of carboxylic acids in transport functions of leaves

### *Osmoregulation and pH-regulation*

It has been shown (Fig. 14.9) that the synthesis of malic acid is required in conjunction with operation of the $H^+$-ATPase of the plasmalemma when it continuously extrudes protons. Acid synthesis is needed for pH-regulation and for production of osmotically active material to maintain turgor. Raven (1977) has stressed that in contrast to algae, aerial leaves of higher plants cannot freely regulate cytoplasmic pH by exchange of $H^+$ or $OH^-$ with an external environment, because the sieve tubes responsible for export of solutes from leaves are themselves symplastic and require regulation of their own pH. Thus, carboxylic acid metabolism is not only necessary in the operation of specific functions like extension growth and movements of stomatal guard cells, but is essential for osmoregulation, pH regulation and electrical charge balance in leaves in general. This is reflected in the high levels of carboxylates such as malate, citrate and isocitrate, oxalate, tartrate (Nierhaus & Kinzel, 1971) often found in leaves; in contrast algae predominantly use inorganic ions as osmotically active material in their vacuoles.

### *Special photosynthetic carbon pathways*

*Crassulacean acid metabolism* (CAM)

Plants performing CAM produce phosphoenolpyruvate (PEP) from starch during the night and synthesize large amounts of malic acid by $CO_2$ dark fixation via PEP-carboxylase (PEP-C). The feedback inhibition of PEP-C by malate, and especially cytoplasmic pH control,

Caption to Fig. 14.10 (*cont.*)
transcellular transport. The reaction scheme aims to demonstrate the need for concerted action of these two modes of transport. PEP, phosphoenolpyruvate; *PEP-C,* phosphoenolpyruvate carboxylase; OAA, oxaloacetate; *Mal*, malate; *MDH*, malate dehydrogenases; *Pyr*, pyruvate; *Mal-enz.*, decarboxylating malic enzyme; *RubP*, ribulosebisphosphate; *RuBP-C*, ribulosebisphosphate carboxylase; *M* = mesophyll, *B* = bundle sheath. From Lüttge (1974) and Lüttge & Higinbotham (1979).

require the accumulation of the malic acid formed, which can amount to 200 μmoles $g^{-1}$ or more, in a separate compartment. Only the vacuoles are large enough. The transport of malic acid into the vacuoles of CAM-leaves is against electrochemical gradients of $H^+$, H-malate$^{1-}$ and malate$^{2-}$. Energetically it is most probable that it is driven by a $2H^+$-ATPase at the tonoplast. This mechanism pumps 2 protons into the vacuole while consuming 1 ATP, with malate$^{2-}$ transport coupled to this $H^+$-pump. The net efflux of malic acid from the vacuoles that occurs during the day is passive. Malate is then decarboxylated, the $CO_2$ obtained is refixed via ribulose-bis-phosphate carboxylase (RuBP-C) and converted to carbohydrate in the Calvin cycle (Lüttge, 1980).

This mechanism allows the plants to close the stomata during the day, when evaporative demand is high, and to open them only during the night for net carbon gain by PEP-C. During the day $CO_2$ for light-dependent reactions of photosynthesis is derived from malate. Thus CAM is a mechanism for adaptation to limited water supply (Osmond, 1978; Kluge & Ting, 1978). Diurnal variations in water relation characteristics accompany the diurnal oscillations of malic acid content (Lüttge & Ball, 1977).

It is possible that CAM has evolved as a specialization from the general mechanisms of osmoregulation and pH regulation mentioned above. Basal carboxylate levels, including especially citrate, isocitrate and some malate, which do not oscillate diurnally, are also observed in leaves of CAM plants (Kluge & Ting, 1978) and a fraction of malate always remains in the tissue at the end of the light phase of the rhythm which is not used up by decarboxylation and photosynthesis (i.e. from 10 up to 40 μmol $g^{-1}$). It is noteworthy that this 'basal level' of malate observed at the end of the light period increases under conditions which suppress the malic acid oscillations of CAM in *Kalanchoe daigremontianum*, as when the light period is extended on long days (Philipps, 1980).

### $C_4$ *photosynthesis*

As with CAM, $C_4$ photosynthesis involves fixation of $CO_2$ by PEP-C and formation of malate prior to chanelling $CO_2$ into the Calvin cycle. Events separated in time in CAM occur simultaneously in $C_4$ photosynthesis but separated in space. Leaves of $C_4$ plants have the characteristic Kranz anatomy, where bundle-sheath cells form layers encircling the vascular strands and mesophyll cells are tightly arranged around them (Fig. 14.10). Malate formation is restricted to the mesophyll cells; decarboxylation of malate and the Calvin cycle operate in the bundle-sheath cells. Depending on the decarboxylation mechanism of malate in the bundle-sheath cells this requires various metabolite shuttles transporting carbon, energy-rich phosphate and reducing power between the two types of cells (Hatch & Osmond, 1976).

## References


Arisz, W. H. (1960). Symplasmatischer Salztransport in *Vallisneria*-Blättern. *Protoplasma*, **46**, 5–62.

Atkinson, M. R., Findlay, G. P., Hope, A. B., Pitman, M. G., Saddler, H. D. W. & West, K. R. (1967). Salt regulation in the mangroves *Rhizophora mucronata* Lam. and *Aegialitis annulata* R.Br. *Australian Journal of Biological Science*, **20**, 589–99.

Bentrup, F. W. (1974). Lichtabhängige Membranpotentiale bei Pflanzen. *Berichte der deutschen botanischen Gesellschaft*, **87**, 515–28.

Bentrup, F. W., Gratz, H. J. & Unbehauen, H. (1973). The membrane potential of *Vallisneria* leaf cells: Evidence for light-dependent proton permeability changes. In *Ion Transport in Plants*, ed. W. P. Anderson, pp. 171–82. London: Academic Press.

Böcher, M., Fischer, E., Ullrich-Eberius, C. I. & Novacky, A. (1980). Effect of fusicoccin on the membrane potential, on the uptake of glucose and glycine, and on the ATP level in *Lemna gibba* G 1. *Plant Science Letters*, **18**, 215–20.

Brinckmann, E. & Lüttge, U. (1974). Lichtabhängige Membranpotentialschwankungen und deren interzelluläre Weiterleitung bei panaschierten Photosynthese-Mutanten von *Oenothera*. *Planta*, **119**, 47–57.

Cho, B.-H. & Komor, E. (1980). The role of potassium in charge compensation for sucrose-proton-symport by cotyledons of *Ricinus communis*. *Plant Science Letters*, **17**, 425–35.

Clarkson, D. T., Sanderson, J. & Scattergood, C. B. (1978). Influence of phosphate-stress on absorption and translocation by various parts of the root system of *Hordeum vulgare* L. (barley). *Planta*, **139**, 47–53.

Cleland, R. E. (1976). Kinetics of hormone-induced $H^+$ excretion. *Plant Physiology*, **58**, 210–13.

Cram, W. J. (1980). A common feature of the uptake of solutes by root parenchyma cells. *Australian Journal of Plant Physiology*, **7**, 41–9.

Cram, W. J. & Pitman, M. G. (1972). The action of abscisic acid on ion uptake and water flow in plant roots. *Australian Journal of Biological Science*, **25**, 1125–32.

Davis, R. F. (1974). Photoinduced changes in electrical potentials and $H^+$ activities of the chloroplast, cytoplasm, and vacuole of *Phaeoceros laevis*. In *Membrane Transport in Plants*, ed. U. Zimmermann & J. Dainty, pp. 195–201. Berlin: Springer-Verlag.

Delrot, S. & Bonnemain, J.-L. (1981). Involvement of protons as a substrate for the sucrose carrier during phloem loading in *Vicia faba* leaves. *Plant Physiology*, **67**, 560–4.

Delrot, S., Despeghel, J.-P. & Bonnemain, J.-L. (1980). Phloem loading in *Vicia faba* leaves: Effect of N-ethylmaleimide and parachloromercuribenzene-sulfonic acid on $H^+$ extrusion, $K^+$ and sucrose uptake. *Planta*, **149**, 144–8.

Denny, P. & Weeks, D. C. (1968). Electrochemical potential gradients of ions in an aquatic angiosperm. *Potamogeton schweinfurthii* (Benn.) *New Phytologist*, **67**, 875–82.

Despeghel, J.-P., Delrot, S., Bonnemain, J.-L. & Besson, J. (1980). Étude du mécanisme de l'absorption des acides aminés par les tissus foliaires du *Vicia faba* L. *Comptes Rendus Hebdomadaires des Séances de l'Académie des Sciences, Paris, Série D*, **290**, 609–14.

Dieffenbach, H., Lüttge, U. & Pitman, M. G. (1980). Release of guttation fluid from passive hydathodes of intact barley plants. II. The effects of abscisic acid and cytokinins. *Annals of Botany*, **45,** 703–12.

Etherton, B. (1970). Effect of indole-3-acetic acid on membrane potentials of oat coleoptile cells. *Plant Physiology*, **45,** 527–8.

Etherton, B. & Rubinstein, B. (1978). Evidence for amino acid-$H^+$ cotransport in oat coleoptiles. *Plant Physiology*, **60,** 933–7.

Falk, H. & Sitte, P. (1963). Zellfeinbau bei Plasmolyse. I. Der Feinbau der *Elodea*-Blattzellen. *Protoplasma*, **57,** 290–303.

Felle, H. & Bentrup, F. W. (1980). Hexose transport and membrane depolarization in *Riccia fluitans. Planta*, **147,** 471–6.

Felle, H., Lühring, H. & Bentrup, F. W. (1979). Serine transport and membrane depolarization in the liverwort *Riccia fluitans. Zeitschrift für Naturforschung*, **34c,** 1222–3.

Geiger, D. R. (1975). Phloem loading. In *Encyclopedia of Plant Physiology*, (new series) vol. 1, *Transport in Plants. I. Phloem Transport*, ed. M. H. Zimmermann & J. A. Milburn, pp. 395–431. Berlin: Springer-Verlag.

Giaquinta, R. (1976). Evidence for phloem loading from the apoplast. Chemical modification of membrane sulfhydryl groups. *Plant Physiology*, **57,** 872–5.

Giaquinta, R. (1980). Mechanism and control of phloem loading of sucrose. *Berichte der deutschen botanischen Gesellschaft*, **93,** 187–201.

Graham, R. P. & Bowling, D. J. F. (1977). Effect of the shoot on the transmembrane potentials of root cortical cells of sunflower. *Journal of Experimental Botany*, **28,** 886–93.

Gunning, B. E. S. (1977). Transfer cells and their roles in transport of solutes in plants. *Science Progress* (*Oxford*), **64,** 539–68.

Gunning, B. E. S. & Robards, A. W. (ed.) (1976). *Intercellular Communication in Plants: Studies on Plasmodesmata*. Berlin: Springer-Verlag.

Guy, M., Reinhold, L. & Rahat, M. (1980). Energization of the sugar transport mechanism in the plasmalemma of isolated mesophyll protoplasts. *Plant Physiology*, **65,** 550–3.

Hartung, W., Ullrich-Eberius, C. I., Lüttge, U., Böcher, M. & Novacky, A. (1980). Effect of abscisic acid on membrane potential and transport of glucose and glycine in *Lemna gibba* G1. *Planta*, **148,** 256–61.

Haschke, H.-P. & Lüttge, U. (1977). Auxin action on $K^+$–$H^+$-exchange and growth, $^{14}CO_2$-fixation and malate accumulation in *Avena* coleoptile segments. In *Regulation of Cell Membrane Activities in Plants*, ed. E. Marrè & O. Ciferri, pp. 243–8. Amsterdam: Elsevier/North-Holland Biomedical Press.

Hatch, M. D. & Osmond, C. B. (1976). Compartmentation and transport in $C_4$ photosynthesis. In *Encyclopedia of Plant Physiology*, (new series) vol. 3, *Transport in Plants. III. Intracellular Interactions and Transport Processes*, ed. C. R. Stocking & U. Heber, pp. 144–84. Berlin: Springer-Verlag.

Heyser, W. (1980). Phloem loading in the maize leaf. *Berichte der deutschen botanischen Gesellschaft*, **93,** 221–8.

Hill, A. E. & Hill, B. S. (1976). Elimination processes by glands: Mineral ions. In *Encyclopedia of Plant Physiology*, (new series) vol. 2, *Transport in Plants. II. Part B. Tissues and Organs*, ed. U. Lüttge & M. G. Pitman, pp. 225–43. Berlin: Springer-Verlag.

Hsiao, T. C. (1976). Stomatal ion transport. In *Encyclopedia of Plant Physiology*, (new series) vol. 2, *Transport in Plants. II. Part B. Tissues and Organs*, ed. U. Lüttge & M. G. Pitman, pp. 195–221. Berlin: Springer-Verlag.

Hutchings, V. M. (1978a). Sucrose and proton cotransport in *Ricinus* cotyledons. I. $H^+$ influx associated with sucrose uptake. *Planta*, **138**, 229–35.

Hutchings, V. M. (1978b). Sucrose and proton cotransport in *Ricinus* cotyledons. II. $H^+$ efflux and associated $K^+$ uptake. *Planta*, **138**, 237–41.

Jacoby, B., Abas, S. & Steinitz, B. (1973). Rubidium and potassium absorption by bean-leaf slices compared to sodium absorption. *Physiologia Plantarum*, **28**, 209–14.

Jeschke, W. D. (1976). Ionic relations of leaf cells. In *Encyclopedia of Plant Physiology*, (new series) vol. 2, *Transport in Plants. II. Part B. Tissues and Organs*, ed. U. Lüttge & M. G. Pitman, pp. 160–94. Berlin: Springer-Verlag.

Jung, K.-D. & Lüttge, U. (1980). Inhibition of sugar and salt elimination by glands with the amino acid analog *p*-fluorophenylalanine. *Annals of Botany*, **45**, 397–401.

Keifer, D. W. (1980). Alteration of cytoplasmic pH in *Chara*, through membrane transport processes . In *Plant Membrane Transport: Current Conceptual Issues*, ed. R. M. Spanswick, W. J. Lucas & J. Dainty, pp. 569–70. Amsterdam: Elsevier/North-Holland Biomedical Press.

Keifer, D. W. & Spanswick, R. M. (1979). Correlation of adenosine triphosphate levels in *Chara corallina* with the activity of the electrogenic pump. *Plant Physiology*, **64**, 165–8.

King, R. W. (1976). Implications for plant growth of the transport of regulatory compounds in phloem and xylem. In *Transport and Transfer Processes in Plants*, ed. I. F. Wardlaw & J. B. Passioura, pp. 415–31. New York: Academic Press.

Kluge, M. & Ting, I. P. (1978). *Crassulacean Acid Metabolism. Analysis of an Ecological Adaptation*. Berlin: Springer-Verlag.

Komor, E., Rotter, M., Waldhauser, J., Martin, E. & Cho, B. H. (1980). Sucrose proton symport for phloem loading in the *Ricinus* seedling. *Berichte der deutschen botanischen Gesellschaft*, **93**, 211–19.

Komor, E. & Tanner, W. (1980). Proton-cotransport of sugars in plants. In *Plant Membrane Transport: Current Conceptual Issues*, ed. R. M. Spanswick, W. J. Lucas & J. Dainty, pp. 247–57. Amsterdam: Elsevier/North-Holland Biomedical Press.

Komor, E., Weber, H. & Tanner, W. (1978). Essential sulfhydryl group in the transport-catalyzing protein of the hexose-proton cotransport system of *Chlorella*. *Plant Physiology*, **61**, 785–6.

Lambers, H. (1979). Efficiency of root respiration in relation to growth rate, morphology and soil composition. *Physiologia Plantarum*, **46**, 194–202.

Läuchli, A. (1979). Regulation des Salztransportes und Salzausschließung in Glycophyten und Halophyten. *Berichte der deutschen botanischen Gesellschaft*, **92**, 87–94.

Lichtner, F. T. & Spanswick, R. M. (1981). Electrogenic sucrose transport in developing soybean cotyledons. *Plant Physiology*, **67**, 869–74.

Löppert, H., Kronberger, W. & Kandeler, R. (1978). Phytochrome-mediated changes in the membrane potential of subepidermal cells of *Lemna paucicostata* 6746. *Planta*, **138**, 133–6.

Lüttge, U. (1973). Photosynthetic $O_2$ evolution and apparent $H^+$ uptake by slices of greening barley and maize leaves in aerobic and anaerobic solutions. *Canadian Journal of Botany*, **51**, 1953–7.

Lüttge, U. (1974). Co-operation of organs in intact higher plants: A review. In *Membrane Transport in Plants*, ed. U. Zimmermann & J. Dainty, pp. 353–62. Berlin: Springer-Verlag.

Lüttge, U. (1977). Nectar composition and membrane transport of sugars and amino acids: A review on the present state of nectar research. *Apidologie*, **8**, 305–19.

Lüttge, U. (1980). Malic acid transport across the tonoplast of *Kalanchoë* leaf cells: Tonoplast biophysics and biochemistry in relation to crassulacean acid metabolism (CAM). In *Plant Membrane Transport: Current Conceptual Issues*, ed. R. M.

Spanswick, W. J. Lucas & J. Dainty, pp. 49–60. Amsterdam: Elsevier/North-Holland Biomedical Press.

Lüttge, U. (1982). Carnivorous plants. In *Encyclopedia of Plant Physiology*, (new series), *Physiological Plant Ecology, Vol. 12 C. Functional Responses to the Chemical and Biological Environment,* ed. O. L. Lange, P. S. Nobel, C. B. Osmond & H. Ziegler. Berlin: Springer-Verlag (In press.)

Lüttge, U. & Ball, E. (1973). Ion uptake by slices from greening etiolated barley and maize leaves. *Plant Science Letters*, **1,** 275–80.

Lüttge, U. & Ball, E. (1976). ATP levels and energy requirements of ion transport in cells of slices of greening barley leaves. *Zeitschrift für Pflanzenphysiologie*, **80,** 50–9.

Lüttge, U. & Ball, E. (1977). Water relation parameters of the CAM plant *Kalanchoë daigremontiana* in relation to diurnal malate oscillations. *Oecologia*, **31,** 85–94.

Lüttge, U. & Ball, E. (1980). $2H^+$ : 1 $malate^{2-}$ stoichiometry during crassulacean acid metabolism is unaffected by lipophilic cations. *Plant, Cell and Environment*, **3,** 195–200.

Lüttge, U., Ball, E. & von Willert, K. (1971). A comparative study of the coupling of ion uptake to light reactions in leaves of higher plant species having the $C_3$- and $C_4$-pathway of photosynthesis. *Zeitschrift für Pflanzenphysiologie*, **65,** 336–50.

Lüttge, U. & Higinbotham, N. (1979). *Transport in Plants*. New York: Springer-Verlag.

Lüttge, U., Jung, K.-D. & Ullrich-Eberius, C. I. (1981). Evidence for amino acid-$H^+$ cotransport in *Lemna gibba* given by effects of fusicoccin. *Zeitschrift für Pflanzenphysiologie*, **102,** 117–25.

Lüttge, U. & Krapf, G. (1969). Die Ultrastruktur der *Nymphaea*-Hydropoten in Zusammenhang mit ihrer Funktion als Salz-transportierende Drüsen. *Cytobiologie*, **1,** 121–31.

Lüttge, U., Pallaghy, C. K. & Osmond, C. B. (1970). Coupling of ion transport in green cells of *Atriplex spongiosa* leaves to energy sources in the light and in the dark. *Journal of Membrane Biology*, **2,** 17–30.

Lüttge, U., Pallaghy, C. K. & von Willert, K. (1971). Microautoradiographic investigations of sulfate uptake by glands and epidermal cells of water lily (*Nymphaea*) leaves with special reference to the effect of poly-L-lysine. *Journal of Membrane Biology*, **4,** 395–407.

Lüttge, U. & Schnepf, E. (1976). Elimination processes by glands: Organic substances. In *Encyclopedia of Plant Physiology,* (new series) vol. 2, *Transport in Plants. II. Part B. Tissues and Organs*, ed. U. Lüttge & M. G. Pitman, pp. 244–77. Berlin: Springer-Verlag.

Lüttge, U. & Zirke, G. (1974). Attempts to measure plasmalemma and tonoplast electropotentials in small cells of the moss *Mnium* using centrifugation techniques. *Journal of Membrane Biology*, **18,** 305–14.

MacRobbie, E. A. C. (1965). The nature of the coupling between light energy and active ion transport in *Nitella translucens. Biochimica et Biophysica Acta*, **94,** 64–73.

Marrè, E. (1979). Fusicoccin, a tool in plant physiology. *Annual Review of Plant Physiology*, **30,** 273–88.

Martin, E. & Komor, E. (1980). Role of phloem in sucrose transport by *Ricinus* cotyledons. *Planta*, **148,** 367–73.

Mitchell, P. (1967). Translocations through natural membranes. *Advances in Enzymology and Related Areas of Molecular Biology*, **29,** 33–87.

Münch, E. (1930). *Die Stoffbewegungen in der Pflanze*. Jena: Gustav Fischer.

Nierhaus, D. & Kinzel, H. (1971). Vergleichende Untersuchungen über die organischen Säuren in Blättern höherer Pflanzen. *Zeitschrift für Pflanzenphysiologie*, **64,** 107–23.

Novacky, A., Fischer, E., Ullrich-Eberius, C. I., Lüttge, U. & Ullrich, W. R. (1978a). Membrane potential changes during transport of glycine as a neutral amino acid and nitrate in *Lemna gibba* G1. *FEBS-Letters*, **88,** 264–7.

Novacky, A., Ullrich-Eberius, C. I. & Lüttge, U. (1978b). Membrane potential changes during transport of hexoses in *Lemna gibba* G1. *Planta*, **138**, 263–70.

Novacky, A., Ullrich-Eberius, C. I. & Lüttge, U. (1980). pH and membrane-potential changes during glucose uptake in *Lemna gibba* G1 and their response to light. *Planta*, **149**, 321–6.

Osmond, C. B. (1968). Ion absorption in *Atriplex* leaf tissue. I. Absorption by mesophyll cells. *Australian Journal of Biological Science*, **21**, 1119–30.

Osmond, C. B. (1978) Crassulacean acid metabolism: A curiosity in context. *Annual Review of Plant Physiology*, **29**, 379–414.

Outlaw, W. H. & Fisher, D. B. (1975). Compartmentation in *Vicia faba* leaves. I. Kinetics of $^{14}C$ in the tissues following pulse labelling. *Plant Physiology*, **55**, 699–703.

Outlaw, W. H., Fisher, D. B. & Christy, A. L. (1975). Compartmentation in *Vicia faba* leaves. II. Kinetics of $^{14}C$-sucrose redistribution among individual tissues following pulse labelling. *Plant Physiology*, **55**, 704–11.

Pate, J. S. (1980). Transport and partitioning of nitrogenous solutes. *Annual Review of Plant Physiology*, **31**, 313–40.

Philipps, R. (1980). Deacidification in a plant with crassulacean acid metabolism associated with anion-cation balance. *Nature*, **287**, 727–8.

Pitman, M. G. (1972). Uptake and transport of ions in barley seedlings. III. Correlation of potassium transport to the shoot with plant growth. *Australian Journal of Biological Science*, **25**, 905–19.

Pitman, M. G. (1977). Ion transport in the xylem. *Annual Review of Plant Physiology*, **28**, 71–88.

Pitman, M. G., Lüttge, U., Läuchli, A. & Ball, E. (1974a). Action of abscisic acid on ion transport as affected by root temperature and nutrient status. *Journal of Experimental Botany*, **25**, 147–55.

Pitman, M. G., Lüttge, U., Läuchli, A. & Ball, E. (1974b). Ion uptake to slices of barley leaves, and regulation of K content in cells of the leaves. *Zeitschrift für Pflanzenphysiologie*, **72**, 75–88.

Pitman, M. G. & Wellfare, D. (1978). Inhibition of ion transport in excised barley roots by abscisic acid; relation to water permeability of the roots. *Journal of Experimental Botany*, **29**, 1125–38.

Prins, H. B. A. (1973). The action spectrum of photosynthesis and the rubidium chloride uptake by leaves of *Vallisneria spiralis*. *Koninklijke Nederlandse Akademie van Wetenschappen, Amsterdam, series C*, **76**, 495–9.

Prins, H. B. A., Harper, J. R. & Higinbotham, N. (1980). Membrane potentials of *Vallisneria* leaf cells and their relation to photosynthesis. *Plant Physiology*, **65**, 1–5.

Prins, H. B. A., Snel, J. F. H., Helder, R. J. & Zanstra, P. E. (1979). Photosynthetic bicarbonate utilization in the aquatic angiosperms *Potamogeton* and *Elodea*. *Hydrobiological Bulletin, Amsterdam*, **13**, 106–11.

Racusen, R. H. & Galston, A. W. (1977). Electrical evidence for rhythmic changes in the cotransport of sucrose and hydrogen ions in *Samanea* pulvini. *Planta*, **135**, 57–62.

Racusen, R. H. & Galston, A. W. (1980). Phytochrome modifies blue-light-induced electrical changes in corn coleoptiles. *Plant Physiology*, **66**, 534–5.

Raschke, K. (1979). Movements of stomata. In *Encyclopedia of Plant Physiology*, (new series) vol. 7, *Physiology of Movements*, ed. W. Haupt & M. E. Feinleib, pp. 383–441. Berlin: Springer-Verlag.

Raschke, K. & Schnabl, H. (1978). Availability of chloride affects the balance between potassium chloride and potassium malate in guard cells of *Vicia faba* L. *Plant Physiology*, **62**, 84–7.

Raven, J. A. (1977). $H^+$ and $Ca^{2+}$ in phloem and symplast: Relation of relative immobility of the ions to the cytoplasmic nature of the transport paths. *New Phytologist*, **79**, 465–80.

Robins, R. J. & Juniper, B. E. (1980). The secretory cycle of *Dionaea muscipula* Ellis. III. The mechanism of release of digestive secretion. *New Phytologist*, **86**, 313–27.

Rona, J.-P., Pitman, M. G., Lüttge, U. & Ball, E. (1980). Electrochemical data on compartmentation into cell wall, cytoplasm, and vacuole of leaf cells in the CAM genus *Kalanchoë*. *Journal of Membrane Biology*, **57**, 25–35.

Satter, R. L. (1979). Leaf movements and tendril curling. In *Encyclopedia of Plant Physiology*, (new series) vol. 7, *Physiology of Movements*, ed. W. Haupt & M. E. Feinleib, pp. 442–84. Berlin: Springer-Verlag.

Schaefer, N., Wildes, R. A. & Pitman, M. G. (1975). Inhibition by *p*-fluorophenylalanine of protein synthesis and of ion transport across roots in barley seedlings. *Australian Journal of Plant Physiology*, **2**, 61–73.

Schnabl, H. & Raschke, K. (1980). Potassium chloride as stomatal osmoticum in *Allium cepa* L., a species devoid of starch in guard cells. *Plant Physiology*, **65**, 88–93.

Schnabl, H. & Ziegler, H. (1977). The mechanism of stomatal movement in *Allium cepa* L. *Planta*, **136**, 37–43.

Smith, F. A. & Raven, J. A. (1979). Intracellular pH and its regulation. *Annual Review of Plant Physiology*, **30**, 289–311.

Smith, R. C. & Epstein, E. (1964). Ion absorption by shoot tisssue: Technique and first findings with excised leaf tissue of corn. *Plant Physiology*, **39**, 338–41.

Spanswick, R. M. (1972). Electrical coupling between cells of higher plants: a direct demonstration of intercellular communication. *Planta*, **102**, 215–27.

Spanswick, R. M. & Miller, A. G. (1977). Measurements of the cytoplasmic pH in *Nitella translucens*. Comparison of values obtained by microelectrodes and weak acid methods. *Plant Physiology*, **59**, 664–6.

Steveninck, R. F. M. van (1976). Effect of hormones and related substances on ion transport. In *Encyclopedia of Plant Physiology*, (new series) vol. 2, *Transport in Plants. II. Part B. Tissues and Organs*, ed. U. Lüttge & M. G. Pitman, pp. 307–42. Berlin: Springer-Verlag.

Stocking, C. R. & Heber, U. (eds.) (1976). *Encyclopedia of Plant Physiology*, (new series) vol. 3, *Transport in Plants. III. Intracellular Interactions and Transport Processes*. Berlin: Springer-Verlag.

Stout, R. G. & Cleland, R. E. (1980). Partial characterization of fusicoccin binding to receptor sites on oat root membranes. *Plant Physiology*, **66**, 353–9.

Sutcliffe, J. F. (1976). Regulation in the whole plant. In *Encyclopedia of Plant Physiology*, (new series) vol. 2, *Transport in Plants. II. Part B. Tissues and Organs*, ed. U. Lüttge & M. G. Pitman, pp. 394–417. Berlin: Springer-Verlag.

Tognoli, L., Beffagna, N., Pesci, P. & Marrè, E. (1979). On the relationship between ATPase activity and FC binding capacity of crude and partially purified microsomal preparations from maize coleoptiles. *Plant Science Letters*, **16**, 1–14.

Ullrich-Eberius, C. I., Lüttge, U. & Neher, L. (1976). $CO_2$-uptake by barley leaf slices as measured by photosynthetic $O_2$-evolution. *Zeitschrift für Pflanzenphysiologie*, **79**, 336–46.

Ullrich-Eberius, C. I., Novacky, A., Fischer, E. & Lüttge, U. (1981). Relationship between energy-dependent phosphate uptake and the electrical membrane potential in *Lemna gibba* G1. *Plant Physiology*, **67**, 797–801.

Volkenburgh, E. van & Cleland, R. E. (1980). Proton excretion and cell expansion in bean leaves. *Planta*, **148**, 273–8.

Walker, N. A. & Smith, F. A. (1975). Intracellular pH in *Chara corallina* measured by DMO distribution. *Plant Science Letters*, **4**, 125–32.

Zurzycki, J. (1968). Changes of the trans-membrane potential of the leaf cell of *Funaria hygrometrica* under influence of light. *Acta Societatis Botanicorum Poloniae*, **37**, 519–31.

# 15

# Physiological responses, metabolic changes and regulation during leaf senescence

H. W. WOOLHOUSE AND G. I. JENKINS

## Introduction

Leaf senescence is a subject embracing many facets of developmental botany, including changing photosynthetic capacity and such physiological responses as photoperiodic sensitivity, respiration and source-sink relations of the leaf. We need to consider the structure and composition of membranes, changes in compartmentation and the altered levels and characteristics of lipids, nucleic acids and proteins which comprise basic elements of the living cells. We are also concerned with the mechanisms by which senescence is regulated in relation to developmental factors, such as flowering and the position of the leaf on the plant, and to the pattern of seasonal variables, such as daylength and temperature.

One aspect of each of these three levels of study is illustrated by reference to photosynthesis, protein turnover and effects of hormones on leaf senescence; before doing this however it is important to sound a warning concerning the limits of generalisation.

Herbaceous plants which occur in habitats such as base-rich alpine meadows, where water, light and nutrients are rarely limiting for growth, tend to be profligate in their turnover of leaves; there is little constraint upon the supply of energy or raw materials to make new ones. In the understorey of a woodland the situation is different: light energy is at a premium and the dictates of economy in energy favour retention of the leaves in a functional state for as long as possible once the energy has been invested in their production. Likewise in nutrient-deficient habitats, such as ombrogenous bogs and forests developed on lateritic substrates in the humid tropics, the 'evergreen' habit of many of the species may represent an adaptation concerned with retention and internal re-cycling of nutrients (Woolhouse, 1967, 1974). The selection pressures in particular habitats have moulded the longevity, morphology and turnover of leaf canopies in many different ways and it would be an unwarranted assumption to suppose therefore that the physiological responses, metabolic changes or regulatory mechanisms involved in leaf senescence are similar from one species to another. The majority of studies of leaf senescence are concentrated on relatively few species and types of leaf; it is unwise therefore to infer too wide a level of generality from such studies.

## Changes in photosynthetic capacity

The rate of photosynthesis decreases during foliar senescence (Šesták & Čatský, 1967; Woolhouse, 1967, 1974, 1982a,b). This is so with a host of species of flowering plants and in leaves with markedly different lifespans and developmental characteristics, but for any given leaf ontogenetic changes in photosynthetic rate are modulated by environmental variables such as nutrient supply (Moss & Peaslee, 1965; Osman & Milthorpe, 1971), water status (Moorby, Munns & Walcott, 1975) and conditions of illumination (Osman & Milthorpe, 1971; Woledge, 1972). Coincident with the decrease in photosynthetic activity of the leaf, its capacity to export photosynthate diminishes and it finally becomes a respiratory burden to the plant, or would do so if it retained the capacity to function as a sink. The extent of the period of maximum photosynthetic rate is of considerable significance, both to the plant in terms of genetic fitness and to the plant breeder as a means of increasing productivity. Thus it is important to understand the causes of the decline in photosynthetic rate. This is not easy because a number of factors may limit photosynthetic activity of a leaf, and the problem is exacerbated by the possibility that their relative effects change during senescence.

### *Photorespiration*

The rate of photosynthesis is frequently presented as the net rate of $CO_2$ uptake in the light per unit leaf area. This is an underestimate of the gross photosynthetic rate because of the simultaneous efflux of $CO_2$ from photorespiration and from any 'dark' respiration continuing in the light. The rate of photorespiratory $CO_2$ efflux is significant and may amount to 20–50% of the net photosynthetic rate in mature leaves of $C_3$ species (Zelitch, 1979). It is therefore necessary to determine to what extent the large decrease in photosynthetic rate during senescence can be accounted for by an increase in the rate of photorespiration. A number of methods have been employed to measure the rate of photorespiration, but unfortunately none is reliable (Ludlow & Jarvis, 1971; Zelitch, 1979). The principal difficulty is that $CO_2$ released by photorespiration and immediately refixed by photosynthesis does not escape from the leaf and therefore cannot be measured; as a result, the rate of photorespiration is underestimated.

Decreases in the rate of photorespiration during senescence have been reported for leaves of a number of species (Mulchi, Volk & Jackson, 1971; Kisaki, Hirabayashi & Yano, 1973; Hodgkinson, 1974; Thomas, Hall & Merrett, 1978; Fraser & Bidwell, 1974; Čatský, Tichá & Solárová, 1976). The data of Thomas *et al.* (1978) for wheat are shown in Fig. 15.1. Here, as with other species studied to date, an increase in photorespiration cannot be responsible for the decline in the net photosynthetic rate; a large decrease in the gross photosynthetic rate remains to be accounted for.

In photorespiration $CO_2$ is released by the oxidation of glycolate formed from glycolate-2-phosphate. The latter is generated in an oxygenase reaction of the enzyme ribulose 1,5-bisphosphate carboxylase-oxygenase (RuBPc-o) which is competitive with the carboxylase reaction:

*carboxylase:* ribulose 1,5-bisphosphate (RuBP) + $CO_2$ + $H_2O \rightarrow$ 2 glycerate 3-phosphate (PGA) + $2H^+$

*oxygenase:* RuBP + $O_2 \rightarrow$ PGA + glycolate 2-phosphate + $2H^+$

The ratio, Ø, of oxygenase/carboxylase activities is dependent on the relative concentrations of $O_2$ and $CO_2$ at the active sites of the enzyme.

A number of observations have raised some interesting points concerning the regulation of photorespiration in senescing leaves. Thomas *et al.* (1978) found that in wheat the rate of photorespiration declined much more slowly than the rate of photosynthesis during senescence, and yet Ø measured for a partially purified RuBPc-o preparation remained constant. Unless the older leaves had some other mechanism of glycolate synthesis, it seems most likely that the explanation lies in the regulation of enzyme activity *in vivo*. An increase in $O_2$ relative to $CO_2$ in the vicinity of the enzyme, as might be occasioned by stomatal closure, could account for these results, but, as discussed below, this explanation may not always be applicable. In lucerne, photosynthesis and photorespiration declined roughly in parallel during senescence, and there was little evidence of stomatal closure (Hodgkinson, 1974). The observation that photosynthesis and photorespiration varied independently in leaves of *Phaseolus vulgaris* of different ages (Fraser & Bidwell, 1974) cannot be explained readily by changes in $O_2$ and $CO_2$ concentrations in the leaf. It is also noted that leaves of certain $C_4$ species appear to develop some $C_3$ characteristics, including a measurable rate of photorespiration, during senescence (Kennedy, 1976; Imai & Murata, 1979).

Fig. 15.1. Gross photosynthesis ($^{14}CO_2$ uptake) (■), net photosynthesis (●), and $CO_2$ evolution in the light (▲, △) during senescence of flag leaves of winter wheat in the field. Closed triangles indicate the difference between $^{14}CO_2$ and $^{12}CO_2$ uptake by the same leaf and open triangles $CO_2$ evolution in the light into $CO_2$-free air. The lines drawn through the points indicate the trends and vertical bars represent the 95% confidence limits of means at each age. Reproduced from Thomas *et al.* (1978).

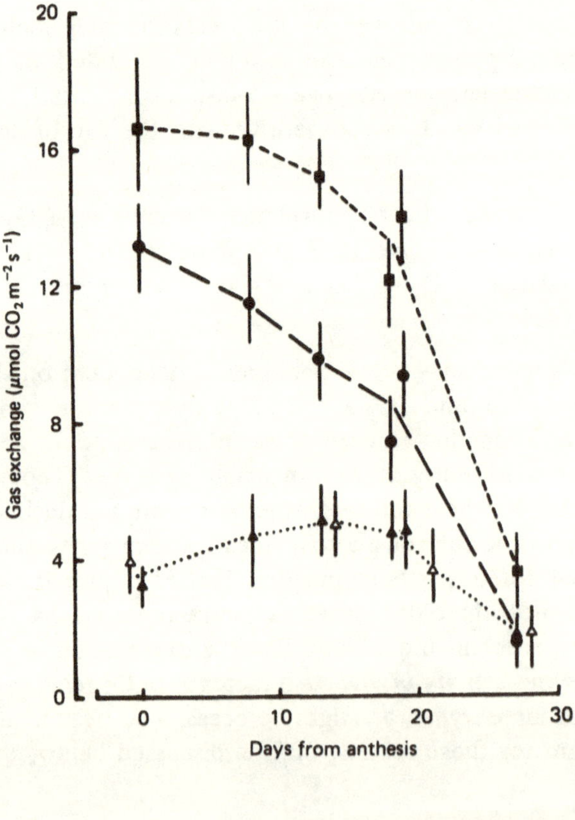

*Leaf conductances for $CO_2$ transfer*

Many workers have attempted to correlate the decrease in rate of photosynthesis during senescence with changes in the various conductances for $CO_2$ transfer across the leaf (see Chapters 10, 12 and 13). The changes observed in stomatal and residual conductances during foliar senescence differ considerably between species. In one group, comprising a few tropical grasses, stomatal conductance, $g'_s$, was found to be very low, and stomatal closure probably limited the rate of photosynthesis in both the mature and senescing leaves (Ludlow & Wilson, 1971; McPherson & Slatyer, 1973). In marked contrast, in a second group (*Perilla frutescens*, Woolhouse, 1968; wheat, Osman & Milthorpe, 1971; *Medicago sativa*, Hodgkinson, 1974), $g'_s$ was reported to be high and to change little during senescence, and the authors concluded that the internal conductance, $g'_i$, and particularly a decreased ability to perform the biochemical reactions of photosynthesis, was the major limiting factor. In the majority of investigations, however, both $g'_s$ and $g'_i$ have been found to change with age, often roughly in parallel (Ludlow & Wilson, 1971; Woodward & Rawson, 1976; Constable & Rawson, 1980; Friedrich & Huffaker, 1980; Fraser & Bidwell, 1974; Čatský *et al.*, 1976). Typical results are shown in Fig. 15.2. In these studies, $g'_s$ remained greater than $g'_i$, and the decrease in the latter was held to be the more likely cause of the decline in the rate of photosynthesis.

Clearly, if stomatal closure was limiting the rate of $CO_2$ uptake in older leaves, the concentration of $CO_2$ in the sub-stomatal cavities and intercellular spaces, $C_i$, would be expected to decrease. It is therefore significant that in recent investigations $C_i$ was found to remain constant during senescence of leaves of *Phaseolus vulgaris* (Davis & McCree, 1978; Goudriaan & van Laar, 1978), sunflower (Rawson & Constable, 1980) and cotton (Constable & Rawson, 1980) while the rate of photosynthesis declined: the data for cotton are shown in Fig. 15.3.

Thus in all but a few species studied to date it seems that changes in $g'_i$ limit photosynthesis during foliar senescence. The question is whether a resistance to $CO_2$ transfer through the mesophyll cells develops in older leaves, or whether the principal limitation is a decrease in the activity of enzymes or electron transport components. Farquhar & von Caemmerer (1981) have concluded that the concentration of $CO_2$ at the carboxylation sites, $C$, is usually only slightly less than $C_i$. This seems to be a reasonable assumption, and unless older leaves develop impervious cell walls or become extremely dehydrated, there is little reason to believe that $C$ would be lower than in young, fully expanded leaves.

A point which needs to be considered here is the cause of the decrease in stomatal aperture in the leaves of many species during senescence. It is possible that largely autonomous changes in, for example, the physical properties of the guard cell walls or the metabolism of the guard cell chloroplasts in older leaves modify the functioning of the stomatal apparatus. Further, since growth regulators such as abscisic acid (Raschke, 1979) and cytokinin (Jewer & Incoll, 1980) are known to affect stomatal action and may be involved in the stomatal mechanism *in vivo*, changes in the levels of these substances in senescing leaves may also be important. However, if stomatal closure during senescence occurs independently of changes in the photosynthetic capacity of the mesophyll chloroplasts, it is a remarkable coincidence that $C_i$ remains constant. It is more likely that the stomata of older leaves continue to operate by some feedback mechanism which serves to keep $C_i$ constant

Fig. 15.2. *Gossypium hirsutum* leaf stomatal (▲, △) and internal (●, ○) conductances to $CO_2$ at 2000 $\mu$mol photons $m^{-2} s^{-1}$ as influenced by leaf age. Solid symbols indicate a winter experiment and open symbols a spring/summer experiment, each point representing the mean of two to eight leaves. The net rate of photosynthesis increased up to a maximum of approximately 25 $\mu$mol $CO_2 m^{-2} s^{-1}$ at day 20, and then decreased to approximately 5 $\mu$mol $CO_2 m^{-2} s^{-1}$ by day 67. Reproduced from Constable & Rawson (1980).

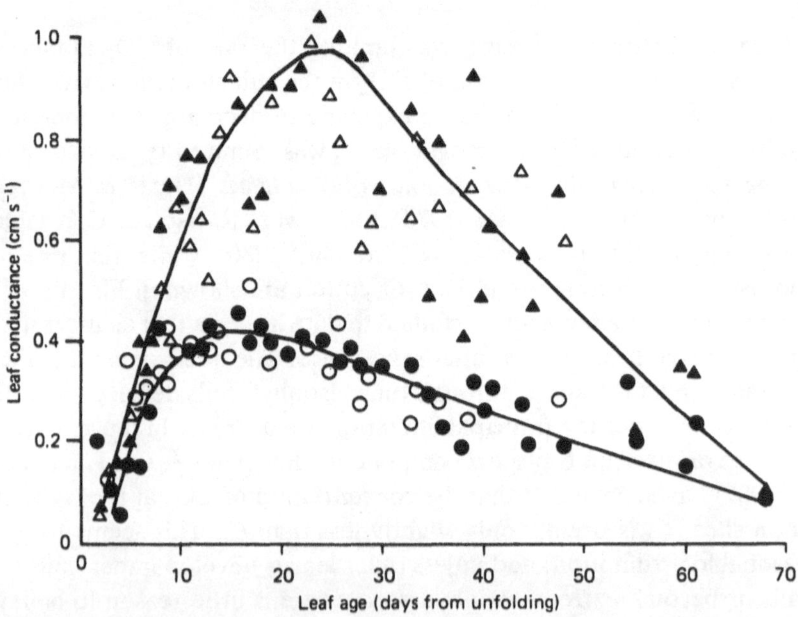

and involves a response to changes in $C_i$ or light or the concentration of certain metabolites in the mesophyll cells (Raschke, 1979; Wong, Cowan & Farquhar, 1979). Thus the decrease in stomatal aperture is probably a consequence of the decline in photosynthetic activity, implying that stomata of older leaves maintain their responsiveness to controlling factors. Those species which show little change in stomatal aperture during senescence and have nearly constant and high values of $g'_s$ may simply have stomata with a minimal response to changes in $C_i$.

### *Enzymes of the reductive pentose phosphate pathway*

Changes in the amounts and activities of enzymes of the reductive pentose phosphate pathway constitute an important potential limitation to the rate of photosynthesis in senescing leaves. In this

Fig. 15.3. Concentration of $CO_2$ in the sub-stomatal cavities and intercellular spaces ($C_i$) of cotton leaves of different ages at 2000 $\mu$mol photons $m^{-2}$ $s^{-1}$ and a volume fraction of $CO_2$ in air of 330 p.p.m. ●, a winter experiment; ○, a spring/summer experiment. Reproduced from Constable & Rawson (1980).

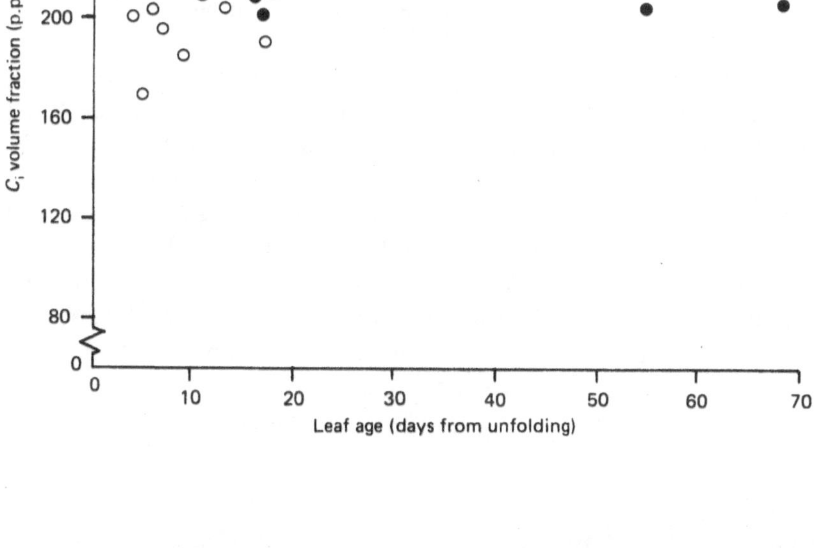

respect regulatory enzymes such as RuBPc-o, fructose bisphosphatase and sedoheptulose bisphosphatase warrant particular consideration. RuBPc-o is present in large amounts in the chloroplast stroma, and in many species it accounts for approximately half of the total soluble protein in the leaves (Kawashima & Wildman, 1970). Following the completion of leaf expansion the amount of this enzyme, and hence the soluble protein content of the leaves, gradually declines, with a concomitant decrease in carboxylase activity per unit leaf area, or per unit fresh weight (Woolhouse, 1967; Hall & Brady, 1977; Hall, Keys & Merrett, 1978; Peoples *et al.*, 1980). It is now realised that the enzyme must be activated by pre-incubation with $CO_2$ and $Mg^{2+}$ prior to assay (Jensen & Bahr, 1977) and where procedures have been employed to ensure complete activation, the specific activity of RuBPc-o has been found either to remain constant (Peoples *et al.*, 1980), or to increase slightly (Friedrich & Huffaker, 1980), during senescence. The earlier reports of a decrease in specific activity may therefore indicate that a larger proportion of the enzyme is in an inactive form in older leaves.

Although the loss of carboxylase activity is often closely correlated with the decline in photosynthetic rate, this does not prove a causal relationship. The important question is whether the carboxylase activity in the leaf is in excess of that required to achieve the observed rates of $CO_2$ fixation. This can be estimated given the gross photosynthetic rate, the amount of enzyme present, and its specific activity *in vivo* (see also Chapter 13). Following Farquhar, von Caemmerer & Berry (1980), the RuBP-saturated specific carboxylase activity of the activated enzyme in the presence of competitive inhibition by oxygen, $v'_e$, is given by:

$$v'_e = v_e \cdot \frac{C}{C + K_c(1 + O/K_o)} \ \mu\text{mol CO}_2 \text{ g enzyme}^{-1}\text{s}^{-1},$$

where $v_e$ is the RuBP and $CO_2$-saturated specific carboxylase activity, $K_c$ and $K_o$ are the Michaelis constants for $CO_2$ and $O_2$ respectively, and $C$ and $O$ the concentrations of $CO_2$ and $O_2$ respectively. If $C_i$ is 10.5 mmol $m^{-3}$ (equivalent to a volume fraction of 235 p.p.m.; Davis & McCree, 1978; Friedrich & Huffaker, 1980; Constable & Rawson, 1980; Fig. 15.3), then $C$ will not usually be less than 6.5 mmol $m^{-3}$. $O$ can be taken as 250 mmol $m^{-3}$, the concentration of $O_2$ in water in equilibrium with air. Data recently reported for a number of $C_3$ plants by Jordan & Ogren (1981) give mean values of $K_c = 12.6$ mmol $m^{-3}$ and $K_o =$ 532 mmol $m^{-3}$ at 25 °C. $v_e$ can be calculated from rates of carboxylation expressed per unit of chlorophyll (chl), which is freshly ruptured spinach chloroplasts approach 1000$\mu$mol $CO_2$ mg $chl^{-1}h^{-1}$ at 20°C (Lilley & Walker, 1975). Given a $Q_{10}$ of 2 this is roughly equivalent to 390$\mu$mol $CO_2$g $chl^{-1}s^{-1}$ at 25°C. In the chloroplasts used in the above experiments, the amount of RuBPc-o g $chl^{-1}$ was somewhat greater than the value of 4–6.5g RuBPc-o g $chl^{-1}$ found for $C_3$ plants by Ku, Schmitt & Edwards (1979) (D. A. Walker, personal communication). If a value of 9 g RuBPc-o g $chl^{-1}$ is used, $v_e$ for spinach at 25 °C is approximately 43 $\mu$mol $CO_2$ g $enzyme^{-1}s^{-1}$. By substitution in the above equation, $v'_e$ is 11 $\mu$mol $CO_2$ g $enzyme^{-1}s^{-1}$. It is not unreasonable to assume that similar activities could be calculated for other $C_3$ species.

This value of $v'_e$ can be used to estimate the maximum potential rates of carboxylation in leaves of different ages (Table 15.1). These estimates are only approximate because a number of the parameters used to calculate $v'_e$ vary between species, and an error is associated with the quantitative determination of RuBPc-o protein. The photosynthetic rates shown for comparison should be adjusted to 25°C; if it is assumed that these rates were light-saturated and that C at the carboxylation sites was approximately 6.5mmol $m^{-3}$, then the data suggest that the enzyme was operating below its maximum capacity *in vivo* in each of the three species. Moreover, during senescence in both wheat and *Perilla*, the rate of photosynthesis declined more rapidly than the amount of enzyme protein. If the specific activity of the enzyme remained constant, then the carboxylase activity must have been limited by some other factor. Clearly, it is necessary to consider the regulation of RuBPc-o activity *in vivo*. The degree of enzyme activation, the conditions in the stroma for catalytic activity, and the concentration of RuBP may all be limiting factors (Jensen & Bahr, 1977). The regeneration of RuBP is itself dependent on the amounts and activities of other enzymes of the reductive pentose phosphate pathway, and on the levels of ATP and NADPH (Walker, 1976; Bassham, 1979).

There is a paucity of information available for other enzymes of the reductive pentose phosphate pathway of senescing leaves (Zima & Šesták, 1979). Batt & Woolhouse (1975) found that in *Perilla* NADPH-glyceraldehyde 3-phosphate dehydrogenase and ribulose 5-phosphate kinase declined in activity during senescence roughly in parallel with RuBPc-o, whereas glycerate 3-phosphate kinase, ribose 5-phosphate isomerase and fructose bisphosphatase remained active for much longer. Further investigation using improved assay procedures is now required. Particular attention should be paid to fructose and

Table 15.1. *Calculated maximum potential carboxylation rates in leaves of different ages*

| Reference | Species and leaf used in experiments | RuBPc-o protein content (g $m^{-2}$) | | Calculated maximum potential carboxylase activity *in vivo* at 25 °C ($\mu$mol $CO_2$ $m^{-2}$ $s^{-1}$) | | Gross photosynthetic rate ($\mu$mol $CO_2$ $m^{-2}$ $s^{-1}$ | | |
|---|---|---|---|---|---|---|---|---|
| | | Mature | Senescent | Mature | Senescent | Mature | Senescent | °C |
| Woolhouse (1967); Woolhouse & Batt (1976) | *Perilla frutescens*; 3rd leaf pair from 45 d (mature) to 70 d (senescent) | 3.3 | 0.75 | 36 | 8 | 5.4[a] | 0.9[a] | 20 |
| Hall *et al.* (1978); N. P. Hall (personal communication) | *Triticum aestivum* (spring variety); flag leaf from 41 d (mature) to 77 d (senescent) | 5.8 | 4.3 | 64 | 47 | 26[a] | 4.0[a] | 26 |
| Friedrich & Huffaker (1980) | *Hordeum vulgare* primary leaves from 12 d (mature) to 22 d (senescent) | 2.4 | 0.64 | 26 | 7 | 5.9 | 1.9 | 24 |

[a] Estimated as 1.25 times the net photosynthetic rate.

sedoheptulose bisphosphatases because the regulation of these enzymes in senescing leaves may be important in controlling the overall rate of photosynthesis.

### *Photosynthetic electron transport*

During senescence the internal membrane system of the chloroplasts becomes increasingly disorganised and disrupted. The thylakoids swell and lose their ordered alignment in the stroma, and eventually the plastids contain little more than membrane vesicles and large plastoglobuli. Over this period large amounts of chlorophyll and lipids of the thylakoid membrane are lost from the leaves. In order to analyse the effects of the degeneration of membrane structure on the capacity of the chloroplasts to carry out electron transport it is necessary to consider both the electron carriers and the reaction-centre and light-harvesting components.

#### *Reaction-centre and light-harvesting chlorophylls*

Few attempts have been made to investigate changes in the number of functional reaction centres during foliar senescence, or to determine if the efficiency of the light-trapping systems remains constant (Šesták, 1977a). Jenkins, Baker & Woolhouse (1981) found that chloroplasts isolated from mature and senescent leaves of *Phaseolus vulgaris* gave very similar light saturation curves for electron transport reactions involving either photosystem one (PSI) or photosystem two (PSII). Assuming that the efficiency of exciton transfer between pigment molecules did not increase during senescence, these results indicate that the average number of chlorophyll molecules able to transfer excitation to each reaction centre was undiminished in older leaves, even though they had lost 75% of their chlorophyll. Analysis of the kinetics of chlorophyll fluorescence emission support this contention (Jenkins, Baker, Bradbury & Woolhouse, 1981). It can be concluded that the reaction centres ceased to function at the same time as, or perhaps before, their antenna chlorophylls were lost from the thylakoid membranes, and that the percentage decrease in the number of functional reaction centres per leaf was at least as great as the percentage decrease in the leaf chlorophyll content. This type of analysis should be extended to examine possible differences in other species. There is, for example, evidence that in some cultivars of tobacco the number of chlorophyll molecules associated with the reaction centres decreases during senescence (Haraguchi & Shimizu, 1970).

Qualitative changes in pigment composition are also found in senescing leaves (Šesták, 1977a), and in some species there appears to be a preferential loss of chlorophylls associated with PSI. Šesták (1969) observed that fewer particles enriched in PSI could be isolated from older leaves of spinach and radish relative to those enriched in PSII, and Šesták & Demeter (1976) reported an increase in the total chl/P700 ratio during the early stages of senescence in leaves of *Phaseolus vulgaris*. More recently, Bricker & Newman (1980) found that the polypeptide of the P700-chlorophyll a-protein complex (P700CP) of PSI decreased in amount relative to the polypeptide of the light-harvesting chlorophyll a/b-protein complex (LHCP) during senescence of soybean cotyledons. The data of Jenkins, Baker & Woolhouse (1981) are consistent with these results, and show that a smaller proportion of the chlorophyll in older primary leaves of *P. vulgaris* was associated with the P700CP. There was also a relative decrease in fluorescence emission characteristic of PSI chlorophylls during senescence. This preferential loss of PSI chlorophylls could be due to the observed breakdown of stromal thylakoids before those of the grana, as reported by Ljubešić (1968), Whatley (1974) and Wolinska (1976).

*Electron transport components*

Unfortunately, studies of electron transport during foliar senescence have been largely confined to measurements of the Hill reaction, often using chloroplasts of a dubious quality. Ferricyanide and 2,6-dichlorophenol indophenol have been employed almost exclusively as the electron acceptors (Šesták, 1977b), whereas it is now known that these compounds can receive electrons from both PSI and PSII depending on the chloroplast preparation (Hauska, 1977). There is also evidence that the site of ferricyanide reduction changes during senescence in *Phaseolus* chloroplasts (Jenkins & Woolhouse, 1981b). The use of these acceptors is now considered inappropriate because it is often uncertain exactly which portion of the electron transport chain is under investigation.

It seems logical that an investigation of the electron transport capacity of chloroplasts from leaves of different ages should initially involve measurements of non-cyclic electron flow. However, until recently, these had been reported only by Haraguchi & Shimizu (1970) and Andersen, Bain, Bishop & Smillie (1972) without any measurements of the rate of photophosphorylation. Jenkins & Woolhouse (1981a) measured coupled, non-cyclic electron transport during senescence of the primary leaves of *Phaseolus vulgaris*, taking particular care to prevent inactivation of the chloroplasts during either isolation or assay of the reaction. Chloroplasts isolated from young, fully expanded leaves gave rates of coupled non-cyclic electron transport from water to the artificial electron acceptor methyl viologen (Fig. 15.4) which compared well with those reported for spinach chloroplasts. Moreover, electron transport was efficiently coupled to photophosphorylation with a mean P/2e ratio of 1.2. During senescence the rate of non-cyclic electron flow per unit chlorophyll declined by 80%, but the P/2e ratio remained constant (Fig. 15.5). The lower activity of chloroplasts from senescent leaves was not the result of an altered sensitivity to components of the isolation and reaction media, and neither was there any evidence of large amounts of inhibitory substances in homogenates of older leaves. The most likely explanation of the results is that a similar decrease in the rate of non-cyclic electron transport occurred *in vivo* during senescence.

Jenkins & Woolhouse (1981b) attempted to identify the site at which electron flow was impaired in chloroplasts from older leaves. Artificial cofactors and inhibitors were employed to measure the activities of particular portions of the electron transport chain (Fig. 15.4). The rates of electron transport through photosystems I and II declined by approximately 25% and 33% respectively during senescence, much less than the decrease of 80% found for non-cyclic electron transport (Table 15.2). It was concluded that the major limitation to non-cyclic electron flow lay at some point between the two photosystems, more specifically

Fig. 15.4. Principal components of the photosynthetic electron transport chain showing the probable sites of inhibition by 3-(3,4-dichlorophenyl)-1,1-dimethylurea (DCMU) and trifluralin, the sites of electron acceptance of oxidised *p*-phenylenediamine ($PD_{ox}$) and methyl viologen (MV), and the site of electron donation of *N,N,Ń,Ń,*-tetramethyl-*p*-phenylenediamine ($TMPD_{red}$). P700 and P680 are the reaction centres, and Fe-S and Q the primary electron acceptors of PSI and PSII respectively. PQ, plastoquinone; $PQH_2$, plastohydroquinone; cyt.f, cytochrome f; PC, plastocyanin; Fd, ferredoxin.

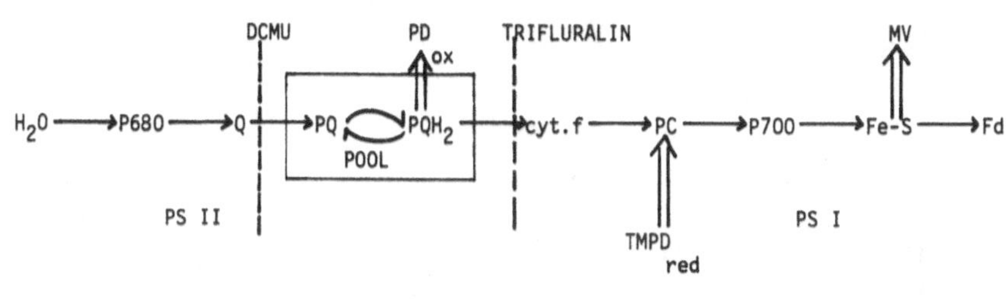

Fig. 15.5. Rates of coupled, non-cyclic electron transport from water to MV, and the P/2e ratios, measured for chloroplasts isolated from primary leaves of *Phaseolus vulgaris* of different ages. Electron transport was measured by monitoring oxygen uptake in the reaction mixtures with an oxygen electrode; photophosphorylation was measured by incorporation of $^{32}P$. Each point refers to one experiment using a chloroplast preparation from a single primary leaf. From Jenkins & Woolhouse (1981a).

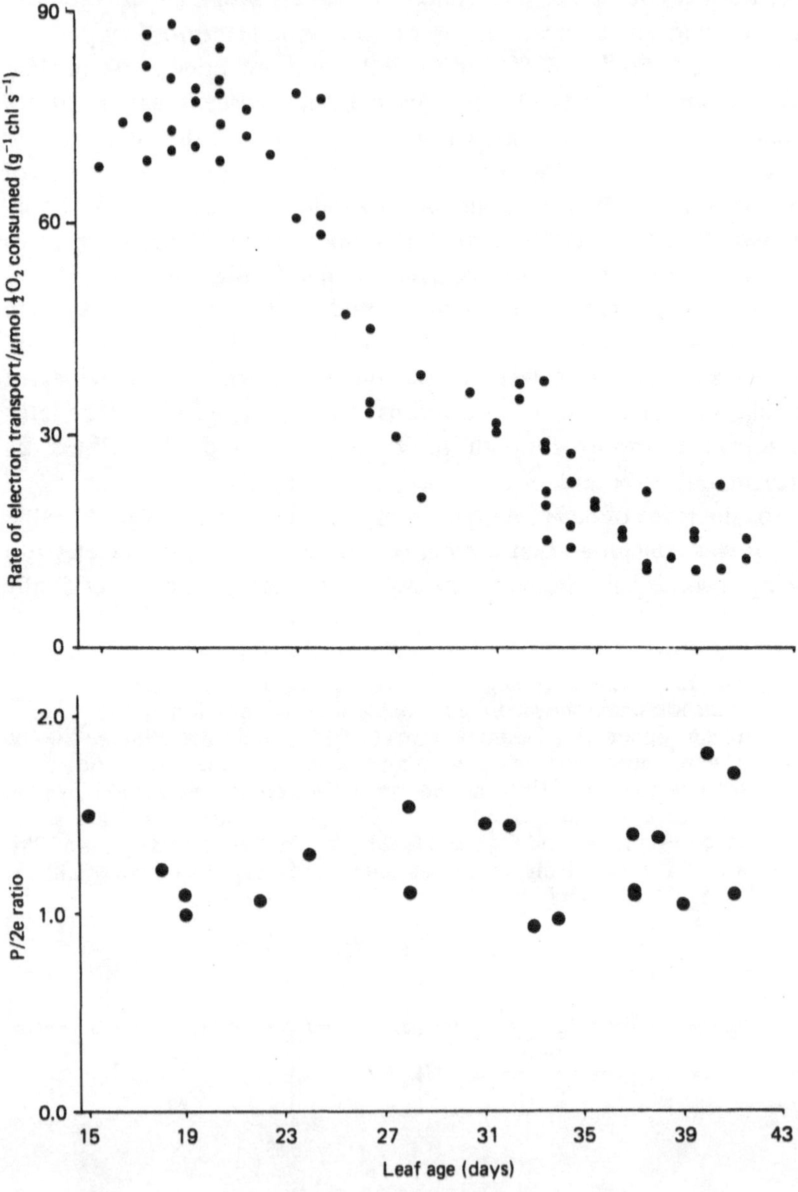

between the acceptor site of oxidised *p*-phenylenediamine ($PD_{ox}$) and the donor site of reduced *N,N,Ń,Ń*-tetramethyl-*p*-phenylenediamine ($TMPD_{red}$). In spinach chloroplasts $PD_{ox}$ is believed to receive electrons from plastohydroquinone (Trebst, 1974), whereas $TMPD_{red}$ donates electrons principally to plastocyanin (Hauska, 1975).

A further indication that an impairment of electron flow between the photosystems developed during senescence has come from analyses of the kinetics of chlorophyll fluorescence emission from intact leaves (Jenkins, Baker, Bradbury & Woolhouse, 1981). Briefly, the results indicate that fewer plastoquinone (PQ) molecules were able to receive electrons from each PSII reaction centre in senescent leaves; the 'pool' of PQ molecules which acts as an electron buffer between the photosystems and interconnects a number of electron transport chains (Siggel, 1976) decreased in size during senescence. Furthermore, the PQ pool remained strongly reduced following illumination of dark-adapted discs taken from senescent leaves, whereas in younger leaves reduction of the pool was followed by reoxidation. This suggests that the rate of electron flow out of the PQ pool to PSI decreased during senescence.

Recent evidence indicates that the sequence and rate of electron transfer is dependent on random lateral diffusion and energetically favourable collisions between the electron transferring compounds and as such is dependent on the physical state of the lipid matrix in which the diffusion occurs (Hackenbrock, 1981). PQ is lipophilic and undoubtedly

Table 15.2. *Rates of photosynthetic electron transport in chloroplasts isolated from primary leaves of* Phaseolus vulgaris *of different ages*

| | Rate of electron transport ($\mu$mol 2 electron equivalents g chl$^{-1}$s$^{-1}$) | | |
|---|---|---|---|
| | young, fully expanded leaves | senescent leaves | % decrease during senescence |
| Coupled, non-cyclic electron transport: water to MV | 76.8 ± 1.3 ($n$ = 22) | 15.0 ± 0.7 ($n$ = 33) | 80 |
| PS I activity: $TMPD_{red}$ to MV + DCMU | 186.0 ± 6.8 ($n$ = 10) | 139.0 ± 5.5 ($n$ = 11) | 25 |
| PS II activity: water to $PD_{ox}$ + trifluralin | 108.0 ± 7.9 ($n$ = 10) | 72.8 ± 5.4 ($n$ = 15) | 33 |

has strong interactions with lipid components of the thylakoid membrane. Alterations in the properties of the lipid bilayer during senescence could therefore impair the function of PQ as an electron and proton carrier or cause it to be released from the membrane. Qualitative changes in thylakoid lipid composition occur during senescence (e.g. Fong & Heath, 1977; Novitskaya *et al.*, 1977), and alterations in the physical state of the membrane lipids have also been reported (McKersie & Thompson, 1978). A detailed study of the effects of these changes on the activity of the electron carriers is now required.

In *Phaseolus vulgaris* the extensive loss of reaction centres during senescence and the decline in rate of electron transport per unit chlorophyll together amount to a decrease of approximately 95% in the capacity of the chloroplast to carry out non-cyclic electron transport (Jenkins & Woolhouse, 1981a; Jenkins, Baker & Woolhouse, 1981). A similar decrease in activity can be calculated for *Nicotiana tabacum* (cv Bright Yellow) using the data of Haraguchi & Shimizu (1970).

### *Regulation of the reductive pentose phosphate pathway*

Photosynthetic electron transport plays an essential role in controlling the activity of several enzymes of the reductive pentose phosphate pathway through the generation of ATP and NADPH and by creating optimal conditions in the stroma for enzyme activation and for the attainment of maximal rates of catalysis (Walker, 1976; Bassham, 1979). During electron transport, protons are transferred from the stroma into the intrathylakoid space and this is partly balanced by a counterflux of $Mg^{2+}$ ions into the stroma. In consequence, the pH of the stroma is raised from approximately pH 7 to pH 8, and the $Mg^{2+}$ concentration is increased by about 1 to 3 mol $m^{-3}$ (Heldt, 1979). These changes are important in modulating the activity of the regulatory enzymes fructose bisphosphatase and sedoheptulose bisphosphatase. Treatments which prevent the increase in stroma pH or $Mg^{2+}$ concentration cause an inhibition of $CO_2$ fixation and an accumulation of fructose and sedoheptulose bisphosphates (Purceld *et al.*, 1978; Flügge, Freisl & Heldt, 1980). Moreover, *in vitro*, the activity of fructose bisphosphatase is raised from zero to near maximum by conditions which approximate those in the stroma during illumination (Baier & Latzko, 1975). The increase in stroma pH and $Mg^{2+}$ concentration also favours the activation of RuBPc-o and provides near optimal conditions for its activity; the physiological significance of the light activation of this enzyme has, however, been questioned (Robinson, McNeil & Walker, 1979).

There is also evidence that electron transport brings about the reductive activation of several enzymes. The electron transport component ferredoxin reduces a soluble protein called thioredoxin which activates the two bisphosphatases (Buchanan, 1980). A membrane-bound reductant is apparently involved in the activation of NADPH-glyceraldehyde 3-phosphate dehydrogenase and ribulose 5-phosphate kinase (Anderson, 1979). The significance of these reductive systems in the regulation of enzyme activity *in vivo* is now the subject of some debate (see, for example, Charles & Halliwell, 1981).

There is little doubt that a decrease in the rate of electron transport during senescence would cause a corresponding decrease in the flux of protons and $Mg^{2+}$ ions across the thylakoid membrane. Although it is difficult to predict how the buffering and cation-binding capacity of the stroma might change during senescence, it is likely that in older leaves the light-induced increase in stroma pH and $Mg^{2+}$ concentration would be diminished. The generation of reductants would also decrease. These changes are likely to affect the activation and activity of key regulatory enzymes, notably RuBPc-o and the bisphosphatases. The effect on the activity of the latter enzymes may be the more important *in vivo*, and in older leaves the activity of RuBPc-o may be limited by the regeneration of RuBP. A decrease in the rate of electron transport during senescence could therefore decrease the rate of carbon fixation independently of changes in the amounts of enzymes in the chloroplasts.

A decrease in the rate of electron transport during senescence would also reduce the levels of ATP and NADPH in the stroma (see Chapter 8), but it is difficult to speculate how the ATP/ADP ratio might change during senescence because other demands for energy in the chloroplast are probably reduced and the total concentration of adenylates is unlikely to remain constant. Measurements of the steady state levels of metabolites in chloroplasts from leaves of different ages are now required to determine whether formation or reduction of PGA, or RuBP regeneration is limiting.

The transport of metabolites between stroma and cytoplasm plays an important role in the regulation of carbon metabolism (see Chapter 13). Although the chloroplast envelope appears to remain intact until a late stage of senescence (Butler & Simon, 1972; Whatley, 1974), it is unknown whether it continues to function effectively as a selective transport barrier. The accumulation of starch which is often found in senescing leaves could be the result of a reduced export of sugars through failure of a transport system. However, since phosphate is translocated out of senescing leaves it is also possible that the export of sugars is limited by a decrease in the availability of phosphate in the cytoplasm.

## Protein turnover and the energy requirement of leaf senescence

The concept of a cycle by which proteins are continuously synthesised and catabolised in the course of metabolism has been gradually developed over the past fifty years. The evidence for such a cycle and the problems of making accurate measurements of turnover on which the operation of such a cycle must rest, are considered elsewhere (Woolhouse, 1982a, b). Here, the ways in which the turnover of constituents may impinge upon and mediate degradative changes in senescing leaves are considered, with special emphasis on membrane-associated changes since these may have much to do with the declining photosynthetic functions.

Changes in the structure and properties of membranes are a universal feature of senescence in plant cells. The evidence comes from electron microscopy (Butler & Simon, 1972), X-ray diffraction studies on isolated membranes (McKersie & Thompson, 1978) and the increasing permeability of cells of senescing tissues.

### *Membrane proteins*

The constituent proteins of thylakoid membranes turn over at very different rates. One of the constituents of the thylakoid is a protein encoded and synthesised within the chloroplast, having a molecular weight of 32 000 (P-32 000); pulse-chase experiments with [$^3$H]-leucine-labelled fronds of *Spirodela* show P-32 000 to be turned over an order of magnitude more rapidly than the other major chloroplast polypeptides (Edelman & Reisfeld, 1980; Mattoo *et al.*, 1981). Nevertheless synthesis of P-32 000 can be detected in chloroplasts from leaves which have ceased the manufacture of RuBP carboxylase.

### *Protein turnover*

Modulation of the level of a particular protein may be achieved by altering the balance of its synthesis and degradation (Fig. 15.6) and this may be altered in favour of synthesis or degradation even when both processes are actually increasing or decreasing in absolute terms. In considering protein turnover we have to consider three sites of transcription – the nucleus, chloroplasts and mitochondria – and three sites of translation – the cytoplasm, chloroplasts and mitochondria. In turn, these imply three sites of turnover (Fig. 15.7) and have important consequences for the pattern of events involved in leaf senescence.

*Control of protein synthesis*

(*a*) *Chloroplasts*. In leaves of *Perilla*, *Phaseolus* and cucumber there is a loss of capacity for chloroplast protein synthesis in the course of senescence, although cytoplasmic protein synthesis is maintained. Declining chloroplastic protein synthesis is associated with a loss of polysomes and the cessation of chloroplast RNA synthesis (Callow & Woolhouse, 1973). The arrest of protein synthesis in chloroplasts of *P. vulgaris*, at the time that leaf expansion is completed, has been traced to

Fig. 15.6. Scheme for the turnover of proteins and other macromolecular constituents of the leaf in which a net loss of constituents may arise from loss of function amongst elements of the biosynthetic system or enhanced activity of enzymes on the catabolic side of the cycle. Some breakdown products are shown as being lost to the system by translocation to other parts of the plant and some are re-cycled through the biosynthetic pathways.

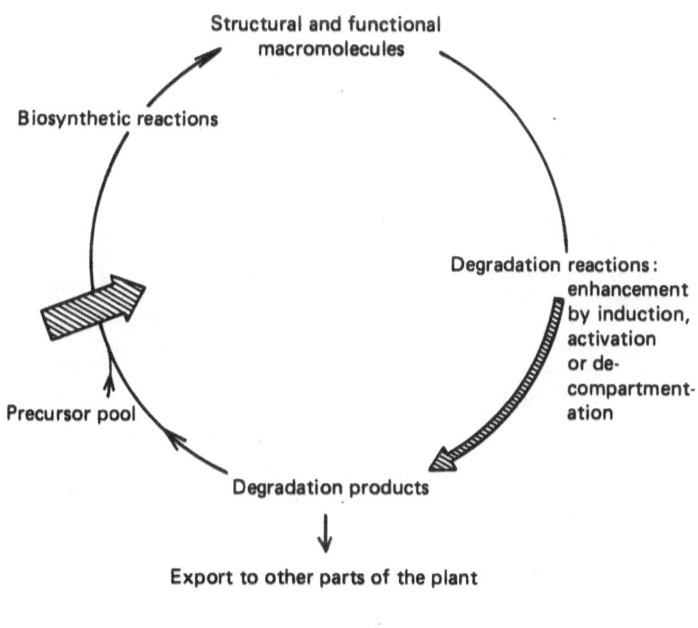

the disappearance of chloroplast RNA polymerase activity. The RNA polymerase found in chloroplasts is encoded in the nucleus and synthesised on cytoplasmic ribosomes and through this key enzyme the nucleus may control the lifespan of the chloroplast population in leaf cells. Further evidence of an over-riding control by the nucleus comes from the work of Yoshida (1961). Leaves of *Elodea densa* were plasmolysed in calcium chloride solution causing some of the cells to separate into two parts, one with and one without a nucleus. In subsequent culture, the chloroplasts of enucleate protoplasts remained green and accumulated starch whilst chloroplasts in the protoplasts containing nuclei disintegrated and were broken down over five days.

(*b*) *Mitochondria*. In the leaves of many plants the rate of

Fig. 15.7. A scheme depicting the three protein cycles which co-exist in leaf cells and aspects of their interrelationships. The numbered steps in each cycle indicate (1) amino acid activation prior to protein synthesis, (2) amino acid incorporation into protein, (3) energy-dependent protein activation prior to degradation (see text) and (4) hydrolytic breakdown of protein. The dotted lines bearing the symbol R represent possible routes along which amino acids or peptides may be transported, thereby linking the substrate pools of the three amino acid cycles, as follows: $R_{o-c}$, organelle to cytoplasm; $R_{c-o}$ cytoplasm to organelle; $R_{c-v}$, cytoplasm to vacuole; $R_{v-c}$, vacuole to cytoplasm and $R_{c-t}$, cytoplasm to the phloem transport system mediating export from the leaf.

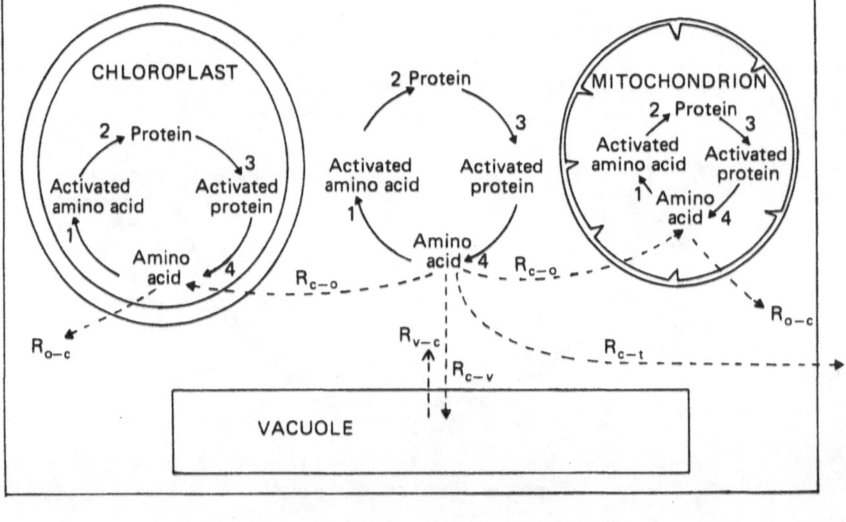

respiration is maintained and may actually increase during yellowing and senescence (James, 1953; Hardwick *et al.*, 1968). This finding is compatible with electron microscope studies which show that the mitochondria of yellowing leaves retain their structural integrity even at an advanced stage of senescence when the chloroplasts are showing extensive degradation and the rate of photosynthesis is declining (Butler & Simon, 1972). It is reasonable to suppose that the sustained mitochondrial activity in senescing leaves serves to fuel vein-loading as the breakdown products of the cells are remobilised, and also the syntheses of lipids, amides and specific enzymes which take place in senescing leaves.

In looking for potential control mechanisms, there are a number of interesting points of difference between the chloroplastic and mitochondrial systems. First, the energy source for chloroplast protein synthesis will become limited as the rates of photosynthetic electron transport and photophosphorylation decrease during senescence, and also as older leaves become shaded by their developing neighbours. Second, there are substantial differences between the protein-synthesising systems of chloroplasts and mitochondria. Both organelles contain immunologically distinct elongation factors, EF-G and EF-Tu, in their protein-synthesising systems but whereas EF-$G_{chl}$ and EF-$Tu_{chl}$ are encoded and synthesised within the plastids, EF-$G_{mit}$ and EF-$Tu_{mit}$ are encoded in the nucleus and synthesised on cytoplasmic ribosomes (Tiboni *et al.*, 1978; Ciferri, Dipasquale & Tiboni, 1979). Thus where there is preferential arrest of metabolism at the chloroplast RNA-polymerase transcription level, there will be a concomitant end to the synthesis of the plastid-encoded elongation factors. If cytoplasmic protein synthesis is maintained at this time it can provide the necessary EF-G and EF-Tu factors and allied constitutes of the mitochondria, since these are encoded in the nucleus and synthesised in the cytoplasm.

*(c) Cytoplasm.* Although the total protein content of leaves declines in the course of senescence (Woolhouse, 1967), there is none-the-less an active synthesis of proteins on 80 S cytoplasmic ribosomes throughout this period (Callow *et al.*, 1972). The RNA content of senescing leaves also declines (Pearson, Thomas & Thomas, 1978; Callow & Woolhouse, 1973) as does the rate of RNA synthesis (Wollgiehn, 1967; Callow *et al.*, 1972). Generally speaking, inhibitors of RNA synthesis, such as actinomycin D, are without effect on leaf senescence, suggesting that senescence is not controlled at the transcriptional level (Thomas & Stoddart, 1980). It may be that relatively

long-lived mRNAs for proteins are associated with leaf senescence and are controlled at the level of processing or translation.

Inhibitors of the ribosome cycle in eukaryotes such as cycloheximide are frequently found to be active inhibitors of leaf senescence, and this has been widely interpreted as evidence of control of senescence at the level of mRNA translation. Such studies undoubtedly afford some interesting pointers but it is essential always to establish the specificity and uniqueness of action of the inhibitor concerned.

To the question of whether the rate of protein synthesis declines and contributes to the net loss of protein which accompanies leaf senescence, the answer is probably yes. However, given that the rate of protein synthesis in the chloroplasts certainly declines (Callow *et al.*, 1972), it is difficult, for technical reasons, to make firm pronouncements concerning the rate of protein synthesis in the cytoplasm. We shall return briefly to the regulation of cytoplasmic protein synthesis in the section where hormonal factors are considered.

### *Control of protein degradation*

A net loss of protein from senescing leaves may be achieved by shutting down the biosynthetic side of the turnover process, operating through the type of control mechanism described above. The contribution of increased rates of degradation to the net loss of protein is now considered.

Table 15.3 lists proteases and other hydrolytic enzymes which have been reported to increase in activity in senescing leaves. It is pertinent to question whether these increases in hydrolase activity are essential for senescence to take place. Frequently, the amount of protease activity detectable before the onset of senescence is more than adequate for the observed rate of breakdown of protein, and the activity during senescence is increased by a factor of two or three only (Waters *et al.*, 1980); a change of this magnitude may be of much less significance than changes of compartmentation which bring enzymes and substrates into contact. Evidence which casts doubt on the significance of measured activities of acid proteases comes from the work of Van Loon *et al.* (1978). Excised oat leaves floated on water in darkness showed the normal yellowing and increase of protease activity; leaves placed in an upright position with their bases in water did not show the increased protease activity but breakdown of protein occurred to the same extent in both sets of leaves.

There are many sites at which there could, in principle, be regulation of levels and activity of proteolytic enzymes (Fig. 15.8). Regulation can be summarised as (i) *de novo* synthesis of proteolytic enzymes (reactions and factors *a* to *l*), (ii) activation and *in situ* regulation (reactions *m* and *n*), and (iii) compartmentation (reactions *o* and *p*). Evidence relating to the events between (*a*) and (*l*) is summarised here and emphasis is concentrated upon new evidence relating to control at the terminal level (*n*)–(*p*). A detailed discussion is given by Woolhouse (1982b).

As already discussed there is no significant evidence for transcriptional control (Fig. 15.8, step *a*) of factors involved in senescence although Grierson (1981) has recently obtained evidence for the synthesis of an mRNA for a $\beta$1–3 glucanase which becomes active during the senescence of tomato fruits.

Message processing (reactions *b* and *c*) may be important in the regulation of protease synthesis in the course of leaf senescence; for instance, cordycepin, an inhibitor of polyadenylation, delays senescence in leaf discs. This category of reactions will also include capping of the mRNA at the 5′-terminus by an inverted 7-methyl guanosine. Capping is known to be a factor in protecting eukaryote mRNAs against exonuclease attack and the cap is involved in initiation of protein synthesis (Hershay, 1980), so that the potential role of such a system in the preservation of long-lived messages for enzymes involved in senescence should be considered.

Table 15.3. *Hydrolytic enzymes reported to show increased activity in senescing leaves (for references, see Woolhouse, 1982a)*

| | |
|---|---|
| Protease | Acid invertase |
| Tobacco | *Lolium temulentum* |
| Mung bean (cotyledons) | Chlorophyllase |
| Pea | Barley and oats |
| Wheat | Esterase |
| Maize | *Festuca pratensis* |
| Barley | Acid phosphatase |
| Oats | *Perilla* |
| *Perilla* | $\beta$-1,3-Glucan hydrolase |
| *Lolium temulentum* | *Nicotiana glutinosa* |
| Ribonuclease | |
| *Rhoeo* | |
| *Phaseolus vulgaris* | |
| Oat | |

The reactions and factors (*d*) to (*k*) refer to the 80 S ribosome cycle in the cytoplasm of the leaf cells, the main components of which have already been considered in the context of regulation of protein synthesis. It is important to emphasise that to date there is only one rigorous demonstration of *de novo* synthesis of an enzyme which increases during leaf senescence (Sacher & Davies, 1974).

Fig. 15.8. Diagram indicating some of the multiple controls which may be involved in the regulation of the increased hydrolytic enzyme activity associated with the later stages of leaf senescence. Reactions and factors (a) to (l) are concerned with aspects of the *de novo* synthesis of the enzymes, reactions (m) and (n) denote *in situ* regulation of the enzymes by processing and secondary modification or release from a zymogen; reaction (o) indicates enzyme regulation by compartmentation; reaction (p) denotes enzyme regulation at the substrate level.

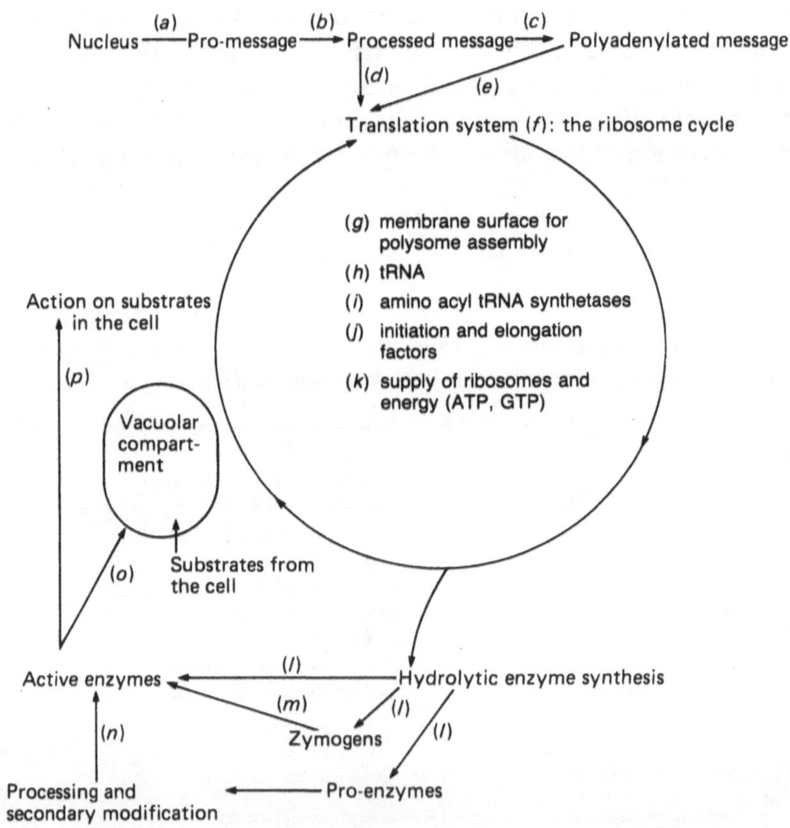

### *Protease regulation by compartmentation and by enzyme activation*

(*a*) *Compartmentation.* The question of whether or not compartmentation is an important factor in the regulation of nucleic acid and protein breakdown in senescing leaves is a matter of controversy at the present time (Woolhouse, 1982a). The central issue concerns the technical problem of breaking open leaf cells and fractionating their constituents without modifying the intracellular distribution of enzymes. Matile (1978) introduced the concept of the vacuole in plant cells as a lysosome-like organelle which serves as a repository for the hydrolytic enzymes. Reviewing the subject and considering the vacuole as an autophagic compartment into which cellular constituents destined for breakdown were transferred, Matile commented:

> The autophagic nature of vacuoles is suggested by numerous ultrastructural observations concerning the presence within vacuoles of structural cytoplasmic material. The bad preservation of such vacuolar inclusions is likely to be due to the action of digestive enzymes. Membrane processes seem to be responsible for the sequestration of cytoplasmic material in vacuoles. On the one hand the engulfment by the invaginating tonoplast results in the formation of intravacuolar vesicles containing a portion of cytoplasm. On the other hand the transfer of cytoplasm into the digestive compartment coincides with the formation of vacuoles: portions of cytoplasm are encircled by components of the endomembrane system which ultimately fuse together and form an autophagic vacuole. These observations suggest that the compartmentation of vacuole hydrolases is never abolished. Yet autophagy cannot be understood unless the morphological observations are complemented with corresponding knowledge about the biochemical events. At the moment we are dealing with a mosaic of somewhat isolated ultrastructural observations of autophagic vacuoles; assessments of turnover reactions, localisations of enzymes and correlations between catabolic processes and hydrolase activities. What is needed is an integrated study of one organism which covers all the relevant aspects.

This comment remains true. More recently Boller & Kende (1979) have provided impressive evidence for the presence of most of the hydrolytic enzymes generally regarded as involved in senescence as being located in the vacuoles in leaves of tobacco. Lin & Wittenbach (1981) used protoplast techniques to achieve separation of intact vacuoles and chloroplasts from leaves of wheat and maize; they concluded that all of the proteolytic activity measured by degradation of ribulose bisphosphate carboxylase, used as a substrate, can be accounted for by that located in the vacuoles. The protease activity associated with the chloroplasts was concluded to be an artefact of contamination. Huffaker & Miller (1978) propose the cytoplasm as a source of endoproteolytic activity in leaves of barley with the chloroplast envelope serving as a barrier between the hydrolase and the chloroplast contents. Heck *et al.* (1981), on the other hand, find 85% of acid proteinase of barley leaves in vacuoles prepared by lysis of protoplasts, which they conclude is 'presumably responsible for the degradation of chloroplastic protein'.

From this conflicting evidence the following points may be noted. Acceptance of the autophagic vacuole as the source of normal protein turnover (Lin & Wittenbach, 1981), diagramatically summarised as Fig. 15.9, is superficially attractive, but raises major difficulties of a logistic nature. Chloroplasts and mitochondria do not undergo simultaneous breakdown in the course of senescence and individual proteins within an organelle turn over at different rates. The autophagic vacuole hypothesis requires that either the proteins with differing half-lives go to the

Fig. 15.9. A simple model depicting the autophagic vacuole concept in relation to proteolysis. Cy, Chl, M, N and V represent the cytoplasm, chloroplast, mitochondria, nucleus and vacuole, respectively. The numbers 1 to 4 represent transport of protein from the organelles to the vacuole. 5 represents the release of amino acids formed by proteolysis in the vacuole. The difficulties in equating a model of this kind with the variety of half-lives of proteins in a given organelle will be evident (see text).

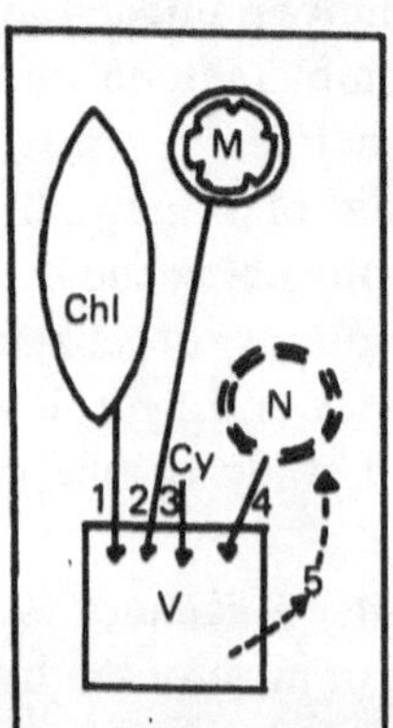

vacuoles at differing rates or the proteases involved in their breakdown come from the vacuoles to the organelles and degrade different proteins at different rates. Protein travelling to the vacuoles raises awkward questions about how they might be removed selectively from the organelles and proteases travelling from the vacuoles to the organelles would require them to exhibit a greater degree of target specificity than has been revealed by *in vitro* assays.

An alternative view is based on the notion that the enzymes of the vacuolar compartment may be primarily concerned with reactions other than turnover, as for example the hypersensitive reactions which are responsible for defence against attack by pests and pathogens (Woolhouse, 1982b). A corollary to this is that the enzymes relevant to the turnover process and the organised phase of senescence may have hitherto been overlooked; a consideration of processes involved in the activation of proteolytic enzymes or their substrates in animal or bacterial cells suggests new avenues of approach to this hypothesis.

(*b*) *Proteins and protease-substrate activation*. The shortcomings of the autophagic vacuole system mean that it cannot provide the sensitive regulatory mechanism which is required for differential turnover of particular proteins. In its stead an alternative mechanism involving a precisely controlled, energy-dependent proteolytic system is now considered.

James (1953) noted that senescence of leaves in darkness was prevented if they were maintained under anaerobic conditions and around the same period it was shown that respiratory inhibitors prevented the breakdown of proteins in liver tissue (Steinberg & Vaughan, 1956). The pioneer work of Linderstrøm-Lang (1950) showed that denatured proteins are much more susceptible to proteolysis than the intact forms. Subsequent work has shown that there is an energy requirement for the degradation of both normal and aberrant proteins in bacteria (Kowit & Goldberg, 1977), reticulocytes (Hershko *et al.*, 1980) and other mammalian cells (Epstein *et al.*, 1975).

It has recently become possible to study this process *in vitro* following the demonstration of ATP-stimulated breakdown of protein in crude extracts of reticulocytes (Hershko *et al.*, 1980) and *E. coli* (Murakami *et al.*, 1979). In extracts of *E.coli* the ATP-dependent system is associated with a membrane fraction which degrades large protein molecules to smaller components which are then hydrolysed to amino acids by soluble proteases which do not require ATP (Voellmy & Goldberg, 1981). The selective breakdown of particular proteins may also involve an ATP-dependent system (Roberts *et al.*, 1978).

Ciechanover *et al.* (1980) extracted a heat-stable polypeptide from reticulocytes of rabbits which was necessary for the ATP-dependent breakdown of protein catalysed by disrupted reticulocytes. The system involved an ATP-dependent formation of a covalent bond between the heat stable peptide and the proteins and the conjugate so formed constitutes the active substrate of the ATP-dependent proteolysis (Hershko *et al.*, 1980). The heat-stable polypeptide is identical with ubiquitin (Wilkinson, Urban & Haas, 1980), a protein of MW 8 500 which occurs in bacteria, fungi, plants and animals. A proposed mechanism for this ATP-regulated degradation of protein is shown in Fig. 15.10. If such a system were to operate in leaves one could envisage that, in each of the protein cycles shown in Fig. 15.7, step 3 would be the

Fig. 15.10. A tentative model for the energy-dependent degradation of proteins in the course of turnover. In reaction 1 an ATP-dependent synthase catalyses the formation of a covalent bond between a group on the activating protein factor ($APF_I$ syn. ubiquitin) and an ε–$NH_2$ group of a lysine residue on the protein molecule destined to be degraded. In reaction 2 a protease now hydrolyses the bound protein to smaller peptides or amino acids leaving a residual lysine residue bonded to the activator protein. In reaction 3 the L-amino lysyl-APF bond is broken; an enzyme of the L-glutamylamine cyclotransferase-type [(5-L-glutamyl)-L-amino acid 5-glutamyltransferase (cyclizing)] may mediate this reaction (Fink *et al.*, 1980). Scheme modified after Hershko *et al.* (1980).

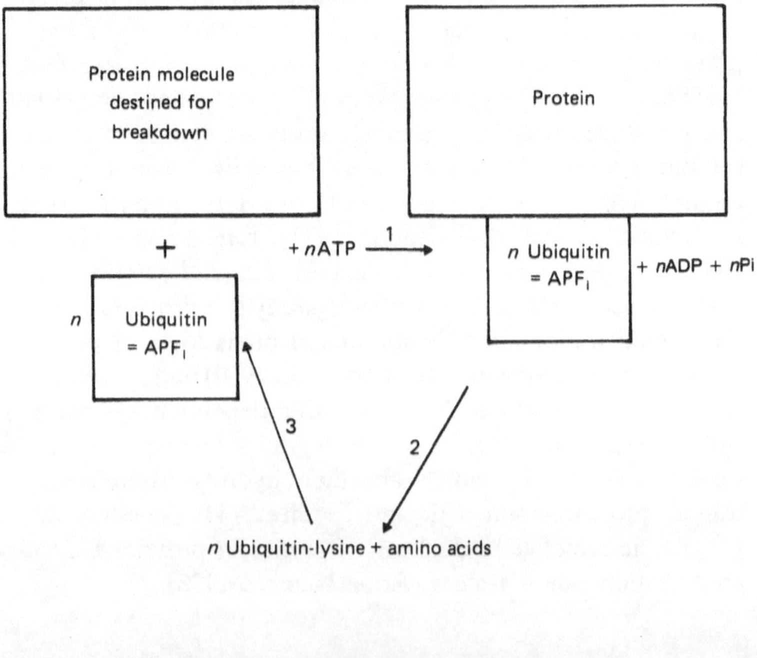

activation of a protein destined for breakdown by the ATP-dependent binding of ubiquitin with step 4 being the subsequent proteolysis. The regulation would occur specifically at stage 3: the ATP-dependent reaction.

The mechanism outlined above is hypothetical but it could explain the respiration-dependence of senescence in leaves. An apparent difficulty in respect of such a scheme is the fact that light, which would be expected to drive the synthesis of ATP, delays leaf senescence. In this connection, however, it is of interest to note that the highly labile P-32 000 becomes stabilised on transfer of leaves to darkness; evidently, there is a light-requirement for the active breakdown of this protein. The motive in developing this discussion is to emphasise that there exist in cells much finer controls for the degradation of specific proteins than are envisaged in the concept of an autophagic vacuole; much more work is needed on the characterisation of proteolytic systems from plant tissues.

## Control mechanisms underlying foliar senescence

### *General considerations*

The environmental and correlative aspects of leaf senescence have been discussed elsewhere (Woolhouse, 1974; Noodén & Leopold, 1978; Thomas & Stoddart, 1980) but some consideration of hormonal aspects, also covered in these reviews, is relevant. Chibnall (1939) found that formation of roots on excised leaves of *Phaseolus vulgaris* led to the arrest of senescence and increased protein synthesis, suggesting that stimuli from the roots may be necessary to maintain the viability of the leaves. Much more is now known about compounds which can control foliar senescence but any clear understanding of the mechanisms involved still seems far distant.

### *Hormone effects*

All five of the known plant hormones or groups of hormones have been implicated in foliar senescence. Cytokinins delay senescence in excised leaves of *Xanthium* and many other species. In leaves of *Prunus* species IAA delays senescence. Gibberellins will retard the senescence of excised leaves of *Taraxacum officinale* but the cytokinins are without effect. In *Rumex* species and *Tropaeolum majus* cytokinins or gibberellins will delay senescence over similar ranges of concentration. In

*Tropaeolum*, the GA content of the leaves decreases in the course of their senescence, providing circumstantial evidence that the effects of particular hormones may be related to their rates of turnover (Woolhouse, 1967, 1974, 1978). Exogenously applied abscisic acid accelerates senescence in leaves of many species and excised leaves of oat incubated in darkness show a five-fold increase in abscisic acid content which may be involved in promoting the rapid senescence which is observed (Gepstein & Thimann, 1980). The aleurone layer shares many attributes of a senescing tissue and it is of interest that abscisic acid exerts an action in this system. Mozer (1980) found that both GA and ABA added alone to barley aleurones induced the appearance of unique mRNAs which are translatable *in vitro*. GA treatment altered *in vivo* protein synthesis by stimulating one set of proteins, the hydrolases, exclusively. ABA induced the synthesis of several new proteins, the functions of which are as yet unknown, and prevented the synthesis of $\alpha$-amylase. Addition of GA and ABA together produced a pattern of protein synthesis resembling that where no hormones were added, but assays of the mRNAs of these aleurones by means of *in vitro* translation showed that the GA-induced mRNAs were still produced. Thus ABA appears to block the translation of the GA-induced messages *in vivo*. It is not yet clear whether this effect of ABA on translation is mediated via the ABA-induced proteins; it would be of particular interest to know whether ABA is operating in a similar manner in senescing leaves.

Although the effect of cytokinins in delaying leaf senescence is well known, their mode of action is still obscure. Studies using cultured cells of *Glycine max*, which are stimulated to growth by cytokinins, suggest a translational level of control. Cytokinin stimulated recruitment of monosomes into polysomes in cultured cells of *Glycine* within 15 minutes of application, apparently by activating existing mRNAs to a form amenable to translation (Tepfer & Foskett, 1978). A similar mechanism could account for the effect of cytokinins in stimulating protein synthesis and thereby delaying senescence in leaf tissues. Another suggested mechanism involves activation of the protein kinases, although a direct effect of cytokinins on phosphorylation of proteins seems unlikely. An initial effect of cytokinin on the intracellular distribution of ions, which in turn alters the pattern of protein synthesis, seems more likely.

The relationship of ethylene to the process of foliar senescence is of particular interest. Although ethylene may have little effect on the rate of senescence of some species, it induces senescence in leaves of tobacco. (Ethylene is produced by senescing leaves of this species.) Excised leaves of evergreens such as *Ilex aquifolium* and *Hedera helix* may be maintained in darkness for many months without undergoing senescence but can be induced to senesce rapidly by exposure to ethylene; if the gas is withdrawn while senescence is incomplete, the process is arrested (H. W. Woolhouse, unpublished). Ethylene biosynthesis from methionine involves *S*-adenosyl methionine (SAM) and 1-aminocyclopropane-1-carboxylic acid (ACC) as intermediates with a recycling of the methylthiol group as a unit into methionine (Adams & Yang, 1979). In excised leaves of tobacco there is a rapid rise in ethylene production prior to the final rapid phase of senescence (Aharoni & Lieberman, 1979), while treatment of attached leaves of *P. vulgaris* with 1 $mol\,m^{-3}$ ACC leads to ethylene production and rapid yellowing and senescence. Auxin treatment of many plant tissues leads to an enhanced production of ethylene (Abeles, 1973) which has been shown to result from increased activity of ACC synthase, the enzyme which converts SAM to ACC (Adams & Yang, 1979). Amino vinyl glycine which inhibits ACC synthase, and $Co^{2+}$ which inhibits the conversion of ACC to ethylene, delay the senescence of excised leaves, and $Ag^{+}$ ions and high $CO_2$ concentrations, which inhibit ethylene action, give rise to similar effects (Aharoni & Lieberman, 1979). Molecular oxygen is needed both for the conversion of ACC to ethylene, and for ethylene action (Beyer, 1976). These results suggest alternative explanations for the delaying of leaf senescence under anaerobic conditions, or at least an additional mechanism to respiratory energy requirements, as a source of this effect.

### *Effects of calcium and polyamines on foliar senescence*

Deficiency of calcium causes a disease syndrome of bulky storage organs, such as apples, brussels sprouts, carrots, cauliflowers and celery. The symptoms include breakdown of cell walls, increased permeability and leakiness of the cell membranes (Simon, 1978), leading to a waterlogged appearance of the tissues with necrotic browning in the later stages; a syndrome which resembles cellular senescence in many of its aspects.

Calcium treatment may delay senescence in excised leaves and delay abscision in petiolar explants of *P. vulgaris* (Noodén & Leopold, 1978). Calcium is used routinely in the pre-treatment of excised roots and discs of storage tissue prior to studies of ion uptake, on the grounds that it diminishes leakiness of the tissues. In these treatments $Ca^{2+}$ has been held to be functioning primarily as a stabiliser of the plasma membranes of the cells. Calcium has recently been found to stimulate production of ACC and ethylene from SAM, leading to the suggestion that plasma membrane stabilisation may be essential to the ACC synthase enzyme system (Lieberman, 1979).

Calmodulin, the calcium-binding protein involved as a 'second messenger' in the regulation of phosphodiesterase enzymes in animals (Wang & Waismann, 1979) has also been identified in plant tissues (Anderson & Cormier, 1978) where it has been implicated in the activation of NAD kinase. These findings should stimulate further study of the role of Ca and Ca-binding proteins in the regulation of metabolic events associated with plant senescence and in those disorders of calcium nutrition which resemble it.

Excised leaves of *Avena sativa* maintained in darkness showed increased RNAase activity after 1 hour, protease activity after 6 hours and chlorophyll degradation after 16 hours. These changes were delayed by treatment with 1–10 mol $m^{-3}$ polyamines, cadaverine, putrescine, spermine and spermidine (Kaur-Sawhney & Galston, 1979). In the light the effect was reversed and the polyamines accelerated senescence. Calcium ions (1–10 mol $m^{-3}$) supplied together with polyamines diminished the polyamine effect, suggesting the possibility of a common ionic attachment mechanism as a basis for $Ca^{2+}$ and polyamine action. These observations raise the intriguing possibility that polyamines may be normal elements of metabolic control in plant tissues just as they are known to be in animals and microorganisms (Bachrach, 1973).

## Conclusion

We see that in respect of the physiology of senescence in leaves much remains to be done, particularly in characterising the changing functions of the photosynthetic apparatus. At the biochemical level our knowledge of the spectrum of proteases, their compartmentation and specificity of action is clearly quite inadequate. The role of lipid metabolism in senescence may be inferred to be important from ultrastructural studies but we are in ignorance of the biochemistry. We have treated the hormonal regulation of senescence but briefly and this may be interpreted as an indication of how little is firmly established. Clearly, there is a long way to go before we can hope to have a thorough understanding of foliar senescence in any of its aspects.

## References

Abeles, F. B. (1973). *Ethylene in Biology*. New York: Academic Press.

Adams, D. O. & Yang, S. F. (1979). Ethylene biosynthesis: identification of 1-aminocyclopropane-1-carboxylic acid as an intermediate in the conversion of methionine to ethylene. *Proceedings of the National Academy of Sciences, USA*, **76**, 170–4.

Aharoni, N. & Lieberman, M. (1979). Ethylene as a regulator of senescence in tobacco leaf discs. *Plant Physiology*, **64**, 801–4.

Andersen, K. S., Bain, J. M., Bishop, D. G. & Smillie, R. M. (1972). Photosystem II activity in agranal bundle sheath chloroplasts from *Zea mays*. *Plant Physiology*, **49**, 461–6.

Anderson, J. M. & Cormier, M. J. (1978). Calcium-dependent regulator of NAD kinase in higher plants. *Biochemical and Biophysical Research Communications*, **84**, 595–602.

Anderson, L. E. (1979). Interaction between photochemistry and activity of enzymes. In *Encylopedia of Plant Physiology*, (new series) vol. 6, *Photosynthesis. II. Photosynthetic Carbon Metabolism and Related Processes*, ed. M. Gibbs & E. Latzko, pp. 271–81. Berlin: Springer-Verlag.

Bachrach, U. (1973). *Function of Naturally Occurring Polyamines*. New York: Academic Press.

Baier, D. & Latzko, E. (1975). Properties and regulation of C-1-fructose-1, 6-diphosphatase from spinach chloroplasts. *Biochimica et Biophysica Acta*, **396**, 141–7.

Bassham, J. A. (1979). The reductive pentose phosphate cycle and its regulation. In *Encylopedia of Plant Physiology*, (new series) vol. 6, *Photosynthesis. II. Photosynthetic Carbon Metabolism and Related Processes*, ed. M. Gibbs & E. Latzko, pp. 9–30. Berlin: Springer-Verlag.

Batt, T. & Woolhouse, H. W. (1975). Changing activities during senescence and sites of synthesis of photosynthetic enzymes in leaves of the labiate, *Perilla frutescens* (L.) Britt. *Journal of Experimental Botany*, **26**, 569–79.

Beyer, E. M. (1976). A potent inhibitor of ethylene action in plants. *Plant Physiology*, **58**, 268–71.

Boller, T. & Kende, H. (1979). Hydrolytic enzymes in the central vacuole of plant cells. *Plant Physiology*, **63**, 1123–32.

Bricker, T. M. & Newman, D. W. (1980). Quantitative changes in the chloroplast thylakoid polypeptide complement during senescence. *Zeitschrift für Pflanzenphysiologie*, **98**, 339–46.

Buchanan, B. B. (1980). Role of light in the regulation of chloroplast enzymes. *Annual Review of Plant Physiology*, **31**, 341–74.

Butler, R. D. & Simon, E. W. (1972). Ultrastructural aspects of senescence in plants. *Advances in Ageing Research*, **4**, 157.

Callow, J. A., Callow, M. E. & Woolhouse, H. W. (1972). *In vitro* protein synthesis, ribosomal RNA synthesis and polyribosomes in senescing leaves of *Perilla*. *Cell Differentiation*, **1**, 79–90.

Callow, J. A. & Woolhouse, H. W. (1973). Changes in nucleic acid metabolism in regreening leaves of *Perilla*. *Journal of Experimental Botany*, **24**, 285–94.

Čatský, J., Tichá, I. & Solárová, J. (1976). Ontogenetic changes in the internal limitations to bean leaf photosynthesis. I. Carbon dioxide exchange and conductance for carbon dioxide transfer. *Photosynthetica*, **10**, 394–402.

Charles, S. A. & Halliwell, B. (1981). Light activation of fructose bisphosphatase in isolated spinach chloroplasts and deactivation by hydrogen peroxide. *Planta*, **151**, 242–6.

Chibnall, A. C. (1939). *Protein Metabolism in the Plant*. New Haven: Yale University.
Ciechanover, A., Heller, H., Elias, S., Haas, A. L. & Hershko, A. (1980). ATP-dependent conjugation of reticulocyte proteins with the polypeptide required for protein degradation. *Proceedings of the National Academy of Sciences*, USA, **77**, 1365–8.
Ciferri, O., Dipasquale, G. & Tiboni, O. (1979). Chloroplast elongation factors are synthesised in the chloroplast. *European Journal of Biochemistry*, **102**, 331–5.
Constable, G. A. & Rawson, H. M. (1980). Effect of leaf position, expansion and age on photosynthesis, transpiration, and water use efficiency of cotton. *Australian Journal of Plant Physiology*, **7**, 89–100.
Davis, S. D. & McCree, K. J. (1978). Photosynthetic rate and diffusion conductance as a function of age in leaves of bean plants. *Crop Science*, **18**, 280–2.
Edelman, M. & Reisfeld, A. (1980). Synthesis, processing and functional probing of P-32000, the major membrane protein translated within the chloroplast. In *Genome Organisation and Expression in Plants*, ed. C. J. Leaver, pp. 353–62. New York: Plenum Press.
Epstein, D., Elias-Bishko, S. & Hershko, A. (1975). Requirement for protein synthesis in the regulation of protein breakdown in cultured Hepatoma cells. *Biochemistry*, **14**, 5199–204.
Farquhar, G. D. & von Caemmerer, S. (1981). Modelling of photosynthetic response to environmental conditions. In *Encyclopedia of Plant Physiology*, (new series) vol. 12B. Berlin: Springer-Verlag (in press).
Farquhar, G. D., von Caemmerer, S. & Berry, J. A. (1980). A biochemical model of photosynthetic $CO_2$ assimilation in leaves of $C_3$ species. *Planta*, **149**, 78–90.
Fink, M. L., Chung, S. I. & Folk, J. E. (1980). $\gamma$-glutamine cyclotransferase: specificity toward *E*-(-L-glutamyl)-L-lysine and related compounds. *Proceedings of the National Academy of Sciences, USA*, **77**, 4564–8.
Flügge, U. I., Freisl, M. & Heldt, H. W. (1980). The mechanisms of the control of carbon fixation by the pH in the chloroplast stroma. *Planta*, **149**, 48–51.
Fong, F. & Heath, R. L. (1977). Age dependent changes in phospholipids and galactolipids in primary bean leaves (*Phaseolus vulgaris*). *Phytochemistry*, **16**, 215–17.
Fraser, D. E. & Bidwell, R. G. S. (1974). Photosynthesis and photorespiration during the ontogeny of the bean plant. *Canadian Journal of Botany*, **52**, 2561–70.
Friedrich, J. & Huffaker, R. C. (1980). Photosynthesis, leaf resistances, and ribulose-1,5-bisphosphate carboxylase degradation in senescing barley leaves. *Plant Physiology*, **65**, 1103–7.
Gepstein, S. & Thimann, K. V. (1980). Changes in the abscisic acid content of oat leaves during senescence. *Proceedings of the National Academy of Sciences, USA*, **77**, 2050–3.
Goudriaan, J. & van Laar, U. H. (1978). Relations between leaf resistance, $CO_2$ concentration and $CO_2$ assimilation in maize, beans, lalang grass, and sunflower. *Photosynthetica*, **12**, 241–9.
Grierson, D. (1981). Control of RNA and enzyme synthesis during ripening. NATO Advanced Study Institute, Sounion Greece (in press).
Hackenbrock, C. R. (1981). Lateral diffusion and electron transfer in the mitochondrial inner membrane. *Trends in Biological Sciences*, **6**, 151–4.
Hall, A. J. & Brady, C. J. (1977). Assimilate source-sink relationships in *Capsicum annuum* L. II. Effects of fruiting and defloration on the photosynthetic capacity and senescence of leaves. *Australian Journal of Plant Physiology*, **4**, 771–83.
Hall, N. P., Keys, A. J. & Merrett, M. S. (1978). Ribulose-1,5-diphosphate carboxylase protein during flag leaf senescence. *Journal of Experimental Botany*, **39**, 31–7.
Haraguchi, N. & Shimizu, S. (1970). Photosynthetic activities in tobacco plants. II.

Relationship between photosynthetic activity and chlorophyll content. *Botanical Magazine*, **83**, 411–18.

Hardwick, K., Wood, M. E. & Woolhouse, H. W. (1968). Photosynthesis and respiration in relation to leaf age in *Perilla frutescens* (L.) Britt. *New Phytologist*, **67**, 79–86.

Hauska, G. A. (1975). The effect of uncouplers on artificial donor shuttles for PSI. In *Proceedings of the 3rd International Congress on Photosynthesis, I*, ed. M. Avron, pp. 689–96. Oxford: Elsevier.

Hauska, G. A. (1977). Artificial acceptors and donors. In *Encylopedia of Plant Physiology*, (new series) vol. 5, ed. A. Trebst & M. Avron, pp. 253–65. Berlin: Springer-Verlag.

Heck, H., Marhnoia, E. & Mahle, P. (1981). Subcellular localisation of acid proteinase in barley mesophyll protoplasts. *Planta*, **151**, 198–200.

Heldt, H. W. (1979). Light-dependent changes of stromal $H^+$, $Mg^{++}$ concentrations controlling $CO_2$ fixation. In *Encylopedia of Plant Physiology*, (new series) vol. 6, *Photosynthesis. II. Photosynthetic Carbon Metabolism and Related Processes*, ed. M. Gibbs & E. Latzko. pp, 202–7. Berlin: Springer-Verlag.

Hershay, J. W. B. (1980). The translational machinery, components and mechanism. In *Cell Biology: A Comprehensive Treatise*, vol. 5, ed. D. M. Prescott & L. Goldstein, pp. 1–68. New York: Academic Press.

Hershko, A., Ciechanover, A., Heller, H., Haas, A. L. & Rose, I. A. (1980). Proposed role of ATP in protein breakdown: conjunction of proteins with multiple chains of the polypeptides of ATP-dependent proteolysis. *Proceedings of the National Academy of Sciences, USA*, **77**, 1783–6.

Hodgkinson, K. C. (1974). Influence of partial defoliation on photosynthesis, photorespiration and transpiration by lucerne leaves of different ages. *Australian Journal of Plant Physiology*, **1**, 561–78.

Huffaker, R. C. & Miller, B. L. (1978). Reutilisation of ribulose bisphosphate carboxylase. In *Photosynthetic Carbon Assimilation*, ed. H. W. Siegelman & G. Hind, pp. 139–52. New York: Plenum.

Imai, K. & Murata, Y. (1979). Changes in apparent photosynthesis, $CO_2$ compensation point and dark respiration of leaves of some *Poaceae* and *Cyperaceae* species with senescence. *Plant Cell Physiology*, **20**, 1653–8.

James, W. O. (1953). *Plant Respiration*. Oxford: Clarendon Press.

Jenkins, G. I., Baker, N. R., Bradbury, M. & Woolhouse, H. W. (1981). Photosynthetic electron transport during senescence of the primary leaves of *Phaseolus vulgaris* (L.). III. Kinetics of chlorophyll fluorescence emission from intact leaves. *Journal of Experimental Botany*, **32**, 999–1008.

Jenkins, G. I., Baker, N. R. & Woolhouse, H. W. (1981). Changes in chlorophyll content and organisation during senescence of primary leaves of *Phaseolus vulgaris* L. in relation to photosynthetic electron transport. *Journal of Experimental Botany*, **32**, 1009–20.

Jenkins, G. I. & Woolhouse, H. W. (1981a). Photosynthetic electron transport during senescence of the primary leaves of *Phaseolus vulgaris* L. I. Non-cyclic electron transport. *Journal of Experimental Botany*, **32**, 467–78.

Jenkins, G. I. & Woolhouse, H. W. (1981b). Photosynthetic electron transport during senescence of the primary leaves of *Phaseolus vulgaris* L. II. The activity of photosystems one and two, and a note on the site of reduction of ferricyanide. *Journal of Experimental Botany*, **32**, 989–97.

Jensen, R. G. & Bahr, R. T. (1977). Ribulose 1,5-bisphosphate carboxylase-oxygenase. *Annual Review of Plant Physiology*, **28**, 379–400.

Jewer, P. C. & Incoll, L. D. (1980). Promotion of stomatal opening in the grass

*Anthephora pubescens* Nees by a range of natural and synthetic cytokinins. *Planta*, **150**, 218–21.

Jordan, D. B. & Ogren, W. L. (1981). Species variation in the specificity of ribulose bisphosphate carboxylase/oxygenase. *Nature*, **291**, 513–15.

Kaur-Sawhney, R. & Galston, A. W. (1979). Interactions of polyamines and light on biochemical processes involved in leaf senescence. *Plant, Cell and Environment*, **2**, 189–96.

Kawashima, N. & Wildman, S. G. (1970). Fraction I Protein. *Annual Review of Plant Physiology*, **21**, 325–58.

Kennedy, R. A. (1976). Relationship between leaf development, carboxylase enzyme activities and photorespiration in the $C_4$-plant *Portulaca oleracea* L. *Planta*, **128**, 149–54.

Kisaki, T., Hirabayashi, S. & Yano, N. (1973). Effect of the age of tobacco leaves on photosynthesis and photorespiration. *Plant Cell Physiology*, **14**, 505–14.

Kowit, S. D. & Goldberg, A. L. (1977). Intermediate steps in the degradation of a specific abnormal protein in *Escherichia coli*. *Journal of Biological Chemistry*, **252**, 8350–7.

Ku, M. S. B., Schmitt, M. R. & Edwards, G. E. (1979). Quantitative determination of RuBP carboxylase oxygenase protein in leaves of several $C_3$ and $C_4$ plants. *Journal of Experimental Botany*, **30**, 89–98.

Lieberman, M. (1979). Biosynthesis and action of ethylene. *Annual Review of Plant Physiology*, **30**, 533–91.

Lilley, R. M. & Walker, D. A. (1975). Carbon dioxide assimilation by leaves, isolated chloroplasts, and ribulose bisphosphate carboxylase from spinach. *Plant Physiology*, **55**, 1087–92.

Lin, W. & Wittenbach, J. A. (1981). Subcellular localisation of proteases in wheat and corn mesophyll protoplasts. *Plant Physiology*, **67**, 969–72.

Linderstrøm-Lang, K. (1950). Structure and enzymatic breakdown of proteins. *Cold Spring Harbor Symposium of Quantitative Biology*, **14**, 117–26.

Ljubešić, N. (1968). Feinbau der chloroplasten während der Vergilbung und Wiederergrüsung der Blätter. *Protoplasma*, **66**, 369–79.

Ludlow, M. M. & Jarvis, P. G. (1971). Methods for measuring photorespiration in leaves. In *Plant Photosynthetic Production; Manual of Methods*, ed. Z. Šestak, J. Čatský & P. G. Jarvis, pp. 294–315. The Hague: Junk.

Ludlow, M. M. & Wilson, G. L. (1971). Photosynthesis of tropical pasture plants. III. Leaf age. *Australian Journal of Biological Sciences*, **24**, 1077–87.

McKersie, B. D. & Thompson, J. E. (1978). Phase behaviour of chloroplast and microsomal membranes during leaf senescence. *Plant Physiology*, **61**, 639–43.

McPherson, H. G. & Slatyer, R. O. (1973). Mechanisms regulating photosynthesis in *Pennisetum typhoides*. *Australian Journal of Biological Sciences*, **26**, 329–39.

Matile, P. (1978). Biochemistry and function of vacuoles. *Annual Review of Plant Physiology*, **29**, 193–213.

Mattoo, A. K., Pick, U., Hoffman-Falk, H. & Edelman, M. (1981). The rapidly metabolized 32000-dalton polypeptide of the chloroplast in the 'proteinaceous shield' regulating photosystem II electron transport and mediating diuron herbicide sensitivity. *Proceedings of the National Academy of Sciences, USA*, **78**, 1572–6.

Moorby, J., Munns, R. & Walcott, J. (1975). Effect of water deficit in photosynthesis and tuber metabolism in potatoes. *Australian Journal of Plant Physiology*, **2**, 323–33.

Moss, D. N. & Peaslee, D. E. (1965). Photosynthesis of maize leaves as affected by age, and nutrient status. *Crop Science*, **5**, 280–1.

Mozer, T. J. (1980). Control of protein synthesis in barley aleurone layers by the plant hormones gibberellic acid and abscisic acid. *Cell*, **20**, 479–85.

Mulchi, C. L., Volk, R. J. & Jackson, W. A. (1971). $O_2$ exchange of illuminated leaves at $CO^2$ compensation. In *Photosynthesis and Photorespiration*, ed. M. D. Hatch, C. B. Osmond & R. O. Slatyer, pp. 35–50. Toronto: Wiley.

Murakami, K., Voellmy, R. & Goldberg, A. L. (1979). Protein degradation is stimulated by ATP in extracts of *Escherichia coli*. *Journal of Biological Chemistry*, **254,** 8194–200.

Noodén, L. D. & Leopold, A. C. (1978). Phytohormones and the endogenous regulation of senescence and abscission. In *Phytohormones and Related Compounds. A Comprehensive Treatise*, vol. II, ed. D. S. Letham, P. B. Goodwin & T. J. V. Higgins. Amsterdam: Elsevier.

Novitskaya, G. V., Rutskaya, L. A. & Molotkovskii, Y. G. (1977). Age changes of lipid composition and activity of the membranes in bean chloroplasts. *Fiziologia Rastenii*, **24,** 35–43.

Osman, A. M. & Milthorpe, F. L. (1971). Photosynthesis of wheat leaves in relation to age, illuminance.and nutrient supply. II. Results. *Photosynthetica*, **5,** 61–70.

Pearson, J. A., Thomas, K. & Thomas, H. (1978). Nucleic acids from leaves of a yellowing and a non-yellowing variety of *Festuca pratensis* Huds. *Planta*, **144,** 85–7.

Peoples, M. B., Beilharz, V. C., Waters, S. P., Simpson, R. J. & Dalling, M. J. (1980). Nitrogen redistribution during grain growth in wheat (*Triticum aestivum* L.). *Planta*, **149,** 241–51.

Purceld, P., Chon, C. J., Portis, A. R., Heldt, H. W. & Heber, U. (1978). The mechanism of the control of carbon fixation by the pH on the chloroplast stroma. Studies with nitrite-mediated proton transfer. *Biochimica et Biophysica Acta*, **501,** 488–98.

Raschke, K. (1979). Movements of stomata. In *Encylopedia of Plant Physiology*, (new series) vol. 7, ed. W. Haupt & M. E. Feinleib, pp. 383–441. Berlin: Springer-Verlag.

Rawson, H. M. & Constable, G. A. (1980). Carbon production of sunflower cultivars in field and control environments. I. Photosynthesis and transpiration of leaves, stems and heads. *Australian Journal of Plant Physiology*, **7,** 555–73.

Roberts, J. W., Roberts, C. W., Craig, N. & Phizicky, E. (1978). Activity of the *E. coli* rec A-gene product. *Cold Spring Harbor Symposium of Quantitative Biology*, **43,** 917–20.

Robinson, S. P., McNeil, P. H. & Walker, D. A. (1979). Ribulose bisphosphate carboxylase – lack of dark inactivation of the enzyme in experiments with protoplasts. *FEBS Letters*, **97,** 296–300.

Sacher, J. A. & Davies, D. D. (1974). Demonstration of *de novo* synthesis of RNase in *Rhoeo* leaf sections by deuterium oxide labelling. *Plant Cell Physiology*, **15,** 157–62.

Šesták, Z. (1969). Ratio of photosystem one and two particles in young and old leaves of spinach and radish. *Photosynthetica*, **3,** 285–7.

Šesták, Z. (1977a). Photosynthetic characteristics during ontogenesis of leaves. I. Chlorophylls. *Photosynthetica*, **11,** 367–448.

Šesták, Z. (1977b). Photosynthetic characteristics during ontogenesis of leaves. 2. Photosystems, components of electron transport chain, and photophosphorylations. *Photosynthetica*, **11,** 449–74.

Šesták, Z. & Čatśky, J. (1967). Sur les relations entre le contenu de chlorophyle et l'activité photosynthetique pendant la croissance et le vieillissement des feuilles. In *Le Chloroplaste*, ed. C. Sironval, pp. 213–62. Paris: Masson.

Šesták, Z. & Demeter, S. (1976). Changes in circular dichroism and P700 content of chloroplasts during leaf ontogenesis. *Photosynthetica*, **10,** 182–7.

Siggel, U. (1976). The function of the plastoquinone as electron and proton carrier in photosynthesis. *Bioelectrochemistry and Bioenergetics*, **3,** 302–18.

Simon, E. W. (1978). The symptoms of calcium deficiency in plants. *New Phytologist*, **80,** 1–15.

Steinberg, D. & Vaughan, M. (1956). Observations on intracellular protein catabolism studied *in vitro*. *Archives of Biochemistry and Biophysics*, **65**, 93–105.

Tepfer, D. A. & Foskett, D. E. (1978). Hormone-mediated translational control of protein synthesis in cultured cells of *Glycine max*. *Developmental Biology*, **62**, 486–92.

Thomas, S. M., Hall, N. P. & Merrett, M. J. (1978). Ribulose 1,5-bisphosphate carboxylase/oxygenase activity and photorespiration during the ageing of flag leaves of wheat. *Journal of Experimental Botany*, **29**, 1161–8.

Thomas, H. & Stoddart, J. L. (1980). Leaf senescence. *Annual Review of Plant Physiology*, **31**, 83–111.

Tiboni, O., Di Pasquale, G. & Cifferi, O. (1978). Purification of the elongation factors present in spinach chloroplasts. *European Journal of Biochemistry*, **92**, 471–7.

Trebst, A. (1974). Energy conservation in photosynthetic electron transport of chloroplasts. *Annual Review of Plant Physiology*, **25**, 423–58.

Van Loon, L. C., Haverkort, A. J. & Lockhurst, G. J. (1978). Changes in protease activity during leaf growth and senescence. *FESPP Abstracts*, **280C**, 544–5.

Voellmy, R. & Goldberg, A. L. (1981). ATP-stimulated endoprotease associated with the cell membrane in *E. coli*. *Nature*, **290**, 419–21.

Walker, D. A. (1976). Regulatory mechanisms in photosynthetic carbon metabolism. *Current Topics in Cell Regulation*, **11**, 203–41.

Wang, J. H. & Waismann, D. M. (1979). Calmodulin and its role in the second messenger system. *Current Topics in Cellular Regulation*, **15**, 47–109.

Waters, S. P., Peoples, M. B., Simpson, R. J. & Dalling, M. J. (1980). Nitrogen redistribution during grain growth in wheat (*Triticum aestivum* L.) I. Peptide hydrolase activity and protein breakdown in the flag leaf, glumes and stem. *Planta*, **148**, 422–8.

Whatley, J. M. (1974). Chloroplast development in primary leaves of *Phaseolus vulgaris*. *New Phytologist*, **73**, 1097–110.

Wilkinson, K. D., Urban, M. K. & Haas, A. (1980). Ubiquitin is the ATP-dependent proteolysis factor I of rabbit reticulocytes. *Journal of Biological Chemistry*, **225**, 7529–32.

Woledge, J. (1972). The effect of shading on the photosynthetic rate and longevity of grass leaves. *Annals of Botany*, **36**, 551–61.

Wolinska, D. (1976). Functional and structural changes in chloroplasts of senescent tobacco leaves. *Acta Societatis Botanicorum Poloniae*, **45**, 341–52.

Wollgiehn, R. (1967). Nucleic acid and protein metabolism of excised leaves. *Symposia of the Society for Experimental Biology*, **21**, 231–46.

Wong, S. C., Cowan, I. R. & Farquhar, G. D. (1979). Stomatal conductance correlates with photosynthetic capacity. *Nature*, **282**, 424–6.

Woodward, R. G. & Rawson, H. M. (1976). Photosynthesis and transpiration in dicotyledonous plants. II. Expanding and senescing leaves of soybean. *Australian Journal of Plant Physiology*, **3**, 257–67.

Woolhouse, H. W. (1967). The nature of senescence in plants. *Symposium of the Society for Experimental Biology*, **21**, 179–230.

Woolhouse, H. W. (1968). Leaf age and mesophyll resistance as factors in the rate of photosynthesis. *Hilger Journal*, **11**, 15–20.

Woolhouse, H. W. (1974). Longevity and senescence in plants. *Science Progress*, **61**, 123–47.

Woolhouse, H. W. (1978). Cellular and metabolic aspects of senescence in higher plants. In *The Biology of Ageing*, ed. J. A. Behnke, C. E. Finch & G. B. Moment, pp. 83–9. New York: Plenum.

Woolhouse, H. W. (1982a). *Biochemical and Molecular Aspects of Leaf Senescence*, ed. H. Smith, (in press).

Woolhouse, H. W. (1982b). The general biology of plant senescence and the role of nucleic acid and protein turnover in the control of senescence processes which are genetically programmed. In *Post Harvest Physiology and Crop Preservation*, NATO Advanced Study Institute, ed. M. Leiberman. New York: Plenum Press.

Woolhouse, H. W. & Batt, T. (1976). The nature and regulation of senescence in plastids. In *Perspectives in Experimental Biology*, vol. 2, ed. N. Sunderland, pp. 163–75. Oxford: Pergamon Press.

Yoshida, Y. (1961). Nuclear control of chloroplast activity in *Elodea* leaf cells. *Protoplasma*, **54,** 476–92.

Zelitch, I. (1979). Photorespiration, studies with whole tissues. In *Encylopedia of Plant Physiology*, (new series) vol. 6, *Photosynthesis. II. Photosynthetic Carbon Metabolism and Related Processes*, ed. M. Gibbs & E. Latzko, pp. 353–67. Berlin: Springer-Verlag.

Zima, J. & Šestak, Z. (1979). Photosynthetic characteristics during ontogenesis of leaves. 4. Carbon fixation pathways, their enzymes and products. *Photosynthetica*, **13,** 83–106.

# 16

## Modelling leaf growth and function

D. A. CHARLES-EDWARDS

The morphology, anatomy and biochemistry of leaves are complex. Each of these leaf attributes differs between leaves from different plant types, and even between leaves from plants of the same type growing in different environments. We need to establish a conceptual framework within which we can order our knowledge about them, and this chapter seeks to show how mathematics provides us with a tool to build one such framework.

Let us suppose that we have identified the main chemical components of a particular leaf process. It is logical that we should next develop a hypothesis about their role *in situ* in the intact leaf and their response to changes in the leaf's environment. Our task is hardly started because we now have to test this hypothesis; that is we have to establish the extent to which it can account for the observed behaviour of the leaf process. If we can translate our hypothesis into a mathematical equation, or set of equations, we may be better able to test it and explore its consequences on other aspects of leaf growth and function. This process of using mathematics to translate (or formalise) a hypothesis is what we commonly call 'mathematical modelling'.

Modelling is only one of many techniques that are available to us in our quest to understand the mechanisms underlying leaf growth and function. Like all other techniques the level of detail employed in any particular study is primarily determined by the objectives of the model. For example, the development of the leaf canopy may be an important determinant of crop productivity, and a predictive model of leaf growth would be of value in this context. Since crop yield, rather than the mechanisms of leaf growth, is of primary interest an empirical (or phenomenological) model of leaf growth may be quite appropriate for this purpose (for example see Monteith & Elston, 1971 and Dennett, Milford & Elston, 1978). In contrast, if the mechanisms determining the rate and/or extent of leaf growth are the main interest a detailed, mechanistic model will be more appropriate (for example see Charles-Edwards, 1979a). (We can define a mechanistic model as one which describes events at one level of organization in terms of the known mechanisms operating at some lower level of organization.) Of course, when we come to construct a mechanistic model we often find that our knowledge about a particular leaf process is inadequate, and this does not allow us to include a mechanistic description of it in our model. We may have sufficient information to include an empirical description of it, but often we can only make reasoned assumptions about it. Our leaf model then becomes an heuristic aid to our research, helping us to establish whether or not those assumptions formalized by our mathematics can account for the observed behaviour of the leaf. If they do not we may be better able to identify the areas where further experimental and/or theoretical work is needed.

The objectives of a modelling programme need to be identified and defined, since they determine the type and level of approach that we take to tackle the problem. One example of this has been given in the preceding paragraph. Another illustrates the simple point that the amount of detail we need to include is determined by the resolution available to us at the level at which we intend to test or apply the model. By 'resolution' I mean the amount of detail we can distinguish at the level at which we are applying our model. We know that the leaf is a heterogeneous assemblage of tissues, each tissue having its own distinct structure and function. Unfortunately our knowledge of the interactions between the different tissues is rudimentary and because of this we are constrained to make the assumption that the leaf can be treated as though it were structurally and functionally homogeneous. This assumption is adequate for most models of leaf growth and function, but we need to be aware of the limitations that it imposes upon them. For example, there are numerous models of photosynthesis by leaves from $C_3$ plants, and the more recent ones are usually based on the dual oxygenase/carboxylase activity of the enzyme ribulose bisphosphate carboxylase. The dual activity of this enzyme dominates the kinetic behaviour of the leaf, to the extent that all models are very similar in their mathematical formulation of leaf photosynthesis (compare the models in Charles-Edwards & Ludwig, 1974, Peisker & Apel, 1975 and Tenhunen, Weber, Yocum & Gates, 1977); each model makes different assumptions about the stoichiometry of the biochemical processes, which leads to subtle mathematical differences between them. However, since they all make Procrustean assumptions about the light and $CO_2$ environments within the leaf, usually that they are uniform throughout the volume of the leaf, it is doubtful that the better or worse fit of one or other of these models to a set of experimental data on leaf photosynthesis could be unambiguously ascribed to its particular assumptions about the biochemical processes of photosynthesis.

Whereas there are numerous mechanistic models of leaf function there are few such models for leaf growth. Most models of leaf growth are empirical phenomenological descriptions, and there can be little doubt that this reflects the conceptual and mathematical difficulties encountered when attempting to model morphological or anatomical changes during leaf growth. In the context of a 'model leaf', homogeneous with respect to structure and function, it is rational to suppose that function is the prime determinant of structure, and simple leaf growth models could be usefully constructed. Models of the anatomical development of the leaf are seriously disadvantaged by our lack of knowledge of the mechanisms of cell differentiation. However, there are several models which have been developed to help examine the effects of changes in leaf anatomy and morphology on photosynthesis (examples can be found in Charles-Edwards & Ludwig, 1975a and Sinclair, Goudriaan & de Wit, 1977).

Because the structure of each model is determined by its objectives, and these differ between models, it is difficult to find any common basis on which the great number of models of leaf growth and functioning can be meaningfully reviewed. I have chosen instead to write an account that reflects my own view of the role of models in helping our understanding of leaf growth and function.

We can take a simple view of the leaf and treat it as an homogeneous agglomeration of individual cells, each with a capacity for growth and division. Thornley (1976) has described an analysis of the growth of a meristem which is relevant in this context. Essentially, he analyses the separate effects of the processes of cell division and cell growth on the gross behaviour of a meristematic tissue. He points out that if the fraction of newly formed cells capable of further division is less then one-half, the proportion of cells capable of division in the tissue as a whole will decline with time until cell division within the tissue ceases. If the extent of growth (size) of the individual cells is determined, it follows that the extent of growth of the tissue as a whole will also be determined. Many observations (cf. Chapter 6) suggest that there is a gradual decrease in the proportions of cells within the leaf capable of further division, but current information does not allow one to distinguish between interrelated features such as constant division rates, all cells capable of dividing at the same size, and differences in rates of cell expansion. Thus, although Thornley's analysis describes the broad features of a limited meristem, and is therefore useful at a general level, it does not attempt to identify the detailed processes involved or the physical or chemical constraints that influence these events, and it is instructive to consider some of them here.

Now, the main function of a leaf is to intercept and absorb solar radiation in the visible waveband incident upon it, and convert this transient, radiant energy into a stable, storable chemical form through the process of photosynthesis. Essentially, the leaf is an array of photocells, the chloroplasts, and in general both the rate and extent of plant growth depend upon the efficiency with which leaves are able to use light energy in the synthesis of new plant material (see Warren Wilson, 1971 and Monteith, 1977). Most plants would therefore benefit if they could maximise their leaf area (increase the size of their array of photocells) for a given investment in new plant material, and it is no casual accident that leaves are often thin laminars, a geometric shape that maximizes the surface area to volume ratio. Unfortunately, thin laminars are not particularly rigid bodies, and this intrinsic deficiency imposes important physical constraints upon their construction and dimensions. Let us consider the elementary physics of a bending beam (for example, see pp. 103–7 in Newman & Searle, 1948). If a rectangular laminar is clamped horizontally along one of its shorter edges, the downward displacement of the unclamped end, $\delta$, is related to the length, $L$, thickness, $h$, and elastic modulus, $Y$, of the laminar according to

$$\delta = 3\varrho L^4/2Yh^2 \tag{16.1}$$

where $\varrho$ is the density of the laminar. Now let us take the displacement, $\delta$, as a simple index of the laminar's rigidity. If $\delta$ is to remain unchanged for a doubling of the laminar length either the thickness must increase four-fold, or the elastic modulus must increase sixteen-fold or there must be some suitable increase in the two together. These constraints are reflected in the assumptions of a recent model for leaf growth (Charles-Edwards, 1979a).

This model was constructed to examine whether the two hypotheses that the rate of synthesis of non-degradable structural material (cell walls) is a function of the rate of water uptake by leaf tissues and that this rate is a function of the amount of osmotically active solute in the leaf (opposed by inelastic forces in the leaf tissues which increase with leaf size) could qualitatively account for the observed growth and metabolic behaviour of real leaves. These hypotheses were advanced to enable the cessation of leaf growth and expansion to be simulated without the introduction of extraneous time dependent functions or mathematical discontinuities which have no physiological basis. The model is capable of describing a number of important features of leaf growth, and indicates that these assumptions may merit further examination.

The rate of uptake of water by the leaf tissues, $\mathrm{d}\theta/\mathrm{d}t$, was written in the form

$$\mathrm{d}\theta/\mathrm{d}t = aM_{\mathrm{L}} - b(M_{\mathrm{S}})^n,\ aM_{\mathrm{L}} > b(M_{\mathrm{S}})^n, \tag{16.2}$$

where $a$, $b$ and $n$ are constants, $M_{\mathrm{L}}$ denotes the amount of labile carbohydrate (sucrose and hexoses) in the leaf and $M_{\mathrm{S}}$ the amount of non-degradable structural material (cell walls). The equation was empirically derived by analogy with published relationships between cell expansion and cell osmotic potential. The inelastic forces opposing tissue water uptake were assumed to increase with leaf size according to $(M_{\mathrm{S}})^n$. Equation (16.1) tells us that the rigidity of a beam can be maintained as its length increases provided its elastic modulus increases according to $a^*(L)^4$ where $L$ is the length and $a^*$ is a constant and there is a compelling similarity between the constraints expressed by this equation (16.1) and the assumption equating $L$ and $M_{\mathrm{S}}$. It suggests that it may be worthwhile attempting to measure the elastic modulus of a leaf during growth.

Traditionally, because we have measured the light flux density incident on the leaf surface rather than the light energy absorbed by the leaf and because of the importance of diffusion processes, the rate of leaf photosynthesis is expressed per unit of leaf area. In contrast to the interception and 'trapping' of light energy, the chemical processes of carbon dioxide assimilation depend upon enzyme and substrate concentrations and are implicitly functions of the metabolic volume of the photosynthetic apparatus. At some point the theoretical analysis of the biochemical processes of photosynthesis as a function of the leaf, cell or organelle volume needs to be reconciled with the experimental measurements of photosynthesis as a function of leaf area. Simply, the dimension of leaf thickness needs to be included. Now, an increase in leaf thickness implies an increase in the metabolic volume of the leaf beneath unit leaf area, and such an increase may have important consequences for the functioning of the leaf photosynthetic apparatus.

There are many models for leaf photosynthesis, some of them also extending to deal with the process of photorespiration. Perhaps the most widely used mechanistic model is one developed by Rabinowitch (1951) which assumes a homogeneous environment throughout the photosynthetic apparatus, that the chemical processes of photosynthesis follow Michaelis-Menten kinetics with the incident light flux density and cellular carbon dioxide concentrations as the substrates for the process, that the movement of carbon dioxide from the bulk air to the photosynthetic sites within the leaf follows simple diffusion kinetics and that the rate of 'dark' respiration is constant (with the respired carbon dioxide being released at the photosynthetic sites). The model gives rise to a non-rectangular hyperbola for the relations between the net rate of leaf photosynthesis and both the light flux density incident on the leaf surface and the ambient carbon dioxide concentration. Its main disadvantage is that it does not describe the dependence of the rate of net photosynthesis on the ambient oxygen concentration. More recent models for leaf photosynthesis and photorespiration overcome this disadvantage (examples are Tenhunen *et al.*, 1977 and Charles-Edwards, 1981).

There is plenty of experimental evidence to suggest that both adaptive and genetic differences in the rate of photosynthesis per unit of leaf area are associated with differences in leaf thickness (see Charles-Edwards, 1978, 1979b). The assimilation of carbon dioxide by a leaf involves the physical movement of gaseous carbon dioxide from the air to the photosynthetic sites as well as the chemical processes of photosynthesis. Changes in leaf thickness might be expected to affect the movement of carbon dioxide in the gaseous phase. The effects of such changes have been subject to both theoretical and experimental study (Charles-Edwards & Ludwig, 1975a; Sinclair *et al.*, 1977; Nobel, 1977; Parkhurst, 1977). There appear to be concomitant changes in the anatomy (stomatal density and distribution) and structure (mesophyll cell size) of the leaf with changes in leaf thickness which tend to minimize the effects of these changes on the photosynthetic functioning of the leaf. Put simply, whereas the rate of photosynthesis per unit leaf area increases with leaf thickness, the rate per unit leaf volume remains almost constant, and function is relatively unaffected by the structural constraints to which the leaf is subject.

Photosynthesis is only one aspect of leaf function. In *Lolium*, for example, there is a correlation between leaf width and the numbers of primary, longitudinal vascular bundles in the leaf (Charles-Edwards, Charles-Edwards & Sant, 1974). This observation implies that with changing leaf size the distance that new assimilate needs to move from the site of assimilation to the site of transport from the leaf remains fairly constant. There are a number of models for translocation of materials from the leaf, and there are comprehensive, physiological accounts of the process of translocation (Canny, 1973; Peel, 1974). Thornley (1976, pp. 50–61) has examined these models and shown that they all lead to one of two mathematical expressions. Either the flux of carbohydrate from the leaf is proportional to the concentration gradient between the leaf and the reservoir (sink), or to the product of the leaf concentration of the carbohydrate and this gradient.

I have dealt mostly with models which treat the leaf as a homogenous collection of tissues, although attempts have been made to study experimentally the concentrations of carbon metabolites in different leaf tissues (for example Outlaw & Fisher, 1975). It has been pointed out previously that in a physically compartmented system it may be preferable for the plant to sacrifice thermodynamic efficiency for kinetic expediency (Charles-Edwards, 1975). Whilst there are clearly problems associated with the physical compartmentation of metabolites it does not seem unreasonable to suppose that phenomenological relationships may exist between the main leaf metabolite fractions averaged over the whole leaf, that is, functional compartmentation may be more dominant than physical compartmentation. For example, it has been observed that when photorespiration in the leaves of $C_3$ plants is suppressed (Ludwig, Charles-Edwards & Withers, 1975), the rate of leaf 'dark' respiration has a different dependence on the rate of leaf photosynthesis. A model for leaf carbon metabolism which presumes functional compartmentation has been advanced to examine this phenomenon (Charles-Edwards & Ludwig, 1975b).

We can use modelling as both an analytical and a heuristic tool; that is, we can use it to help us to analyse experimental observation, distinguish casual and causal relationships and to advance hypotheses. If our analysis of leaf functioning is well founded, models can be used to predict, or simulate, how real leaves would behave in particular situations. Mathematics is a powerful and necessary tool in our quest to understand about leaves. A model provides a conceptual, mathematical framework within which we can start to order the large amount of information we have about leaf growth and functioning. The constraints upon us when we start to develop a model are two-fold. Firstly, we need to determine the objectives of the modelling exercise at the outset so that our model is appropriate to those objectives. Secondly, we must be aware of the amount of detail that we can meaningfully distinguish at the level at which we intend to apply the model. It seems trivial to remark that it is little use to build a model for translocation when we want to know about respiration, or that a model of leaf growth rooted in molecular orbital theory might not be particularly helpful to us, but less extreme examples of inappropriate modelling exercises are not uncommon. Equally inappropriate is the criticism of many experimentalists that models are based on too many assumptions. A model *is* a formal representation of assumption. To many model builders working alongside experimental physiologists the words of Hilaire Belloc are only too relevant:

> A scientist who ought to know
> Assures me that it must be so.
> So let us never, never doubt
> What nobody is sure about.

## References

Canny, M. J. (1973). *Phloem Translocation.* Cambridge University Press.

Charles-Edwards, D. A. (1975). Efficiency and expediency in plant growth. *Annals of Botany,* **39,** 161–2.

Charles-Edwards, D. A. (1978). An analysis of the photosynthesis and productivity of vegetative crops in the U.K. *Annals of Botany*, **42,** 717–31.

Charles-Edwards, D. A. (1979a). A model for leaf growth. *Annals of Botany*, **44,** 523–35.

Charles-Edwards, D. A. (1979b). Photosynthesis and crop growth. In *Photosynthesis and Plant Development*, ed. R. Marcelle, H. Clijsters & M. Van Pooke. The Hague: W. Junk.

Charles-Edwards, D. A. (1981). *The Mathematics of Photosynthesis and Productivity.* London: Academic Press.

Charles-Edwards, D. A., Charles-Edwards, J & Sant, F. I. (1974). Leaf photosynthetic activity in six temperate grass varieties grown in contrasting light and temperature environments. *Journal of Experimental Botany*, **25,** 87, 715–24.

Charles-Edwards, D. A. & Ludwig, L. J. (1974). A model for leaf photosynthesis by $C_3$ plant species. *Annals of Botany*, **38,** 921–30.

Charles-Edwards, D. A. & Ludwig, L. J. (1975a). The basis of expression of leaf photosynthetic activities. In *Environmental and Biological Control of Photosynthesis*, ed. R. Marcelle. The Hague: W. Junk.

Charles-Edwards, D. A. & Ludwig, C. J. (1975b). A model of leaf carbon metabolism. *Annals of Botany*, **39,** 819–29.

Dennett, M. D., Milford, J. R. & Elston, J. (1978). The effect of temperature on the relative leaf growth rate of crops of *Vicia Faba* L. *Agricultural Meteorology*, **19,** 505–14.

Ho, L. M. & Thornley, J. M. M. (1978). Energy requirements for assimilate translocation from mature tomato leaves. *Annals of Botany*, **42,** 481–3.

Ludwig, L. J., Charles-Edwards, D. A. & Withers, A. C. (1975). Tomato leaf photosynthesis and respiration in various light and carbon dioxide environments. In *Environmental and Biological Control of Photosynthesis*, ed. R. Marcelle. The Hague: W. Junk.

Milthorpe, F. L. & Newton, P. (1963). Studies on the expansion of the leaf surface. *Journal of Experimental Botany*, **14,** 483–95.

Monteith, J. L. (1977). Climate and the efficiency of crop production in Britain. *Philosophical Transactions of the Royal Society of London*, **B281,** 277–94.

Monteith, J. L. & Elston, J. (1971). Microclimatology and crop production. In *Potential Crop Production*, ed. P. F. Waring & J. P. Cooper. London:Heinemann.

Newman, F. H. & Searle, U. H. L. (1948). *The General Properties of Matter*. London: Arnold.

Nobel, P. S. (1977). Internal leaf area and cellular $CO_2$ resistance: photosynthetic implications of variations with growth conditions and plant species. *Physiologia Plantarum*, **40,** 137–44.

Outlaw, W. H. & Fisher, D. B. (1975). Compartmentation in *Vicia faba* leaves. I. Kinetics of $^{14}C$ in the tissues following pulse labelling. *Plant Physiology*, **55,** 699–703.

Parkhurst, D. F. (1977). A three-dimensional model for $CO_2$ uptake by continuously distributed mesophyll in leaves. *Journal of Theoretical Biology*, **67,** 471–88.

Peel, A. J. (1974). *Transport of Nutrients in Plants*. London: Butterworths.

Peisker, M. & Apel, P. (1975). Influence of oxygen on photosynthesis and photorespiration in leaves of *Triticum aestivum* L. *Photosynthetica*, **9,** 16–23.

Rabinowitch, E. I. (1951). *Photosynthesis and Related Processes*, vol. 2, part 1. New York: Interscience.

Sinclair, T. R., Goudriaan, J. & de Wit, C. T. (1977). Mesophyll resistance to $CO_2$ compensation concentration in leaf photosynthesis models. *Photosynthetica*, **11,** 1, 56–65.

Tenhunen, J. D., Weber, J. A., Yocum, C. S. & Gates, D. M. (1977). Development of a photosynthesis model with an emphasis on ecological applications. 2. Analysis of a data set describing the 'PM' surface. *Oecologia*, **26,** 2, 101–19.

Thornley, J. H. M. (1976). *Mathematical Models in Plant Physiology*. London: Academic Press.

Warren Wilson, J. W. (1971). Maximum yield potential. In *Transition from Extensive to Intensive Agriculture with Fertilizers*. Proceedings of the 7th Colloquium of the International Potash Institute. Berne: IPI.

# 17

# Performance and productivity of foliage in the field

J. L. MONTEITH AND J. ELSTON

## Subject and object

This final chapter deals with foliage rather than with individual leaves, drawing together three themes which are related to previous contributions. The first is the nature of the environment to which foliage is exposed in the field. The second is the development of foliage as a light-gathering system in a plant community, closely linked to the development of a root system gathering water and nutrients. The third is the relation between the interception of light by foliage, and the associated rate of dry matter production and eventually of yield.

Although these themes are clearly linked, it has been difficult to treat them consistently. When considering the microclimate of leaves in the field, we shall point out that natural environments are characterised by variability both in time and in space which cannot be fully reproduced in the laboratory or in growth rooms. On the other hand, when we attempt to relate the growth of a plant community to irradiance and to leaf photosynthesis, we shall stress that simplification is the key to progress and to understanding. As Charles-Edwards indicates (Chapter 16), there is no way in which all the physiological information discussed in previous chapters can be incorporated in some gigantic simulation model of plant growth. Instead, we shall try to show how the performance of individual leaves is linked to the productivity of foliage in a community, drawing attention to parts of the story which are vague or unconvincing because of the lack of experimental evidence.

**Microclimate and epiclimate**

The environment of a young shoot emerging from the soil is one of extreme variability. Large vertical gradients of temperature exist in bright sunshine, particularly when the surface is dry, so that leaf or stem tissue growing close to the surface can be 15 to 20 °C above the contemporary screen temperature (Geiger, 1965; Stoller & Wax, 1973). At night, temperature gradients reverse because of radiative heat loss and the surface can then become 5 °C cooler than the air at screen height. So a diurnal variation in the conventional screen climate of, say, ± 10 °C can be associated with a variation of ± 20 °C in the temperature of a seedling.

Seedlings may also be exposed to severe water stress if the soil surface is not rewetted by rain or irrigation. Survival then depends on achieving the right balance between the development of a system for collecting light energy and $CO_2$ above the ground and a mutually dependent system for collecting water and nutrients below ground. The survival strategies of seedlings in natural environments have been little studied, at least in terms of physical processes, despite their obvious agricultural significance in the semi-arid tropics where poor establishment is one of the main causes of low yield.

Once established, a seedling can develop a shoot and leaf system which shades the ground beneath it, creating a more equable climate for the root system and reducing the loss of water by direct evaporation from the soil surface. However, an isolated plant such as a shrub or a tree must then come to terms with substantial spatial variability in the radiative environment of its leaves. When the sun is shining in the late morning or early afternoon, outer leaves on the south side of a structure (northern hemisphere) will be much more strongly irradiated and will often have a faster photosynthesis rate than corresponding north-facing leaves, or foliage within the crown. Yet trees commonly achieve a symmetrical form with minor irregularities unrelated to the sun's path, presumably because internal controls effectively move assimilate from sunlit to shaded parts of the foliage. When asymmetry exists, it is often a consequence of another climatic element – wind.

Horizontal and vertical gradients of irradiance and windspeed may produce physiologically significant differences of leaf temperature (Rackham, 1974). The temperature differences shown in Fig. 17.1 represent an extreme case but are a salutary reminder of the contrast between microclimates which plants experience in the field and those in growth rooms where ventilation is designed to minimise temperature gradients.

The distribution of microclimatic elements is somewhat different in a crop or plant community where individuals of the same age and size are more or less uniformly spaced. There is often a very marked vertical gradient of irradiance with the uppermost leaves receiving radiation directly from the sun and sky and the lowest leaves receiving mainly diffuse light which has been transmitted or reflected by other elements of the foliage (Szeicz, 1974; Ross, 1973). Because leaves absorb strongly

Fig. 17.1. Temperature (°C) of leaves, flowers and roots of *Sempervivum montanum*, growing in the Alps at 2200 m, air temperature 22 °C, 1300 to 1400 local time (from Larcher, 1980).

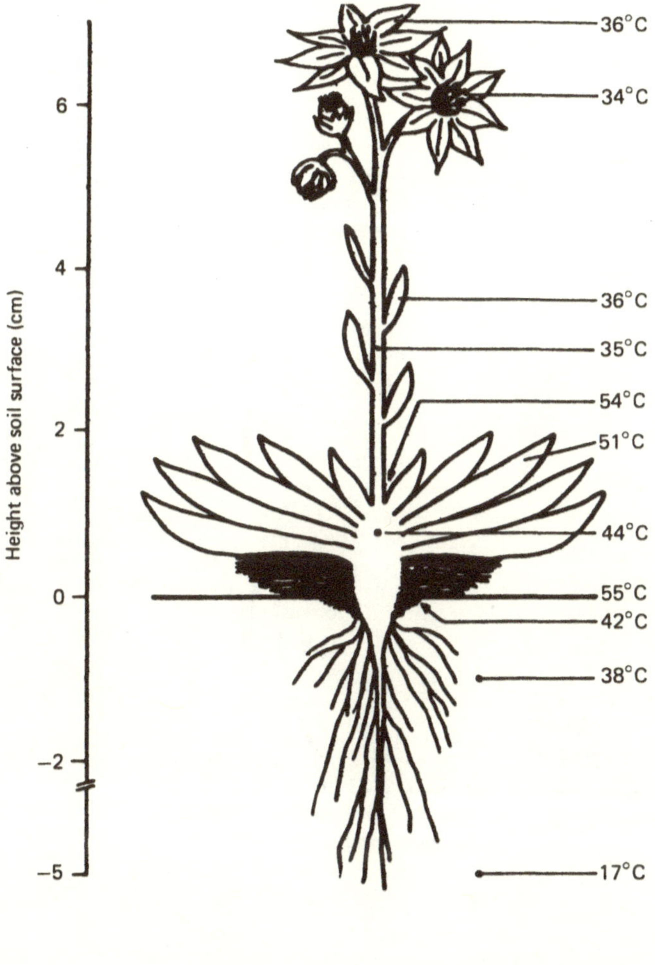

in the visible region of the spectrum but not in the near infra-red, shade light makes little contribution to photosynthesis but contains a strong spectral signal because the far red : red ratio is much larger than in direct sunlight (Fig. 17.2). The signal is perceived by the pigment phytochrome and appears to stimulate stem and leaf extension in some species (Smith, 1981a).

The distinction between sunlit and shaded leaves is rather artificial: it is common for only part of a leaf to be exposed within a sunfleck while the rest is in shade. Virtually nothing is known about how this spatial

Fig. 17.2. Spectral quantum flux density of solar radiation above a wheat canopy and at ground level below the canopy (measurements by M. G. Holmes, from Smith, 1981b).

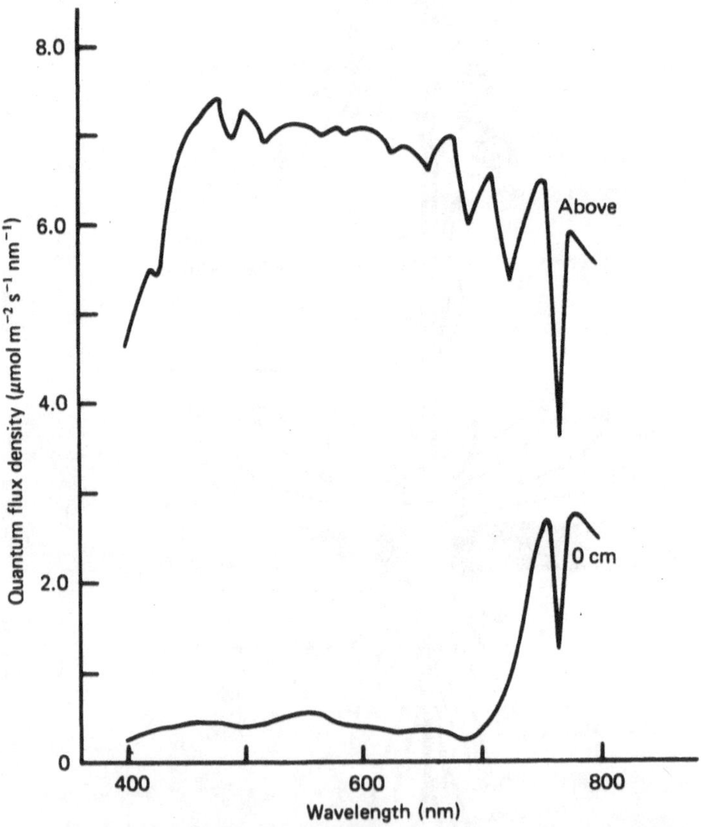

variation, changing continuously through the day, affects the mean rate of $CO_2$ and water vapour exchange per unit leaf area.

The irregular distribution of radiant energy is the main reason for the complex epiclimate of leaves – the physical state of the leaf surface (Monteith, 1981). Even when a leaf is uniformly irradiated, the difference between air temperature and tissue temperature at any point depends partly on the local thickness of the aerodynamic boundary layer and partly on the local diffusion resistance of stomata. In general, the temperature difference is least around the edges of a leaf where the boundary layer is thin (Grace, 1981) and is greatest along the mid-rib (Monteith, 1973). However, this pattern can be distorted by fluctuations in windspeed and direction and by fluttering as well as by the distribution of sunflecks.

Because of the cooling effect of transpiration, the mean temperature of a leaf freely supplied with water rarely departs from the temperature of the ambient air by more than ± 2 °C even in bright sunshine; but when stomata are closed, the mean excess temperature can rise to between 5 and 10 °C (Monteith, 1981).

An extensive community of uniform vegetation such as an arable crop creates a microclimate with significant horizontal as well as vertical features (Oke, 1978). For example, when a crop is surrounded by dry fallow the soil surface will be warmer than the foliage during the day (Davenport & Hudson, 1967). Both air temperature and foliage temperature decrease downwind from the leading edge of the stand and there is a corresponding increase of humidity. The most extreme changes occur when crops growing in otherwise dry areas are irrigated. Changes of temperature and of saturation deficit of the type illustrated by Fig. 17.3 are likely to have substantial systematic effects on the transpiration rate and photosynthesis rate of individual leaves and therefore on yield.

In summary, leaves on isolated plants and in communities are exposed to microclimates which are extremely variable both in time and in space and the epiclimates of leaf surfaces are even more variable. Individual aspects of this variability such as temperature fluctuations can be simulated in controlled environments but not the ensemble. Field measurements are essential for exploring the response to extreme conditions, e.g. the combination of high temperature and water stress. When the environment is not extreme, plants are good integrators, responding, for example, to daily mean temperature and to daily totals of radiation.

**Life and death**

*Principles*

Most green plants depend on leaves to intercept radiant energy and to absorb $CO_2$ so that expansion of the leaf surface and growth of the whole plant are intimately related. The growth of any healthy plant is governed by a complex system of checks and balances which ensure that the amount of assimilate allocated to leaf growth matches the growth of roots and of the structures needed to support leaves.

In physiological studies of single leaves, exchanges of water vapour and carbon dioxide are usually expressed on the basis of leaf area. In the field, the contribution from individual leaves is less important than the transpiration and assimilation per unit of field area. To link leaf area to field area, Watson (1947) introduced the concept of leaf area index ($L$) and this simple logical step had profound implications for the progress of crop ecology as a quantitative study. Subsequent developments in crop physiology, agronomy and pasture management demonstrated the economic usefulness of the concept; and it provides an essential link

Fig. 17.3. Changes of air temperature (●), vapour pressure (○), and vapour pressure deficit (■) across 300 m of irrigated cotton down-wind of dry fallow in the Sudan Gezira (from Davenport & Hudson, 1967).

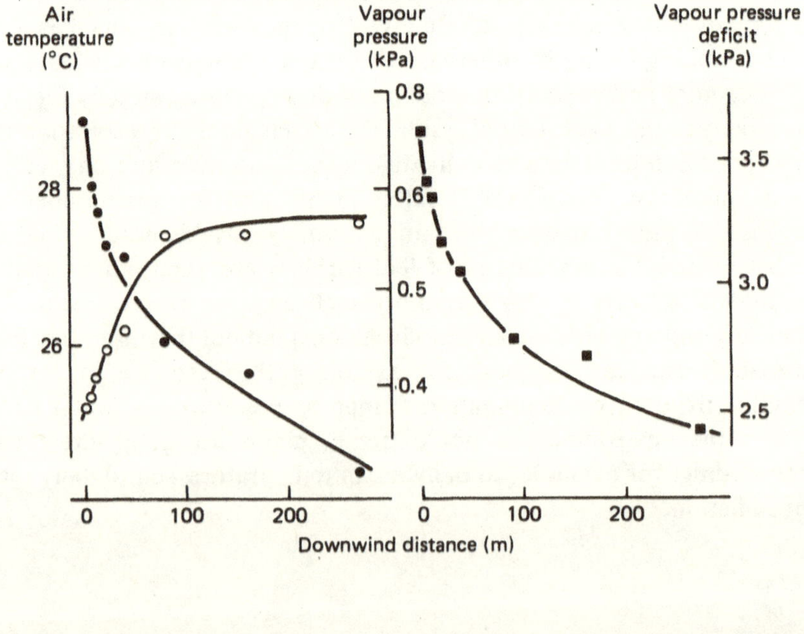

between the field measurements described in this chapter and the laboratory work described in previous chapters.

Leaf area index can be treated as the product of the number of leaves per unit of ground area and their average size. During the early stages of growth in an arable crop, both these quantities increase with time so that $L$ increases rapidly. At least in a favourable environment, the density of sowing usually ensures that within a few weeks there is enough foliage to intercept more than 90% of the incident radiation. At this stage the leaf area index has a maximum value $L_m$ and tends to decline thereafter for a number of reasons which we now explore.

In the first place, it appears that the lowest leaves of most crop species cannot survive when the average irradiance is between one and two orders of magnitude less than full daylight. The average irradiance at the level of the lowest leaves can be expressed as a fraction $\exp(-KL)$ of the irradiance at the top of the canopy, where $K$ is an attenuation coefficient depending on the spectral properties and the geometrical arrangement of the foliage. In the limit, $K = 1$ for a canopy of opaque, horizontal leaves, randomly distributed, but for most arable crops $K$ is between 0.4 and 0.8. Hayashi & Ito (1962) showed that $L_m$ was inversely correlated with $K$ in 14 varieties of rice, and the correlation can be demonstrated by plotting maximum $L$ against $1/K$ (Fig. 17.4) so that straight lines passing through the origin represent a constant value of $KL_m$ and therefore a constant fraction of transmitted light. The open squares on the graph are for rice (after Hayashi & Ito) and the full squares are for 12 other arable crops as reported in the literature. It is clear that the correlation of $L_m$ with $1/K$ is not confined to rice; it appears to be a general rule that species with more erect leaves tend to achieve larger values of $L_m$. Moreover, the variation of $K$ within varieties of one species (*Oryza sativa*) is comparable with the variability found in a whole range of species.

The interpretation of Fig. 17.4 in terms of the conservative nature of $KL_m$ is more consistent with experience than the concept of an 'optimum' value of $L$ advanced by Donald (1961), Black (1963) and others. They claimed that plant stands could achieve supra-optimal values of $L$ at which the crop growth rate was less than the value at $L_m$ because of the respiratory load of the lowest leaves which made little or no contribution to photosynthesis. Their evidence was based on experiments where the apparent decline of crop growth rate with increasing $L$ was probably a consequence of the loss of dead leaves and it is now widely accepted that crop stands rarely, if ever, develop *equilibrium* levels of $L_m$ beyond the value at which crop growth rate is maximal. *Transient* values of $L$ above $L_m$ were clearly demonstrated by Hiroi & Monsi (1966) in densely sown stands of sunflower. The behaviour of these stands is compatible with a maximum stable leaf area index in full sunlight of about 6, but in shade corresponding to 23% of full sunlight, $L_m$ was about 2.

Apart from the inhibitory effects of shading, the decline of $L$ in arable crops is a consequence of senescence and therefore of the complex internal factors discussed in Chapter 15. In determinate, annual species, the decline of leaf area is often associated with a decrease in the average photosynthetic efficiency of leaves, because new leaves are not produced after flowering and because the growth of seeds may require synthesis of proteins from precursors originating from protein breakdown in leaves.

The longevity of a leaf canopy in an annual crop therefore depends on the rate at which complete ground cover is achieved (leading to say 95% interception of incident light), on the maintenance of a maximum leaf

Fig. 17.4. Maximum leaf area index of crop stands plotted against $1/K$ (for total solar radiation). Open squares, rice crops (Hayashi & Ito, 1962); closed squares, artichoke, barley, casava, corncockle, flax, kale, lucerne, radish, sorghum, sunflower, *Vicia* bean and white clover (published values). Dashed lines indicate constant transmission of radiation.

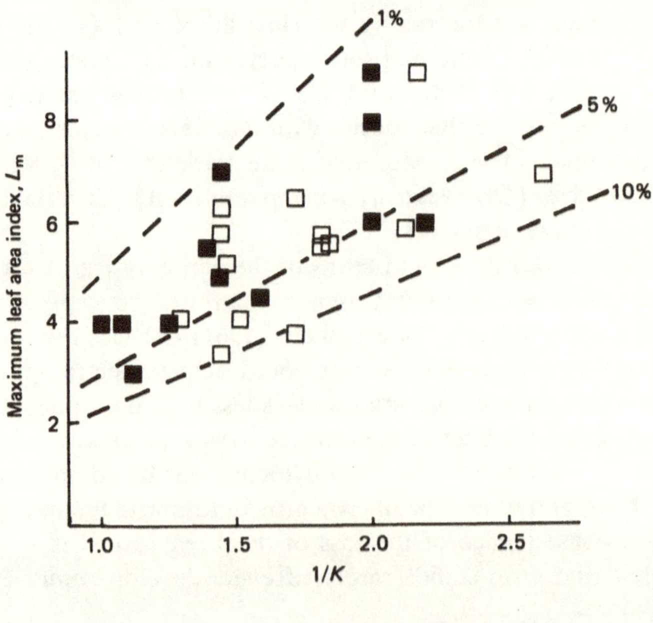

area index determined by the architecture of the canopy, and by the subsequent senescence and death of leaves. Each stage depends on the interaction of endogenous controls and environmental factors, mainly temperature and water supply. The significance of these factors has already been discussed in terms of studies in controlled environments but we now cite some relevant evidence from the field.

### *Practice*

Many studies in controlled environments have shown that rates of initiation, appearance, and expansion of leaves are strongly dependent on temperature (Chapter 7). In contrast, very few systematic measurements were reported from the field until Williams & Biddiscombe (1965) demonstrated that the rate of extension of grass leaves was closely correlated with temperature. Later laboratory and field work by Watts (1972) and by Peacock (1975) identified the meristem as the site of temperature perception in *Zea mays* and *Lolium perenne* respectively. Gallagher (1979) and Baker (1979) went a step further by demonstrating that the rate of leaf initiation in barley and wheat is strongly correlated with the prevailing daily mean air temperature which was close to the mean meristem temperature for autumn and spring sown crops. For reasons discussed in the last section, mean air and mean meristem temperature are likely to differ significantly in more extreme environments.

Temperature thus has a substantial effect on the area of leaves, as well as upon their rate of initiation. For example, Dennett, Elston & Milford (1979) found that, at daily irradiances of less than about 19.5 MJ $m^{-2}$, the final area to which *Vicia* leaves grew increased with temperature. At higher irradiances, however, increasing temperature tended to reduce final leaf area. The difference in final area was produced by differences in the duration of growth rather than in the mean growth rate. In cereals, the rate of leaf extension is clearly affected by temperature and the effects are larger than perturbations produced in the field by changes in water potential (Fig. 17.5) (Gallagher & Biscoe, 1979) or by carbohydrate concentration (Kemp & Blacklow, 1980). However, low concentration of available carbohydrate must be expected to limit extension. Wenkert, Lemon & Sinclair (1978) found that leaf water status hardly affected the rate of elongation of soyabean leaves in the field, except during pod-filling at adjacent nodes. The rate of elongation was always sensitive to variation in temperature.

The relatively small response to a substantial drop in the total potential is consistent with the type of adjustment described in Chapter 11.

Legg and his colleagues (1979) examined the effects of drought on growth and yield in barley in the field. Leaf area, light interception, stomatal resistance and the rate of photosynthesis of individual leaves and ears were all measured. The largest effects of drought on yield came from effects of water stress on leaf area and hence on light interception (whose significance is discussed later). Differences in stomatal resistance and the rate of photosynthesis were comparatively unimportant.

Karamanos (1976), using *Vicia faba*, and Rawson, Constable & Howe

Fig. 17.5. Mean values of extension rate for leaf 8 in a barley crop (○); and total leaf water potential (△) as functions of air temperature on 1 June 1972. Numbers refer to start of 2 hour period for mean values. Dashed line AB is expected response with no water stress (from Gallagher & Biscoe, 1979).

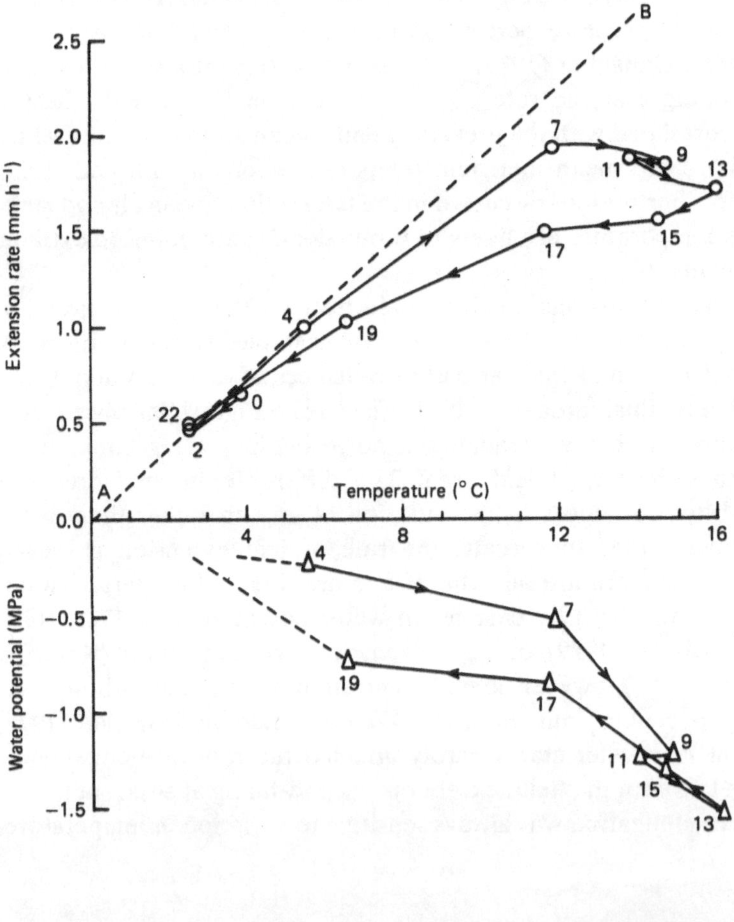

(1980), studying several cultivars of sunflower, found that the mean rate of expansion in leaf area was more substantially affected by water stress than was the duration of expansion. Similarly nitrogen fertilisers, which operate primarily by increasing leaf area, increase the rate of expansion rather than the duration of expansion of individual leaves of sugar beet (Fig. 17.6; K. Birkby, unpublished).

In general, temperature seems to have the largest effect on the *duration* of growth. Water, nitrogen and salinity seem to have their largest effects on the mean *rate* of expansion, rather than on the duration of growth.

Both the leaf area of a plant at any instant and the leaf area index of a community are net quantities – the balance established by the processes of leaf initiation, appearance and expansion which add to leaf area and leaf death which subtracts from the area. This statement is almost trivial but is worth making to emphasise the point that the processes which govern the production and death of leaves have rarely, if ever, been studied together in the field. Little is known about the environmental control of the complex set of interrelated changes in a senescent leaf (Chapter 15) and although death is simpler to assess, it is often unrecorded.

Fig. 17.6. Increase in length of sugar beet leaves, including petiole, plotted against time from noon on 16 July, 1981, with subsequent noons indicated by vertical markers. Upper, 0 kg N ha$^{-1}$; lower, 150 kg N ha$^{-1}$ (K. Birkby, unpublished).

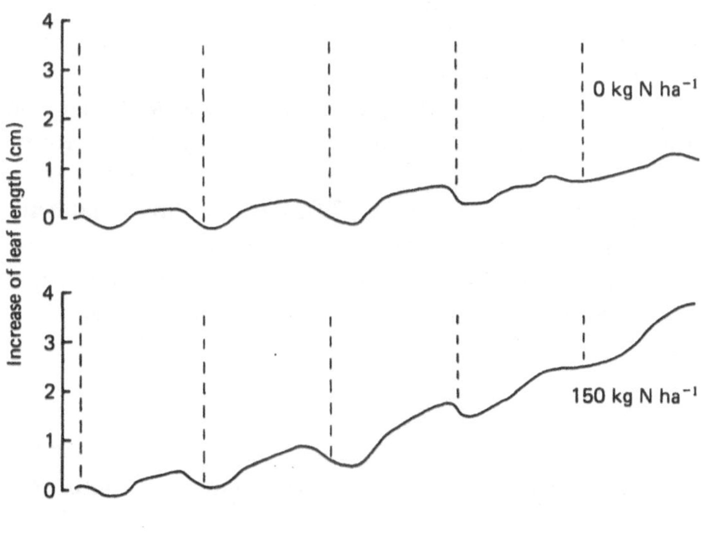

The timing of the death of a leaf or of the start of the death of a set of leaves appear to be related to a number of apparently unrelated factors. Fukai & Silsbury (1976) found that death started at about day 28 for subterranean clover grown at four different temperatures and with different leaf numbers, leaf area indices and dry weights. Finch-Savage (1976), using crops of *Vicia* beans, found that death started slowly at a date which varied between crops and years and reached an appreciable rate only at the maximum dry weight of the seed, i.e. at about 0.85 of total crop life. Littleton *et al.* (1979) found that the time to the start of leaf death in cowpeas was inversely related to the mean air temperature over the period and that a linear regression accounted for about 90% of the variation in time. All three of these species are legumes and the last two are grain legumes.

Fukai & Silsbury (1976) clearly showed the way in which leaf area and leaf area index depend upon the balance between leaf appearance and leaf death in subterranean clover. In their measurements, the rate of leaf appearance exceeded the rate of leaf death at 15 °C and leaf area index was an increasing function of time for the whole of their period of measurement. At 25 °C and 30 °C the rate of leaf death exceeded the rate of leaf appearance after a time; leaf area index reached a maximum value and then began to decline.

Finch-Savage (1976) showed that the rate of leaf death in *Vicia* beans depends upon temperature and upon irradiance, vapour pressure deficit and water stress, both in field crops and in growth room experiments (Fig. 17.7). Fukai & Silsbury (1976) found a strong dependence upon temperature in subterranean clover. Disease is probably important in determining the longevity of individual leaves in many field crops (Elston, Harkness & McDonald, 1976).

The overall consequences of the physiological processes which determine the rates of appearance, expansion and death and the duration of expansion can be summarised diagrammatically for leaf area index (Finch-Savage, 1976; Littleton *et al.*, 1979). Fig 17.8 shows that leaf area index at any moment consists of the accumulated leaf area index less the loss of area from death. Both curves depend upon the numbers appearing or dying and on the areas of the individual leaves.

## Variables and parameters

We began this chapter by emphasising the complexity and variability of the environment in which given plants grow and we have

Fig. 17.7. Number of dead leaves as a function of time after sowing for *Vicia* beans grown at Sonning Farm, University of Reading, 1975. Sowing dates: 1, 30 April; 2, 26 May; 3, 12 June; D, no irrigation; W, irrigated to maintain leaf water potential above −0.4 MPa. Note that death rate accelerated at maximum seed weight but thereafter was slower in successive sowings because of decreasing temperature during autumn. (After Finch-Savage, 1976.)

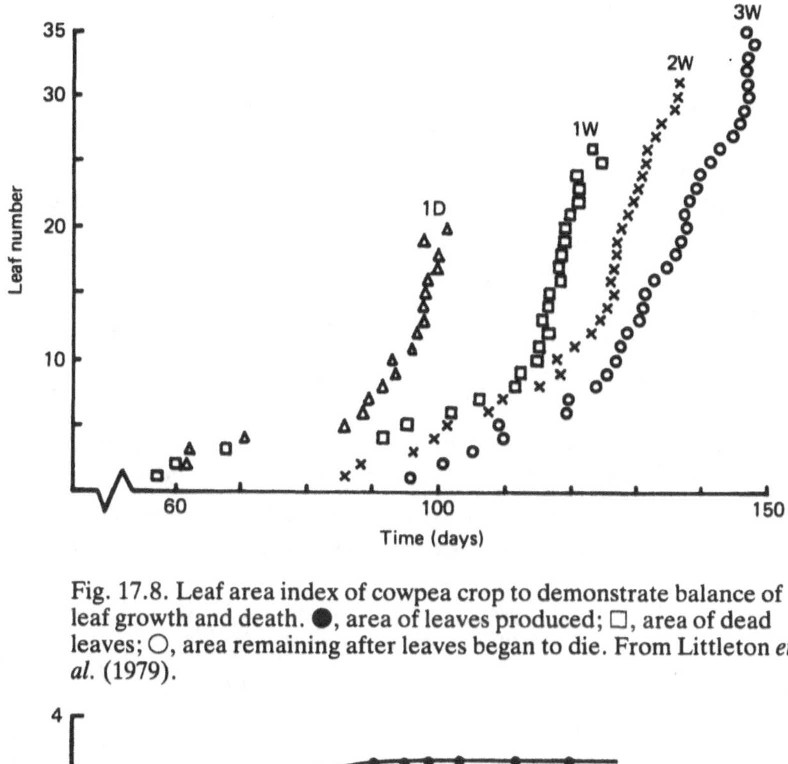


Fig. 17.8. Leaf area index of cowpea crop to demonstrate balance of leaf growth and death. ●, area of leaves produced; □, area of dead leaves; ○, area remaining after leaves began to die. From Littleton *et al.* (1979).

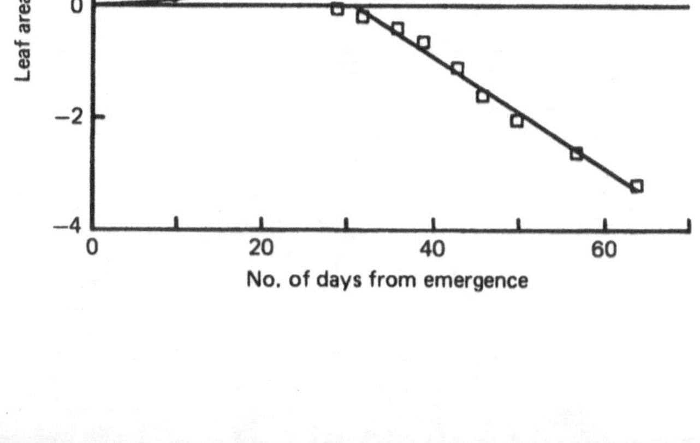

reviewed briefly some of the ways in which leaves respond to that environment. Our treatment has been largely descriptive with the excuse that mathematical models of such a complex system are beyond the scope of this book. However, progress in the interpretation of field measurements needs an analytical framework and several elements of this structure will now be described.

At the outset, it is important to distinguish the variables and the parameters of the system. The two terms are often confused in everyday speech (especially by politicians!) and the distinction is not always clear-cut. By 'variables' we mean elements of the environment or physiological quantities normally changing in time or in space, but sometimes constant during periods of observation. 'Parameters' are constants in a mathematical context but in ecology they can be defined less rigorously as conservative quantities, changing slowly with time or altering little between species. Progress in analysing the behaviour of any ecological system depends on identifying the conservative elements which provide a frame of reference against which variables can be measured.

What conservative quantities are associated with leaf production and with plant growth in the field? We look first at the relation between the amount of dry matter produced by crops giving record yields and the length of the growing season for each crop (Fig. 17.9). The conservative

Fig. 17.9. Total dry weight at harvest for very high yielding crops in relation to growth period. ■, $C_4$; □, $C_3$. (After Monteith, 1978.)

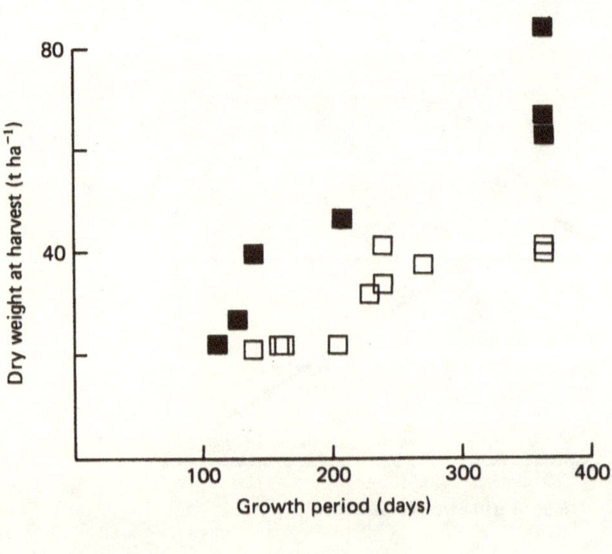

quantity in the analysis is the mean rate of dry matter production given by the slopes of the two fitted lines, viz. 13.0 ± 1.6 g $m^{-2}$ $day^{-1}$ for $C_3$ species and 22.0 ± 3.6 g $m^{-2}$ $day^{-2}$ for $C_4$ species (Monteith, 1978). The corresponding variable is the time from sowing to harvest determined essentially by the longevity of leaves. Fig. 17.9 implies that the factors which determine the development and death of leaves are much more significant discriminants of yield than the photosynthetic capacity of leaves, notwithstanding the multiplicity of internal and external factors governing rate of leaf photosynthesis, reviewed in Chapters 12 and 13. Differences of yield are a result of differences in the *duration* of canopies rather than the *rate* at which they produce dry matter. This conclusion is supported by a comparison of the maximum growth rates of arable crops in the Netherlands (Sibma, 1968) and by a similar comparison of vegetable crops in Britain (Greenwood *et al.*, 1977). These two studies showed that when crops in Western Europe have developed enough foliage to cover the ground, they are likely to produce dry matter at rates between 16 and 20 g $m^{-2}$ $day^{-1}$ provided they are not affected by disease or by drought.

The inference that growth rate is much more conservative than growth duration emerged long ago from traditional growth analysis. Watson (1952) identified 'leaf area duration' rather than 'net assimilation rate' as a major factor determining differences of yield between species in the same season and for the same species over a number of seasons.

Unfortunately, neither leaf area duration (the time integral of $L$) nor net assimilation rate (rate of dry matter production per unit of $L$) are constant enough to be treated as parameters of growth. The quest for such parameters made progress when it was found that the amount of dry matter accumulated by arable crops was almost proportional to the amount of radiation intercepted by their foliage. The constant of proportionality – dry matter per unit of energy – is a measure of the photosynthetic efficiency of a crop and is a useful parameter in several senses.

In the first place, the efficiency of a crop stand is often nearly constant during vegetative growth as long as there is a continuous production of new leaf tissue to offset the senescence of older leaves as they become increasingly shaded (Fig. 17.10). Second, evidence, so far limited to a few species only, suggests that efficiency may be relatively insensitive to temperature (recent work on millet at Nottingham University), to water stress (Legg *et al.*, 1979, working on barley) and to nitrogen supply (Fig. 17.10

and measurements on sugar beet by P. V. Biscoe and colleagues). Third, efficiency appears to be conservative for different varieties of the same species and even for different species, at least within the main $C_3$ and $C_4$ groups (Gallagher & Biscoe, 1978; Monteith, 1977). In Table 17.1, which contains a characteristic set of values, there are at least two reasons why the figures for cereals and rape should be lower than for sugar beet and potatoes. Harvest of the root crops included a large

Fig. 17.10. Dry weight of tops in stands of barley from sequential harvests as function of accumulated solar radiation (0.4–3 $\mu$m) intercepted by foliage. Time of ear emergence shown by arrows. Dashed line fitted to final harvests – 1.22 g $MJ^{-1}$. Other lines fitted by eye. (From Gallagher, 1976.)

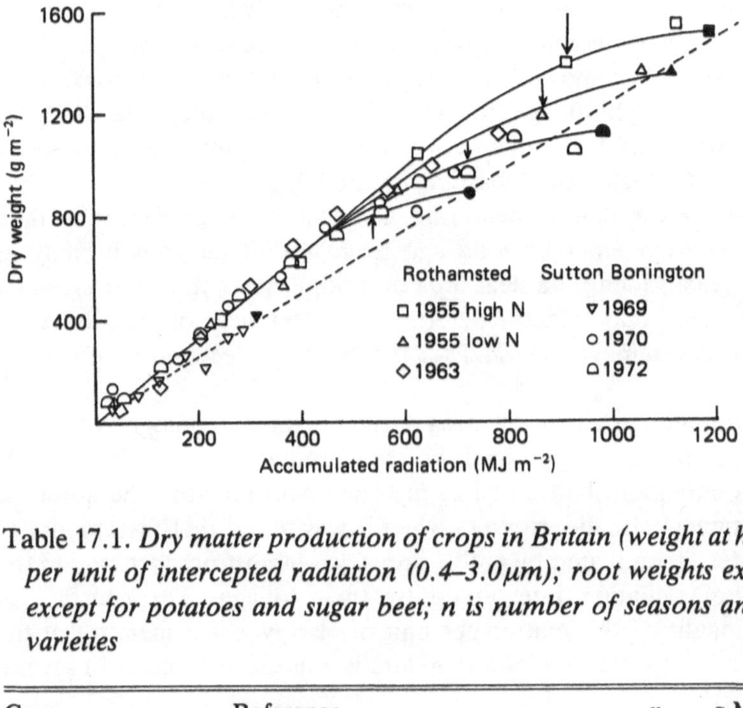


Table 17.1. *Dry matter production of crops in Britain (weight at harvest) per unit of intercepted radiation (0.4–3.0 $\mu$m); root weights excluded except for potatoes and sugar beet; n is number of seasons and/or varieties*

| Crop | Reference | *n* | g $MJ^{-1}$ |
|---|---|---|---|
| Barley | Gallagher & Biscoe (1978) | 6 | 1.1–1.3 |
| Winter and spring wheat | Gallagher & Biscoe (1978) | 9 | 0.9–1.2 |
| Potatoes | Scott & Allen (1978) | 15 | 1.2–1.5 |
| Sugar beet | Scott & Allen (1978) | 13 | 1.2–1.5 |
| Oil seed rape | Mendham, Shipway & Scott (1981) | 16 | 1.0–1.2 |

fraction of the root dry weight but root weights were excluded from the cereal and rape analysis. Moreover, it is likely that the decline in photosynthetic efficiency towards the end of the growing season occurs later in indeterminate species than in determinate species which stop producing new leaves when they flower.

If it is legitimate to treat as constant the efficiency with which a crop intercepts radiation, several conclusions follow (Monteith, 1981). In the first place, the fraction of intercepted radiation is nearly proportional to $(1 - \exp(-KL))$ so that the amount of dry matter production by a crop will depend on the time integral of this quantity rather than on the time integral of $L$ (i.e. the leaf area duration). Moreover, it can readily be shown that net assimilation rate cannot be a conservative quantity throughout the life of a crop because it is nearly proportional to $(1 - \exp(-KL))/L$. This function is independent of $L$ (and equal to $K$) when $L$ is very small but it is inversely proportional to $L$ during the main period of growth when the denominator is close to unity. Finally, if the ratio of leaf area to plant dry weight is assigned an appropriate mean value, it is possible to derive an expression for the dry matter which a crop fails to produce (and the corresponding 'lost' time) during the period when the canopy is incomplete. Because the leaf area ratio increases with temperature in many species, the loss of dry matter decreases with temperature.

## Conclusions

Clearly the growth and yield of a crop are determined primarily by the amount of radiation intercepted during the growing season, which, in turn, is primarily determined by the timing, size and duration of the canopy. In agriculture, field operations usually affect leaf area index rather than the actual rate of photosynthesis per unit leaf area. For example, sowing rate and date, rate of nitrogen fertilizer applied, irrigation treatment, many insecticide and fungicide applications and some soil cultivation treatments have their main effects on leaf area index. Economically, we are concerned with the marginal response to an increase in leaf area index produced by variation in some field operation. Agronomically, we need to find the major factors controlling the growth and duration of the canopy and to match canopy duration and size to the pattern of irradiance.

However, whole plant and crop physiologists have traditionally studied the process of photosynthesis, which has been seen as the base for studies of growth. The invention of the infra-red gas analyser, by Luft, made these quantitative studies much easier as well as more precise. A recent survey of an abstracting journal showed that papers about effects of the environment on photosynthesis exceed those on leaf growth by about 3 to 1. Most of the work is laboratory based and relatively few studies have been made of photosynthesis in field crops.

Field measurements of leaf initiation, growth and death are essential if our understanding, and control, of growth and yield is to improve.

## References

Baker, C. K. (1979). The Environmental Control of Development in Winter Wheat. Ph. D. thesis, University of Nottingham.

Black, J. N. (1963). The interrelationship of solar radiation and leaf area index in determining the rate of dry matter production of swards of subterranean clover (*Trifolium subterraneum* L.). *Australian Journal of Agricultural Research*, **14,** 20–38.

Davenport, D. C. & Hudson, J. P. (1967). Changes in evaporation rates along a 17-km transect in the Sudan Gezira. *Agricultural Meteorology*, **4,** 339–52.

Dennett, M. D., Elston, J. & Milford, J. R. (1979). The effect of temperature on the growth of individual leaves of *Vicia faba* L. in the field. *Annals of Botany*, **43,** 197–208.

Donald, C. M. (1961). Competition for light in crops and pastures. In *Symposia of the Society for Experimental Biology XV: Mechanisms in Biological Competition*, pp. 282–313. Cambridge University Press.

Elston, J., Harkness, C. & McDonald, D. (1976). The effects of Cercospora leaf disease on the growth of groundnuts (*Arachis hypogaea*) in Nigeria. *Annals of Applied Biology*, **83,** 39–51.

Finch-Savage, W. E. (1976). A Study of Leaf Senescence and Death in Crops of *Vicia faba* L. Ph.D. thesis, University of Reading.

Fukai, S. & Silsbury, J. H. (1976). Responses of subterranean clover communities to temperature. I. Dry matter production and plant morphogenesis. *Australian Journal of Plant Physiology*, **3,** 527–43.

Gallagher, J. N. (1976). The Growth of Cereals in Relation to Weather. Ph.D. thesis, University of Nottingham.

Gallagher, J. N. (1979). Field studies of cereal leaf growth. I. Initiation and expansion in relation to temperature and ontogeny. *Journal of Experimental Botany*, **30,** 625–36.

Gallagher, J. N. & Biscoe, P. V. (1978). Radiation absorption, growth and yield of cereals. *Journal of Agricultural Science*, **91,** 47–60.

Gallagher, J. N. & Biscoe, P. V. (1979). Field studies of cereal leaf growth. III. Barley leaf extension in relation to temperature, irradiance, and water potential. *Journal of Experimental Botany*, **30,** 645–55.

Geiger, R. (1965). *The Climate Near the Ground*. Cambridge, Massachusetts: Harvard University Press.

Grace, J. (1981). Some effects of wind on plants. In *Plants and their Atmospheric Environment*, ed. J. Grace, E. D. Ford & P. G. Jarvis, pp. 31–56. Oxford: Blackwell Scientific Publications.

Greenwood, D. J., Cleaver, T. J., Loquens, S. H. M. & Niendorf, K. B. (1977). Relationship between plant weight and growing period for vegetable crops in the United Kingdom. *Annals of Botany*, **41,** 987–97.

Hayashi, K. & Ito, H. (1962). Studies on form of plant in rice varieties with particular reference to efficiency in utilizing sunlight. I. The significance of extinction coefficient in rice plant communities. *Proceedings at the Crop Science Society of Japan*, **30**, 329–33.

Hiroi, T. & Monsi, M. (1966). Dry matter economy of *Helianthus annuus* communities grown at varying densities and light intensities. *Journal of the Faculty of Science, Tokyo University*, **111**, 241–85.

Karamanos, A. J. (1976). An Analysis of the Effect of Water Stress on Leaf Area Growth in *Vicia faba* L. in the Field. Ph.D. thesis, University of Reading.

Kemp, D. R. & Blacklow, W. M. (1980). Diurnal extension rates of wheat leaves in relation to temperature and carbohydrate concentration of the extension zone. *Journal of Experimental Botany*, **31**, 821–8.

Larcher, W. (1980). *Physiological Plant Ecology*, second edition. Berlin: Springer-Verlag.

Legg, B. J., Day, W., Lawlor, D. W. & Parkinson, K. J. (1979). The effects of drought on barley growth: models and measurements showing the relative importance of leaf area and photosynthetic rate. *Journal of Agricultural Science*, **92**, 703–16.

Littleton, E. J., Dennett, M. D., Elston, J. & Monteith, J. L. (1979). The growth and development of cowpeas (*Vigna unguiculata*) under tropical field conditions. I. Leaf area. *Journal of Agricultural Science*, **93**, 291–307.

Mendham, N. J., Shipway, P. A. & Scott, R. K. (1981). The effect of delayed sowing and weather on growth, development and yield of winter oil-seed rape (*Brassica rapus*). *Journal of Agricultural Science*, **96**, 389–416.

Monteith, J. L. (1973). *Principles of Environmental Physics*. London: Edward Arnold.

Monteith, J. L. (1977). Climate and the efficiency of crop production in Britain. *Philosophical Transactions of the Royal Society of London*, **B281**, 277–94.

Monteith, J. L. (1978). Reassessment of maximum growth rates for $C_3$ and $C_4$ crops. *Experimental Agriculture*, **14**, 1–5.

Monteith, J. L. (1981). Coupling of plants to the atmosphere. In *Plants and their Atmospheric Environment*, ed. J. Grace, E. D. Ford & P. G. Jarvis, pp. 1–29. Oxford: Blackwell Scientific Publications.

Oke, T. (1978). *Boundary Layer Climates*. London: Methuen.

Peacock, J. M. (1975). Temperature and leaf growth in *Lolium perenne*. II. The site of temperature perception. *Journal of Applied Ecology*, **12**, 115–23.

Rackham, O. (1974). Temperatures of plant communities as measured by pyrometric and other methods. In *Light as an Ecological Factor*, II, ed. G. C. Evans, R. Bainbridge & O. Rackham, pp. 423–49. Oxford: Blackwell Scientific Publications.

Rawson, H. M., Constable, G. A. & Howe, G. N. (1980). Carbon production of sunflower cultivars in field and controlled environments. II. Leaf growth. *Australian Journal of Plant Physiology*, **7**, 575–86.

Ross, J. (1973). Radiative transfer in plant communities. *Vegetation and Atmosphere*, vol. 1, ed. J. L. Monteith, pp. 13–55. London: Academic Press.

Scott, R. K. & Allan, E. J. (1978). Crop physiological aspects of importance to maximum yields – potatoes and sugar beet. In *Maximising Yields of Crops*, ADAS-ARC Symposium, pp. 25–30. London: HMSO.

Sibma, L. (1968). Growth of closed crop surfaces in The Netherlands. *Netherlands Journal of Agricultural Science*, **16**, 211–16.

Smith, H. (1981a). Light quality as an ecological factor. In *Plants and their Atmospheric Environment*, ed. J. Grace, E. D. Ford & P. G. Jarvis, pp. 93–610. Oxford: Blackwell Scientific Publications.

Smith, H. (1981b). Adaption to shade. In *Physiological Processes Limiting Plant Productivity*, ed. C. B. Johnson, pp. 159–73. London: Butterworths.

Stoller, E. W. & Wax, L. M. (1973). Temperature variation in the surface layers of an agricultural soil. *Weed Research*, **13**, 273–82.

Szeicz, G. (1974). Solar radiation in canopies. *Journal of Applied Ecology*, **11**, 1117–56.

Watson, D. J. (1947). Comparative physiological studies on the growth of field crops. I. Variation in net assimilation rate and leaf area between species and varieties, and within and between years. *Annals of Botany*, **11**, 41–76.
Watson, D. J. (1952). The physiological basis of variation in yield. *Advances in Agronomy*, **4**, 101–45.
Watts, W. R. (1972). Leaf extension in *Zea mays. Journal of Experimental Botany*, **23**, 704–21.
Wenkert, W., Lemon, E. R. & Sinclair, T. R. (1978). Leaf elongation and turgor pressure in field-grown soybean. *Agronomy Journal*, **70**, 761–4.
Williams, C. N. & Biddiscombe, E. F. (1965). Extension growth of grass tillers in the field. *Australian Journal of Agricultural Research*, **16**, 14–22.

# Summary and Discussion

D. GEIGER and F. L. MILTHORPE

The presentations in this section revealed a pervading issue – our inability to describe conditions and determine rates of processes in organelles, cells, or tissues at particular locations within the leaf. Although the study of isolated organelles, cells and tissues has provided much data, it has failed generally to help us in understanding differences in rates of processes or in states at specific sites in the intact leaf. Better methods are needed to determine parameters such as resistance to carbon dioxide transfer, rates of solute transfer, levels of a hormone or of water potential, and turgor gradients at a particular place in the leaf. The existence of this gulf between functions of the intact leaf and of its components is a frustration, but also offers a challenge.

Lüttge pointed out several difficulties in relating water relations of cells and tissues to parameters describing the water status of the intact leaf. Pressure probe measurements give good data on water status but the necessary cell volume measurements are not very reliable. Methods under development may help solve this problem. The lack of homogeneity of leaf tissues makes it difficult to interpret bulk parameters measured in whole leaves. Several persons referred to the problems that arise from substituting into equations values taken from sources that are inappropriate. For instance, values of hydraulic conductivity differ by a factor of $10^4$ for different membranes. Likewise, values for parameters measured on bulk tissue may not be appropriate for calculations of cell parameters. A major point of Barlow's presentation, i.e. that stomatal conductance is correlated with turgor pressure to a higher degree than to leaf water potential, sparked considerable discussion. The difficulty of knowing water potential and turgor for individual cells was mentioned above. Other issues include solute transfer between guard and epidermal cells, abscisic acid (ABA) level in relation to turgor of leaf cells, and water vapour gradient or flux. Determinations of conditions at specific sites within the leaf are needed; as emphasized by Radford, the technique used and time over which measurements are made greatly influence the results.

As discussed in Part I, we must pay heed to the fourth dimension (time) in studying the physiological response of leaves. The importance of previous growth history was emphasized by presenters and discussants. Plants grown under cycles of uniform conditions in environmentally controlled cabinets respond differently from those exposed to the wider range of conditions found in the field. Time-lapse studies of leaf growth presented by Terry demonstrated the different response to water stress upon subsequent repeated exposures. The nature of persistent after-effects need to be studied. Also, in spite of several studies, there is still the question about how long enhancement of photosynthesis continues following carbon dioxide enrichment of the atmosphere. Studies of acclimation, adaptation and after-effects need to be extended over inconveniently long periods. A case in point is the persistent inhibition of photosynthesis for up to several days after the leaf is exposed to water stress. Some workers conclude that, even after water has been restored and ABA levels drop, stomates do not open as widely because the former high level of ABA resets the stomatal response to carbon dioxide. Another proposal to explain the carry-over is that photo-inhibition occurs following exposure of the leaf cells to low carbon dioxide during water stress; this idea requires experimental evidence as it has practical implications for yield and may help shed light on the role of photorespiration under stress conditions.

Preparation in advance of stress in crassulacean acid metabolism (CAM) plants is found to be induced by photoperiod. No other examples of anticipation of seasonal stress were offered. Adaptation to light, water and temperature stress seems to be triggered by the onset of the stress itself. Ludlow stated that new leaves are adapted to the conditions under which they developed while older leaves undergo structural changes and become adapted.

Terry questioned the interpretation usually given to the saturation of photosynthetic carbon fixation per unit area at high photon flux density. He stressed the fact that the lack of increase in rate is a result of saturating the available photosynthetic units in the thylakoids. It was pointed out by several that increasing carbon dioxide concentration in the atmosphere causes a marked increase in carbon fixation rate; this supports the interpretation that carbon dioxide availability is limiting. The concept of co-limiting factors was introduced by Terry as an alternative to Blackman's Law of the Minimum. He noted that, when photosynthetic unit density is increased by supplying additional iron, the rate of carbon fixation increases over the former saturation level. The occurrence of the topic of photosynthetic unit density several times in the discussion periods and the inconclusive outcome of the exchanges indicates the need for further clarification. There seem to be more things occuring than can be accommodated by the present model. Several discussants raised the point that the size of the photosynthetic unit may be able to change rapidly with changing conditions; further, individual units may be capable of communicating with each other. The characteristics are reasonable given the fluid nature of the lipids of the thylakoid membranes. A change in the nature of properties of these lipids under various conditions may explain changes in chloroplast function. A note of caution was sounded for studies of the size and organization of photosynthetic units. It is possible that loss of activity of some reaction centres during isolation or treatment may lead to apparent changes in size of the units. It was suggested that more work was needed on the role of lipid composition, especially in relation to the ability of chloroplasts to adapt rapidly to internal and external factors.

Although the importance of proton co-transport in solute transport in higher plants is gaining support, there are aspects which need further clarification. The role of the carriers and proton co-transport in processes such as ion or sugar secretion, and in efflux from mesophyll into the apoplast needs to be investigated; details need to be added to our present models for these processes. The view was offered that onset of phloem loading in leaves which were formerly sinks results from an increase in the electrochemical gradient of protons. We need to move beyond the general description of proton co-transport to study its specialization in specific applications such as glands, nectaries, ion secretion into xylem and transfer in sink organs.

Isolated protoplasts are being used increasingly in transport studies in cells. Bentrup expressed the caution that the inability of these cells to develop turgor appears to change their electrical and transport properties. Autotrophically grown cell suspensions appear to be superior, giving some realistic characteristics. It should be noted that each time an organ like a leaf is disassembled to a lower level of organization, interactions are lost. Some features can be studied better in isolated cells but the loss of characteristics and functions needs to be kept in mind when interpreting and generalizing from results.

An aspect which needs greater attention is the ability of leaves to control allocation of fixed carbon. Several speakers emphasized that we were arbitrarily featuring the leaf in isolation to focus our discussion. The role of the leaf in allocating carbon for its own growth, for storage pools more readily available for respiration and for export should be kept in mind. Intercommunication and interdependence, stressed in the section on leaf initiation, is certainly of great importance when considering the function and significance of the mature leaf.

Woolhouse emphasized that he had discussed three aspects only of senescence; many others were certainly involved. He, as earlier contributors, had taken photosynthesis as the major physiological process. Terry queried whether the supply of ATP and NADPH consequent on the slowing down of electron transport was of itself limiting. Woolhouse pointed out that there appeared to be circumstances where electrons, normally available for $CO_2$ reduction, were selectively diverted to activate enzymes. Huffaker wondered if the contention of excess RuBPcase activity held in the late stages of senescence as there was a close correlation between photosynthesis and RuBPcase activity. Woolhouse reiterated his belief in respect of bean and was supported, in respect of wheat, by Brady who also pointed out that the protein-synthesizing system in the chloroplast was maintained. In this there appeared to be real species differences as it was maintained in chinese cabbage but not *Perilla*. Gifford asked if there was any explanation for why the plant should invest such a large proportion of its soluble protein in RuBPcase if this enzyme was 5–10 fold in excess. Woolhouse thought, as with pyrenoids in some algae, it could be an important storage protein.

The inevitable controversy between experimentalists and modellers followed Charles-Edward's contribution. For example, Woolhouse maintained that few models in plant physiology were helpful; they were useful only if predictive and at the frontier of experimentation. However, he agreed that in the field of electrochemical aspects of ion transport models had been fruitful; it was the biochemical aspects which were unsatisfactory. The need for continuing dialogue between experimenters and modellers was emphasized by several speakers as was the need for more biologists to obtain an adequate background in mathematics. Charles-Edwards emphasized that a model was simply a formal mathematical statement of assumptions about a system; even if not predictive, a heuristic objective could help understand systems (supported by Silk). Pleas had been made throughout the Symposium for more and better data but this approach was limited unless the observations could be fitted into an acceptable and testable framework; a model which included structural and physical as well as biochemical components allowed this and could remove much of the observed variation. A recurrent example in this discussion concerned the limited expansion of a leaf: this reduces to the limited expansion of single cells (increase in wall deposition, decreased elasticity, changes in turgor, ?) and, perhaps, more importantly, to the issue of why cells fail to continue to divide. Harte pointed out that, despite lack of satisfactory explanations, the time when leaves stopped growing could be reasonably predicted.

Monteith brought a new perspective to the discussion, pointing out that the interception of light and, hence, the development of the leaf surface appeared to play a much larger part in determining productivity than did differences in rates of photosynthesis per unit area to which so much attention was paid in earlier papers. However, this latter theme was continued by Nasyrov and colleagues (communicated), who maintained that the most important task in improving productivity lay in understanding the interaction between photosynthesis and photorespiration (see also *Annual Review of Plant Physiology*, **29**, 215–37, 1978). They believed that photorespiration was not necessarily a wasteful process and that the glycolate pathway was important in nitrate reduction, formation of amino acids and in maintaining the Calvin cycle under stress; the aim should be to inhibit the oxygenase activity of RuBPCase/Oase whilst maintaining the glycolate pathway. It should be possible to obtain forms of $C_3$ plants with high activity of PEP carboxylase in the cytoplasm by experimental mutagenesis followed by screening; such should allow enhanced transport of $CO_2$ into the chloroplast and increased net photosynthesis.

Monteith, as did Biscoe and Ritchie, placed a great deal of reliance on 'thermal time', an approach which had helped greatly in simplifying and integrating observations. A note of caution is in order: this concept requires a linear response between the attribute studied and temperature and falls down if temperatures exceed the optimum for an appreciable period of time. Experimenters (and modellers) in applying this concept need to ensure that the basic assumptions are met. Biscoe presented data showing that rates of expansion varied with leaf position and nitrogen supply but that final size was related most closely to duration of growth; his observations suggested that many of the differences were due to differences in nitrogen uptake. Ritchie, from several years' experience in modelling leaf production, maintained that the leaf area index could be readily simulated taking into account the linear relations between expansion rate and the reciprocal of growth duration with temperature, the linear relation of $A_{max}$ with duration and the curvilinear relation between $A_{max}$ of a given leaf and $A_{max}$ of the next oldest leaf. Starck pointed out that observations from her laboratory indicated that, in respect of grain filling, cereals were divided into three groups: wheat in which most of the assimilate for the grain came from current photosynthesis in the (flag) leaf; barley and oats in which it came from current photosynthesis by the ear; and rye in which it came from stored photosynthate in the culm. These observations lead from the initiation, development and functioning of a leaf out into wider issues of crop growth, aspects which could not be covered in this symposium.

# Index

For EU product safety concerns, contact us at Calle de José Abascal, 56–1°, 28003 Madrid, Spain or eugpsr@cambridge.org.

www.ingramcontent.com/pod-product-compliance
Lightning Source LLC
LaVergne TN
LVHW090254210926
848480LV00001B/21

* 9 7 8 0 5 2 1 2 7 9 2 8 4 *